*Edited by*
*Matthias Dehmer,*
*Kurt Varmuza,*
*and Danail Bonchev*

**Statistical Modelling of Molecular Descriptors in QSAR/QSPR**

*Titles of the series*
*"Quantitative and Network Biology"*

**Volume 1**

Dehmer, M., Emmert-Streib, F., Graber, A., Salvador, A. (eds.)

**Applied Statistics for Network Biology**

**Methods in Systems Biology**

2011

ISBN: 978-3-527-32750-8

Schneider, G., Baringhaus, K.-H.

**Molecular Design**

**Concepts and Applications**

2008

ISBN: 978-3-527-31432-4

*Related Titles*

Wang, B.

**Drug Design of Zinc-Enzyme Inhibitors**

**Functional, Structural, and Disease Applications**

2009

ISBN: 978-0-470-27500-9

Todeschini, R., Consonni, V.

**Molecular Descriptors for Chemoinformatics**

**Volume I: Alphabetical Listing / Volume II: Appendices, References**

2009

ISBN: 978-3-527-31852-0

Hinchliffe, A.

**Molecular Modelling for Beginners**

2009

ISBN: 978-0-470-51314-9

*Quantitative and Network Biology*
*Series Editors M. Dehmer and F. Emmert-Streib*
*Volume 2*

# Statistical Modelling of Molecular Descriptors in QSAR/QSPR

*Edited by Matthias Dehmer,*
*Kurt Varmuza,*
*and Danail Bonchev*

WILEY-BLACKWELL

**The Editors**

***Matthias Dehmer***
UMIT
Institut für Bioinformatik und Translationale Forschung
Eduard Wallnöfer Zentrum 1
6060 Hall/Tyrol
Austria

***Kurt Varmuza***
Technische Universität Wien
Institut für Verfahrenstechnik, Umwelttechnik und Technische Biowissenschaften
Getreidemarkt 9/166
1060 Wien
Austria

***Danail Bonchev***
Virginia Commonwealth University
Biological Complexity Center
1015 Floyd Avenue, #3132
Richmond, VA 23284-2030
USA



**Library of Congress Card No.:** applied for

**British Library Cataloguing-in-Publication Data**
A catalogue record for this book is available from the British Library.

**Bibliographic information published by the Deutsche Nationalbibliothek**
The Deutsche Nationalbibliothek lists this publication in the Deutsche Nationalbibliografie; detailed bibliographic data are available on the Internet at http://dnb.d-nb.de.



Wiley-Blackwell is an imprint of John Wiley & Sons, formed by the merger of Wiley's global Scientific, Technical, and Medical business with Blackwell Publishing.



**Typesetting** Thomson Digital, Noida, India
**Printing** Strauss GmbH, Mörlenbach
**Cover Design** Schulz Grafik-Design, Fußgönheim

Printed in the Federal Republic of Germany
Printed on acid-free paper

**Print ISBN:** 978-3-527-32434-7
**ePDF ISBN:** 978-3-527-64502-2
**oBook ISBN:** 978-3-527-64512-1
**ePub ISBN:** 978-3-527-64501-5
**mobi ISBN:** 978-3-527-64503-9

# Contents

# Preface

Molecular descriptors have been applied extensively in, for example, bioinformatics, network biology structure-oriented drug design, medicinal chemistry, chemometrics, chemical graph theory, and mathematical chemistry. Also, their positive impact in quantitative structure–activity relationship/quantitative structure–property relationship (QSAR/QSPR) has been demonstrated and important subgroups of descriptors such as topological indices have been explored. The book *Statistical Modeling of Molecular Descriptors in QSAR/QSPR* presents theoretical and practical results toward the statistical analysis and modeling of molecular descriptors. An intriguing and important field of activity for applying the results discussed in this book is QSAR and QSPR. Particularly the contributors put the emphasis on employing statistical methods for modeling data generated by using molecular descriptors. In this sense, the major goal of the book is to advocate and promote a combination of statistical, information-theoretic, and data analysis techniques to meet the challenge of designing empirical models by using molecular descriptors. Importantly, some of these contributions demonstrate the ability of molecular descriptors for predicting physicochemical or even toxic properties of chemicals successfully. Also, mathematical properties of molecular and topological descriptors are investigated.

We would like to sketch the idea of choosing the book cover in brief. Note that it has been inspired by a short NASA report from April 1995 tries to demonstrate the complexity of QSAR/QSPR in a multivariate setting. The authors of this report, D.A. Noever, R.J. Cronise, and R.A. Relwani, exposed spiders to substances with different toxicity and claimed that the changes in the spider webs reflect the degree of toxicity. For caffeine – the molecule shown on the book cover – the spiders produced only unstructured webs instead of rather symmetrical, radial webs as shown in the background of the cover.

From a statistical point of view, one regrets that no estimations of the reproducibility are given in the report and obviously no further literature exists dealing with this subject; although the original report has been cited frequently. From a point of view of QSAR one may doubt that the toxic effect on spiders can be easily translated to explain toxic effect on other animals or even humans. Furthermore, the effect is not really surprising considering well-known effects of drugs and ethanol when it comes to humans. When speculating, one may be seduced to look for relationships between the networks describing chemical structures and the networks of distorted spider webs.

A different approach is the crucial idea on which the book and its contributions is based: Starting from a molecular structure, a set of descriptors is calculated, for example, information-theoretic indices by using Shannon's entropy as indicated by the cover figure. Hence, a set of chemical structures can be thereby represented by a matrix where each row corresponds to a structure. Typically, multivariate data analysis methods can be applied to such data to generate empirical models that relate a property of substances to the molecular descriptors derived from the chemical structures. Essential for such empirical models is a careful and cautious evaluation of the performance – otherwise one might quickly run into speculation and circular reasoning. In this context, we hope that the book may help to avoid this and also might be stimulating for understanding the mentioned problems more deeply.

Exemplarily, the topics we are going to tackle in this book range from modeling molecular descriptors, studying statistical and information-theoretic techniques, multivariate data analysis, and machine learning techniques for QSAR and QSPR. The book is intended for researchers, graduate, and advanced undergraduate students in the interdisciplinary fields such as biostatistics, bioinformatics, chemistry, chemometrics, mathematical chemistry, molecular medicine, medical informatics, network biology, and systems biology. Each chapter is comprehensively presented, accessible not only to researchers from this field but also to advanced undergraduate or graduate students.

Many colleagues, whether consciously or unconsciously, have provided us with input, help and support before and during the preparation of the present book. In particular, we would like to thank Maria and Gheorghe Duca, Frank Emmert-Streib, Boris Furtula, Ivan Gutman, Armin Graber, Martin Grabner, D. D. Lozovanu, Alexei Levitchi, Alexander Mehler, Abbe Mowshowitz, Arcady Mushegian, Andrei Perjan, Ricardo de Matos Simoes, Fred Sobik, Dongxiao Zhu and apologize to all who have not been named mistakenly. Matthias Dehmer thanks his wife Jana. Also, we would like to thank our editors Andreas Sendtko and Gregor Cicchetti from Wiley-VCH who have been always available and helpful and we are grateful to Frank Emmert-Streib for fruitful discussions. Last but not least, Matthias Dehmer and Kurt Varmuza thank the Austrian Science Funds for supporting this work (project P22029-N13).

Finally, we hope this book helps to spread out the enthusiasm and joy we have for this field and to inspire people regarding their own practical or theoretical research problems.

Hall/Tyrol, Vienna and Richmond, January 2012
*Matthias Dehmer*
*Kurt Varmuza*
*Danail Bonchev*

# List of Contributors

***Dimitris K. Agrafiotis***
Johnson & Johnson Pharmaceutical
Research & Development, LLC
Welsh & McKean Roads
Spring House, PA 19477
USA

***Jürgen Bajorath***
Rheinische Friedrich-Wilhelms-
Universität
B-IT (Bonn-Aachen International
Center for Information Technology)
Department of Life Science Informatics
Dahlmannstr. 2
53113 Bonn
Germany

***Curt M. Breneman***
Rensselaer Polytechnic Institute
Department of Chemistry and Chemical
Biology
110 8th Street
Troy, NY 12180
USA

***Eduardo A. Castro***
Universidad de Buenos Aires
Facultad de Farmacia y Bioquímica
PRALIB (UBA-CONICET)
Junín 956
C1113AAD Buenos Aires
Argentina

***Yiyu Cheng***
Zhejiang University
College of Pharmaceutical Sciences
388 Yuhangtang Road
Hangzhou, Zhejiang 310058
China

***Timothy Clark***
Friedrich-Alexander-Universität
Erlangen – Nürnberg
Computer-Chemie-Centrum
Nägelsbachstrasse 25
91052 Erlangen
Germany

and

University of Portsmouth
Center for Molecular Design
Mercantile House
Portsmouth PO1 2EG
UK

***Viviana Consonni***
University of Milano – Bicocca
Milano Chemometrics & QSAR
Research Group
P.za della Scienza 1
20126 Milano
Italy

***Hong Fang***
ICF International at FDA's National Center for Toxicological Research
3900 NCTR Road
Jefferson, AR 72079
USA

***Grzegorz Fic***
Rzeszow University of Technology
Faculty of Chemistry
Department of Physical Chemistry and Computer Chemistry
Al. Powstancow Warszawy 6
35-959 Rzeszow
Poland

***Gonzalo Cerruela García***
University of Córdoba
Department of Computing and Numerical Analysis
Campus de Rabanales
Albert Einstein Building
14071 Córdoba
Spain

***Weigong Ge***
U.S. Food and Drug Administration
National Center for Toxicological Research
Center for Bioinformatics
Division of Systems Biology
3900 NCTR Road
Jefferson, AR 72079
USA

***Miguel Ángel Gómez-Nieto***
University of Córdoba
Department of Computing and Numerical Analysis
Campus de Rabanales
Albert Einstein Building
14071 Córdoba
Spain

***Daniel J. Graham***
Department of Chemistry
Loyola University Chicago
6525 North Sheridan Road
Chicago, IL 60631
USA

***Huixiao Hong***
Center for Bioinformatics
Division of Systems Biology
National Center for Toxicological Research
U.S. Food and Drug Administration
3900 NCTR Road, Building 5,
Room 5C-109A
Jefferson, AR 72079
USA

***Bögel Horst***
Martin-Luther-University
Department of Chemistry
Kurt-Mothes-Str. 2
06122 Halle
Germany

***Tao-Wei Huang***
Rensselaer Polytechnic Institute
Department of Chemistry and Chemical Biology
110 8th Street
Troy, NY 12180
USA

***Christian Kramer***
Novartis Pharma AG
Novartis Institutes for BioMedical Research
Forum 1
Novartis Campus
4056 Basel
Switzerland

***Michael Krein***
Rensselaer Polytechnic Institute
Department of Chemistry and Chemical Biology
110 8th Street
Troy, NY 12180
USA

***Fabian López-Vallejo***
Computational Chemistry
Torrey Pines Institute for Molecular Studies
11350 SW Village Parkway
Port St. Lucie, FL 34987
USA

***Irene Luque Ruiz***
University of Córdoba
Department of Computing and Numerical Analysis
Campus de Rabanales
Albert Einstein Building
14071 Córdoba
Spain

***George Maroulis***
University of Patras
Department of Chemistry
26500 Patras
Greece

***José L. Medina-Franco***
Computational Chemistry
Torrey Pines Institute for Molecular Studies
11350 SW Village Parkway
Port St. Lucie, FL 34987
USA

***Andrew G. Mercader***
Instituto de Investigaciones Fisicoquímicas Teóricas y Aplicadas (INIFTA, UNLP, CCT La Plata-CONICET)
Diag. 113 y 64, Sucursal 4, C.C. 16
1900 La Plata
Argentina

and

Universidad de Buenos Aires
Facultad de Farmacia y Bioquímica
PRALIB (UBA-CONICET)
Junín 956
C1113AAD Buenos Aires
Argentina

***Lisa Morkowchuk***
Rensselaer Polytechnic Institute
Department of Chemistry and Chemical Biology
110 8th Street
Troy, NY 12180
USA

***Koh-Hei Nitta***
Kanazawa University
Institute of Science and Engineering
Division of Natural System
Kanazawa 920-1192
Japan

***Grażyna Nowak***
Rzeszow University of Technology
Faculty of Chemistry
Department of Physical Chemistry and Computer Chemistry
Al. Powstancow Warszawy 6
35-959 Rzeszow
Poland

***Roger Perkins***
U.S. Food and Drug Administration
National Center for Toxicological Research
Center for Bioinformatics
Division of Systems Biology
3900 NCTR Road
Jefferson, AR 72079
USA

***Feng Qian***
ICF International at FDA's National Center for Toxicological Research
3900 NCTR Road
Jefferson, AR 72079
USA

***Dominic P. Searson***
Newcastle University
School of Chemical Engineering and Advanced Materials
Newcastle upon Tyne NE1 7RU
UK

***Leming Shi***
U.S. Food and Drug Administration
National Center for Toxicological Research
Center for Bioinformatics
Division of Systems Biology
3900 NCTR Road
Jefferson, AR 72079
USA

***Svetoslav Slavov***
U.S. Food and Drug Administration
National Center for Toxicological Research
Center for Bioinformatics
Division of Systems Biology
3900 NCTR Road
Jefferson, AR 72079
USA

***Zhenqiang Su***
ICF International at FDA's National Center for Toxicological Research
3900 NCTR Road
Jefferson, AR 72079
USA

***Roberto Todeschini***
University of Milano – Bicocca
Milano Chemometrics & QSAR Research Group
P.za della Scienza 1
20126 Milano
Italy

***Weida Tong***
U.S. Food and Drug Administration
National Center for Toxicological Research
Center for Bioinformatics
Division of Systems Biology
3900 NCTR Road
Jefferson, AR 72079
USA

***Martin Vogt***
Rheinische Friedrich-Wilhelms-Universität
B-IT (Bonn-Aachen International Center for Information Technology)
Department of Life Science Informatics
Dahlmannstr. 2
53113 Bonn
Germany

***Yap Chun Wei***
National University of Singapore
Faculty of Science
Department of Pharmacy
18 Science Drive 4
Singapore 117543
Singapore

**_Mark J. Willis_**
Newcastle University
School of Chemical Engineering and Advanced Materials
Newcastle upon Tyne NE1 7RU
UK

**_Allen Wright_**
Newcastle University
School of Chemical Engineering and Advanced Materials
Newcastle upon Tyne NE1 7RU
UK

**_Liew Chin Yee_**
National University of Singapore
Faculty of Science
Department of Pharmacy
18 Science Drive 4
Singapore 117543
Singapore

**_Austin B. Yongye_**
Computational Chemistry
Torrey Pines Institute for Molecular Studies
11350 SW Village Parkway
Port St. Lucie, FL 34987
USA

# 1
# Current Modeling Methods Used in QSAR/QSPR

*Liew Chin Yee and Yap Chun Wei*

## 1.1
## Introduction

A drug company has to ensure the quality, safety, and efficacy of a marketed drug by subjecting the drug to a variety of tests [1]. Therefore, drug development is a time-consuming and expensive process. From the initial stage of target discovery, development often takes an average of 12 years [2] and was estimated to cost USD868 million per marketed drug [3]. This high cost and lengthy process is due to the high risk of drug development failure. It was estimated that only 11% of the drugs that completed developmental stage were approved by the US or European regulators [4]. In year 2000, it was found that 10% of attrition during drug development was contributed by poor pharmacokinetic and bioavailability, while in the clinical stage, 30% of attrition was due to lack of efficacy and another 30% was caused by toxicity or clinical safety issues [4]. Thus, it will be useful to predict these failures prior to the clinical stage in order to reduce drug development costs. It was claimed that savings of USD100 million in development costs per drug could be attained with 10% prediction improvement [5]. Therefore, various methods, such as *in vitro*, *in vivo*, or *in silico* methods, are being used early in the drug development stage to filter out potential failures. An example of an *in silico* method is quantitative structure–activity relationship (QSAR) models, which can be used to understand drug action, design new compounds, and screen chemical libraries [6–9]. Recently, the European Chemicals Legislation, Registration, Evaluation and Authorisation of Chemicals (REACH) suggested the use of *in silico* methods as reliable toxicological risk assessment [10, 11].

QSARs, or quantitative structure–property relationships (QSPRs), are mathematical models that attempt to relate the structure-derived features of a compound to its biological or physicochemical activity. Similarly, quantitative structure–toxicity relationship (QSTR) or quantitative structure–pharmacokinetic relationship (QSPkR) is used when the modeling applies on toxicological or pharmacokinetic systems. QSAR (also QSPR, QSTR, and QSPkR) works on the assumption that structurally similar compounds have similar activities. Therefore, these methods have predictive and

*Statistical Modelling of Molecular Descriptors in QSAR/QSPR*. First Edition.
Edited by M. Dehmer, K. Varmuza, and D. Bonchev

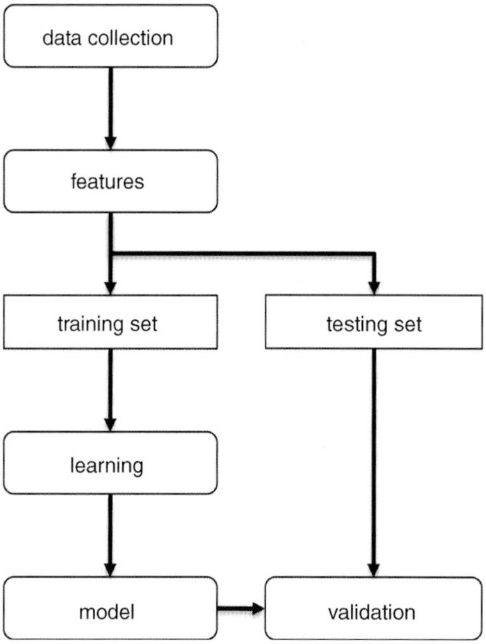


**Figure 1.1** General workflow of developing a QSAR model.

diagnostic abilities. They can be used to predict the biological activity (e.g., $IC_{50}$) or class (e.g., inhibitor versus noninhibitors) of compounds before the actual biological testing. They can also be used in the analysis of structural characteristics that can give rise to the properties of interest.

As illustrated in Figure 1.1, developing QSAR models starts with the collection of data for the property of interest while taking into consideration the quality of the data. It is necessary to exclude low-quality data as they will lower the quality of the model. Following that, representation of the collected molecules is done through the use of features, namely molecular descriptors, which describes important information of the molecules. There are many types of molecular descriptors but not all will be useful for a particular modeling task. Thus, uninformative or redundant molecular descriptors should be removed before the modeling process. Subsequently, for tuning and validation of the QSAR model, the full data set is divided into a training set and a testing set prior to learning.

During the learning process, various modeling methods like multiple linear regression, logistic regression, and machine learning methods are used to build models that describe the empirical relationship between the structure and property of interest. The optimal model is obtained by searching for the optimal modeling parameters and feature subset simultaneously. This finalized model built from the optimal parameters will then undergo validation with a testing set to ensure that the model is appropriate and useful.

This chapter gives an introduction to the algorithm of the various modeling methods that have been commonly used in constructing QSAR models. We have used most of these methods in developing QSAR models for various pharmacodynamic, pharmacokinetic, and toxicological properties [12–16]. Even though our research have found that models developed using more complex modeling methods like the newer machine learning methods frequently outperform those developed using traditional statistical methods, it is essential to have a good foundation of all these methods. This is because no method is useful for all QSAR problems and the principle of parsimony states that we should use the simplest method that provides the desired performance level. This is to prevent overfitting of the data, which can lead to a loss in generalizability. Data collection, data processing, computation and selection of features, and model validation have been thoroughly reviewed elsewhere [17–22], so they are not described here. Software that is available for QSARs development will be discussed.

## 1.2 Modeling Methods

In general, methods for constructing QSAR can be divided into two groups: methods for regression problems or classification problems. The methods are organized into the two groups in the following section.

### 1.2.1 Methods for Regression Problems

#### 1.2.1.1 Multiple Linear Regression

Multiple linear regression (MLR) is one of the most fundamental and common modeling method for regression QSAR. Recent application of MLR in QSAR or QSPR includes prediction for luteinizing hormone-releasing hormone antagonists [23], 5-HT6 receptor ligands [24], interleukin-1 receptor-associated kinase 4 inhibitors [25], potencies of endocrine disruptors [26], and chlorine demand by organic molecules [27]. MLR is favored for its simplicity and ease of interpretation as the model assumes a linear relationship between the compound's property, $\hat{y}$, and its feature vector, denoted **X**, which is usually the molecular descriptors. Thus, with the notion of X, the property of an unknown compound can be predicted by the fitted model. The following equation represents a general expression of a MLR model:

$$\hat{y} = \beta_0 + \beta_1 X_1 + \beta_2 X_2 + \cdots + \beta_k X_k$$

where $\beta_0$ is the model constant, $X_1, \ldots, X_k$ are molecular descriptors with their corresponding coefficients $\beta_1, \ldots, \beta_k$ (for molecular descriptors 1 through $k$). These coefficients can be obtained through the use of estimators like least-squares method which minimizes the sum of squared residuals.

The size of the coefficients may reveal the degree of influence of the corresponding molecular descriptors on the target property. In addition, a positive coefficient suggests that the corresponding molecular descriptor contributes positively to the target property, while a negative coefficient suggests negative contribution. However, these interpretations may not be accurate as collinear descriptors have the potential to influence the coefficients such that erroneous values may be assigned. Thus, the molecular descriptors in the model should be independent of each other and the number of instances for model building should be at least five times the number of descriptors used [28]. In addition, the assumption of a linear relationship makes MLR less suitable to model complex problems like toxicity, where multiple mechanisms may interplay to elicit a toxic response. Nonetheless, MLR has been used for modeling toxicity systems [29–33].

To date, MLR remains in use with enhancements or in combination with feature selection to improve its performance. Examples of enhancements are: the use of independent component analysis – MLR in QPSR of aqueous solubility [34], local lazy regression [35], retro-regression applied on boiling points of nonanes [36], ensemble feature selection [37], and other feature selection methods like genetic algorithm, ridge regression, partial least-squares method, pair-correlation method, forward selection, and best subset selection in the application of MLR [38–42].

#### 1.2.1.2 **Partial Least Squares**

Partial least squares (PLSs) assumes a linear relationship between feature vector, **X**, and target property, $\hat{y}$, but unlike MLR, PLS is more appropriate when the number of features greatly exceed the number of samples and when features are highly collinear [43]. It is to note that advancement has brought about methods like quadratic-PLS and kernel-PLS for nonlinear systems, multiway-PLS, unfolding-PLS, hierarchical-PLS, three-block bifocal PLS, and so on. These will not be discussed here and interested readers can refer to the review by Hasegawa *et al.* [44].

PLS works on the assumption that the examined system is subjected to the influence of just a few causal factors, termed *latent factors* or *latent variables* [43, 45]. PLS avoids the problem of collinear features by extracting these latent factors that can explain the variations of the molecular descriptors while simultaneously models the response of the target property.

PLS was also interpreted as the initialism for “Projection to Latent Structure” [45]. As illustrated in Figure 1.2, the latent factors can be estimated through X-scores and Y-scores, which are extracted from the molecular descriptors and desired compound properties, respectively. Subsequently, the X-scores are used to predict the Y-scores, which in turn can be used to predict the compound properties. The number of latent factors used in PLS is an important consideration for QSAR modeling, and it is usually obtained through the use of cross-validation methods like *n*-fold cross-validation and leave-one-out methods, where a portion of the samples is used as training set, while the other portion is set aside as testing set to validate the model that was built from the training set.

PLS has been applied on various QSAR studies like toxicity of quaternary ammonium compounds on *Chlorella vulgaris* [46], angiotensin II AT1 receptor antagonists [47], CYP11B2 binding affinity and CYP11B2/CYP11B1 selectivity [48], toxicity to *Daphnia magna* [49], and nonpeptide HIV-1 protease inhibitors prediction [50]. PLS is also used as an analysis method in the popular 3D-QSAR technique,

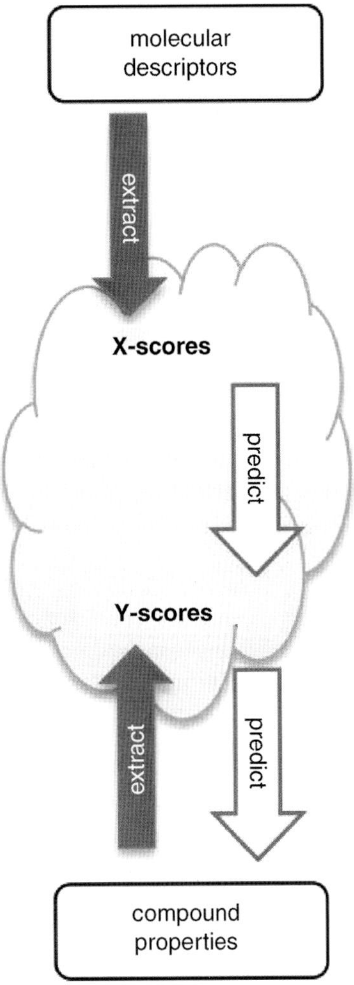


**Figure 1.2** Extraction of latent factors from molecular descriptors and compound properties.

Comparative Molecular Field Analysis (CoMFA) that is available in the SYBYL software [51]. In CoMFA, the molecular descriptors are obtained from the magnitude of the steric and electrostatic field of molecules, which are sampled at regular interval when the molecules are aligned to a common substructure. As a result, a large number and correlated descriptors may be produced from a small training sample. Hence, PLS is applied to reduce the number of descriptors to make them more suitable for further analysis.

#### 1.2.1.3 **Feedforward Backpropagation Neural Network**

Artificial neural network (ANN) attempts to imitate a biological neural network and is inspired from the structure, processing, and learning method of a biological brain.

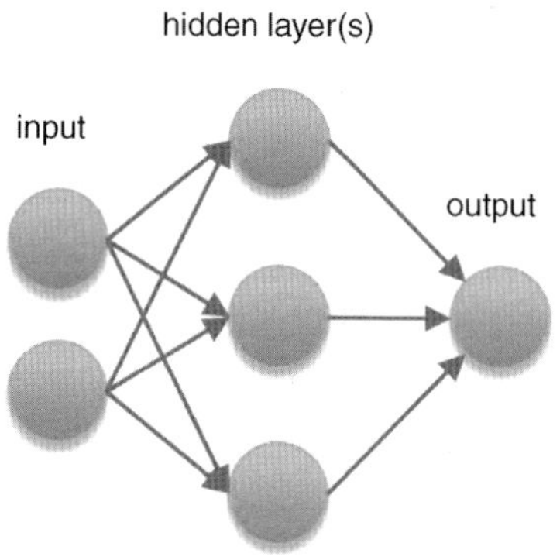


**Figure 1.3** A simple structure showing the three layers of an artificial neural network.

It is a network of processing elements (akin to neurons) with weighted connections between them. A typical artificial neural network consists of three or more layers: an input layer, hidden layer(s), and an output layer as shown in Figure 1.3. In training, ANN adapts the weight of the connections until it approximates the input–output relationship of the training data. For model building, the number of hidden layers and the number of elements in the hidden layers is commonly optimized [52], although one hidden layer with large number of elements is generally sufficient to approximate most functions [53]. Nonetheless, it is not trivial to find an optimal topology that can generalize the data well; many rounds of training are usually required, which makes building an ANN model a time-consuming process.

A feedforward neural network was the first and possibly the simplest type of ANN. The input layer of the network represents molecular descriptors and the output layer represents the target properties of compounds. The network is feedforward because the elements in one layer are only connected to the elements in the next layer and the information moves forward from one layer to another toward the output layer. In the case of fully connected neural network, each element in the hidden layer receives information from all elements in the previous layer. Subsequently, an activation function, which is commonly linear or sigmoidal, will transform the input before forwarding the information to the next layer (if any) and eventually to the output layer.

The learning of a feedforward neural network can be done through a variety of techniques and one of the most popular methods is through backpropagating errors [54]. In backpropagation, the output of the network (from forward phase) is compared with the actual compound property to calculate a predefined error-function. This calculated value is then feedback (backward phase) into the network, allowing the algorithm to re-adjust the weights of the connections, which were randomly assigned initially, with the aim to minimize the error. The approximation improves as the errors converge after numerous training cycles. However, the model may run into the problem of overfitting and thus incapable of predicting the property of unknown compounds that are sufficiently different from the training set. Therefore, there are various methods such as the use of a validation set or techniques like early stopping, pruning, and weight decay to minimize the risk of overfitting [55–58].

Examples of application of feedforward backpropagation neural network in QSAR studies are toxic effect on fathead minnows [59, 60], calcium channel antagonist activity [61], alpha adrenoreceptors agonists [62], air to water partitioning for organic pesticides [63], aldose reductase inhibitors [64], antinociceptive activity [65], and boiling points [66]. Neural network is used in QSAR because of its ability to approximate any target function and also it can handle redundant descriptors well, as their weights can be learned and reduced to insignificant levels [67]. However, it is not easy to optimize the best network topology. Furthermore, parameters like learning rate and momentum needs to be defined by the user, thus the lack of automation makes the process rather time consuming. In addition, if the network is not optimized, undersized network may not approximate the relationship between the descriptors and target property well. Conversely if the network is oversized, overfitting may occur. Other disadvantages of neural network includes its susceptibility to noisy data which can be overcome with the use of a validation set during training, and also hard to interpret connection weights which makes optimization of compound structures difficult for medicinal chemists. Nonetheless, it is to note that a few studies have attempted to interpret neural network for QSAR studies with success [68–70], indicating that neural network is still a useful tool for QSAR studies.

#### 1.2.1.4 General Regression Neural Network

One of the difficulty of building a neural network is the lack of automation in the selection of parameters and network topology. Coupled with many iterations that may be needed by the backpropagation method to converge to an acceptable error, model building is usually a time-consuming process [71]. To overcome these disadvantages, Specht introduced a one-pass neural network learning algorithm as an alternative to increase the training speed. With the implementation of the one-pass algorithm, user-defined parameters have been reduced to a minimal, of which optimization of the network mainly involves adjusting the sigma, σ, of the estimation kernel [71].

The one-pass algorithm was first implemented in probabilistic neural network for use in classification problems. Following that, the general regression neural network (GRNN) was introduced for estimation of regression problems [72]. It is noted that GRNN was rediscovered by Schløler [73, 74] a year later, and it is related to the kernel regression invented by Nadaraya [75] and Watson [76] independently [71]. GRNN was used in QSAR studies of CCR2 inhibitors [77], HIV-1 reverse transcriptase inhibitors [78], drug total clearance [79], estrogenic activity [80], phytoestrogen binding to estrogen receptors [81], aqueous solubility of nitrogen- and oxygen-containing small organic molecules [82], and QSAR on effect of substitution on the phenyl ring of orally active β-lactam inhibitor [83].

For target property, *y*, let X represents a value of molecular descriptor *x*, and *f(x,y)* represents the joint probability density function (PDF) of *x* and *y*. The prediction of the target property can then be obtained by the conditional expected value of *y* given by X [71, 72, 82]:

$$E[y|X] = \frac{\int_{-\infty}^{-\infty} yf(X, y)dy}{\int_{-\infty}^{-\infty} f(X, y)dy}$$

The joint probability density function, $f(x,y)$, symbolizes the relationship between the target property and molecular descriptors, and it is usually not known [72]. Therefore, the value is commonly estimated from the training data using techniques like Parzen's nonparametric estimator [84]:

$$g(x) = \frac{1}{n\sigma}\sum_{i=1}^{n} W\left(\frac{x-x_i}{\sigma}\right)$$

where $n$ is the set cardinality, $\sigma$ is a smoothing parameter that defines the kernel width, $W$ is a weight function, and $(x - x_i)$ is the distance between a given instance and an instance in the training data. Cacoullos has expanded Parzen's nonparametric estimator for the multivariate case [85] and becomes

$$g(x_1, \ldots, x_p) = \frac{1}{n\sigma_1 \ldots \sigma_p}\sum_{i=1}^{n} W\left(\frac{x_1-x_{1,i}}{\sigma_1}, \ldots, \frac{x_p-x_{p,i}}{\sigma_p}\right)$$

For the weight function, $W$, a commonly used function is the Gaussian kernel. Updating the equation for PDF gives the following:

$$g(x) = \frac{1}{n}\sum_{i=1}^{n} \exp\left(-\sum_{j=1}^{p}\left(\frac{x_j-x_{j,i}}{\sigma_j}\right)^2\right)$$

There are two types of models defined by the number of sigma used: single-sigma models and multisigma models where an individual $\sigma$ for each molecular descriptor is used. Multisigma models are suitable for cases which the descriptors are of different nature and importance. In general, these models are able to perform considerably better than the corresponding single-sigma models [82]. In single-sigma model, a single $\sigma$ value is used to simplify the equation. Single-sigma model is preferred for cases where the descriptors have similar importance because reasonable models can still be obtained with faster computation speed [86].

$$\hat{y} = \frac{\sum_{i=1}^{n} y_i \exp(-D(\mathrm{x}, \mathrm{x}_i))}{\sum_{i=1}^{n} \exp(-D(\mathrm{x}, \mathrm{x}_i))}$$

The above is the basic equation of GRNN obtained from substituting Parzen's nonparametric estimator for $f(\mathrm{x},y)$, where the distance function, $D(x,x_i)$, can be the squared weighted Euclidean distance:

$$D(\mathrm{x}, \mathrm{x}_i) = \sum_{j=1}^{p}\left(\frac{x_j-x_{j,i}}{\sigma_j}\right)^2$$

GRNN can be visualized as a network with four layers as shown in Figure 1.4. The first layer is the input layer where each molecular descriptor forms an element in the layer. From the input layer, scaled information is fed to the pattern layer. The pattern layer contains elements that represent each training compounds. In this layer, each of these pattern elements calculates a distance measure between the input compound and the training compound that it represents, and then further processes it with

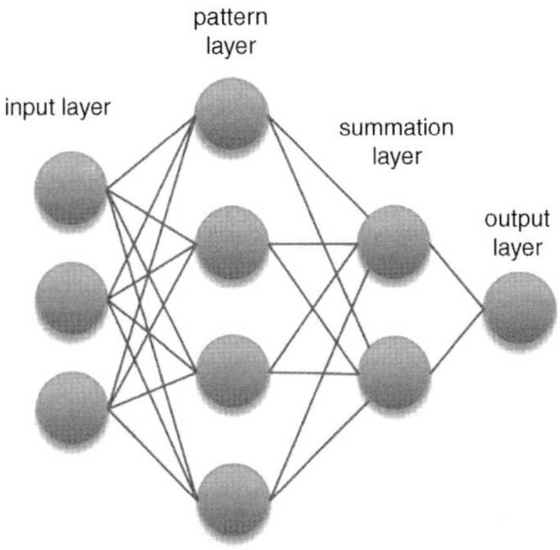


**Figure 1.4** GRNN is a neural network with four layers.

Parzen's nonparametric estimator. Each element of the pattern layer is connected to precisely two elements in the summation layer where the numerator and denominator of the basic GRNN equation are calculated. At the final step, the prediction for the target property is obtained by division in the sole node of the output layer.

#### 1.2.1.5 **Gaussian Processes**

Gaussian process (GP) is a generalization of the Gaussian probability distribution to an infinitely large pool of possible functions [8, 87, 88] and it is formally defined as a collection of random variables, any finite number of which have a joint Gaussian distribution [87]. GP is one of the kernel methods for model development and have been applied in QSAR modeling of ADMET and aqueous solubility [89–91], and hERG inhibitors [92].

Consider target property $y$ with molecular descriptors $x$, in modeling, we would like to find a function $f(x)$ that can relate the observations of $x$ with the observed $y$. One can assume a linear basis function (or any other basis functions) so that the parameters of the function $f(x)$ can be determined from the observed data set. However, just like multiple linear regression, the restriction to a linear basis function will result in a poorly modeled function if the underlying relationship between the target property and molecular descriptors is nonlinear. Therefore, with Bayesian setting as an alternative, one may include all possible types of functions that may be derived, and assign a prior probability to each of them such that more plausible functions will have higher probabilities, and to use this knowledge in making a weighted decision. However, the number of possible functions may be countless and thus impractical to compute. Therefore, a Gaussian process can be introduce to "generalize" them – an inference is made from these functions such that "hyper"-parameters, that is, a collection of random variables, may be derived or refined and made closer to the observations as more data points are introduced. Since GP is a nonparametric model, we need not worry about the linear or nonlinear underlying relationship [8, 87, 88].

A Gaussian process is completely specified by its mean function $m(\mathrm{x})$ and covariance function $\mathrm{cov}(f(\mathrm{x}), f(\mathrm{x}')) = k(\mathrm{x}, \mathrm{x}')$ [87], thus written as

$$f(\mathrm{x}) \sim GP(m(\mathrm{x}), k(\mathrm{x}, x'))$$

where the mean function is usually taken as zero. The covariance function is sometimes called the kernel trick, and similar to that used in support vector machines. The learning process involves optimizing the properties for this covariance function. Some covariance functions that may be used are:

$$\text{squared exponential}: \quad k(\mathrm{x}_i, \mathrm{x}_j) = \exp\left(-\frac{|x_i - x_j|^2}{2l^2}\right)$$

$$\gamma\text{-exponential}: \quad k(\mathrm{x}_i, \mathrm{x}_j) = \exp\left(-\left(\frac{|x_i - x_j|^{\gamma}}{l}\right)\right)$$

where $l$ is the characteristic length scale. In the case of squared exponential, sometimes known as radial basis function (RBF), the inverse of $l$ determines the degree of influence of the molecular descriptors. When $l$ had a large value, the covariance will become independent of the molecular descriptor, thus rendering the descriptor irrelevant [87].

To predict the property of an unknown compound $\mathrm{x}_*$, the GP model assigns probabilistic outcomes, that is, a predicted mean:

$$\bar{f}(\mathrm{x}_*) = \sum_{i=1}^{n} \alpha_i k(\mathrm{x}_i, \mathrm{x}_*)$$

and a standard deviation for the unknown compound [8]:

$$\mathrm{std} f(\mathrm{x}_*) = \left(k(\mathrm{x}_*, \mathrm{x}_*) - \sum_{i=1}^{N}\sum_{j=1}^{N} k(\mathrm{x}_i, \mathrm{x}_*) k(\mathrm{x}_j, \mathrm{x}_*) H_{ij}\right)^{\frac{1}{2}}$$

where the coefficient $\alpha = (\alpha_1, \ldots, \alpha_N)$ is obtained by solving a system of linear equations

$$(K + \sigma^2 I)\alpha = y$$

where $I$ is the unit matrix, $y = (y_1, \ldots, y_N)$, and $H_{ij}$ are the elements of the matrix $H = (K + \sigma^2 I)^{-1}$. For the mathematical details, please refer to the chapter by Rasmussen and Williams [93] and also Schwaighofer *et al.* [89].

## 1.2.2 Methods for Classification Problems

### 1.2.2.1 Logistic Regression

Logistic regression (LR) is similar to linear regression in many ways. LR is used to model the probability of the occurrence of some event as a linear function of a set of predictors. For example, the relationship between categorical target property (usually a property with binary outcomes like inhibitor/noninhibitor) and a set of molecular descriptors. The following equation calculates the probability:

$$\hat{y} = \frac{1}{1 + e^{-(\beta_0 + \beta_1 X_1 + \beta_2 X_2 + \cdots + \beta_k X_k)}}$$

where $\beta_0$ is the model's intercept, $X_1, \ldots, X_k$ are molecular descriptors with their corresponding regression coefficients $\beta_1, \ldots, \beta_k$ (for molecular descriptors 1 through $k$).

Given an unknown compound, LR calculates the probability that the compound belongs to a certain target property. For example, in predicting whether an unknown compound is toxic or nontoxic, LR tries to estimate the probability of the compound being a toxic substance. If the calculated $\hat{y}$ is $>0.5$, then it is more probable that the compound is toxic. Conversely, if $\hat{y} < 0.5$, then the compound is more probable to be nontoxic.

Similar to multiple linear regression, the regression coefficients in LR can describe the influence of a molecular descriptor on the outcome of the prediction. When the coefficient has a large value, it shows that the molecular descriptor strongly affect the probability of the outcome, whereas a zero value coefficient shows that the molecular descriptor has no influence on the outcome probability. Likewise, the sign of the coefficients affects the probability as well, that is, a positive coefficient increases the probability of an outcome, while a negative coefficient will result in the opposite. From our experience, LR usually cannot handle high dimensional data well, especially if the data set is large (e.g., more than 60 000 compounds) with large class imbalance (active: nonactive = 1: 68), which can be common in a training set meant for virtual screening. Such LR models commonly have increased computational time and do not perform well [15].

Applications of LR in QSAR studies includes modeling of nucleosides against amastigotes of *Leishmania donovani* [94], skin sensitization prediction [95–97], *Tetrahymena pyriformis* toxicity [14], classification of antibacterial activity [98], and sediment toxicity prediction [99].

#### 1.2.2.2 **Linear Discriminant Analysis**

Linear discriminant analysis (LDA) is commonly used for classification problems and also for dimensionality reduction. It works on data that has categorical target properties and molecular descriptors that are continuous variables. LDA tries to find a hyperplane that can best separate different classes of a target property. The hyperplane is defined by a linear discriminant function which is a linear combination of molecular descriptors:

$$L = \sum_{i=1}^{k} w_i x_i$$

where $L$ is the discriminant score, molecular descriptors $x_i, \ldots, x_k$ and their corresponding weights, $w_i, \ldots, w_k$. As depicted in Figure 1.5, the discriminant function, $L$, represents the function where a set of data points is projected onto. This function is

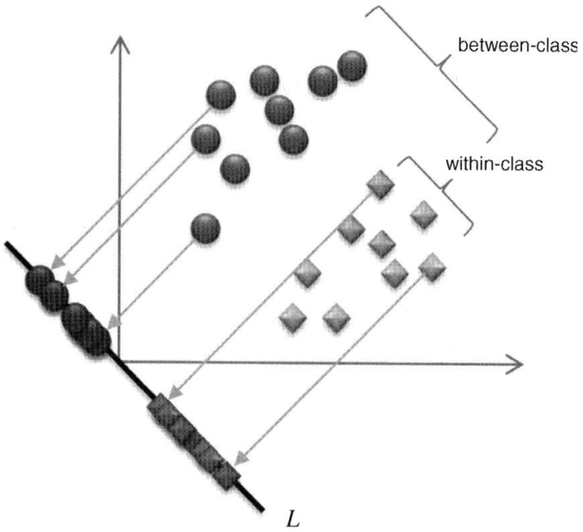


**Figure 1.5** Distribution of a two-class samples on a transformed axis, $L$.

obtained through optimizing a set of weights, $w_i$, that maximizes the ratio of the between-class to the within-class variance to obtain the greatest class separation.

The class of an unknown compound is assigned based on its $L$ discriminant score falling below or above a threshold $C$. When two classes are of equal cardinality and have similar distributions, a possible value for $C$ is the average of the mean discriminant scores from both classes:

$$C_{\text{threshold}} = \frac{\mu_{L1} + \mu_{L2}}{2}$$

Application of LDA in QSAR includes: prediction of passive blood-brain partitioning [100], mutagenicity [101], prediction of antiparasitic drugs against different parasite species [102], prediction of antitrypanosomal compounds [103], aquatic toxicity of pesticide [104], telomerase inhibitory activity [105], trichomonacidals prediction [106], and QSAR for compounds with antiproliferative activity [107].

#### 1.2.2.3 **Decision Tree and Random Forest**

A decision tree (DT) is a structure with hierarchical arrangement of nodes and branches. A DT has three types of nodes: a root node, internal nodes, and leaf nodes. A root node does not have any incoming branches, while an internal node has one incoming branch and two or more outgoing branches. Lastly, the leaf nodes, also known as terminal nodes, have one incoming branch and no outgoing branches. Each leaf node is assigned with a target property, while a nonleaf node (root or internal node) is assigned with a molecular descriptor that becomes a test condition which branches out into groups of differing characteristics.

The classification of an unknown compound is based on the leaf node that it reaches after going through a series of questions (nodes) and answers (deciding

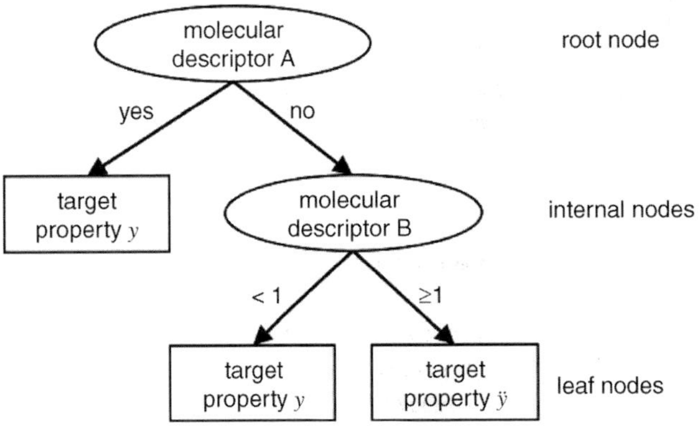


**Figure 1.6** Decision tree has three types of nodes.

which branches to take), starting with the first question from the root node. In the example in Figure 1.6, an unknown compound will be classified with target property *y*, if it fulfilled a certain condition for molecular descriptor *A*. Otherwise, molecular descriptor *B* of the unknown compound is checked at the next step. If the value is less than 1, the unknown compound will be labeled with target property *y*. If not, the unknown will be given the label of target property $\ddot{y}$.

A decision tree is constructed by systematically subdividing the information within a training set with rules and relationships. With a given set of descriptors, many possible variations of trees may be constructed and they may have varying accuracies. Nonetheless, there are algorithms such as the Hunt's algorithm that can be used to induce a decision tree [67]. The algorithms frequently use a recursive greedy heuristic to select which descriptors to split the training data. In the first step of Hunt's algorithm, it examines if all instances in the training set belong to one class *y*. If so, a leaf node *y* will be created and all these training cases will be associated with this node. If not, a molecular descriptor with a certain threshold is chosen to split the samples into smaller subsets, where each of these subsets form a new child node, and the process repeats from the first step until all training cases are associated with a leaf node. The threshold of molecular descriptors that specify the best split can be determined using measures like misclassification error, entropy, and Gini index that enables comparison of "impurities" in the parent node and child nodes; the child nodes should have less impurity than the parent node, therefore, the greater is the impurity difference, the better is the selected threshold for splitting the samples.

Decision trees have the advantage of easy interpretation especially if they are small, and the performance of the decision tree is not so easily affected by unnecessary descriptors. It has been applied on QSAR of cytochrome P450 activities [108], peptide–protein-binding affinity [109], catalysts discovery [110], and in a study of substrates, inhibitors, and inducers of P-glycoprotein [111]. However, a potential drawback of decision tree is its susceptibility to model overfitting due to lack of data or the presence of mislabeled training instances. To overcome the problem of over-fitting, methods such as pruning, cross-validation or random forest may be used.

Pruning works by preventing the construction of an excessively complicated tree that flawlessly fits the whole data set, of which mislabeled data may be present. On the other hand, random forest (RF) uses consensus classification to reduce the problem of overfitting while improving the accuracy [112–114]. The algorithm works by growing many decision trees, thus, collectively known as a "forest" that makes a final prediction based on the majority prediction from each of the trees. To construct each tree, a training sample of size $D$ is selected at random with replacement from an original data with size $D$. Using the new training sample, a tree is grown with randomly selected descriptors and it is not pruned. RF is easy to use as the user only need to fix two parameters: the number of trees in the forest and the number of descriptors in each trees. It was recommended that a large number of trees should be grown and the number of descriptors to be taken from the square root of the total descriptors [115].

RF can handle large number of training data and descriptors. Besides classifying an unknown compound, it can be extended for unsupervised clustering and outlier detection [114]. RF can also be used to infer the influence of the descriptors in a classification task and also to estimate missing data. It was found that RF is less affected by noisy data or data with many weak inputs [114]. Although it is claimed that RF does not overfit, it was shown that the performance of RF can be influenced by imbalanced data set or small sample size and also by the number of trees and features selected [116, 117]. Therefore, the parameters for RF should be carefully selected and the optimization may be carried out through the use of cross-validation [117].

#### 1.2.2.4 *k*-Nearest Neighbor

*k*-nearest neighbor (kNN) is a type of lazy learner whereby it delays the learning of the training data until it is needed to classify an unknown sample. It is useful for QSAR studies because QSAR works on the assumption of compounds with similar structure should have similar activities [118]. kNN has been applied on QSAR studies of binding affinity and receptor subtype selectivity of human 5HT1E and 5HT1F receptor-ligands [119], anti-HIV activity of Isatin analogs [120], inhibitors of γ-amino butyric acid transaminase [121], T-helper-type-2 cells receptor antagonist [122], selective cyclooxygenase-2 inhibitors [123], and geranylgeranyltransferase-I inhibitors [124].

kNN works by measuring the distance between the unknown compound and every compound in the training set and then classifies a test compound by searching for the training compounds that are similar in characteristics to the unknown compound. There are various types of distance measures that may be used; two of the common ones are the Euclidean distance

$$d(\mathrm{p},\mathrm{q}) = \sqrt{\sum_{i=1}^{n} (p_i - q_i)^2}$$

and the Manhattan distance

$$d(p,q) = \sqrt{\sum_{i=1}^{n} |p_i - q_i|}$$

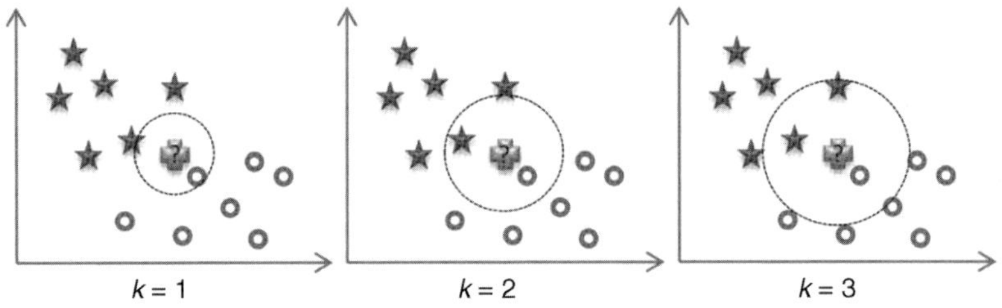


**Figure 1.7** Classification of the unknown compound changes when $k$ is different.

where $n$ is the number of molecular descriptors, and $p_i$ and $q_i$ is the $i$th descriptor for compounds $p$ and $q$, respectively. The class of the unknown compound is then determined by the majority of the class of its $k$ neighbor(s). The number of neighbors, $k$, is a user-defined integer that needs to be optimized as it will affect the performance of the model (Figure 1.7). Misclassification can occur if the $k$ is too small or too large. When dealing with binary classification problems, an odd number $k$ is usually chosen to reduce the ambiguity of the predictions. We find that $k$ of 3 is usually sufficient to produce good results and it is frequently used in kNN-related research. Also, since kNN relies on measuring distances between compounds, the chemical descriptors should be scaled to avoid descriptors with large magnitudes from adversely influencing the kNN model.

#### 1.2.2.5 **Probabilistic Neural Network**

Probabilistic neural network (PNN) is similar to the general regression neural network, but it is used in classification problems [125, 126]. PNN is effective in nonlinear mapping, pattern recognition, estimation of target property and likelihood ratios [127]. PNN has been applied on prediction of acute toxicity to *Daphnia magna* and fathead minnow [128, 129], *Tetrahymena pyriformis* toxicity [14], anti-HIV activity of compounds [130], and QSAR of soluble epoxide hydrolase inhibitors [82]. PNN works on the basis of nonparametric estimators of conditional probability density functions and the Bayes strategy to minimize expected risk [127]. Similar to GRNN, the probability density function for each target property for a univariate case can be estimated by the Parzen's nonparametric estimator.

In a binary classification problem where a compound belongs to either class $\theta_i$ or $\theta_j$, PNN predicts the class of unknown compounds based on a set of molecular descriptors, $x$, through the use of Bayes decision rule:

$$\begin{aligned} d(x) &= \theta_i \quad \text{if} \quad h_i\, c_i f_i(x) > h_j\, c_j f_j(x) \\ d(x) &= \theta_j \quad \text{if} \quad h_i\, c_i f_i(x) < h_j\, c_j f_j(x) \end{aligned}$$

where class $\theta_i$ and $\theta_j$ have the probability density function of $f_i(x)$and $f_j(x)$, prior probabilities of $h_i$ and $h_j$, and costs of misclassification of $c_i$ and $c_j$, respectively. The prior probabilities and misclassification cost are treated as being equal in most applications. In classification tasks, an unknown compound is predicted with target property $\theta_j$ if the product of all three variables is greater for target property $\theta_j$ than for any other data with class $\theta_i$ not equal to $\theta_j$.

The network architecture for PNN is similar to GRNN, but unlike GRNN which has two elements in the summation layer, PNN will have the same number of elements as the number of target properties. Within each summation element, the estimated probability density function for the corresponding target property is obtained by summing all the acquired inputs from the pattern layer. This information is then passed on to the single element in the output layer where the probability density functions will be evaluated and the class of an unknown compound is assigned with the target property with the highest value.

Like GRNN, PNN has the advantage of fast training for classification problems with small sample size and target properties because it uses the one-pass neural network learning algorithm. Also, PNN is able to handle sparse samples and learns only the necessary information from outliers, thus making it robust and less susceptible to outliers [131]. Memory requirement increases with the increase of data size [131], which may cause a problem if computational resource is limited. Therefore, PNN is usually recommended for problems of small to moderately large sample sizes. Alternatively, clustering methods should be used to reduce the number of elements required in the network before modeling [127].

#### 1.2.2.6 **Support Vector Machine**

Support vector machine (SVM) is based on the structural risk minimization principle from statistical learning theory [132] and it is probably one of the most well-known kernel methods for model development [133]. It is a classifier that is less affected by duplicated data and has lower risk of model overfitting [67]. SVM has become very popular in recent years with its applications in various pattern recognition fields like bioinformatics, medical, economics, and cheminformatics [15, 16, 134–140].

In binary classification of linearly separable data, SVM tries to build a maximal margin hyperplane to separate one class of compounds from the other class as illustrated in Figure 1.8. The hyperplane, also known as the decision boundary, is

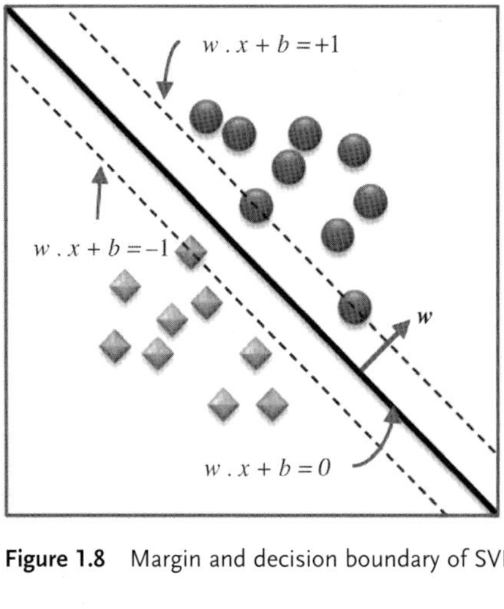


**Figure 1.8** Margin and decision boundary of SVM in linearly separable case.

built on the basis of the data points called support vectors and can be represented by the following:

$$w \cdot x + b = 0$$

The parameters $w$ and $b$ are estimated during learning and they must satisfy the following conditions:

$$\begin{array}{lll} \text{Class } 1: & w \cdot x_i + b \geq +1, & \text{if } y_i = +1 \\ \text{Class } -1: & w \cdot x_i + b \leq -1, & \text{if } y_i = -1 \end{array}$$

and at the same time maximizing the margin by minimizing the following function:

$$f(w) = \frac{\|w\|^2}{2}$$

where $y_i$ is the class label and $x_i$ is a vector of molecular descriptors for compound $i$, $w$ is a vector perpendicular to the hyperplane, and $\|w\|^2$ is the Euclidean norm of $w$. With optimized parameters $w$ and $b$, an unknown compound with vector $x$ can be classified by

$$\hat{y} = \text{sign}[(w \cdot x) + b]$$

The unknown compound is classified as Class 1 if $\hat{y} > 0$ and classified as Class $-1$ when $\hat{y} < 0$.

Optimization of the parameters of the hyperplane can also be viewed as a convex optimization problem where the margins should be constructed furthest from each of the convex hulls as shown in Figure 1.9. Using the Lagrange multiplier, the function for the optimization problem becomes

$$L = \sum_{i=1}^{l} \alpha_i - \frac{1}{2} \sum_{i,j} \alpha_i \alpha_j y_i y_j (\mathrm{x}_i \cdot \mathrm{x}_j)$$

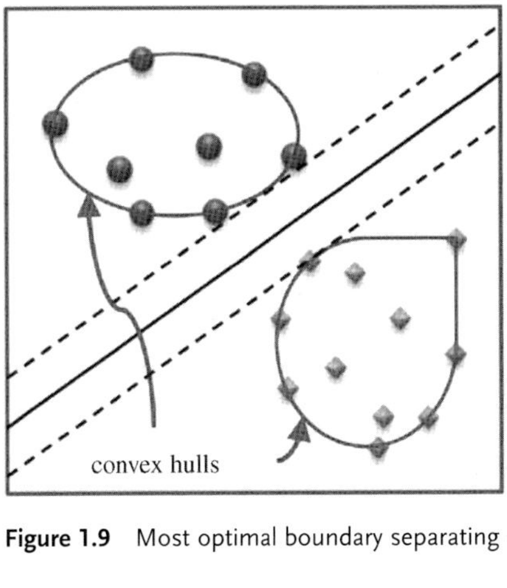


**Figure 1.9** Most optimal boundary separating closest points in the convex hulls.

where the parameters $\alpha_i$ are called the Lagrange multipliers and it is obtained by maximizing the expression. Subsequently, feasible solutions for $w$ and $b$ can be obtained once $\alpha_i$ are found. In nonlinearly separable data, SVM uses a kernel function, $k(\mathrm{x}_i, \mathrm{x}_j)$, to map the vectors into a higher dimensional feature space to make them linearly separable. Classification of an unknown compound, $\mathrm{x}_j$, is performed using

$$\hat{y} = \mathrm{sign}\left(\sum_{i=1}^{l} \alpha_i^0\, y_i\, k(\mathrm{x}_i, \mathrm{x}_j) + b\right)$$

where $l$ is the number of support vectors. The unknown compound is classified as a positive or negative compound if $\hat{y} > 0$ or $\hat{y} < 0$, respectively. $\alpha_i^0$ and $b$ are estimated by maximizing the Lagrangian expression

$$\sum_{i=1}^{l} \alpha_i - \frac{1}{2}\sum_{i=1}^{l}\sum_{j=1}^{l} \alpha_i\, \alpha_j\, y_i\, y_j\, k(\mathrm{x}_i, \mathrm{x}_j)$$

under the following conditions:

$$0 \leq \alpha_i \leq C$$
$$\sum_{i=1}^{l} \alpha_i\, y_i = 0$$

where $C$ is a penalty for training errors. Some common kernel function, $k(\mathrm{x}_i, \mathrm{x}_j)$, that may be used are

Polynomial kernel: $k(\mathrm{x}_i, \mathrm{x}_j) = (\mathrm{x}_i \cdot \mathrm{x}_j)^d$

Gaussian radial basis function(rbf): $k(\mathrm{x}_i, \mathrm{x}_j) = \exp\left(-\frac{\|\mathrm{x}_i - \mathrm{x}_j\|^2}{2\sigma^2}\right)$

We have used SVM in many of our QSAR models and it has been shown to perform well on many problems and is robust even when there is redundant and overlapping data [14–16]. Another advantage of SVM is that it is relatively simple to use as there are only a few user-defined parameters. For example, if the Gaussian rbf kernel is selected, the user will only need to fine-tune the parameters for $C$ and $\sigma$. Furthermore, the final results of SVM are reproducible and stable, unlike those of methods like neural networks, which may change from run to run because of the random initialization of the weights [133].

## 1.3 Software for QSAR Development

There are many commercial or free software available for QSAR development. These include specialized software for drawing chemical structures, interconverting chemical file formats, generating 3D structures, calculating chemical descriptors, developing QSAR models, and general-purpose software that have all the necessary components for QSAR development. A good website for QSAR resources is the Cheminformatics and QSAR Society website (http://www.qsar.org/). There are lists of software, data sets, and resources pertaining to QSAR in the website.

### 1.3.1
### Structure Drawing or File Conversion

**ChemDraw** (http://www.cambridgesoft.com/software/ChemDraw/)

ChemDraw is a commercial software for chemical structure drawing and editing. It may be packaged with other programs such as ChemDraw ActiveX/Plugin Viewer, Chem3D, ChemBioFinder, and ChemNMR which enhance the functionality of the program. Besides drawing and editing of chemical structures, the program offers integration of the drawn structure into Microsoft Office documents, conversion of structure from name (or name from structure), 13C and 1H NMR prediction, query of online databases, and many other features [141].

**ACD/ChemSketch** (http://www.acdlabs.com/resources/freeware/chemsketch/)

ACD/ChemSketch is a software for drawing of chemical structures that comes with other functionalities such as calculation of molecular properties, 2D and 3D structure cleaning, structure naming, and prediction of log*P*. The software is available in two versions: the commercial and freeware version. The freeware version does not include ACD/Dictionary, technical support, ACD/Lab extension for ChemDraw, and the function to search files by structure [142].

**Open Babel** (http://openbabel.org/)

Conversion of files at different stages of QSAR development may be necessary to satisfy the input requirements of various software. The file conversion can be easily done by using software like Open Babel. Open Babel is an open-source program that enables users to search, convert files, analyze or store data from molecular modeling projects [143]. Open Babel can convert over 90 chemical file formats, and it also has compounds preprocessing functionality like "adding hydrogen bond," "convert dative bonds," and "generate 3D coordinates."

### 1.3.2
### 3D Structure Generation

**CORINA** (http://www.molecular-networks.com/products/corina)

CORINA is one of the commercial software offered by Molecular Networks. It is used for generating three-dimensional structure of small- and medium-sized compounds, necessary as a preprocessing step prior to calculation of 3D molecular descriptors or structure-based docking studies. CORINA can be used as a component in Accelrys Pipeline Pilot or on its own through a Java-based graphical user interface or command line interface which supports batch processing [144].

**Concord** (http://tripos.com/index.php?family = modules,SimplePage,,,&page = sybyl_concord)

Concord is available as one of SYBYL applications. It is a commercial software that converts 2D inputs into 3D structures rapidly. The main benefits of Concord includes the variety of built-in geometry optimization options and its capability of handling inputs and outputs of common industry-standard formats [145].

**Frog** (http://bioserv.rpbs.jussieu.fr/Help/Frog-Help.html)

Frog is an online tool for 3D conformation generation from 1D or 2D information using Merck molecular force field [146]. It is accessible as one of the web services in the Mobyle@RPBS (http://mobyle.rpbs.univ-paris-diderot.fr/cgi-bin/portal.py) website. Frog accepts compound structures in the form of SMILES or SDF restricted to 5000 compounds per submission. A newer version, Frog 2, is able to accept 3D information as input to generate multiconformations. Frog is able to process structures with common atoms only and ions in the input file must be removed first [147].

**smi23d** (http://www.chembiogrid.org/projects/proj_ws_all.html)

smi23d is an open-source program that can be downloaded and compiled for use in Windows or Linux [148, 149]. The program generates 3D structures from SMILES string. It is also accessible via a web service called REST, hosted by the Indiana University. To use the service, one would simply need to append the SMILES string at the end of this URL: "http://cheminfov.informatics.indiana.edu/rest/thread/d3.py/SMILES/" for conversion into 3D structure.

### 1.3.3 Descriptor Calculation

**ADRIANA.Code** (http://www.molecular-networks.com/products/adrianacode)

ADRIANA.Code is one of the commercial software offered by Molecular Networks for computing molecular descriptors. Similar to CORINA, ADRIANA.Code can be used as a component in Accelrys Pipeline Pilot or on its own through a graphical user interface or command line interface. The descriptors calculated include physicochemical property descriptors, shape- and size-related descriptors, autocorrelation of 2D interatomic distance distributions, autocorrelation or radial distribution functions of 3D interatomic distance distributions weighted and autocorrelation of distances between surface points [150].

**Dragon** (http://www.talete.mi.it/products/dragon_description.htm)

Dragon is a commercial software for the computation of molecular descriptors [151]. It also has a free-for-use web-based version called the E-Dragon (http://www.vcclab.org/lab/edragon/) which uses Dragon version 5.4 [152]. It is to note that the features of E-Dragon is more restricted compared to the commercial version as every job submission can only handle up to a maximum of 149 compounds with a maximum of 150 atoms per compound. Currently, Dragon version 5.5 can compute 3224 molecular descriptors which are divided into 22 blocks. These blocks include constitutional or topological descriptors, walk and path counts, connectivity or information indices, 2D autocorrelations, BCUT descriptors, topological charges indices, 3D-MoRSE descriptors, WHIM descriptors, GETAWAY descriptors, functional group counts, 2D frequency fingerprints and so on. Dragon can work in both Windows and Linux, and it also has simple functions for conducting preliminary graphical and statistical analysis of descriptors, for example, histograms, Pareto plots, and 2D and 3D scatter plots.

**Molconn-Z** (http://www.edusoft-lc.com/molconn/)

Molconn-Z is a commercial software for molecular descriptor calculation that works on multiple platforms, for example, Windows, Mac OS X, and Linux. It calculates molecular connectivity chi indices, kappa shape indices, electrotopological state indices, topological indices, counts of subgraphs, and vertex eccentricities [153].

**PaDEL-Descriptor** (http://padel.nus.edu.sg/software/padeldescriptor/)

We released the first version of PaDEL-Descriptor in 2008 and have recently updated it to version 2.0. PaDEL-Descriptor is an open-source Java-based software developed using the Chemistry Development Kit for the calculation of molecular descriptors and fingerprints. Currently, it can calculate 797 descriptors and 10 types of fingerprints which includes 1D, 2D, and 3D descriptors, for example, atom-type electrotopological state descriptors, McGowan volume, molecular linear free energy relation descriptors, ring counts, WHIM, Petitjean shape index, count of chemical substructures identified by Laggner, and binary fingerprints and count of chemical substructures identified by Klekota and Roth [154]. PaDEL-Descriptor works as a standalone program and also available as a Java Web Start version. It has a graphical user interface, a command line interface, and can also be used as an extension to RapidMiner. The program also has some compound preprocessing capabilities like "remove salt," "add hydrogen," and "convert to 3D" [155].

### 1.3.4 Modeling

**KNIME** (http://www.knime.org/)

Konstanz Information Miner (KNIME) is an open-source platform with pipelining ability for data integration, processing, analysis, and exploration [156]. Modules for data preprocessing, modeling, visualization, and others, are organized as "nodes" which allows the users to create data flow by connecting these nodes. There are more than 100 processing nodes in the KNIME base version. It also integrates modules from WEKA and has a plugin that allow execution of R scripts. In addition, it has chemistry nodes based on the Chemistry Development Kit (CDK) which enables the calculation of molecular properties and fingerprints. User customized nodes can be implemented in KNIME easily. This enables organizations such as Tripos [157], ChemAxon [158], and Schrödinger [159] to offer their commercial tools as KNIME extensions (nodes).

**RapidMiner** (http://rapid-i.com/)

RapidMiner is an open-source system with a large collection of algorithms for data analysis and model development. There are more than 500 operators for data processing, model development, evaluation, and visualization, and it also integrates another modeling library, WEKA [160]. It has the "Optimize Parameters" operator which allows semiautomation of parameters searching. The software is able to run on major platforms like Windows, Linux, and Mac OS X. Users are able to visualize the modeling workflow in the form of an intuitive process interface and users also have the option of adding their own algorithms in the form of extensions, written in Java, into RapidMiner easily.

**WEKA** (http://www.cs.waikato.ac.nz/ml/weka/)

WEKA has a rich compilation of modeling methods and tools for data preprocessing, classification, regression, clustering, and visualization, which are organized into different sections in the WEKA Explorer [161]. It is an open-source software that can run on major platforms like Windows, Mac OS X, and Linux. The WEKA Explorer is used for most modeling tasks. Alternatively, the WEKA KnowledgeFlow which is the graphical front of the software can be used to allow the user to see the flow of the data processing or modeling. WEKA is a flexible software as new analysis methods can be added easily with users own implementation of algorithms or downloads from the "WEKA-related Projects" in the WEKA website.

**Orange** (http://www.ailab.si/orange/)

Orange is a free program that offers tools for some simple data preparation, evaluation, visualization, classification, regression, and clustering. Each of these functions are available as widgets and the user will need to connect these widgets in a flow for data analyses, similar to WEKA KnowledgeFlow and KNIME. It has the "Select Data" widgets which allow easy manipulation and filtering of data. Although Orange does not have widgets for automated parameter optimization and has fewer operators compared to programs like WEKA or RapidMiner, the variety is suffice for many modeling tasks [162].

**TANAGRA** (http://eric.univ-lyon2.fr/~ricco/tanagra/en/tanagra.html)

TANAGRA is an open-source software containing tools for data analysis, statistics, modeling, and database exploration [163]. Some of these tools are applications for feature selection, feature construction, for example, principal component analysis, descriptive statistics, *t*-test, various clustering algorithms, and modeling methods like multiple linear regression, regression trees, SVM, random forest, naive Bayes classifier, and decision trees. The website of TANAGRA contains a collection of comprehensive tutorials describing the usage of these applications.

**MATLAB** (http://www.mathworks.com/products/matlab/)

MATLAB is a commercial software that provides an interactive system for algorithm development, data visualization, data analysis, and numeric computation with wide application in image processing, financial analysis, computational biology, and so on. Data can be analyzed easily with ready-to-use functions, but users are also allowed to customize some of these tools or add their own algorithms for use. It also has functions to integrate MATLAB-based algorithms with external applications and languages such as Microsoft Excel, Java, and C++. This enables developed QSAR models to be easily distributed as stand-alone programs or software modules.

**R** (http://www.r-project.org/)

R is a free software environment for graphical and statistical analysis that can run on Windows, Linux, and Mac OS X. It has a variety of statistical tools like linear or nonlinear modeling, classical statistical tests, time-series analysis, classification, and clustering. R is extensible as user can add more utilities by creating their own functions or via add-on "packages" that comes with the distribution of R or downloadable through the CRAN websites. Some packages of interests are *kernlab* which provides algorithms for kernel-based machine learning methods such as kernel PCA; *klar* which provides various visualization and classification functions like support vector machine; *RWeka* which is the R interface to WEKA; *nnet* and *tree* for their application in single-layer neural network and classification and regression trees modeling. More information on packages related to modeling is available in the R website under Task View for Machine Learning and Statistical Learning [164].

### 1.3.5 General purpose

**SYBYL** (http://www.tripos.com/) SYBYL, as a basic program, provides a wide range of molecular modeling tools which includes tools in structure building, optimization, and comparison (and visualization) of structures and related data. It also comes with a broad selection of force fields that can be used in compound analysis [51]. One of the widely used QSAR techniques, Comparative Molecular Field Analysis (CoMFA), can be found as an integrated module in SYBYL. Besides ligand-based design, users may choose to integrate other SYBYL applications for receptor-based design, structural biology, library design or cheminformatics.

**Discovery Studio** (http://accelrys.com/products/discovery-studio/)Accelrys Discovery Studio contains a suite of applications for optimizing the drug discovery process. It has applications for properties examination, leads identification and candidate's optimization. For example, other than structure-based design, simulation, QSAR and library design tools, it also contains the predictive ADME and toxicology application which help in identifying undesirable compounds early in the discovery process. It also has the Protein Modeling and Sequence Analysis, and Biopolymer Building tools to allow better understanding of the biological function of the targets. Automation of routine tasks can be done through the use of Pipeline Pilot or scripting in Perl. Discovery Studio is also integrated with web resources like the protein data bank (PDB) and PubChem.

**MOE: Molecular Operating Environment** (http://www.chemcomp.com/software.htm) MOE provides drug discovery software suites for structured-based design, pharmacophore discovery, protein and antibody modeling, molecular modeling and simulations, cheminformatics and (HTS) QSAR, medicinal chemistry applications, and methods developments and deployment. The Cheminformatics and (HTS) QSAR suite comes with pipeline tools to process SD files and calculation of over 600 molecular descriptors, model building, similarity searching and combinatorial library design. Customizable functions can be added through their built-in scripting language, Scientific Vector Language (SVL), that comes with over 1000 specific functions for chemical structure manipulation and analyses [165]. MOE is operating systems independent and is also adapted into MOE/batch and MOE/web that enables usage in batch mode (nongraphical interface) and web interface.

**CODESSA™** (http://www.codessa-pro.com/index.htm)CODESSA is a commercial software that combines various mathematical and computational systems to build QSAR models. CODESSA is capable of calculating a range of molecular descriptors based on the chemical compounds' three-dimensional structure and/or quantum-chemical wave function. Some of the molecular descriptors calculated are constitutional, topological, geometrical, electrostatic, charged surface area, quantum-chemical, molecular-orbital related, and thermodynamic descriptors. CODESSA is also used for developing models, and cluster analysis of molecular descriptors (or data). It also has tools for model interpretation and compound property prediction from its chemical structure [166]. The use of CODESSA can be integrated with AMPAC™ where the quantum mechanical information produced by AMPAC can be used to calculate molecular descriptor [167].

## 1.4 Conclusion

Table 1.1 shows the number of published studies from two abstract and citation databases for the period of June 2005 to June 2010. These studies were found by PubMed and Scopus when the different data analysis methods were used as keywords together with "qsar" in the searches limited to title, abstract, and keywords. Take note that the list is nonexhaustive and the results from the two databases are not exclusive from each other.

From Table 1.1, it is clear that classical (simple) modeling methods like MLR and PLS remained as popular choices for QSAR studies. SVM is an increasingly popular method and this may be attributed to its ease of use and generally superior performance in comparative studies [168–171]. However, its usage is still lower than those of MLR and PLS. This may be due to the fact that MLR or PLS models are simple to interpret. There are also many new modeling methods introduced in recent years and they may potentially be used for QSAR studies. These new methods include local lazy regression [35], retro-regression [36], rotated general regression neural network [71], kernel partial least squares and unfolding partial least squares [44].

It is important to note that there is no particular method that is ideal for all problems [91, 169, 172]. The choice of an algorithm should be based on the nature of

**Table 1.1** Number of published studies in PubMed and Scopus with different algorithms as keywords.

| Data analysis methods | No. of published studies | | |
|---|---|---|---|
| | PubMed | Scopus | Rank |
| Gaussian process | 7 | 16 | 8 |
| *k*-Nearest neighbor | 38 | 56 | 5 |
| Linear discriminant analysis | 51 | 77 | 4 |
| Logistic regression | 9 | 15 | 9 |
| Multiple linear regression | 218 | 447 | 1 |
| Neural network: feedforward backpropagation | 3 | 10 | 11 |
| Neural network: general regression | 2 | 13 | 10 |
| Neural network: probabilistic | 4 | 6 | 12 |
| Partial least squares | 241 | 409 | 2 |
| Support vector machines | 69 | 125 | 3 |
| Trees: decision tree | 20 | 37 | 6 |
| Trees: random forest | 14 | 26 | 7 |

the data, and also whether the goal is to build a predictive or diagnostic model. Classification methods are suitable for the prediction of compound class (active versus nonactive) when the availability of activity information is limited. On the other hand, regression methods are suitable for quantitative (activity values e.g., $IC_{50}$ or $EC_{50}$) prediction when sufficient activity information of compounds possessing the same property is available.

Some potential obstacles in model development include insufficient data size (e.g., lack of report of weak or nonactive compounds, or proprietary compounds), lack of curation of data, lack of standardized procedures that facilitate sharing of data or models, toxicological modeling which is inherently a complicated problem [9, 19, 173–176]. Although these issues present challenges in model development and much research is needed to address them, our research and those of other QSAR modelers have shown that both classification and regression methods have been effective in the predictions of a diverse set of compounds and of various physicochemical, pharmacodynamic, pharmacokinetic, and toxicological properties so far [174, 177].

Despite all the successful stories in QSAR modeling, in our opinion, computational methods may not and should not completely replace conventional *in vitro* or *in vivo* testing methods, but should be further developed as an important and essential complementary tool in the drug development process. Moreover, there are still many research opportunities in the areas of data harmonization, model applicability domain, model validation, inclusion of nonstructural information in modeling, similarity or pharmacophore research, consensus modeling and many more [19, 22, 175, 178–183].

Lastly, we recommend current and future modelers to subscribe to the "Organization for Economic Cooperation and Development (OECD) Principles for the Validation of QSAR in developing models for scientific or regulatory use. All QSAR models developed should be presented with the following information [184]:

- a defined endpoint;
- an unambiguous algorithm;
- a defined domain of applicability;
- appropriate measures of goodness-of-fit, robustness and predictivity; and
- a mechanistic interpretation if possible.

In essence, the guidelines demonstrate the importance of a transparent validation process and reliable QSAR for its acceptance in regulatory context. The readers are to refer to the guideline document, available on the OECD website (http://www.oecd.org), for a comprehensive explanation and interpretation of the guidelines.

## References


1 Snodin, D.J. (2002) *Toxicology Letters*, **127**, 161–168.

2 Kraljevic, S., Stambrook, P.J., and Pavelic, K. (2004) *EMBO Reports*, **5**, 837–842.

3 Adams, C.P. and Brantner, V.V. (2006) *Health Aff (Millwood)*, **25**, 420–428.

4 Kola, I. and Landis, J. (2004) *Nature Reviews. Drug Discovery*, **3**, 711–716.

5 Critical Path Opportunities Reports > Challenges and Opportunities Report – March (2004) http://www.fda.gov/ScienceResearch/SpecialTopics/CriticalPathInitiative/CriticalPathOpportunitiesReports/ucm077262.htm#f15.

6 Yap, C.W., Xue, Y., Li, Z.R., and Chen, Y.Z. (2006) *Current Topics in Medicinal Chemistry*, **6**, 1593–1607.

7 Guido, R.V., Oliva, G., and Andricopulo, A.D. (2008) *Current Medicinal Chemistry*, **15**, 37–46.

8 Schwaighofer, A., Schroeter, T., Mika, S., and Blanchard, G. (2009) *Combinatorial Chemistry and High Throughput Screening*, **12**, 453–468.

9 Valerio, L.G. Jr. (2009) *Toxicology and Applied Pharmacology*, **241**, 356–370.

10 Worth, A.P., Bassan, A., De Bruijn, J., Gallegos Saliner, A., Netzeva, T., Patlewicz, G., Pavan, M., Tsakovska, I., and Eisenreich, S. (2007) *SAR and QSAR in Environmental Research*, **18**, 111–125.

11 Lilienblum, W., Dekant, W., Foth, H., Gebel, T., Hengstler, J., Kahl, R., Kramer, P.J., Schweinfurth, H., and Wollin, K.M. (2008) *Archives of Toxicology*, **82**, 211–236.

12 Yap, C.W. and Chen, Y.Z. (2005) *Journal of Chemical Information and Modeling*, **45**, 982–992.

13 Yap, C.W. and Chen, Y.Z. (2005) *Journal of Pharmaceutical Sciences*, **94**, 153–168.

14 Xue, Y., Li, H., Ung, C.Y., Yap, C.W., and Chen, Y.Z. (2006) *Chemical Research in Toxicology*, **19**, 1030–1039.

15 Liew, C.Y., Ma, X.H., Liu, X., and Yap, C.W. (2009) *Journal of Chemical Information and Modeling*, **49**, 877–885.

16 Liew, C.Y., Ma, X.H., and Yap, C.W. (2010) *Journal of Computer-Aided Molecular Design*, **24**, 131–141.

17 Todeschini, R. and Consonni, V. (2000) *Handbook of Molecular Descriptors*, vol. 11, Wiley-VCH, Weinheim.

18 Guyon, I. and Elisseeff, A. (2003) *The Journal of Machine Learning Research*, **3**, 1157–1182.

19 Tropsha, A., Gramatica, P., and Gombar, Vijay K. (2003) *The QSAR & Combinatorial Science*, **22**, 69–77.

20 Schultz, T.W., Netzeva, T.I., and Cronin, M.T. (2003) *SAR and QSAR in Environmental Research*, **14**, 59–81.

21 Scior, T., Medina-Franco, J.L., Do, Q.T., Martinez-Mayorga, K., Yunes Rojas, J.A.,

and Bernard, P. (2009) *Current Medicinal Chemistry*, **16**, 4297–4313.

22 Sprous, D.G., Palmer, R.K., Swanson, J.T., and Lawless, M. (2010) *Current Topics in Medicinal Chemistry*, **10**, 619–637.

23 Fernández, M. and Caballero, J. (2007) *Journal of Molecular Modeling*, **13**, 465–476.

24 Goodarzi, M., Freitas, M.P., and Ghasemi, N. (2010) *European Journal of Medicinal Chemistry*, **45**, 3911–3915.

25 Pourbasheer, E., Riahi, S., Ganjali, M.R., and Norouzi, P. (2010) *Journal of Enzyme Inhibition and Medicinal Chemistry*, vol 25, 844–853.

26 Papa, E., Kovarich, S., and Gramatica, P. (2010) *Chemical Research in Toxicology*, **23**, 946–954.

27 Luilo, G.B. and Cabaniss, S.E. (2010) *Environmental Science & Technology*, **44**, 2503–2508.

28 Topliss, J.G. and Edwards, R.P. (1979) *Journal of Medicinal Chemistry*, **22**, 1238–1244.

29 Benigni, R., Giuliani, A., Franke, R., and Gruska, A. (2000) *Chemical Reviews*, **100**, 3697–3714.

30 Cash, G.G. (2001) *Mutation Research*, **491**, 31–37.

31 Abraham, M.H., Hassanisadi, M., Jalali-Heravi, M., Ghafourian, T., Cain, W.S., and Cometto-Muniz, J.E. (2003) *Toxicological Sciences*, **76**, 384–391.

32 Patlewicz, G., Basketter, D.A., Smith Pease, C.K., Wilson, K., Wright, Z.M., Roberts, D.W., Bernard, G., Gimenez Arnau, E., and Lepoittevin, J.-P. (2004) *Contact Dermatitis*, **50**, 91–97.

33 Papa, E., Villa, F., and Gramatica, P. (2005) *Journal of Chemical Information and Modeling*, **45**, 1256–1266.

34 Hiromasa, K., Masamoto, A., and Kimito, F. (2008) *Journal of Chemical Information and Modeling*, **48**, 534–541.

35 Guha, R., Dutta, D., Jurs, P.C., and Chen, T. (2006) *Journal of Chemical Information and Modeling*, **46**, 1836–1847.

36 Randić, M. (2001) *Journal of Chemical Information and Computer Sciences*, **41**, 602–606.

37 Dutta, D., Guha, R., Wild, D., and Chen, T. (2007) *Journal of Chemical Information and Modeling*, **47**, 989–997.

38 Farkas, O. and Heberger, K. (2005) *Journal of Chemical Information and Modeling*, **45**, 339–346.

39 Rebehmed, J., Barbault, F., Teixeira, C., and Maurel, F. (2008) *Journal of Computer-Aided Molecular Design*, **22**, 831–841.

40 Riahi, S., Ganjali, M., Pourbasheer, E., and Norouzi, P. (2008) *Chromatographia*, **67**, 917–922.

41 Gharagheizi, F. (2008) *The QSAR and Combinatorial Science*, **27**, 165–170.

42 Yuan, Y., Zhang, R., Hu, R., and Ruan, X. (2009) *European Journal of Medicinal Chemistry*, **44**, 25–34.

43 An Introduction to Partial Least Squares Regression. http://www.ats.ucla.edu/stat/sas/library/pls.pdf.

44 Hasegawa, K. and Funatsu, K. (2010) *Current Computer-Aided Drug Design*, **6**, 103–127.

45 Wold, S., Sjöström, M., and Eriksson, L. (2001) *Chemometrics and Intelligent Laboratory Systems*, **58**, 109–130.

46 Zhu, M., Ge, F., Zhu, R., Wang, X., and Zheng, X. (2010) *Chemosphere*, **80**, 46–52.

47 Paliwal, S.K., Pal, M., and Siddiqui, A.A. (2010) *Medicinal Chemistry Research*, **19**, 475–489.

48 Roy, P.P. and Roy, K. (2009) *Journal of Enzyme Inhibition and Medicinal Chemistry*, **25**, 354–369.

49 Kar, S. and Roy, K. (2010) *Journal of Hazardous Materials*, **177**, 344–351.

50 Deeb, O. and Goodarzi, M. (2010) *Chemical Biology and Drug Design*, **75**, 506–514.

51 Tripos:: SYBYL-X. http://www.tripos.com/.

52 Ritchie, M.D., White, B.C., Parker, J.S., Hahn, L.W., and Moore, J.H. (2003) *BMC Bioinformatics*, **4**, 28.

53 Hornik, K., Stinchcombe, M., and White, H. (1989) *Neural Networks*, **2**, 359–366.

54 Rumelhart, D.E., Hinton, G.E., and Williams, R.J. (1986) *Nature*, **323**, 533–536.

55 Mundie, D.B. and Massengill, L.W. (1991) *IEEE Transactions on Neural Networks*, **2**, 168–170.

56 Krogh, A. and Hertz, J.A. (1992) *Journal of Physics A –Mathematical and General*, **25**, 1135–1147.

57 Reed, R. (1993) *IEEE Transactions on Neural Networks*, **4**, 740–747.
58 Prechelt, L. (1998) *Neural Networks: Tricks of the Trade*, Springer, Berlin/Heidelberg, pp. 553.
59 Kaiser, K.L.E., Niculescu, S.P., and Schüürmann, G. (1997) *Water Quality Research Journal of Canada*, **32**, 637–657.
60 Cui, X., Wang, Z., Zhang, Z., Yuan, X., and Harrington, P.D.B. (2008) Proceedings – 4th International Conference on Natural Computation, ICNC 2008, 2: pp. 134–138.
61 Hemmateenejad, B., Akhond, M., Miri, R., and Shamsipur, M. (2003) *Journal of Chemical Information and Computer Sciences*, **43**, 1328–1334.
62 González-Arjona, D., López-Pérez, G., and Gustavo González, A. (2002) *Talanta*, **56**, 79–90.
63 Goodarzi, M., Ortiz, E.V., Coelho, L.d.S., and Duchowicz, P.R. (2010) *Atmospheric Environment*, **44**, 3179–3186.
64 Patra, J.C. and Singh, O. (2009) *Journal of Computational Chemistry*, **30**, 2494–2508.
65 Ramírez-Galicia, G., Garduño-Juárez, R., Hemmateenejad, B., Deeb, O., Deciga-Campos, M., and Moctezuma-Eugenio, J.C. (2007) *Chemical Biology and Drug Design*, **70**, 53–64.
66 Nohair, M., Mallouk, N., Benmarzouk, M., and Mohssine, E.M. (2009) *Chemical Product and Process Modeling*, **4**, 1–22.
67 Tan, P.-N., Steinbach, M., and Kumar, V. (2005) *Introduction to Data Mining*, Addison-Wesley, New York.
68 Baskin, I.I., Ait, A.O., Halberstam, N.M., Palyulin, V.A., and Zefirov, N.S. (2002) *SAR and QSAR in Environmental Research*, **13**, 35–41.
69 Guha, R. and Jurs, P.C. (2005) *Journal of Chemical Information and Modeling*, **45**, 800–806.
70 Guha, R. (2008) *Journal of Computer-Aided Molecular Design*, **22**, 857–871.
71 Gholamrezaei, M. and Ghorbanian, K. (2007) International Joint Conference on Neural Networks, 2007 IJCNN 2007, pp. 1959–1964.
72 Specht, D.F. (1991) *IEEE Transactions on Neural Networks*, **2**, 568–576.
73 Schløler, H. and Hartmann, U. (1992) *Neural Networks*, **5**, 903–909.
74 Specht, D.F. (1993) *Neural Networks*, **6**, 1033–1034.
75 Nadaraya, E.A. (1964) *Theory of Probability and its Applications*, **9**, 141–142.
76 Watson, G.S. (1964) *Sankhyā: The Indian Journal of Statistics. Series A*, **26**, 359–372.
77 Arkan, E., Shahlaei, M., Pourhossein, A., Fakhri, K., and Fassihi, A. (2010) *European Journal of Medicinal Chemistry*, **45**, 3394–3406.
78 Hu, R., Doucet, J.P., Delamar, M., and Zhang, R. (2009) *European Journal of Medicinal Chemistry*, **44**, 2158–2171.
79 Yap, C.W., Li, Z.R., and Chen, Y.Z. (2006) *Journal of Molecular Graphics and Modelling*, **24**, 383–395.
80 Ji, L., Wang, X., Luo, S., Qin, L., Yang, X., Liu, S., and Wang, L. (2008) *Science in China. Series B, Chemistry, Life Sciences & Earth Sciences*, **51**, 677–683.
81 Agatonovic-Kustrin, S. and Turner, J.V. (2006) *Letters in Drug Design and Discovery*, **3**, 436–442.
82 Mosier, P.D. and Jurs, P.C. (2002) *Journal of Chemical Information and Computer Sciences*, **42**, 1460–1470.
83 Mager, P.P. and Reinhardt, R. (2002) *Molecular Simulation*, **28**, 287–294.
84 Parzen, E. (1962) *The Annals of Mathematical Statistics*, **33**, 1065–1076.
85 Cacoullos, T. (1966) *AnISM*, **18**, 179–189.
86 Masters, T. (1995) *Advanced Algorithms for Neural Networks: A C++ Sourcebook*, John Wiley & Sons, Inc, New York.
87 Rasmussen, C.E. and Williams, C.K.I. (2006) *Gaussian Processes for Machine Learning*, MIT Press, Cambridge, MA
88 Seeger, M. (2004) *International Journal of Neural Systems*, **14**, 69–106.
89 Schwaighofer, A., Schroeter, T., Mika, S., Laub, J., ter Laak, A., Sülzle, D., Ganzer, U., Heinrich, N., and Müller, K.-R. (2007) *Journal of Chemical Information and Modeling*, **47**, 407–424.
90 Obrezanova, O., Gola, J.M.R., Champness, E.J., and Segall, M.D. (2008) *Journal of Computer-Aided Molecular Design*, **22**, 431–440.
91 Obrezanova, O. and Segall, M.D. (2010) *Journal of Chemical Information and Modeling*, **50**, 1053–1061.
92 Hansen, K., Rathke, F., Schroeter, T., Rast, G., Fox, T., Kriegl, J.M., and Mika, S.

(2009) *Journal of Chemical Information and Modeling*, **49**, 1486–1496.

93 Rasmussen, C.E. and Williams, C.K.I. (2006) *Gaussian Processes for Machine Learning*, MIT Press, Cambridge, MA.

94 Oliveira, K.M.G. and Takahata, Y. (2008) *The QSAR & Combinatorial Science*, **27**, 1020–1027.

95 Fedorowicz, A., Zheng, L., Singh, H., and Demchuk, E. (2004) *International Journal of Molecular Sciences*, **5**, 56–66.

96 Li, Y., Pan, D., Liu, J., Kern, P.S., Gerberick, G.F., Hopfinger, A.J., and Tseng, Y.J. (2007) *Toxicological Sciences*, **99**, 532–544.

97 Liu, J., Kern, P.S., Gerberick, G.F., Santos-Filho, O.A., Esposito, E.X., Hopfinger, A.J., and Tseng, Y.J. (2008) *Journal of Computer-Aided Molecular Design*, **22**, 345–366.

98 Cronin, M.T.D., Aptula, A.O., Dearden, J.C., Duffy, J.C., Netzeva, T.I., Patel, H., Rowe, P.H., Schultz, T.W., Worth, A.P., Voutzoulidis, K., and Schüürmann, G. (2002) *Journal of Chemical Information and Computer Sciences*, **42**, 869–878.

99 Lee, J.H., Landrum, P.F., Field, L.J., and Koh, C.H. (2001) *Environmental Toxicology and Chemistry/SETAC*, **20**, 2102–2113.

100 Vilar, S., Chakrabarti, M., and Costanzi, S. (2010) *Journal of Molecular Graphics and Modelling*, **28**, 899–903.

101 Pérez-Garrido, A., Helguera, A.M., Rodríguez, F.G., and Cordeiro, M.N.D.S. (2010) *Dental Materials*, **26**, 397–415.

102 Prado-Prado, F.J., García-Mera, X., and González-Díaz, H. (2010) *Bioorganic and Medicinal Chemistry*, **18**, 2225–2231.

103 Castillo-Garit, J.A., Vega, M.C., Rolon, M., Marrero-Ponce, Y., Kouznetsov, V.V., Torres, D.F.A., Gómez-Barrio, A., Bello, A.A., Montero, A., Torrens, F., and Pérez-Giménez, F. (2010) *European Journal of Pharmaceutical Sciences*, **39**, 30–36.

104 Wang, G., Li, Y., Liu, X., and Wang, Y. (2009) *The QSAR and Combinatorial Science*, **28**, 1418–1431.

105 Castillo-González, D., Cabrera-Pérez, M.A., Pérez-González, M., Helguera, A.M., and Durán-Martínez, A. (2009) *European Journal of Medicinal Chemistry*, **44**, 4826–4840.

106 Rivera-Barroto, O.M., Marrero-Ponce, Y., Meneses-Marcel, A., Escario, J.A., Barrio, A.G., Arán, V.J., Alho, M.A.M., Pereira, D.M., Nogal, J.J., Torrens, F., Ibarra-Velarde, F., Montenegro, Y.V., Huesca-Guillén, A., Rivera, N., and Vogel, C. (2009) *The QSAR and Combinatorial Science*, **28**, 9–26.

107 Saíz-Urra, L., Pérez-Castillo, Y., González, M.P., Ruiz, R.M., Cordeiro, M.N.D.S., Rodríguez-Borges, J.E., and García-Mera, X. (2009) *The QSAR and Combinatorial Science*, **28**, 98–110.

108 Hammann, F., Gutmann, H., Baumann, U., Helma, C., and Drewe, J. (2009) *Molecular Pharmacology*, **6**, 1920–1926.

109 Ivanciuc, O. (2009) *Current Proteomics*, **6**, 289–302.

110 Wang, X.Z., Perston, B., Yang, Y., Lin, T., and Darr, J.A. (2009) *Chemical Engineering Research and Design*, **87**, 1420–1429.

111 Hammann, F., Gutmann, H., Jecklin, U., Maunz, A., Helma, C., and Drewe, J. (2009) *Current Drug Metabolism*, **10**, 339–346.

112 Amit, Y. and Geman, D. (1997) *Neural Computation*, **9**, 1545–1588.

113 Ho, T.K. (1998) *ITPAM*, **20**, 832–844.

114 Breiman, L. (2001) *MLear*, **45**, 5–32.

115 Larivière, B. and Van Den Poel, D. (2005) *Expert Systems with Applications*, **29**, 472–484.

116 Machine Learning Benchmarks and Random Forest Regression. http://escholarship.org/uc/item/35x3v9t4.

117 Statnikov, A., Wang, L., and Aliferis, C.F. (2008) *BMC Bioinformatics*, **9**, 319.

118 Kubinyi, H. (1998) *Perspectives in Drug Discovery and Design*, **9–11**, 225–252.

119 Wang, X.S., Tang, H., Golbraikh, A., and Tropsha, A. (2008) *Journal of Chemical Information and Modeling*, **48**, 997–1013.

120 Pawar, V., Lokwani, D., Bhandari, S., Mitra, D., Sabde, S., Bothara, K., and Madgulkar, A. (2010) *Bioorganic and Medicinal Chemistry*, **18**, 3198–3211.

121 Bansal, S.K., Sinha, B.N., and Khosa, R.L. (2010) *Medicinal Chemistry Research*, vol 20, 549–553.

122 Jain, A. and Agrawal, R.K. (2009) 2nd International Conference on Biomedical and Pharmaceutical Engineering, ICBPE 2009, 1–4.

123 Bhandari, S., Bothara, K., Pawar, V., Lokwani, D., and Devale, T. (2009) *Internet Electronic Journal of Molecular Design*, **8**, 14–28.

124 Peterson, Y.K., Wang, X.S., Casey, P.J., and Tropsha, A. (2009) *Journal of Medicinal Chemistry*, **52**, 4210–4220.

125 Specht, D.F. (1990) *Neural Networks*, **3**, 109–118.

126 Specht, D.F. (1988) *IEEE International Conference on Neural Networks*, **1**, 525–532.

127 Specht, D.F. and Romsdahl, H. (1994) IEEE International Conference on Neural Networks, 1994 IEEE World Congress on Computational Intelligence, 2, pp. 1203–1208.

128 Niculescu, S.P., Atkinson, A., Hammond, G., and Lewis, M. (2004) *SAR and QSAR in Environmental Research*, **15**, 293–309.

129 Niculescu, S.P., Lewis, M.A., and Tigner, J. (2008) *SAR and QSAR in Environmental Research*, **19**, 735–750.

130 Vilar, S., Santana, L., and Uriarte, E. (2006) *Journal of Medicinal Chemistry*, **49**, 1118–1124.

131 Niculescu, S.P. (2003) *Journal of Molecular Structure: THEOCHEM*, **622**, 71–83.

132 Vapnik, V. (1995) *The Nature of Statistical Learning Theory*, Springer, New York, London.

133 Bennett, K.P. and Campbell, C. (2000) *SIGKDD Explorations Newsletter*, **2**, 1–13.

134 Kim, H.S. and Sohn, S.Y. (2010) *European Journal of Operational Research*, **201**, 838–846.

135 Zuluaga, M.A., Magnin, I.E., Hernández Hoyos, M., Delgado Leyton, E.J.F., Lozano, F., and Orkisz, M. (2010) *International Journal of Computer Assisted Radiology and Surgery*, vol 6, 163–174.

136 Fernandez, M., Ahmad, S., and Sarai, A. (2010) *Journal of Chemical Information and Modeling*, **50**, 1179–1188.

137 Shen, J., Cheng, F., Xu, Y., Li, W., and Tang, Y. (2010) *Journal of Chemical Information and Modeling*, **50**, 1034–1041.

138 Conforti, D. and Guido, R. (2010) *Computers and Operations Research*, **37**, 1389–1394.

139 Basu, S., Das, N., Sarkar, R., Kundu, M., Nasipuri, M., and Kumar Basu, D. (2010) *Pattern Recognition*, **43**, 3507–3521.

140 Khorrami, H. and Moavenian, M. (2010) *Expert Systems with Applications*, **37**, 5751–5757.

141 CambridgeSoft Desktop Software – ChemDraw (Windows/Mac). http://www.cambridgesoft.com/.

142 ACD/Labs.com:: Freeware:: ACD/ChemSketch. http://www.acdlabs.com/resources/freeware/chemsketch/.

143 Open Babel: About. http://openbabel.org/wiki/Open_Babel:About.

144 CORINA: Generation of 3D coordinates. http://www.molecular-networks.com/software/corina/index.html.

145 Tripos:: Concord. http://tripos.com/index.php?family=modules, SimplePage,,,&page=sybyl_concord.

146 Leite, T.B., Gomes, D., Miteva, M.A., Chomilier, J., Villoutreix, B.O., and Tuffery, P. (2007) *Nucleic Acids Research*, **35**, W568–572.

147 Frog: FRee Online druG conformation generation. http://bioserv.rpbs.jussieu.fr/Help/Frog-Help.html.

148 3D Conformer Generation – Metabolomics Fiehn Lab. http://fiehnlab.ucdavis.edu/staff/kind/ChemoInformatics/Concepts/3D-conformer.

149 Simple 3D Conformer Generation with Smi23D. http://depth-first.com/articles/2007/12/12/simple-3d-conformer-generation-with-smi23d.

150 ADRIANA.Code. http://www.molecular-networks.com/products/adrianacode.

151 Talete – Dragon. http://www.talete.mi.it/products/dragon_description.htm.

152 E-Dragon Software. http://www.vcclab.org/lab/edragon/.

153 Molconn-Z. http://www.edusoft-lc.com/molconn/.

154 Klekota, J. and Roth, F.P. (2008) *Bioinformatics (Oxford, England)*, **24**, 2518–2525.

155 PaDEL-Descriptor. http://padel.nus.edu.sg/software/padeldescriptor/index.html.

156 KNIME | Konstanz Information Miner. http://www.knime.org/.

157 Tripos:: Tripos Chemistry Extension for KNIME. http://tripos.com/index.php?

family=modules,SimplePage, TCE_Knime.

158 Integration nodes≪ ChemAxon – toolkits and desktop applications for cheminformatics. http://www.chemaxon.com/about/our-partners/integration-nodes/.

159 Schrödinger – Products Guides – KNIME http://www.schrodinger.com/products/14/8/.

160 Mierswa, I., Wurst, M., Klinkenberg, R., Scholz, M., and Euler, T. (2006) KDD '06: Proceedings of the 12th ACM SIGKDD International Conference on Knowledge Discovery and Data Mining, pp. 935–940.

161 Hall, M., Frank, E., Holmes, G., Pfahringer, B., Reutemann, P., and Witten, I.H. (2009) *SIGKDD Explorations Newsletter*, **11**, 10–18.

162 Orange – Data Mining Fruitful & Fun. http://www.ailab.si/orange/.

163 Rakotomalala, R. (2005) Actes de EGC'2005, RNTI-E-3, 2, 697–702.

164 The R Project for Statistical Computing. http://www.r-project.org/.

165 MOE – Methods Development and Deployment. http://www.chemcomp.com/software-mdd.htm.

166 Codessa. http://www.codessa-pro.com/index.htm.

167 AMPAC, Semichem Inc. http://www.semichem.com/ampac/default.php.

168 Czermiński, R., Yasri, A., and Hartsough, D. (2001) *Quantitative Structure–Activity Relationships*, **20**, 227–240.

169 Brace, C.L., Melville, J.L., Pickett, S.D., and Hirst, J.D. (2007) *Journal of Chemical Information and Modeling*, **47**, 219–227.

170 Xue, Y., Yang, X.G., Chen, D., Wang, M., and Chen, Y.Z. (2009) *Journal of Computational Chemistry*, **30**, 1202–1211.

171 Darnag, R., Mostapha Mazouz, E.L., Schmitzer, A., Villemin, D., Jarid, A., and Cherqaoui, D. (2010) *European Journal of Medicinal Chemistry*, **45**, 1590–1597.

172 Plewczynski, D., Spieser, S.A.H., and Koch, U. (2006) *Journal of Chemical Information and Modeling*, **46**, 1098–1106.

173 Benigni, R. and Giuliani, A. (2003) *Bioinformatics (Oxford, England)*, **19**, 1194–1200.

174 Yap, C.W., Xue, Y., Li, H., Li, Z.R., Ung, C.Y., Han, L.Y., Zheng, C.J., Cao, Z.W., and Chen, Y.Z. (2006) *Mini Reviews in Medicinal Chemistry*, **6**, 449–459.

175 Tropsha, A. and Golbraikh, A. (2007) *Current Pharmaceutical Design*, **13**, 3494–3504.

176 Yang, S.Y. (2010) *Drug Discovery Today*, **15**, 444–450.

177 Yap, C.W., Li, H., Ji, Z.L., and Chen, Y.Z. (2007) *Mini Reviews in Medicinal Chemistry*, **7**, 1097–1107.

178 Sedykh, A., Zhu, H., Tang, H., Zhang, L., Richard, A., Rusyn, I., and Tropsha, A. (2010) *Environmental Health Perspectives*, vol 119, 364–370.

179 Lessigiarska, I., Worth, A.P., Netzeva, T.I., Dearden, J.C., and Cronin, M.T. (2006) *Chemosphere*, **65**, 1878–1887.

180 Verma, J., Khedkar, V.M., and Coutinho, E.C. (2010) *Current Topics in Medicinal Chemistry*, **10**, 95–115.

181 Kruhlak, N.L., Contrera, J.F., Benz, R.D., and Matthews, E.J. (2007) *Advanced Drug Delivery Reviews*, **59**, 43–55.

182 Dearden, J.C. (2003) *Journal of Computer-Aided Molecular Design*, **17**, 119–127.

183 Wawer, M., Lounkine, E., Wassermann, A.M., and Bajorath, J. (2010) *Drug Discovery Today*, **15**, 630–639.

184 OECD, *Guidance Document on the Validation of (Quantitative) Structure–Activity Relationships Models*, Organisation for Economic Co-operation and Development (2007).

# 2
# Developing Best Practices for Descriptor-Based Property Prediction: Appropriate Matching of Datasets, Descriptors, Methods, and Expectations

*Michael Krein, Tao-wei Huang, Lisa Morkowchuk, Dimitris K. Agrafiotis, and Curt M. Breneman*

## 2.1
## Introduction

For well over 100 years, chemists have explored the relationship between the chemical structure and biological activity, and dreamed of predicting them as well as other measurable properties. The first description of a relationship between composition and activity [1] was based on observations of correlation between specific molecular features and observable physiochemical properties [2]. With some data tabulation, it was found that structure–activity relationships could be used to quantify chemical intuition: For a small change in the molecular structure, a corresponding small change in activity could be explained by analyzing regular changes the numerical representations of molecular structure. The power inherent in this type of relationship quickly became obvious, and increased in importance with the quick tabulation abilities of computers. The reductionist qualities of quantitative structure–activity relationships (QSARs) have resulted in both praise and condemnation for the discipline throughout its existence [3–5]. Without debating the philosophical validity of reductionist views, a more practical approach is to understand how and when QSARs are applicable to relevant problems. As discussed below, there are many choices to make when matching available data with types of chemical descriptors and machine learning methodologies (Figure 2.1). Inherent in these choices are decisions that affect the level of difficulty and computational effort needed to develop a model and to establish its domain of applicability – a crucial element for managing end-user expectations of model performance. Most models are constructed using methods that project or compress information into a simpler form, consequently representing a compromise between mode interpretability and predictive power. For any nontrivial QSAR the importance of good chemical descriptors cannot be overstated – even the most capable machine learning methodology cannot extract "signal" from descriptor variance that is not monotonically related to the endpoint of interest. This is the essential "Tao of building QSARs," where the ultimate goal is to construct chemically meaningful, validated models. Achievement of this goal relies

*Statistical Modelling of Molecular Descriptors in QSAR/QSPR*. First Edition.
Edited by M. Dehmer, K. Varmuza, and D. Bonchev

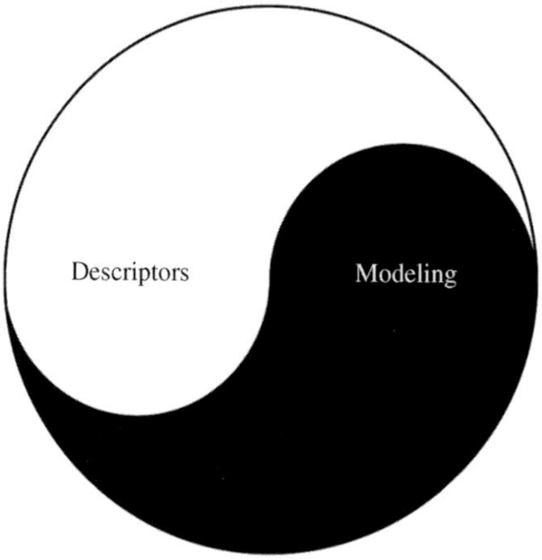


**Figure 2.1** A synergy between problem-relevant chemical descriptors and problem-appropriate validated modeling methodology maximizes model performance.

on a making a series of balanced, rational decisions in dataset evaluation, descriptor selection, and choice of the modeling method.

In order to make intelligent choices for the initial development of models, a set of "best practice" guidelines can be useful. These considerations begin with an analysis of the quality of the data, selection of appropriate descriptors, and consideration of the importance of model validation and interpretation. Years of hard-earned experience have emphasized that careful use of specific methodologies and modeling strategies are required to establish robust models. While "best practice" principles provide a foundation for building usable QSARs, there is no silver bullet: application of the best modeling practices will not atone for inadequate data, and vice versa. A combination of diligence and data is ultimately required.

### 2.1.1
### Posing the Question

The first step in building a valid QSAR is to formulate a question that can be answered within the framework of statistical learning methods. This may seem trivial, but the quality and quantity of available experimental data often dictates what is possible: *there is no "one-size-fits-all" approach* [6]. Most problems can be thought of as one of three broad types: classification, regression, or ranking. Classification models assign molecules into one of a given number of categories such as "toxic" or "nontoxic;" they are especially suited to discrete high throughput assay results. Regression modeling has a different goal: to find a function mapping the descriptors onto the (usually continuous) response. These models are more appropriate for modeling dose-response or kinetic data. Predictions based on regression models will thus be real-valued estimates of the activity. This information can also be used in ranking models, where the focus is to prioritize one structure over another for a specific application.

### 2.1.2 Validating the Models

Regardless of the underlying relationship established within a QSAR framework, statistical models built with empirical data are meaningless without proper validation: A lack of rigor in defining the capability of a model to make reliable predictions will inevitably lead to wrong conclusions about underlying mechanisms and produce faulty predictions. The failures of QSARs are well noted, and blind usage of poorly understood QSAR models is little more than numerology. At a bare minimum, QSAR models must be able to fit known data with fidelity as quantified by valid statistical measures. For instance, typical measures of goodness-of-fit for regression models are the root mean squared error (RMSE) and Pearson's product-moment correlation coefficient squared ($R^2$), as calculated by the equations

$$\mathrm{RMSE} = \sqrt{\sum\nolimits_{i=1}^{n} (y_i - \hat{y}_i)2n}$$

$$r^2 = \left( \frac{\sum_{i=1}^{n} (\hat{y}_i - \bar{\hat{y}})(y_i - \bar{y})}{\sqrt{\sum_{i=1}^{n} (\hat{y}_i - \bar{\hat{y}})^2} \sqrt{\sum_{i=1}^{n} (y_i - \bar{y})^2}} \right)^2$$

In each equation, $\hat{y}_i$ is the predicted value, $\bar{\hat{y}}$ is the average predicted value across all samples, $y_i$ is the actual value, and $\bar{y}$ is the average actual value. For classification models, measures such as the balanced error rate (BER) [7], area under the curve (AUC) of a receiver operating characteristic (ROC) curve [8], and the adjusted rand index (ARI) [9] are commonly used; for ranking, Spearman's rank correlation coefficient and the Kendall tau rank correlation coefficient are the most widespread [10]. While models must possess high $R^2$ (close to 1) and a low RMSE, these conditions alone do not validate the models or the assumptions upon which they were built. Model robustness may be tested with data internal to the original model or with external data; bootstrapping and cross-validation are the most popular internal validation methods. Several extensive reviews of both methodologies have already been published [11, 12]. It is critical to note that both methods are based on sampling the original data: Performance estimation is based on QSARs established on a fraction of the total data and assessed on another fraction of the data. Obviously, the reported estimations will depend on the composition and size of the data fractions and aggregate statistics (the performance of the models averaged over successive rounds of validation). It has been argued that internal validation alone is insufficient to guarantee model performance: *internal validation is only a performance estimate.* The real demonstrated utility of a model is its ability to make reliably accurate predictions in the real world [13]. Where external experimental data within the applicability domain of the model is available, model performance should be assessed against that data as a "gold standard" of validation.

### 2.1.3
### Interpreting the Models

Attempting to interpret validated QSARs can often provide insights into the physical basis of activity or potential mechanisms of action, and may help to rationalize the existence of observed outliers. Appropriate model interpretation may also provide a direction forward by shedding light on important aspects of the problem domain. The extent to which models can be interpreted depends upon many factors, but some level of interpretability is necessary in order to connect modeling results with chemical intuition [15]. One of the ways that QSAR models can be interpreted is by mapping the smoothness of the relationship between the molecular structure and activity in the context of their relative similarities in chemical space. As multiple mechanisms of activity can confound both model building and interpretation, attention must be paid to not overinterpret models that lack clear relationships between physically based descriptors and observed properties [3]. Another consideration is that all QSAR training datasets will only represent specific samplings of chemical space, and will therefore not be able to inform statistical models about behavior outside their domains of applicability, that is, the chemical space in which the QSAR model is validated. The applicability domain is defined as the physicochemical, structural, or biological space spanned by the known data available for model training where the model is likely to make reasonable predictions [14].

## 2.2
## Leveraging Experimental Data and Understanding their Limitations

The development of QSAR models begins and ends with experimental data, including the activities of the molecules to be modeled. A critical assumption in building QSAR models is that the measured activity is based on a single mode of action: one reaction, one pathway, one endpoint [4]. Obviously, strict adherence to standardized experimental conditions and protocols is preferred, and datasets culled from multiple sources can be used but must be treated with extreme caution. Beyond this, if one follows the "Tao of QSAR building" one strives for balance in dataset creation: balance in size, molecular classes, and activities or endpoint values.

The size of a dataset is an important attribute, in which there needs to be enough information present to form a substantive relationship between structure and function. Too little available data means that no substantive relationship can be formed; too much data and the relationship formed may be uninterpretable. Depending on the dynamic range of the response, a lower limit of 20 compounds is required to train a robust QSAR model in the best of conditions, and ideally several hundred cases should be available [15]. Common machine learning methods intended to be rigorously validated become computationally problematic when dealing with more than several thousand compounds, and more importantly, each compound in such a large dataset contributes insignificantly to the total information available in the model. As such, it is reasonable to pare down the number of compounds available to model by utilizing a diversity-sampling scheme. As the desired domain of applicability broadens, the likelihood increases that more than one mechanism or mode of action is reflected in observed responses. Diversity analysis should be performed [16] to strategically subsample the data to best represent the structural classes or motifs represented by the molecules in the dataset. Alternatively, multiple models can be built, where each model represents a specific molecular class–activity relationship [17, 18]. Techniques such as local lazy regression [19] may be used to capture these "neighborhood" effects implicitly, at the expense of model interpretability. If a global QSAR model is desired, extreme care must be taken to validate its performance over a wide range of chemical space, since most QSAR models are most effective within a limited range.

A balanced treatment of the reported molecular activities is also important; experimental errors should be documented and be a small fraction of the total response (there should be a high signal-to-noise ratio in the experiment) [12]. If the activity collected describes classes, such as active versus inactive molecules, the classes should be approximately balanced. Unbalanced datasets must be treated by appropriate methodologies so as to not bias against the minority class(es) [20]. If the activity reported is real, that is, it can take on any value from a continuous distribution and would thus be amenable to a regression analysis, an even distribution of values throughout the activity range is ideal: Large gaps between activities will bias regression models just as unbalanced classes will bias classification models. Apropos to ranking models, ranked responses should be unambiguous for the above-stated reasons [15].

## 2.3
## Descriptors: The Lexicon of QSARs

Defining what substances are and how one substance relates to another has been debated long before the field of cheminformatics came into being – it is only with the supposition that small changes in structure are proportional to changes in a measured activity that QSAR exists. The plethora of available chemical descriptors is diverse, since they can be derived from chemical constituency, topology, conformation, graph-theoretic views of structure, electronic properties including interpretations of electron density, and many other features. All of these approaches to descriptor calculation necessarily capture only a part of all molecular information, therefore limiting the ultimate accuracy of the QSARs built from them. Exhaustive reviews of descriptor classes are well covered elsewhere [21]. There are currently thousands of descriptors available through commercial and academic software packages – clearly, there are many different ways to quantify chemical structure. The relative relevance of these manifold interpretations of chemical structure is thus problem dependent, and should be chosen accordingly.

### 2.3.1
### Classical QSAR Descriptors and Uses

The physicochemical molecular descriptors behind the QSAR/QSPR concept were first popularized by Hammett, and were widely used and further modified in the late 1950s and early 1960s. Hammett described linear free energy relationships by the equation

$$\log K/K_0 = \sigma\varrho$$

where $K$ is an equilibrium constant and $K_0$ is the reference equilibrium constant (where the R-group is hydrogen). $\sigma$ is the substituent constant, which describes the effect of the R-group, and $\varrho$ is the reaction constant, characteristic of a particular reaction type [22]. Systematic alteration of R-groups allows quantitative explanation of substituent effects, such as acidity and reactivity. To account for the steric effects of *ortho* substituents, Taft introduced the sensitivity factor $\delta$ in the otherwise familiar form:

$$\log K/K_{CH_3} = \sigma^*\varrho^* + \delta E_S$$

noting that the reference reaction is with R = methyl and that $E_S$ is the steric substituent constant [23]. In an effort to relate these formulas to the drug design, Hansch added the lipophilicity parameter $\pi$ to the now-familiar QSAR equation:

$$\log(1/C) = -k\pi + k'\pi^2 + \sigma\varrho + k''$$

where C is the concentration of a drug needed to achieve a desired biological activity. Here, $k$, $k'$, $\varrho$, and $k''$ are the regression coefficients fit from experimental data [24]. To approximate molecular partitioning at *in vivo* aqueous-membrane interfaces, the lipophilicity parameter $\pi$ represents equilibrium molecular partitioning $P$ between water and 1-octanol, compared to the R = H case:

$$\pi_R = \log P_R - \log P_H$$

As QSARs became increasingly popular, the term *Hansch analysis* came to mean using the above multiple regression equation to quantitatively describe the relationship between biological activity and global or substituent parameters of a series of compounds. Successful uses today highlight the explanative power of marrying physically interpretable descriptors with linear modeling methods [25, 26].

### 2.3.2
### Experimentally Derived Descriptors

The partition coefficient $P$ above is one example of an experimentally derived descriptor introduced early in the history of QSAR. Indeed, "sensing" structure through the results of experiments is an intuitive approach to descriptor generation.

#### 2.3.2.1 **Biodescriptors**

Recent advances in the high-throughput experimental design have enriched the availability and depth of this descriptor space. With a considerable thrust from the National Institute of Health, the *Molecular Libraries Initiative* has brought assay results and methodologies into the public domain [27]. Parallel advances in computational power have enabled researchers to take advantage [28] of these multiple high-throughput assays, which are now employed to describe both on- and off-target drug effects by creating an activity profile for each compound tested. Such a profile captures information about ligand/binding site complex interactions and relationships that go beyond representations of independent ligand or binding site structure. As with any other descriptors, bioactivity profiles can be clustered for visual inspection and used for prediction of activity (Figure 2.2).

In this manner, Fliri *et al.* have created molecular property descriptors from the percent inhibition values of a broad spectrum of *in vitro* assays [29, 30]. For a set of structurally diverse compounds, a total of 92 assays were employed that represent a cross-section of the druggable proteome.

Hawkins *et al.* modeled toxicity through the combined use of classical chemical descriptors and biological descriptors created from information in 2D electrophoresis gels [31]. The resulting models described spot abundances of expressed proteome exposed to 14 halocarbons and a control. New information was found in the biodescriptors, and it was suggested that further analysis could lead to additional insight into the mechanism of hepatotoxicity.

Balasubramanian *et al.* created complex graphs from 2D gel experiments. Charge and mass were represented as an atomic adjacency matrix, decomposition of this matrix yielded novel biodescriptors that characterized the proteomics maps [32].

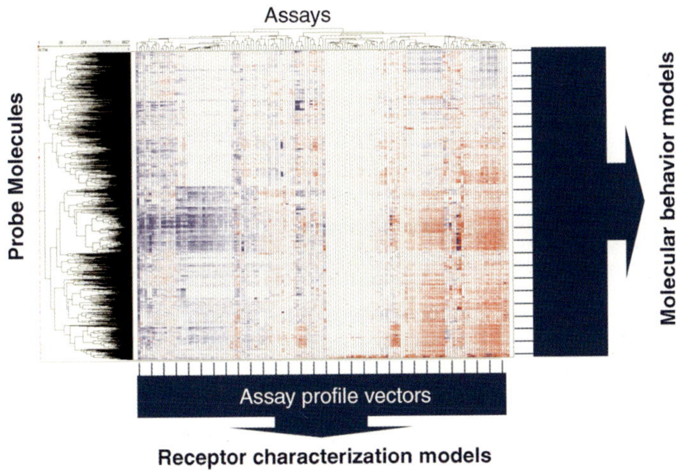


**Figure 2.2** Bioactivity profiles report a molecule's activity over a wide spectrum of biological assays. Clustered results demonstrate structure–activity relationships.

#### 2.3.2.2 Descriptors from Spectroscopy/Spectrometry and Microscopy

Experimentally derived descriptors need not be confined to the biological domain. Investigations using IR [33] and NMR spectra [34] as QSAR descriptors have yielded mixed results, but offer the potential of physical interpretation. Further, the concept of broadly applicable experimental descriptors emerges: Experimental descriptors may contain information about time and length scales not accessible to computational descriptions of the atomic-level structure. This becomes crucial when modeling multiscale systems. The relevance of spectroscopic data bridges multiple disciplines on multiple time domains: Materials, biological, and chemical systems can be observed in real time with high resolution (Figure 2.3).

In a similar fashion, images taken from the analysis of AFM, SEM, or TEM imagery can be thought of as a bridge between length scales. Relevant descriptors of structural defects, wetting, surface chemistry, morphology, and percolation are important in describing physical and electrical properties of materials systems within a QSAR framework [35]. Work has progressed in this area, but there is significant room for improvement, and linking these descriptors to particular informatics applications remains a tantalizing goal.

### 2.3.3 0D, 1D and 2D Computational Descriptors

In their earliest forms, chemical descriptors were numerical characterizations derived from experimental determinations of structure. Moving beyond experimental descriptions and subsequent parameterizations, Kier *et al.* introduced the concept of descriptors as mathematical characterizations of the molecular structure [36]. Like their experimental counterparts, mathematical descriptors have a rich history, diversity, and range of complexity. One-dimensional representations based on atom counts and linear sequences are quick to compute and thus are (historically) valuable in high-throughput processes, such as their use as filters early in drug design pipelines [37]. Slightly more complex are 2D descriptors, which encode the topology of the molecular graph. Bond counts, branching measures, and graph theoretic representations of molecular structure are common descriptors of this type. Information theoretic measures such as entropy of the molecular graph have been well

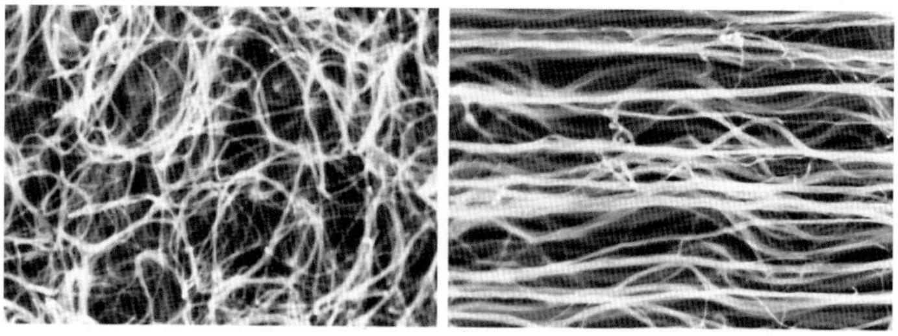

**Figure 2.3** SEM image of "wavy" versus "aligned" carbon nanotube forests – morphology significantly affects material properties.

studied, and describe molecular complexity. Extensive reviews of the development and use of such descriptors are available [21, 38–40].

Two-dimensional representations of local atomic environments are better known as fragment descriptors [41]; 2D descriptors are typically computed in sets, and are represented as arrays of binary, integer, or real numbers. In most cases, they encode the presence or absence of specific structural patterns in the target molecule, and are classified by three broad criteria: dimensionality (fixed-length or variable-length), resolution (binary or nonbinary), and uniqueness (hashed or nonhashed). Fixed-length descriptors have the same dimensionality for every molecule (e.g., the number of occurrences of a finite number of predefined fragments), while variable-length descriptors capture patterns of higher diversity that may exist in one molecule but not in another. Binary descriptors, which are often referred to as fingerprints, encode just the presence or absence of a particular feature, whereas nonbinary descriptors capture the actual counts or more complex invariants of the molecular graph expressed as real numbers. Finally, hashed fingerprints employ a special mathematical function (referred to as the hash function) that maps a large and potentially variable-sized set of patterns into a lower dimensional fixed-length array. Since this can lead to collisions (where multiple distinct patterns map onto the same position in the array), the hash functions are designed so that the distribution of hash values is as even as possible. Common fingerprint types include Daylight, Unity, MDL MACCS keys, extended-connectivity fingerprints, and molprint 2D fingerprints [42].

### 2.3.4
### 3D Descriptors and Beyond

Going beyond atomic connectivity, 3D descriptors represent geometry-dependent molecular features, that is, they can encode conformational information. The practical use of such information is easily exemplified in the century-old idea of representing "lock and key" or "induced fit" enzyme–substrate binding via pharmacophores: Molecular frameworks that carry essential features responsible for a drug's biological activity. A 3D molecular fingerprint representation of pharmacophores would thus indicate the presence or absence of groups of atoms that define these features, for example, aromatic, lipophilic, hydrogen bond donor/acceptor, and so on [43]. Such groups can be defined by topological (in 2D) or geometric (in 3D) distance, based on whole molecules or molecule fragments, such as the popular FEPOPS descriptors [44]. Other popular descriptors are WHIM [45], 3D MoRSE [46], and GETAWAY [47] descriptors.

Although in principle of high quality, the descriptors above often suffer from incomplete sampling of conformational space and fail to capture biologically relevant noncovalent interactions. By treating noncovalent interactions as steric and electrostatic forces, molecular mechanics force fields often sufficiently capture these biomolecular properties. CoMFA descriptors are generated by sampling the steric and electrostatic fields around aligned sets of ligands in their estimated bioactive conformations [48]. Alignment of the molecules is necessary for the descriptors of one molecule to have any relationship with the descriptors of another, and thus fidelity of the descriptors depend on the quality of molecular alignment and its resemblance to the actual bioactive conformation. Cramer *et al.* created topomer descriptors [49, 50] via a deterministic protocol for generating invariant 3D representations of molecular fragments from their 2D topologies – a potential solution to the alignment issues of CoMFA descriptors. Also demonstrating the relevance of a molecule's shape are ultrafast shape recognition (USR) descriptors [51]: high-throughput molecular shape moments with respect to a small set of well-defined points within a molecule. They have the advantage of providing an alignment-free, compact-shape profile of a molecule, and their successful use in virtual screening and similarity searching applications has been demonstrated [52, 53].

### 2.3.5 Local Molecular Surface Property Descriptors

Clark *et al.* have argued that descriptors based on local molecular surface properties that indirectly encode constituency are more likely to be able to "scaffold hop," that is, acknowledge that different molecular scaffolds could lead to the same global properties, thus leading to more widely applicable QSAR/QSPR models [54]. Common surface descriptors are representations of polarity and solvent accessibility. Descriptors of this type are often used in the modeling of ADMET properties (absorption, distribution, metabolism, excretion, and toxicity) crucial to *in-silico* drug discovery [37, 55]. However, representation and sensitivity of these types of descriptors with consideration of the solvation environment is an ongoing and important body of work necessary [56–58] to describe molecular partitioning in different solvents and transport through biological interfaces. Through the combination of shape and force field-derived properties Das *et al.* created property-encoded shape distributions (PESDs) descriptors [59]. These high-throughput fingerprint descriptors are completely alignment-free, and useful in classifying protein-binding sites.

### 2.3.6 Quantum Chemical Descriptors

In principle, the electron density of a system contains all molecular information [60, 61], thus, descriptor representations of the electron density have the potential to be invaluable to QSAR modeling. Electron density distribution-derived quantum chemical descriptors are applicable to a wide variety of small molecules now that QM calculations of drug-like molecules can be routinely performed on commodity hardware. However, the exponential scaling of such calculations with the number of electrons in a system precludes routine use for large biomolecules without clever tricks [62] (Figure 2.4).

Breneman *et al.* have created electron-density-derived descriptors applicable for use in high throughput methodologies and on large biomolecules [63, 64]. Among these, RECON descriptors employ an atoms-in-molecules-based approach that allows rapid computation of molecular electronic properties from precomputed

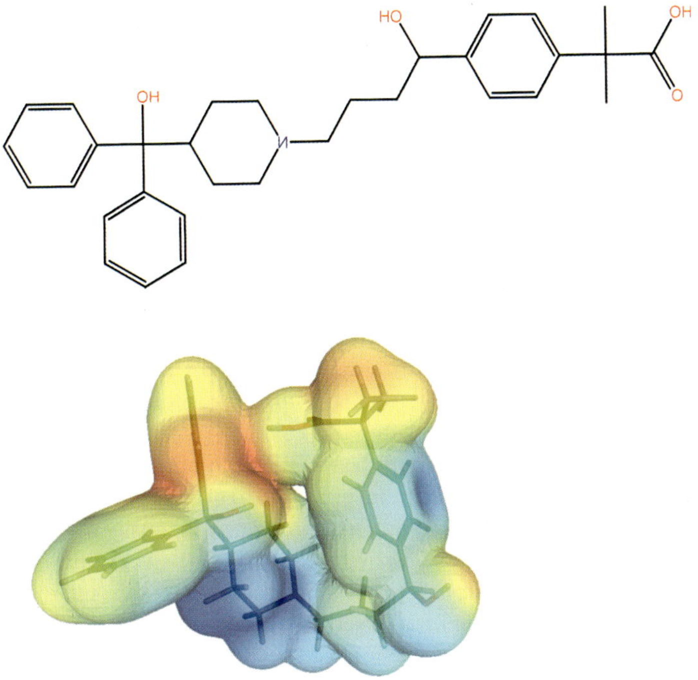


**Figure 2.4** 2D drawn representation and energy-minimized conformation of popular antihistamine Fexofenadine. Here, the electrostatic potential is encoded on the 0.002 electrons/bohr3 isodensity surface of that conformation.

*ab initio* atomic charge density fragments. These atomic fragments satisfy Bader's virial partitioning scheme and thus are approximately additive and transferable from one molecule to another. Calculation time of RECON descriptors therefore scales linearly with both molecular and database size.

Molecular shape information can be incorporated by property autocorrelations binned by the distance *Rxy* between atom pairs *x*, *y* for each property *P*, over all atoms *n*:

$$A(Rxy) = (1/n)\Sigma x, yPxPy$$

Additionally, RECON property moment descriptors can be calculated and treated as an extension of USR shape moments [65]. Such descriptors are useful in high throughput scenarios, but work on the assumption that properties are transferrable and additive, and that the molecules of interest are well represented by the library of pregenerated atomic fragments.

Trading off speed for more accurate representations of density-derived properties, property-encoded surface translator (PEST) descriptors [66] employ *ab intio* density calculations and charges. PEST descriptors, an extension of the Zauhar shape signatures method [67], rely on a ray-bouncing algorithm to sample a property-encoded 3D molecular surface; they are histogram or wavelet encoded representations of the 3D molecular surface. Complementary to RECON, PEST, and PESD descriptors are the GRIND descriptors by Cruciani *et al.* [68, 69] GRIND descriptors encode molecular interaction field distributions in the form of correlograms rather than capturing molecular surface information.

## 2.4
## Machine Learning Methods: The Grammar of QSARs

A methodology may fall naturally from the molecular descriptors used: From Hammett to Hansch, *linear free energy relationships* (LFERs) took the form of linear regression of one to several descriptors. Classical multiple linear regression (MLR) has the ability to quickly produce robust QSAR models that are easily interpretable, as each coefficient in the model indicates the importance of the respective descriptor. Such models assume that descriptors are orthogonal to each other, and for good reliability in fitting; there should be a minimum of a 5: 1 ratio of experimental data to descriptors [70]. Unless a descriptor's scaling is appropriately manually modified, only linear relationships will be captured. The first successful QSARs were the careful pairing of MLR with a small number of relatively uncorrelated physically interpretable descriptors over an analogous series of molecules. This is a perfectly legitimate usage of the method, and MLR continues to be used successfully today [18].

As with the availability of thousands of descriptors, a variety of popular machine learning methods have become available to cheminformaticians. In keeping with *the Tao of QSAR building*, there will be a balance between model predictivity and model interpretability. The goal is model parsimony: The preferred models are the simplest models that faithfully represent the relationships inherent in the data. To accomplish this goal, methods must sift through the multitude of available descriptors and identify those that have a causal relationship to the biological or physical process under investigation. This can be done in an unsupervised or supervised way.

### 2.4.1
### Principal Component Analysis

The first methodologies described in this section are mappings from a high-dimensional descriptor space to a lower dimensional one (Figure 2.5). A downward projection of the molecules in descriptor space allows direct visualization of the molecular relationships within that space. Principal component analysis (PCA) is an orthogonal linear transformation used for dimensionality reduction [71]. In PCA, descriptors are transformed to a new coordinate system such that the greatest variance lies on the first coordinate (the first principal component axis), the second greatest variance on the second coordinate, and so on.

These dimensions (principal components) are thus composed of linear combinations of the original descriptors. Interpretation of such projections is again relatively straightforward, as the importance of the linear combinations of descriptors can be quickly retrieved and visualized.

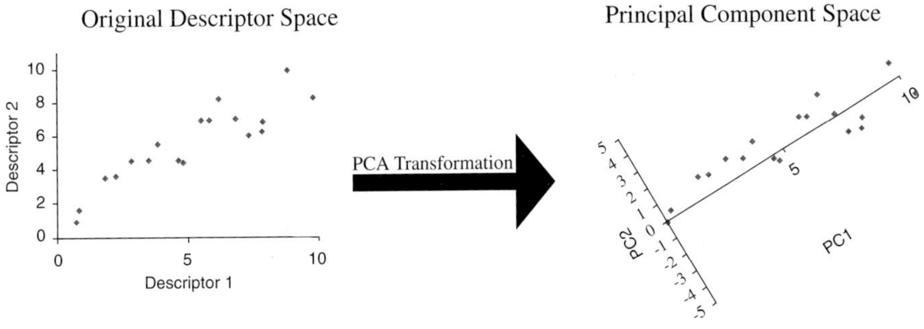


**Figure 2.5** Principal component analysis transforms data so as to represent maximal variance in the data. Principal component 1 (PC 1) is the linear combination of Descriptor 1 and Descriptor 2 with the highest variance.

### 2.4.2
### Factor Analysis

Though less frequently used in QSAR, factor analysis is a closely related technique that attempts to extract coherent subsets of variables that are relatively independent from one another [72]. It is often the case that the variable we are interested in is not directly observable, but is implicitly captured in other quantities that we measure or compute. Factor analysis is an attempt to explain the correlations between variables in the form of underlying factors (called *latent variables*), which are themselves not directly observable, but which are thought to be representative of the underlying process that has created these correlations.

Just as with principal components, factors are linear combinations of original variables. On the surface, factor analysis and PCA are very similar. Both rely on an eigenvalue analysis of the covariance matrix, and both use linear combinations of variables to explain a set of observations. However, in PCA the quantities of interest are the observed variables themselves; the combination of these variables is simply a means for simplifying their analysis and interpretation. Conversely, in factor analysis the observed variables are of little intrinsic value; what is of interest is the underlying factors.

### 2.4.3
### Multidimensional Scaling, Stochastic Proximity Embedding, and Other Nonlinear Dimensionality Reduction Methods

Of course, most biological processes are highly complex, and linear relationships are not always sufficient to explain the underlying phenomena. The variables that we measure and/or compute are often highly correlated, leading to observations that lie on or close to a low-dimensional nonlinear manifold. The dimensionality and geometry of that manifold are implicitly captured in the relationships between the data points, especially in the similarities between them. Consequently, many nonlinear manifold learning approaches try to embed the data points in a low-dimensional space so as to best preserve these similarities. Two common techniques for performing this embedding are multidimensional scaling (MDS) [73] and nonlinear mapping (NLM) [74], which use a steepest descent algorithm to minimize a so-called stress function that measures the deviation between the original (dis)similarities and the distances on the low-dimensional map. Unfortunately, these methods scale at least quadratically with the number of points, and can only be applied to datasets of the moderate size. A more recent method known as stochastic proximity embedding (SPE) [75, 76] uses a linearly scaling self-organizing algorithm that is much faster to execute and trivial to implement, and allows the processing of datasets orders of magnitude larger than those accessible to MDS or NLM. More importantly, it can be combined with other learning algorithms such as neural networks to effectively encode the mapping in an analytical function, which can then be used to map additional objects virtually instantaneously [77–79].

Still, while these methods are nonlinear, their naïve application does not guarantee that the intrinsic dimensions will be identified. The reason is that conventional similarity measures such as the Euclidean distance tend to underestimate the proximity of points on a nonlinear manifold and lead to erroneous embeddings. To properly reconstruct the manifold, one needs to preserve geodesic distances, that is, the lengths of the shortest paths between two points along the surface of the manifold itself. The problem is that these distances are not known *a priori*. The ISOMAP method addresses this limitation by estimating the geodesic distances and using classical MDS to find the optimum low-dimensional configuration [80]. The problem of scaling, however, is exacerbated because the estimation of the geodesic distances is even worse than quadratic. A similar problem plagues locally linear embedding (LLE) [81], a related approach that produces globally ordered maps by constructing locally linear relationships between the data points.

The isometric variant of SPE (ISPE) [82, 83] solves this problem without requiring estimation of geodesic distances, by simply utilizing the fact that the geodesic distance is always greater than or equal to the input proximity if the latter is a true metric. The details of the algorithm and some illustrative applications are described in the original publications and a recent review [84].

### 2.4.4 Clustering

Visualization is an important tool in analyzing, interpreting, and rationalizing data. Clustering data into several categories can show class relationships, class deficiencies, and bias in dataset selection. Several algorithms are available to cluster QSAR data, and extensive reviews have been published [85–87]. One of the most widely used clustering techniques is $k$-means clustering, where data are separated into an arbitrary $k$ classes. First, a random $k$ cases are chosen from the dataset and designated as the centers of the $k$ clusters. Remaining cases are then assigned to those clusters based on a distance measure, such as the Euclidean distance. The center of each cluster is then recalculated as the average of all cases with membership in that cluster. All cases are then reconsidered for membership in each class based on distances from the newly computed centers. This iterative process continues until either no case is classified to a different cluster, or until the number of iterations reaches some preset maximum. Based on the random nature of the algorithm, it is likely that several seedings will yield qualitatively different clusters, though general class relationships should be preserved. Thus, *k*-means clustering is often a popular way to understand relationships between data [88].

### 2.4.5 Partial Least Squares (PLS)

Partial least squares regression (PLS) is commonly used in cheminformatics problems and can be thought of as an extension of principle component regression (PCR), based on PCA [71, 89–91]. PLS determines a linear regression model by finding a direction vector in the descriptor space that maximizes the explained variance in the response. The result is a modeling methodology that is robust with respect to collinear descriptors and easy to control, as the number of descriptor linear combinations (*latent variables*) is the only model parameter to optimize, an advantage over more complicated methods.

### 2.4.6 *k*-Nearest Neighbors (kNN)

Nearest neighbor, or *k-nearest-neighbor* modeling is a simple and intuitive way to predict a molecule's class [92], property [93], or rank [94] (Figure 2.6). In this method,

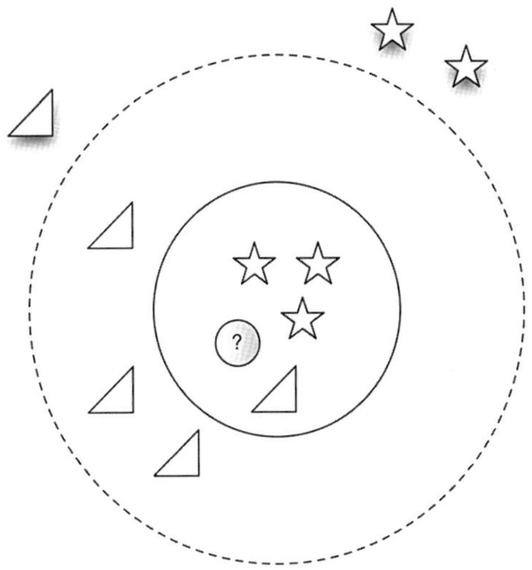


**Figure 2.6** Example of *k*-NN classification. The test sample (circle with a question mark) should be classified either as a star or a triangle. If $k = 4$ it is classified as a star because there are three stars and one triangle inside the inner circle. If $k = 7$ it is classified as a triangle (four triangles vs. three stars inside the outer circle).

a new case is compared to known cases based on a distance measure. The $k$ closest (nearest) cases determine the predicted molecule's class membership via un-weighted or weighted voting [95].

Regression modeling via kNN is likewise straightforward – the predicted activity is a weighted or un-weighted average of the $k$ nearest known neighbors' activities. Although intuitive, $k$-nearest-neighbor modeling does have limitations. First, as only $k$ neighbors are used to predict a new case, this method is extremely sensitive to noisy data. A single misclassified training datum could cause a new case to be predicted incorrectly [5]. By extension, irrelevant descriptors will likewise lead to spurious predictions. And, of course, the predicted value can never be lower or greater than the minimum and maximum activity in the training set – put differently, a new case can never be better or worse than an existing one.

### 2.4.7 Neural Networks

Artificial neural networks map descriptors to activity through a series of weighted connections, or synapses, between neurons (Figure 2.7). Synapses both carry and manipulate information via an often nonlinear transfer function [10]. Basic neural networks have three layers: the input layer of neurons, at least one hidden layer of neurons, and an output layer of neurons. More complex neural networks will have additional layers of neurons to capture additional subtlety in molecular relationships.

Neural networks are trained iteratively, where each period of training is referred to as an epoch. Training involves the minimization of the error between the known activity and the activity predicted by the neural net. As neural nets are highly

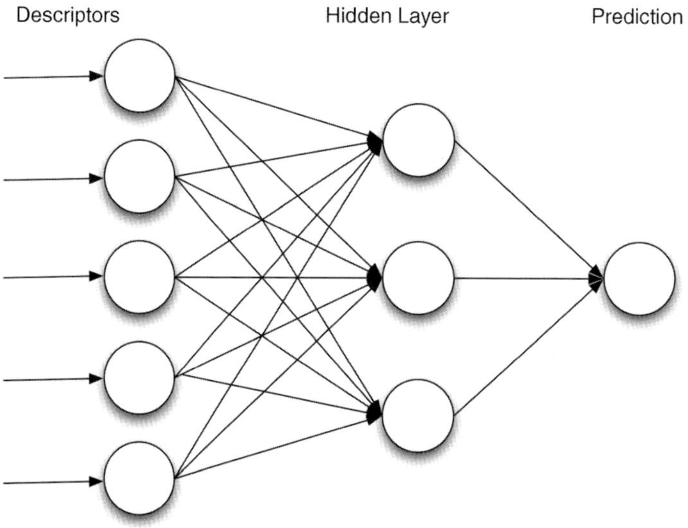


**Figure 2.7** Simplified view of an artificial neural network.

nonlinear and have the potential to minimize error over time, they are incredibly powerful models, but are very easy to overtrain [96]. Methodologies such as early stopping are employed to prevent overtraining [97]. In early stopping, the original QSAR dataset is partitioned into a training and validation set; the neural network is formed on the training set, performance is evaluated based on error of the validation set, and when performance no longer improves, the training of the neural network is stopped. Beyond susceptibility to overtraining, the complex nature of neural networks makes interpretation very difficult. It is possible to read back neuron weights in toy neural networks but as the problem size grows, this quickly becomes impossible. Indirect interpretation strategies such as sensitivity analysis are one approach; neural networks still are methods that favor accuracy over interpretability [98].

### 2.4.8 Ensemble Models

Neural networks are known to be unstable, in that slight changes in the training set and/or training parameters can lead to significant changes in their predictive performance. Research has shown that the accuracy of classification or regression methods can be significantly improved through aggregation of individual predictors. The so-called ensemble techniques, such as bagging [99], boosting [100], and stacking [101], combine multiple models to achieve better predictivity than could be obtained from any of the constituent models. Obviously, combining the output of multiple predictors is useful only if there is disagreement between them, so much of the work in this field has been devoted to methods for introducing diversity into the model ensemble. Ensemble techniques (also referred to as consensus or data fusion techniques) are becoming increasingly prevalent in chemoinformatics, where they are used not only in QSAR and QSPR modeling [102–105] but also in similarity searching [106–109], and protein-ligand docking and scoring [110, 111].

### 2.4.9 Decision Trees and Random Forests

Decision trees map chemical descriptors to a classification, regression, or ranking result by creating a hierarchy based on descriptor values [112] (Figure 2.8). Conclusions are reached in these trees by following logical branches that are composed of nodes, where leaves represent classifications based on a single descriptor and branches represent a path of decisions leading to an answer.

Given the visual nature of the models and simple decisions at each level of the tree, an understanding and interpretation of decision trees is straightforward. Additionally, trees are insensitive to descriptor scaling and relative importance, and can handle categorical and numerical data. While decision trees have been found to be both robust and efficient, optimizing them to behave as such is no small feat [113]. Learning an optimal tree is NP-complete, so heuristics are employed to create decision trees for all but the simplest problems. There is also a significant danger in overfitting, which can be addressed by "pruning." The most common strategy is to

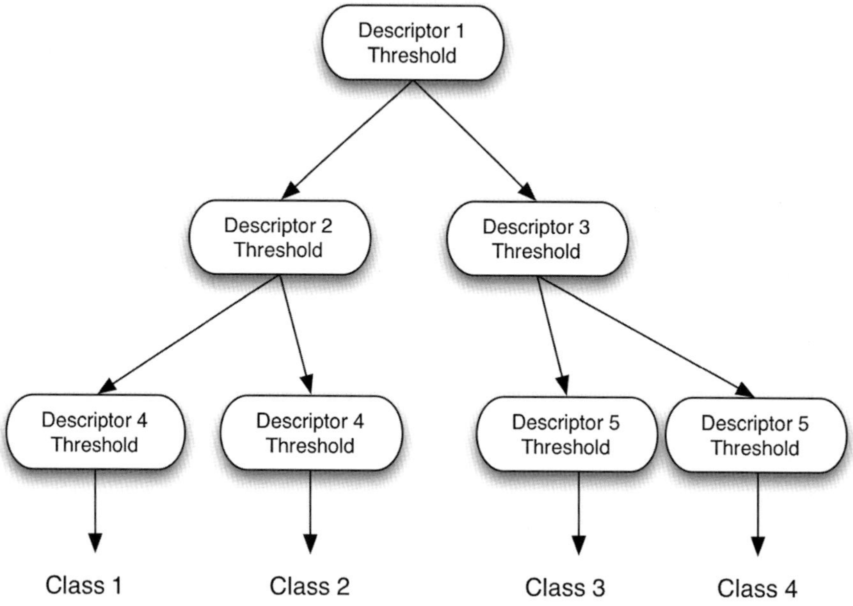


**Figure 2.8** Simplified decision tree for molecular classification.

evaluate every possible subtree, working from the leaves backward, for possible replacement by a terminal node. Nontrivial heuristics including genetic algorithms [114] have also been demonstrated as useful pruning strategies [115, 116].

Random forests [117] are a consensus approach to decision trees, where many decision trees are created based on random subsets of both descriptors and cases [118, 119]. Like neural networks, random forests have the ability to create a highly accurate model, but only within a narrow applicability domain. They suffer from the same interpretation problems as neural networks, as multiple trees confound interpretation. Like simple decision trees, random forests can easily be visualized and can be effective in classifications, regressions, and rankings. Nonoptimal forests are quick to build, can handle large datasets, and as many random subset trees are evaluated, can produce unbiased estimates of generalization error.

### 2.4.10
### Kernel Methods

Kernel methods are a class of algorithms that approach fitting data to a response by mapping the data into a high-dimensional feature space, an approach known as the kernel trick [120–122]. In that high-dimensional space, a variety of relationships become trivial to establish. The kernel trick is efficiently achieved by computing inner products of pairs of data in descriptor space, without explicit calculation of the coordinates of the higher dimensional space [123]. Additionally, as an implicit mapping is established, fitting error is independent of the original number of descriptors. While this makes kernel methods less susceptible to overfitting with respect to the number of descriptors, removing irrelevant descriptors can still improve the model quality [124]. Commonly used kernel methods in QSAR include linear discriminant analysis (LDA), support vector machines (SVMs), and kernel partial least squares (KPLS). Linear discriminant analysis (LDA) is used in classification problems by identifying the hyperplane in descriptor space that, when the data are projected onto it, maximizes the ratio of the distances between class centers to the distances between the farthest points of the same class [125]. Threshold constants can be established to correct for bias in unbalanced data or to bias toward certain conditions, for example, to minimize false negative predictions.

Support vector machines (SVMs) are a powerful general approach to modeling (Figure 2.9). Originally conceived to solve pattern recognition problems [120], SVMs are based on the need to balance training error and the capacity (flexibility) of the model used to fit the data. For a classification problem, this translates to establishing a hyperplane that is farthest from frontier molecules of any class (the functional margin). In general, a wider margin leads to a lower chance of misclassification, and a lower error. Note that the model depends solely on the frontier molecules and the established hyperplane [126].

SVMs resist overfitting the data, producing models that generalize well (have a wider applicability domain). They have been extended beyond classification problems and applied successfully to a wide range of chemical problems in classification, regression, and ranking [127, 128].

Kernel partial least squares regression works by exploiting the kernel trick within a PLS framework, though Bennett *et al.* have demonstrated similarity in performance to other kernel methods with the added benefit of robust and simple model optimization [129]. Other than the choice of kernel, the only parameter that affects the model's fitting ability is the number of latent variables.

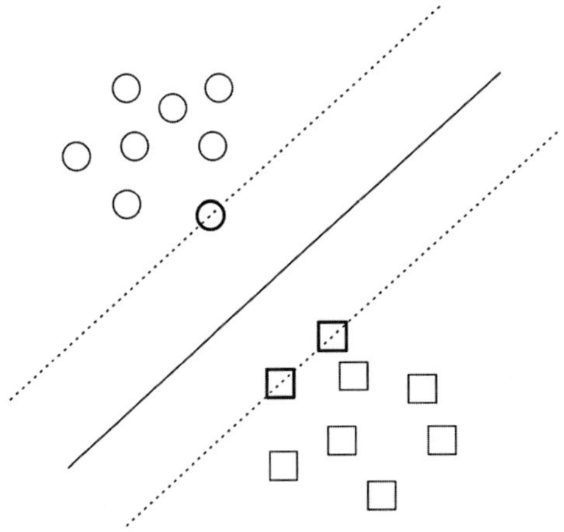

**Figure 2.9** A support vector machine trained with data belonging to two classes. Shown are the maximum-margin hyperplane (solid line) and margins (dashed lines). Samples on the margin are called the support vectors.

### 2.4.11
### Ranking Methods

Recently, ranking methods have received much attention in the field of virtual screening, where the goal is to identify and prioritize screening hits for lead optimization. As an accurate prediction of potency throughout the range of compounds is not the same problem as a top-rank ordering, models tuned for compound ranking inherently behave differently than regression models [130]. Ranking models allow users to select a number of most potent compounds for virtual screening, while classification methods can only be applied for differentiating fixed boundaries of compound potency [131].

Geppert *et al.* utilized SVM-based ranking for hit identification in virtual screening and found that it outperforms standard similarity methods using calculated 2D fingerprints.

Agarwal *et al.* developed the RankSVM and Bipartite RankSVM methods and applied them to various QSAR datasets [132]. They found that these ranking algorithms outperform standard SVM methods based on various ranking metrics such as bipartite ranking error and Spearman correlation coefficient. The essential improvement of their ranking algorithms was achieved through modifying the empirical risk term of the loss function for minimizing ranking errors.

In the above QSAR studies, standard classification and regression methods still could be used for ranking. However, in problems such as Cytochrome P450 (CYP) regioselectivity prediction, the goal is to predict which atoms or sites within a molecule would be subject to metabolism by a given CYP. Only a partial ranking of sites is known, so regression methods cannot be applied. On the other hand, classification models cannot be directly applied either, since the ranking information of atoms across different molecules is unknown. To address this problem, Bergeron *et al.* have developed the multiple instance ranking (MIRank) algorithm [133]. Similar to Agarwal's approach, their method was designed specifically for minimizing ranking error of atoms within each molecule and they have reported good model performance over a diverse dataset.

## 2.5
## Defining Modeling Strategies: Putting It All Together

General strategies for successful modeling exist for all three major types of QSAR modeling. Treatment of the original data is always an important consideration. Detection and treatment of experimental outliers is a complex, challenging, and controversial area of research in chemistry [3–5]. There exist numerous opinions on the matter; if the goal is to focus on a specific structure–activity relationship, classical QSAR models have been shown to benefit from "dividing and conquering" outliers, so as to produce several simple and very specific models [134]. Other applications to drug design and discovery suggest that outliers in the chemical space improve the applicability domain of the model and should be left in the QSAR dataset [98].

Multiple strategies exist for outlier detection, based on activity as well as the chemical structure [134, 135].

Once experimental data is acquired, chemical structures must be digitized. Depending on the goal of the QSAR, appropriate protonation states may need to be assigned. While structure repositories such as PubChem are convenient, structural representations are not always faithful and thus are a source of error in QSAR models [136]. An awareness of this situation is necessary and includes removal of disconnected salts, as well as representation of the appropriate tautomer form of the compound. Once drawn structures are satisfactorily represented, descriptor generation may proceed. The easiest solution, mass descriptor generation, is a trap for the prospective modeler: Prefabricated descriptor collections may be uninterpretable and or have little physical meaning attached to them [3]. Deliberate descriptor choice via hypothesis formation is initially slower but generally more successful [12]. The modeler should focus on descriptors that are believed to be relevant in capturing mechanistic detail; by selecting these descriptors, there is a positive bias toward interpretability. If the hypothesis has been proven incorrect by the resulting QSAR models, alternate hypotheses may be explored.

If there is sufficient data available, an external test set should be generated by splitting the original data [16] (or ideally, additional data should be gathered to form an external test set). This test set should be chosen within the domain of applicability of the model, and if data is plentiful, a retrospective analysis of blind test set predictions will allow a better picture of the true domain of the model's applicability. In order to prevent bias toward this test set, it should not be used to influence preprocessing techniques, modeling parameter selection, or model training: Application of the above techniques should be applied blindly to the test set [12].

Before QSAR model building occurs, data preprocessing is usually required. This entails data scaling and the removal of collinear descriptors (Figure 2.10). Preprocessing decisions are based on the constraints of the data and the modeling methodology in the context of past experience. For instance, descriptors must usually be scaled to fall within a range of zero to one for use in a neural network model [137]. For other modeling methods such as support vector machines, standardizing the data ensures that the models are not biased toward large-valued descriptors [138]. Before or after scaling, the removal of collinear descriptors beyond a chosen co-correlation threshold is highly desired; such descriptors effectively contribute nothing to the QSAR model as they contain the same information [89, 139]. Inclusion may increase instability in modeling methods such as MLR, and interpretation of a QSAR built from collinear descriptors is confounding: Which descriptor describes the mechanism?

Building a robust and useful QSAR involves multiple checks of self-consistency known as *model validation* [140]. A common development cycle of a QSAR is presented below, where model building is wrapped within internal validation techniques.

Cross-validation is the most commonly used technique for internal validation and parameter selection [11]. Over multiple rounds of model building, different divisions of data are iteratively withheld from model learning and are used strictly for

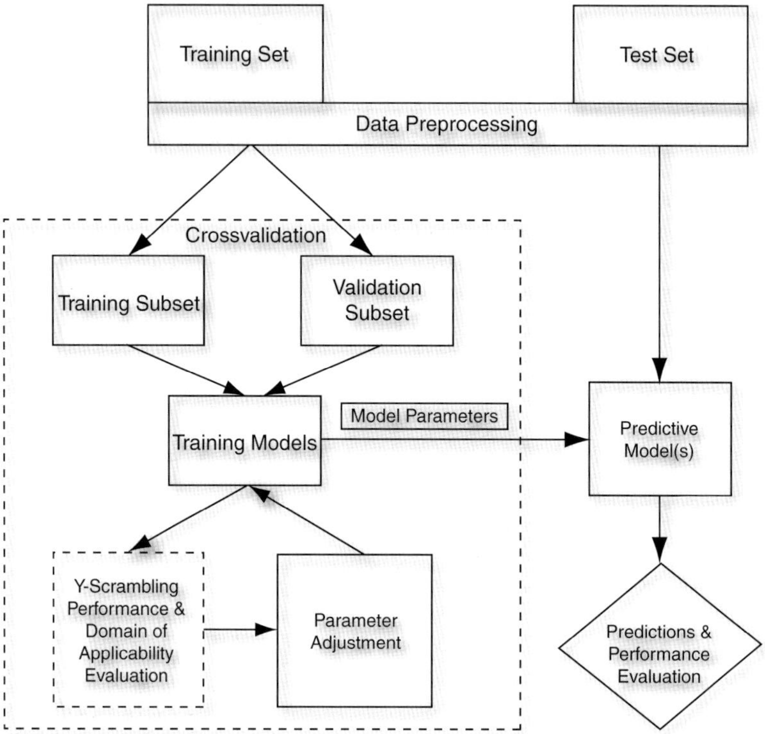


**Figure 2.10** A validated QSAR workflow. The external test set data does not influence overall data scaling and model creation.

performance estimation. Successive rounds of cross-validation may be used to optimize model parameters to control degrees of model flexibility or nonlinearity, or to choose a subset of descriptors that maximizes model performance. Leave-one-out cross-validation (LOO-CV) takes this to the logical extreme, where a single molecule is withheld for performance evaluation, and successive models are built to estimate performance over the entire dataset. As datasets do not provide uniform coverage of chemical space, performance estimates using LOO-CV are often exaggerated and should be used with caution, or not at all [13]. A more robust performance estimation method is leave-N-out cross-validation, where multiple molecules are withheld for evaluation estimation. Depending on the heterogeneity and size of the dataset, it has been suggested that up to 30% of the data be withheld for cross-validation [11, 141]. Similar in spirit but different in execution, bootstrapped validation methodologies rely on random subset selection for model training and for performance estimation [142, 143]. Since random selections are made, particular molecules may unevenly influence a bootstrapped QSAR model, but if multiple rounds of bootstrapping take place, aggregate statistics converge quickly. Bootstrapping is preferred in the treatment of large datasets, where multiple rounds of leave-N-out cross-validation result in an explosion in the number of QSAR models that need to be evaluated and the amount of computational effort involved.

Model validation is not adequate to ensure that relationships found are real. *This is a critical observation* – it is always possible that when combined properly, specific sets of descriptors are capable of predicting *any* endpoint. Use of y-scrambling as a test of method susceptibility to overfitting should be strongly considered as a means of testing the robustness of a particular modeling approach [15, 144]. In this technique, the activity vector is shuffled, and a new, "fake" QSAR model is built using the original, unscrambled chemical descriptors, and the shuffled response. This process occurs several times, and for the majority of cases, it is expected that these scrambled, fake models will exhibit low performance compared to the real model, as described by model performance metrics (Figure 2.11).

Thus, y-scrambling examines the assumption that there is only one plausible physically interpretable QSAR relationship being modeled [145]. If many or all of the scrambled QSAR models built using a specific set of descriptors and a given machine learning method appear to have relatively high performance, it implies that QSAR models based on the given modeling method and descriptors demonstrate many plausible hypotheses and a lack of differentiation between them. Realistically, this means the modeling methodology and descriptors form an inappropriate QSAR: there may be too many descriptors, too flexible a model, or some combination thereof. A conservative observation would be that these QSAR models are not robust, meaning the domain to which they apply is smaller than other, more general models.

This domain, the applicability domain, defines the region of trustworthiness of predictions, and can be described in terms of the descriptors within the model [142]. As interpolation is more reliable than extrapolation, new molecules whose activities are to be predicted should ideally exist within the space defined by the training set molecules. Outside that space, predictions are unreliable. An applicability domain

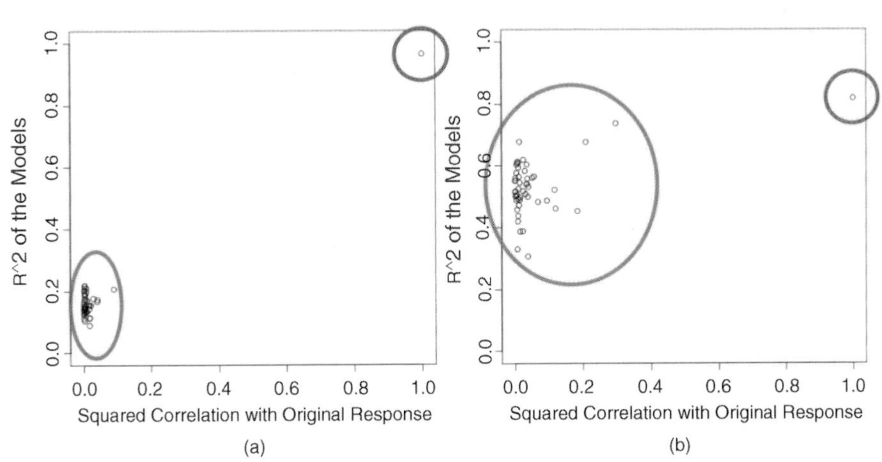


**Figure 2.11** Comparison of y-scrambling results. (a) It indicates robust modeling, while (b) suggests a distinct possibility of model overtraining.

can be defined by the descriptor range, distance, geometry, or probability density distribution, and there is much debate over which method best estimates prediction reliability. As different applicability domain measures suggest different applicability domains, several measures should be used in consensus to highlight similarities and differences in the representation of the domain of applicability [146, 147].

Consensus approaches to modeling and model validation have lead to interesting observations in QSAR model predictivity and methodology effectiveness. Tetko *et al.* used mixtures of Gaussian distributions as well as statistical tests to evaluate model applicability domains with respect to their ability to discriminate compounds with small and large prediction errors [148]. Their analysis was performed on 12 varied QSAR models of aqueous toxicity against *Tetrahymena. pyriformis*. They concluded that the standard deviation of a molecule's prediction in the ensemble provided the best estimation of the accuracy of predictions. A larger standard deviation corresponds to a larger disagreement between models, and a lower expected prediction accuracy.

Das *et al.* have shown a clear domain of applicability in the prediction of protein-binding affinity using a combination of PESD descriptors and cross-validated SVM modeling [149]. Chi-squared distances to the training data were correlated with ligand-binding prediction accuracy; as distance from the training set increased, the prediction accuracy decreased in a linear and predictable manner.

Identification of "activity cliffs," locations where structure–activity relationships break down, is critical for a gaining an understanding of model reliability in specific domains [135]. These cliffs arise due to inadequate representation of the activity by descriptors and/or an inadequate capacity of a model to learn the true relationship between structure and activity, such as an induced fit mechanism. Effective use of activity cliff representations, such as the structure–activity landscape index (SALI) [150] and structure–activity relationship index (SARI) [151], is a new but promising technique for analyzing the reliability of QSAR models [152, 153].

## 2.6 Conclusions

QSAR models should be based on testable hypotheses; mechanisms of action may be uncertain, but they should be plausible. Statistical diagnostics from modeling may indicate the underlying QSAR assumption to be invalid. At this point, it would be advisable to consider reasons for the failure of the hypothesis rather than forcing a model to fit the data. Apropos to this, the experimental data gathered should be as free from error as possible, and the amount and sources of error should be known. The models built from this data should not be expected or desired to return more accurate predictions than the error in the data will allow; model validation strategies are employed to ensure that the model is robust over the range of chemical space upon which the model is built. Understanding this space is critical in knowing when the model can provide meaningful predictions. So as to understand this space, deliberate and plausible descriptor choices must be made, in much the same way that an initial QSAR hypothesis must be made. Rational choices in descriptor preprocessing chosen with respect to the modeling method used lead to better model performance and a clearer interpretation of the QSAR. Identifying the performance of the QSAR over an appropriate range of molecules is a goal that can lead to the exploration of plausible mechanisms of action and a greater understanding of the underlying chemistry. Realizing that those choices leading up to QSAR model use should be rational and balanced, QSAR modeling is transformed into a journey to better understand the significance and context of the relationships between the data.

## References


1 Brown, A.C. and Fraser, T.R. (1868) On the connection between chemical constitution and physiological action; with special reference to the physiological action of the salts of the ammonium bases derived from strychnia, Brucia, Thebaia, Codeia, Morphia, and Nicotia. *Journal of Anatomy and Physiology*, **2** (2), 224–242.

2 Meyer, H.H. (1899) Zur theorie der alkoholnarkose: I. Mittheilung. Welche eigenschaft der anaesthetika bedingt ihre narkotische wirkung? *Archives of Experimental Pathology and Pharmacology*, **42**, 109–118.

3 Cronin, M.T.D. and Schultz, T.W. (2003) Pitfalls in QSAR. *Journal of Molecular Structure: THEOCHEM*, **622** (1–2), 39–51.

4 Zvinavashe, E., Murk, A.J., and Rietjens, I.M.C.M. (2008) Promises and pitfalls of quantitative structure–activity relationship approaches for predicting metabolism and toxicity. *Chemical Research in Toxicology*, **21** (12), 2229–2236.

5 Scior, T., Medina-Franco, J.L., Do, Q.T., Martinez-Mayorga, K., Yunes Rojas, J.A., and Bernard, P. (2009) How to recognize and workaround pitfalls in QSAR studies: A critical review. *Current Medicinal Chemistry*, **16**, 4297–4313.

6 Doweyko, A. (2008) QSAR: Dead or alive? *Journal of Computer-Aided Molecular Design*, **22** (2), 81–89.

7 Hochberg, Y. and Tamhane, A.C. (2008) *Frontmatter*, John Wiley & Sons, Inc, New York.

8 Jin, H. and Ling, C.X. (2005) Using AUC and accuracy in evaluating learning algorithms. *Knowledge and Data Engineering, IEEE Transactions on*, **17** (3), 299–310.

9 Steffen, A., Kogej, T., Tyrchan, C., and Engkvist, O. (2009) Comparison of molecular fingerprint methods on the basis of biological profile data. *Journal of Chemical Information and Modeling*, **49** (2), 338–347.

10 King, R., Hirst, J., and Sternberg, M. (1993) New approaches to QSAR: Neural networks and machine learning. *Perspectives in Drug Discovery and Design*, **1** (2), 279–290.

11 Gramatica, P. (2007) Principles of QSAR models validation: Internal and external. *QSAR & Combinatorial Science*, **26** (5), 694–701.

12 Dearden, J.C., Cronin, M.T.D., and Kaiser, K.L.E. (2009) How not to develop a quantitative structure–activity or structure–property relationship (QSAR/QSPR). *SAR and QSAR in Environmental Research*, **20** (3/4), 241–266.

13 Golbraikh, A. and Tropsha, A. (2002) Beware of q2. *Journal of Molecular Graphics and Modelling*, **20** (4), 269–276.

14 Weaver, S. and Gleeson, M.P. (2008) The importance of the domain of applicability in QSAR modeling. *Journal of Molecular Graphics and Modelling*, **26** (8), 1315–1326.

15 Tropsha, A. (2010) Best practices for QSAR model development, validation, and exploitation. *Molecular Informatics*, **29** (6–7), 476–488.

16 Golbraikh, A. and Tropsha, A. (2002) Predictive QSAR modeling based on diversity sampling of experimental datasets for the training and test set selection. *Journal of Computer-Aided Molecular Design*, **16** (5), 357–369.

17 Baroni, M., Clementi, S., Cruciani, G., Kettaneh-Wold, N., and Wold, S. (1993) D-optimal designs in QSAR. *Quantitative Structure–Activity Relationships*, **12** (3), 225–231.

18 Hansch, C. and Leo, A. (1995) Exploring QSAR: Fundamentals and applications in chemistry and biology. *American Chemical Society*, **1**, 1–557.

19 Guha, R., Dutta, D., Jurs, P.C., and Chen, T. (2006) Local lazy regression: Making use of the neighborhood to improve QSAR predictions. *Journal of Chemical Information and Modeling*, **46** (4), 1836–1847.

20 Chen, J.J., Tsai, C.A., Young, J.F., and Kodell, R.L. (2005) Classification ensembles for unbalanced class sizes in predictive toxicology. *SAR and QSAR in Environmental Research*, **16** (6), 517–529.

21 Todeschini, R. and Consonni, V. (2000) *Handbook of Molecular Descriptors*, vol. 11, Wiley-VCH, Weinheim

22 Hammett, L.P. (1937) The effect of structure upon the reactions of organic compounds. Benzene derivatives. *Journal of the American Chemical Society*, **59** (1), 96–103.

23 Taft, R.W. (1952) Polar and steric substituent constants for aliphatic and o-benzoate groups from rates of esterification and hydrolysis of esters. *Journal of the American Chemical Society*, **74**, 3120–3128.

24 Hansch, C., Muir, R.M., Fujita, T., Maloney, P.P., Geiger, F., and Streich, M. (1963) The correlation of biological activity of plant growth regulators and chloromycetin derivatives with Hammett constants and partition coefficients. *Journal of the American Chemical Society*, **85**, 2817–2824.

25 Hansch, C., Kurup, A., Garg, R., and Gao, H. (2001) Chem-bioinformatics and QSAR: A review of QSAR lacking positive hydrophobic terms. *Chemical Reviews*, **101** (3), 619–672.

26 Garg, R., Gupta, S.P., Gao, H., Babu, M.S., Debnath, A.K., and Hansch, C. (1999) Comparative QSAR studies on anti-HIV drugs. *Chemical Reviews*, **99**, 3525–3601.

27 Guha, R. and Schürer, S. (2008) Utilizing high throughput screening data for predictive toxicology models: Protocols and application to MLSCN assays. *Journal of Computer-Aided Molecular Design*, **22** (6), 367–384.

28 Chen, B. and Wild, D.J. (2010) PubChem BioAssays as a data source for predictive models. *Journal of Molecular Graphics and Modelling*, **28** (5), 420–426.

29 Fliri, A.F., Loging, W.T., Thadeio, P.F., and Volkmann, R.A. (2005) Biospectra analysis: Model proteome characterization for linking molecular structure and biological response. *Journal of Medicinal Chemistry*, **48**, 6918–6925.

30 Fliri, A.F., Loging, W.T., Thadeio, P.F., and Volkmann, R.A. (2005) Biological spectra analysis: Linking biological activity profiles to molecular structure. *Proceedings of the National Academy of Sciences of the United States of America*, **102** (2), 261–266.

31 Hawkins, D.M., Basak, S.C., Kraker, J., Geiss, K.T., and Witzmann, F.A. (2006) Combining chemodescriptors and biodescriptors in quantitative structure–activity relationship modeling. *Journal of Chemical Information and Modeling*, **46** (1), 9–16.

32 Balasubramanian, K., Khokhani, K., and Basak, S.C. (2006) Complex graph matrix representations and characterizations of proteomic maps and chemically induced changes to proteomes. *Journal of Proteome Research*, **5**, 1133–1142.

33 Benigni, R., Giuliani, A., and Passerini, L. (2001) Infrared spectra as chemical descriptors for QSAR models. *Journal of Chemical Information and Computer Sciences*, **41** (3), 727–730.

34 Willighagen, E.L., Denissen, H.M.G.W., Wehrens, R., and Buydens, L.M.C. (2006) On the use of $^{1}H$ and $^{13}C$ 1D NMR spectra as QSPR descriptors. *Journal of Chemical Information and Modeling*, **46** (2), 487–494.

35 Puzyn, T., Leszczynski, J., and Cronin, M.T. (2009) Recent advances in QSAR studies: methods and applications, in *Recent Advances in QSAR Studies: Methods and Applications*, vol. 1 (eds T. Puzyn, J. Leszczynski, and M.T. Cronin), Springer, Berlin, pp. 383–403.

36 Kier, L.B., Murray, W.J., Randic, M., and Hall, L.H. (1975) Molecular connectivity: I. Relationship to nonspecific local anaesthesia. *Journal of Pharmacological Sciences*, **64**, 1971–1974.

37 Dearden, J.C. (2007) *In silico* prediction of ADMET properties: How far have we come? *Expert Opinion on Drug Metabolism & Toxicology*, **3** (5), 635–639.

38 Randic, M. (2001) The connectivity index 25 years after. *Journal of Molecular Graphics and Modelling*, **20** (1), 19–35.

39 Bonchev, D. (1983) *Information Theoretic Indices for Characterization of Chemical Structures*, Research Studies Press, Chichester, UK.

40 Dehmer, M. and Mowshowitz, A. (2011) A history of graph entropy measures. *Information Sciences*, **181** (1), 57–78.

41 Zefirov, N.S. and Palyulin, V.A. (2002) Fragmental approach in QSPR. *Journal of Chemical Information and Computer Sciences*, **42**, 1112–1122.

42 Nikolova, N. and Jaworska, J. (2003) Approaches to measure chemical similarity – a review. *The QSAR and Combinatorial Science*, **22**, 1006–1026.

43 Nettles, J.H., Jenkins, J.L., Bender, A., Deng, Z., Davies, J.W., and Glick, M. (2006) Bridging chemical and biological space: "Target Fishing" using 2D and 3D molecular descriptors. *Journal of Medicinal Chemistry*, **49** (23), 6802–6810.

44 Nettles, J.H., Jenkins, J.L., Williams, C., Clark, A.M., Bender, A., Deng, Z., Davies, J.W., and Glick, M. (2007) Flexible 3D pharmacophores as descriptors of dynamic biological space. *Journal of Molecular Graphics and Modelling*, **26** (3), 622.

45 Todeschini, R. and Gramatica, P. (1997) SD-modelling and prediction by WHIM descriptors: Part 5. Theory development and chemical meaning of WHIM descriptors. *Quantitative Structure–Activity Relationships*, **16** (2), 113–119.

46 Schuur, J.H., Selzer, P., and Gasteiger, J. (1996) The coding of the three-dimensional structure of molecules by molecular transforms and its application to structure–spectra correlations and studies of biological activity. *Journal of Chemical Information and Computer Sciences*, **36** (2), 334–344.

47 Consonni, V., Todeschini, R., and Pavan, M. (2002) Structure/response correlations and similarity/diversity analysis by GETAWAY descriptors: Part 1. Theory of the novel 3D molecular descriptors. *Journal of Chemical Information and Modeling*, **42**, 682–692.

48 Cramer, R.D., Patterson, D.E., and Bunce, J.D. (1988) Comparative molecular field analysis (CoMFA): Part 1. Effect of shape on binding of steroids to carrier proteins. *Journal of the American Chemical Society*, **110**, 5959–5967.

49 Cramer, R.D., Jilek, R.J., and Andrews, K.M. (2002) dbtop: topomer similarity searching of conventional structure databases. *Journal of Molecular Graphics & Modelling*, **20**, 447–462.

50 Jilek, R.J. and Cramer, R.D. (2004) Topomers: a validated protocol for their self-consistent generation. *Journal of Chemical Information and Computer Sciences*, **44**, 1221–1227.

51 Ballester, P.J. and Richards, W.G. (2007) Ultrafast shape recognition to search compound databases for similar molecular shapes. *Journal of Computational Chemistry*, **28**, 1711–1723.

52 Ballester, P.J. and Richards, W.G. (2007) Ultrafast shape recognition for similarity search in molecular databases. *Proceedings of the Royal Society of London. Series A, Mathematical and Physical Sciences*, **463**, 1307–1321.

53 Ballester, P.J., Finn, P.W., and Richards, W.G. (2009) Ultrafast shape recognition: Evaluating a new ligand-based virtual screening technology. *Journal of Molecular Graphics & Modelling*, **27**, 836–845.

54 Clark, T. (2004) QSAR and QSPR based solely on surface properties? *Journal of Molecular Graphics & Modelling*, **22**, 519–525.

55 Liu, R., Sun, H., and So, S.-S. (2001) Development of quantitative structure–property relationship models for early ADME evaluation in drug discovery: Part 2. Blood–brain barrier penetration. *Journal of Chemical Information and Computer Sciences*, **41** (6), 1623–1632.

56 Jover, J., Bosque, R., and Sales, J. (2008) QSPR prediction of p$K_a$ for benzoic acids in different solvents. *QSAR and Combinatorial Science*, **27** (5), 563–581.

57 Tulp, I., Sild, S., and Maran, U. (2009) Relationship between structure and permeability in artificial membranes: Theoretical whole molecule descriptors in development of QSAR models. *QSAR and Combinatorial Science*, **28** (8), 811–814.

58 Ioakimidis, L., Thoukydidis, L., Mirza, A., Naeem, S., and Reynisson, J. (2008) Benchmarking the reliability of QikProp: Correlation between experimental and predicted values. *QSAR and Combinatorial Science*, **27** (4), 445–456.

59 Das, S., Kokardekar, A., and Breneman, C.M. (2009) Rapid comparison of protein binding site surfaces with property encoded shape distributions. *Journal of Chemical Information and Modeling*, **49** (12), 2863–2872.

60 Mezey, P.G. (1993) *Shape in Chemistry*, Wiley-VCH, New York.

61 Mezey, P.G. (1999) The holographic electron density theorem and quantum similarity measures. *Molecular Physics*, **96** (2), 169–178.

62 Walker, P.D. and Mezey, P.G. (1994) *Ab initio* quality electron densities for proteins: A MEDLA approach. *Journal of the American Chemical Society*, **116**, 12022–12032.

63 Breneman, C.M., Thompson, T.R., Rhem, M., and Dung, M. (1995) Electron density modeling of large systems using the transferable atom equivalent method. *Computers and Chemistry*, **19** (3), 161.

64 Whitehead, C.E., Breneman, C.M., Sukumar, N., and Ryan, M.D. (2003) Transferable atom equivalent multi-centered multipole expansion method. *Journal of Computational Chemistry*, **24**, 512–529.

65 Sukumar, N., Krein, M., and Breneman, C.M. (2008) Bio- and chem-informatics: Where do the twain meet? *Current Opinion in Drug Discovery & Development*, **11** (3), 311–319.

66 Breneman, C.M., Sundling, C.M., Sukumar, N., Shen, L., Katt, W.P., and Embrechts, M.J. (2003) New developments in PEST shape/property hybrid descriptors. *Journal of Computer-Aided Molecular Design*, **17**, 231–240.

67 Zauhar, R.J., Moyna, G., Tian, L., Li, Z., and Welsh, W.J. (2003) Shape signatures, a new approach to computer-aided ligand- and receptor-based drug design. *Journal of Medicinal Chemistry*, **46**, 5674–5690.

68 Pastor, M., Cruciani, G., McLay, I., Pickett, S., and Clementi, S. (2000) GRid-INdependent descriptors (GRIND): A novel class of alignment-independent three-dimensional molecular descriptors. *Journal of Medicinal Chemistry*, **43**, 3233–3243.

69 Cruciani, G., Crivori, P., Carrupt, P.A., and Testa, B. (2000) Molecular fields in quantitative structure–permeation relationships: The VolSurf approach. *Journal of Molecular Structure: THEOCHEM*, **503** (1–2), 17.

70 Topliss, J.G. and Edwards, R.P. (1979) Chance factors in studies of quantitative–structure property relationships. *Journal of Medicinal Chemistry*, **22**, 1238–1244.

71 Eriksson, L., Andersson, P., Johansson, E., and Tysklind, M. (2006) Megavariate analysis of environmental QSAR data: Part I. A basic framework founded on principal component analysis (PCA), partial least squares (PLS), and statistical molecular design (SMD). *Molecular Diversity*, **10** (2), 169–186.

72 Cooley, W., Lohnes, P., and Analysis, M.D. (1971) *Multivariate Data Analysis*, Wiley, New York.

73 Borg, I. and Groenen, P.J.F. (1997) *Modern Multidimensional Scaling: Theory and Applications*, Springer, New York.

74 Sammon, J.W. (1969) A nonlinear mapping for data structure analysis. *IEEE Transactions on Computers*, **18**, 401–409.

75 Agrafiotis, D.K. (2003) Stochastic proximity embedding. *Journal of Computational Chemistry*, **24** (10), 1215–1221.

76 Rassokhin, D.N. and Agrafiotis, D.K. (2003) A modified update rule for stochastic proximity embedding. *Journal of Molecular Graphics and Modelling*, **22** (2), 133–140.

77 Agrafiotis, D.K. and Lobanov, V.S. (2000) Nonlinear mapping networks. *Journal of Chemical Information and Computer Sciences*, **40** (6), 1356–1362.

78 Rassokhin, D.N., Lobanov, V.S., and Agrafiotis, D.K. (2001) Nonlinear mapping of massive data sets by fuzzy clustering and neural networks. *Journal of Computational Chemistry*, **22** (4), 373–386.

79 Agrafiotis, D.K., Rassokhin, D.N., and Lobanov, V.S. (2001) Multidimensional scaling and visualization of large molecular similarity tables. *Journal of Computational Chemistry*, **22** (5), 488–500.

80 Tenenbaum, J.B., Silva, V.d., and Langford, J.C. (2000) A global geometric framework for nonlinear dimensionality reduction. *Science*, **290** (5500), 2319–2323.

81 Roweis, S.T. and Saul, L.K. (2000) Nonlinear dimensionality reduction by locally linear embedding. *Science*, **290** (5500), 2323–2326.

82 Agrafiotis, D.K. and Xu, H. (2002) A self-organizing principle for learning nonlinear manifolds. *Proceedings of the National Academy of Sciences of the United States of America*, **99** (25), 15869–15872.

83 Agrafiotis, D.K. and Xu, H. (2003) A geodesic framework for analyzing molecular similarities. *Journal of Chemical Information and Computer Sciences*, **43** (2), 475–484.

84 Agrafiotis, D.K., Xu, H., Zhu, F., Bandyopadhyay, D., and Liu, P. (2010) Stochastic proximity embedding: Methods and applications. *Molecular Informatics*, **29** (11), 758–770.

85 Stanforth, R.W., Kolossov, E., and Mirkin, B. (2007) A measure of domain of applicability for QSAR modelling based on intelligent *K*-means clustering. *QSAR & Combinatorial Science*, **26** (7), 837–844.

86 Kerzic, D., Blazic, B.J., and Batagelj, V. (1994) Comparison of three different approaches to the property prediction problem. *Journal of Chemical Information and Computer Science*, **34**, 391–394.

87 Bennett, K., Demiriz, A., and Embrechts, M. (1999) Semi-Supervised Clustering Using Genetic Algorithms, Artificial Neural Networks in Engineering.

88 Rose, V.S., Croall, I.F., and Macfie, H.J.H. (1991) An application of unsupervised neural network methodology Kohonen topology-preserving mapping to QSAR analysis. *Quantitative Structure–Activity Relationships*, **10** (1), 6–15.

89 Wold, S., Ruhe, A., Wold, H., and Dunn, W.J. III (1984) The collinearity problem in linear regression. The partial least squares (PLS) approach to generalized inverses. *SIAM Journal of Scientific Statistical Computing*, **5** (3), 735.

90 Hoskuldson, A. (1988) PLS regression methods. *Journal of Chemometrics*, **2**, 211.

91 Geladi, P. (1988) Notes on the history and nature of partial least-squares (PLS) modelling. *Journal of Chemometrics*, **2**, 231.

92 Kauffman, G.W. and Jurs, P.C. (2001) QSAR and k-nearest neighbor classification analysis of selective cyclooxygenase-2 inhibitors using topologically-based numerical descriptors. *Journal of Chemical Information and Computer Sciences*, **41** (6), 1553–1560.

93 Konovalov, D.A., Coomans, D., Deconinck, E., and Vander Heyden, Y. (2007) Benchmarking of QSAR models for blood–brain barrier permeation. *Journal of Chemical Information and Modeling*, **47** (4), 1648–1656.

94 Votano, J.R., Parham, M., Hall, L.H., Kier, L.B., Oloff, S., Tropsha, A., Xie, Q., and Tong, W. (2004) Three new consensus QSAR models for the prediction of Ames genotoxicity. *Mutagenesis*, **19** (5), 365–377.

95 Patterson, D.E., Cramer, R.D., Ferguson, A.M., Clark, R.D., and Weinberger, L.E. (1996) Neighborhood behavior: A useful concept for validation of "molecular diversity" descriptors. *Journal of Medicinal Chemistry*, **39** (16), 3049–3059.

96 Huuskonen, J., Salo, M., and Jyrki, Taskinen, (1997) Neural network modeling for estimation of the aqueous solubility of structurally related drugs. *Journal of Pharmaceutical Sciences*, **86** (4), 450–454.

97 Livingstone, D.J., Manallack, D.T., and Tetko, I.V. (1997) Data modelling with neural networks: Advantages and limitations. *Journal of Computer-Aided Molecular Design*, **11** (2), 135–142.

98 Guha, R. (2008) On the interpretation and interpretability of quantitative structure–activity relationship models. *Journal of Computer-Aided Molecular Design*, **22** (12), 857–871.

99 Breiman, L. (1996) Bagging predictors. *Machine Learning*, **24** (2), 123–140.

100 Freund, Y. and Schapire, R.E. (1995) A decision-theoretic generalization of on-line learning and an application to boosting, in *Proceedings of the Second European Conference on Computational Learning Theory*, Springer, Berlin, pp. 23–37.

101 Breiman, L. (1996) Stacked regressions. *Machine Learning*, **24** (1), 49–64.

102 Agrafiotis, D.K., Cedeño, W., and Lobanov, V.S. (2002) On the use of neural network ensembles in QSAR and QSPR. *Journal of Chemical Information and Computer Sciences*, **42** (4), 903–911.

103 Mattioni, B.E., Kauffman, G.W., Jurs, P.C., Custer, L.L., Durham, S.K., and Pearl, G.M. (2003) Predicting the genotoxicity of secondary and aromatic amines using data subsetting to generate a model ensemble. *Journal of Chemical Information and Computer Sciences*, **43** (3), 949–963.

104 Merkwirth, C., Mauser, H., Schulz-Gasch, T., Roche, O., Stahl, M., and Lengauer, T. (2004) Ensemble methods for classification in cheminformatics. *Journal of Chemical Information and Computer Sciences*, **44** (6), 1971–1978.

105 Seierstad, M. and Agrafiotis, D.K. (2006) A QSAR model of hERG binding using a large, diverse, and internally consistent training set. *Chemical Biology & Drug Design*, **67** (4), 284–296.

106 Ginn, C.M.R., Willett, P., and Bradshaw, J. (2000) Combination of molecular similarity measures using data fusion. *Perspectives in Drug Discovery and Design*, **20**, 1–16.

107 Salim, N., Holliday, J., and Willett, P. (2002) Combination of fingerprint-based similarity coefficients using data fusion. *Journal of Chemical Information and Computer Sciences*, **43** (2), 435–442.

108 Baurin, N., Mozziconacci, J.-C., Arnoult, E., Chavatte, P., Marot, C., and Morin-Allory, L. (2003) 2D QSAR consensus prediction for high-throughput virtual screening. An application to COX-2 inhibition modeling and screening of the NCI database. *Journal of Chemical Information and Computer Sciences*, **44** (1), 276–285.

109 Baber, J.C., Shirley, W.A., Gao, Y., and Feher, M. (2005) The use of consensus scoring in ligand-based virtual screening. *Journal of Chemical Information and Modeling*, **46** (1), 277–288.

110 Charifson, P.S., Corkery, J.J., Murcko, M.A., and Walters, W.P. (1999) Consensus scoring: A method for obtaining improved hit rates from docking databases of three-dimensional structures into proteins. *Journal of Medicinal Chemistry*, **42** (25), 5100–5109.

111 Paul, N. and Rognan, D. (2002) ConsDock: A new program for the consensus analysis of protein–ligand interactions. *Proteins: Structure, Function, and Bioinformatics*, **47** (4), 521–533.

112 Bruce, C.L., Melville, J.L., Pickett, S.D., and Hirst, J.D. (2007) Contemporary QSAR classifiers compared. *Journal of Chemical Information and Modeling*, **47** (1), 219–227.

113 Myles, A.J., Feudale, R.N., Liu, Y., Woody, N.A., and Brown, S.D. (2004) An introduction to decision tree modeling. *Journal of Chemometrics*, **18** (6), 275–285.

114 Carvalho, D.R. and Freitas, A.A. (2004) A hybrid decision tree/genetic algorithm method for data mining. *Information Sciences*, **163** (1–3), 13–35.

115 Dudek, A.Z., Arodz, T., and Galvez, J. (2006) Computational methods in developing quantitative structure–activity relationships (QSAR): A review. *Combinatorial Chemistry and High Throughput Screening*, **9**, 213–228.

116 Hou, T., Wang, J., Zhang, W., and Xu, X. (2006) ADME evaluation in drug discovery: Part 7. Prediction of oral absorption by correlation and classification. *Journal of Chemical Information and Modeling*, **47** (1), 208–218.

117 Breiman, L. (2001) Random Forests. *Machine Learning*, **45** (1), 5–32.

118 Svetnik, V., Liaw, A., Tong, C., Culberson, J.C., Sheridan, R.P., and Feuston, B.P. (2003) Random forest: A classification and regression tool for compound classification and QSAR modeling. *Journal of Chemical Information and Computer Sciences*, **43** (6), 1947–1958.

119 Tong, W., Hong, H., Fang, H., Xie, Q., and Perkins, R. (2003) Decision forest: combining the predictions of multiple independent decision tree models. *Journal of Chemical Information and Computer Sciences*, **43** (2), 525–531.

120 Cortes, C. and Vapnik, V.N. (1995) Support-vector networks. *Machine Learning*, **20**, 273–297.

121 Scholkopf, B., Smola, A.J., Williamson, R.C., and Bartlett, P.L. (2000) New support vector algorithms. *Neural Computation*, **12**, 1207–1245.

122 Rosipal, R. and Trejo, L.J. (2001) Kernel partial least squares regression in reproducing kernel Hilbert space. *Machine Learning Resources*, **2**, 97–123.

123 Bennett, K. and Campbell, C. (2000) Support vector machines: Hype or hallelujah. *SIGKDD Explorations*, **2** (2), 1–13.

124 Embrechts, M.J., Arciniegas, F.A., Ozdemir, M., Breneman, C.M., Bennett, K.P., and Lockwood, L. (2001) Bagging Neural Network Sensitivity Analysis for Feature Reduction in QSAR Problems. 2001 INNS–IEEE International Joint Conference on Neural Networks, Washington, DC, IEEE Press, Washington, DC, pp. 2478–2482.

125 Bakken, G.A. and Jurs, P.C. (2000) Classification of multidrug-resistance reversal agents using structure-based descriptors and linear discriminant analysis. *Journal of Medicinal Chemistry*, **43** (23), 4534–4541.

126 Bennett, K. and Demiriz, A. (2000) Optimization approaches to semi-supervised learning, in *Applications and Algorithms of Complementarity* (eds M.C. Ferris, O.L. Mangasarian, and J.S. Pang), Kluwer Academic, Boston.

127 Burbidge, R., Trotter, M., Buxton, B., and Holden, S. (2001) Drug design by machine learning: Support vector machines for pharmaceutical data analysis. *Computational Chemistry*, **26**, 5–14.

128 Czerminski, R., Yasri, A., and Hartsough, D. (2001) Use of support vector machine in pattern classification: Application to QSAR studies. *Quantitative Structure–Activity Relationships*, **20**, 227–240.

129 Bennett, K.P. and Embrechts, M.J. (2003) An optimization perspective on partial least squares, in *Advances in Learning Theory: Methods, Models and Applications*, vol. 190 (eds J.A.K. Suykens, G. Horvath, S. Basu, C. Micchelli, and J. Vandewalle), IOS Press, Amsterdam, pp. 227–250.

130 Ryan, D., McLellan, M., and Breneman, C.M. (2008) QSAR model stability: How much information is in the data? American Chemical Society National Meeting, New Orleans, LA.

131 Geppert, H., Horváth, T., Gärtner, T., Wrobel, S., and Bajorath, J. (2008) Support-vector-machine-based ranking significantly improves the effectiveness of similarity searching using 2D fingerprints and multiple reference compounds. *Journal of Chemical Information and Modeling*, **48** (4), 742–746.

132 Agarwal, S., Dugar, D., and Sengupta, S. (2010) Ranking chemical structures for drug discovery: A new machine learning approach. *Journal of Chemical Information and Modeling*, **50** (5), 716–731.

133 Bergeron, C., Zaretzki, J., Breneman, C., and Bennett, K.P. (2008) Multiple instance ranking. Proceedings of the 25th International Conference on Machine Learning, ACM, Helsinki, Finland, pp. 48–55.

134 Verma, R.P. and Hansch, C. (2005) An approach toward the problem of outliers in QSAR. *Bioorganic and Medicinal Chemistry*, **13** (15), 4597–4621.

135 Maggiora, G.M. (2006) On outliers and activity cliffs – why QSAR often disappoints. *Journal of Chemical Information and Modeling*, **46** (4), 1535.

136 Fourches, D., Muratov, E., and Tropsha, A. (2010) Trust, but verify: On the importance of chemical structure curation in cheminformatics and QSAR modeling research. *Journal of Chemical Information and Modeling*, **50** (7), 1189–1204.

137 Embrechts, M.J., Robert Kewley, J., and Breneman, C. (1998) Computationally

intelligent data mining for the automated design and discovery of novel pharmaceuticals, in *Smart Engineering Systems: Neural Networks, Fuzzy Logic, Evolutionary Programming, Data Mining and Rough Sets, November 1–4, 1998* (eds C.H. Dagli, M. Akay, A.L. Buczak, O. Ersoy, and B.R. Fernandex), ASME Press, St. Louis, MO, pp. 397–403.

**138** Mazzatorta, P., Benfenati, E., Neagu, D., and Gini, G. (2002) The importance of scaling in data mining for toxicity prediction. *Journal of Chemical Information and Computer Sciences*, **42** (5), 1250–1255.

**139** Wold, S. and Dunn, W.J. (1983) Multivariate quantitative structure–activity relationships (QSAR): Conditions for their applicability. *Journal of Chemical Information and Computer Sciences*, **23** (1), 6–13.

**140** Tropsha, A., Gramatica, P., and Gombar, V.K. (2003) The importance of being earnest: Validation is the absolute essential for successful application and interpretation of QSPR models. *QSAR and Combinatorial Science*, **22** (1), 69–77.

**141** Roy, P.P., Leonard, J.T., and Roy, K. (2008) Exploring the impact of size of training sets for the development of predictive QSAR models. *Chemometrics and Intelligent Laboratory Systems*, **90** (1), 31–42.

**142** Liu, H., Papa, E., and Gramatica, P. (2006) QSAR prediction of estrogen activity for a large set of diverse chemicals under the guidance of OECD principles. *Chemical Research in Toxicology*, **19** (11), 1540–1548.

**143** Cramer, R.D., Bunce, J.D., Patterson, D.E., and Frank, I.E. (1988) Crossvalidation, bootstrapping, and partial least squares compared with multiple regression in conventional QSAR studies. *Quantitative Structure–Activity Relationships*, **7** (1), 18–25.

**144** Rücker, C., Rücker, G., and Meringer, M. (2007) y-randomization and its variants in QSPR/QSAR. *Journal of Chemical Information and Modeling*, **47** (6), 2345–2357.

**145** Clark, R. and Fox, P. (2004) Statistical variation in progressive scrambling. *Journal of Computer-Aided Molecular Design*, **18** (7), 563–576.

**146** Jaworska, J., Nikolova-Jeliazkova, N., and Aldenberg, T. (2005) QSAR applicability domain estimation by projection of the training set in descriptor space: a review. *Alternatives to Laboratory Animals*, **33** (5), 445–459.

**147** Dimitrov, S., Dimitrova, G., Pavlov, T., Dimitrova, N., Patlewicz, G., Niemela, J., and Mekenyan, O. (2005) A stepwise approach for defining the applicability domain of SAR and QSAR models. *Journal of Chemical Information and Modeling*, **45** (4), 839–849.

**148** Tetko, I.V., Sushko, I., Pandey, A.K., Zhu, H., Tropsha, A., Papa, E., Öberg, T., Todeschini, R., Fourches, D., and Varnek, A. (2008) Critical assessment of QSAR models of environmental toxicity against *Tetrahymena pyriformis*: Focusing on applicability domain and overfitting by variable selection. *Journal of Chemical Information and Modeling*, **48** (9), 1733–1746.

**149** Das, S., Krein, M.P., and Breneman, C.M. (2010) Binding affinity prediction with property-encoded shape distribution signatures. *Journal of Chemical Information and Modeling*, **50** (2), 298–308.

**150** Guha, R. and Van Drie, J.H. (2008) Structure–activity landscape index: Identifying and quantifying activity cliffs. *Journal of Chemical Information and Modeling*, **48**, 646–658.

**151** Peltason, L. and Bajorath, J. (2007) SAR index: quantifying the nature of structure–activity relationships. *Journal of Medicinal Chemistry*, **50** (23), 5571–5578.

**152** Peltason, L., Iyer, P., and Bajorath, J. (2010) Rationalizing three-dimensional activity landscapes and the influence of molecular representations on landscape topology and the formation of activity cliffs. *Journal of Chemical Information and Modeling*, **50** (6), 1021–1033.

**153** Bajorath, J., Peltason, L., Wawer, M., Guha, R., Lajiness, M.S., and Van Drie, J.H. (2009) Navigating structure–activity landscapes. *Drug Discovery Today*, **14** (13–14), 698–705.

# 3
# Mold$^2$ Molecular Descriptors for QSAR

*Huixiao Hong, Svetoslav Slavov, Weigong Ge, Feng Qian, Zhenqiang Su, Hong Fang, Yiyu Cheng, Roger Perkins, Leming Shi, and Weida Tong*

## 3.1
## Background

### 3.1.1
### History of QSAR

The physical, chemical, and biological properties of compounds are uniquely determined by their structure and vary with it in a systematic way. Some of simple properties such as polarizability, dipole moments, heat of formation, and so on can be obtained directly by using quantum-mechanical calculations. However, more complex properties such as partition coefficients, retention indices, toxicological endpoints, pharmacological activity, and so on require deeper exploration of the chemical structure to determine their relationship. The chemical structure itself can be expressed in various ways: by defining the position and the type of the building blocks (atoms or functional groups), by using electron density maps or isopotential surfaces, by describing the connectivity/topology, or by calculating any alternative set of parameters representing the structure. The numerical value characteristics for a given structure are known as molecular descriptors. These can be obtained either experimentally (Hammett substituent constants, octanol/water partition coefficients, Ostwald solubility coefficients, etc.) or computationally (Kier and Hall indices, partial positively or negatively charged surface areas, electrostatic potentials, etc.). When expressed quantitatively, the relationship between the molecular structure and the property of interest is known as a quantitative structure–activity relationship (QSAR), and "A" is normally taken to be biological activity. The value of QSAR is self-evident, recognizing that when a difficult to calculate or measure property or activity is known for a set of chemicals with varying structure, a correlation can be constructed from which to readily calculate that property or activity for other not yet synthesized or tested chemicals.

*Statistical Modelling of Molecular Descriptors in QSAR/QSPR*. First Edition.
Edited by M. Dehmer, K. Varmuza, and D. Bonchev

Despite being coined recently, the QSAR has its roots in the nineteenth century, when Crum-Brown and Frazer proposed the following very simple equation for the curare-like paralyzing properties of a set of quaternised strychnines:

$$\Phi = f(C) \tag{3.1}$$

In this equation $f$ is a measure of the biological activity and $C$ represents relevant structural features characterizing the quaternising group [1]. Shortly thereafter, Richardson [2] showed that the toxicity effect of ethers and alcohols could be related to their water solubility by a reciprocal function. Richet [3] discovered the relationship between the narcotic effect of alcohols and their molecular weight, and Meyer [4] and Overton [5] reported QSARs describing the narcotic effect of a series of diverse organic compounds as a function of their corresponding oil/water partition coefficients. In the 1930s, chemists focused their attention on the relationship between the chemical structure and the equilibrium or rate constants of chemical reactions [6]. Among all studies, the work of Hammett, who invented a scale of electronic effects of substituents, attracted the most significant attention. In his equation,

$$\varrho^{*}\sigma_x = \log K_X - \log K_H \tag{3.2}$$

where $K_X$ and $K_H$ represent the "equilibrium constants" for a reaction involving, respectively, an $X$ substituted compound and its unsubstituted parent, $\varrho$ is a reaction constant that is characteristic of a particular reaction, and $\sigma_x$ is the substituent constant for the substituent $X$. The Hammett equation enabled the creation of $\sigma$ scales useful for characterization of different types of electronic effects [7].

In the 1960s Corwin Hansch proposed a hydrophobicity model system based on octanol/water partition coefficients (log$P$) [8]. However, his most significant contribution to the emerging QSAR field was the supposition that the use of only one variable (such as log$P$) might be inadequate to explain a compound's potency. The generalized form of his equation now known as "Hansch equation" [9] is

$$\text{Log}\,1/C = a\pi + b\pi^2 + c\sigma + dE_S + \text{constant} \tag{3.3}$$

The terms in Equation 3.3 representing the hydrophobic and electronic substituent are the constants $\pi$ and $\sigma$, respectively, and the steric substituent is the constant $E_S$ [10].

In 1979, Cramer and Milne first attempted to compare a set of congeners by aligning them in space and consecutively mapping their molecular fields to a 3D grid [11]. Several important events facilitated the broader application of this approach: (i) in 1986, Svante Wold proposed the use of the partial least-squares (PLS) analysis, as opposed to principal component analysis (PCA), to correlate the field values with the activity; (ii) this key work was published in the prestigious *Journal of the American Chemical Society* [12] and the method was officially named as Comparative Molecular Field Analysis (CoMFA); (iii) appropriate software became commercially available [13].

The increased computational power available in the late 1990s enabled simulations of the ligands (drugs) binding to receptors. Unsurprisingly, the new powerful method was called “molecular docking.” The first successful “blind” docking was published in 1996 [14] for TEM-1 Beta-lactamase to the Beta-lactamase inhibitor protein.

### 3.1.2
### Introduction to QSAR

As understood in the field of computational chemistry, QSAR is a rational approach for revealing the relationship between the structure expressed in terms of empirical or calculated molecular descriptors and the property or the biological/pharmacological activity of a series of compounds. Once the functional form of the dependence on the chemical structure is known, it is straightforward to predict how each structural change affects the activity/property. This approach might also enable important structural features to be identified, providing physicochemical insights for synthesizing molecules more likely to possess improved characteristics. For example, features increasing a desired bioactivity and/or decreasing an undesirable one might be revealed.

The so-called classical or two-dimensional QSAR relates the activities of a series of compounds to a set of independent variables encoding certain structural features. In addition to the lipophilicity, polarizability, and electronic properties, steric and topological parameters are also frequently used to describe the size, flexibility, and the “information” richness of the molecules. In some cases, indicator variables have been used to differentiate racemates and active enantiomers [15, 16].

Instead of dealing with scalar quantities (2D-QSAR descriptors), the 3D-QSAR utilizes steric, electrostatic, and hydrophobic fields. This method assumes that when “looking” at the ligand (drug molecule), the binding site does not see individual atoms and chemical bonds. From a far distance, the binding site would “feel” only the electrostatic potential of the molecule, while at a closer distance it would “feel” the relatively solid body of the molecule with its unique charge distribution pattern at the solvent-accessible surface area.

Molecular docking is an alternative QSAR approach that does not rely on the calculation of molecular descriptors. Rather, the process by which the ligand and the receptor bind together in three-dimensional space is directly simulated. Importantly, the free energy of the bound system can be estimated and then used to identify ligands that are most likely to interact favorably with a particular receptor.

The primary steps of QSAR are generation, validation, and, if possible, a physicochemical interpretation of the predictive models. The model generation procedure involves identifying informative descriptors extracted from the initial descriptor pool that explain a sufficient amount of variance in the experimental data. Statistical and data mining methods that are used include decision forest (DF) [17–19]; consensus analysis (CANCON) [20]; stepwise multiple linear regression (SMLR) [21]; principal components regression (PCR) [22]; partial least squares (PLS) [23]; genetic function approximation (GFA) [24]; artificial neural networks (ANN) [25]; support vector machine [26]; and Random Forest [27].

Once a QSAR model is developed, it has to be subjected to a validation procedure in order to estimate its power of prediction. The two main validation categories are internal and external validation. Popular internal validation methods include the leave-one-out cross-validation [28], *k*-fold cross-validation [28], bootstrapping [29], and scrambling [30]. Special care must be taken to minimize or avoid the bias in internal cross-validation due to leakage of information from the training set. External validation relies on a procedure where the chemical structures selected to form the external test set are different and specifically excluded from those used to train the model. The QSAR model developed using only the chemicals from the training set is then applied to the "unknown" external test set to verify the predictive ability of the model. True external validation is generally considered to yield a more accurate assessment than cross-validation.

One of the most challenging aspects of QSAR is the interpretation of the structure–activity relationship information that is encoded in the model. It has to be emphasized that the perceived value of the model increases if its physical/chemical/pharmacological significance can be explained. Unless the equation is very simple, the interpretation of even a conventional QSAR model might be a complicated task for the following reasons: the regression coefficients are not always directly interpretable because they may represent a complex picture of two or more superimposed structural trends buried in the model; some structural descriptors might be derived by using purely mathematical constructs without a direct physicochemical interpretation; and some descriptors might be surrogate measures for structural features characterized not accurately enough by another descriptor with a more intuitive physicochemical meaning.

The ultimate goal of almost every QSAR model is the successful prediction of compounds with desired characteristics. In this endeavor, the 3D-QSAR models provide more useful insights and accurate predictions since the exact location of the positively and negatively contributing substituents are known. However, the 2D-QSAR models may achieve similar levels of accuracy if fragment descriptors are used. The predictions should then be analyzed, and if inaccurate predictions were made, the reasons need to be further explored.

### 3.1.3
### Molecular Descriptors: Bridge for QSAR

In mathematics, a QSAR model, either qualitative or quantitative, is a mathematical function that can be used to describe the relationship between structures of chemicals and their biological activities. In other words, a QSAR model is a mathematical function that can be used to determine the biological function of a chemical from its structure. While a model that can be used to predict a compound's biological activity directly from its molecular structure (the red path in Figure 3.1) is an ideal solution, it is usually very difficult and may not be attainable. Therefore, to construct a QSAR model, an indirect approach is usually used to tackle this problem. First, a set of molecular descriptors that encode chemical structure information are calculated for compounds. Then a mathematical (or statistical) model is developed to correlate the biological activity of interest with

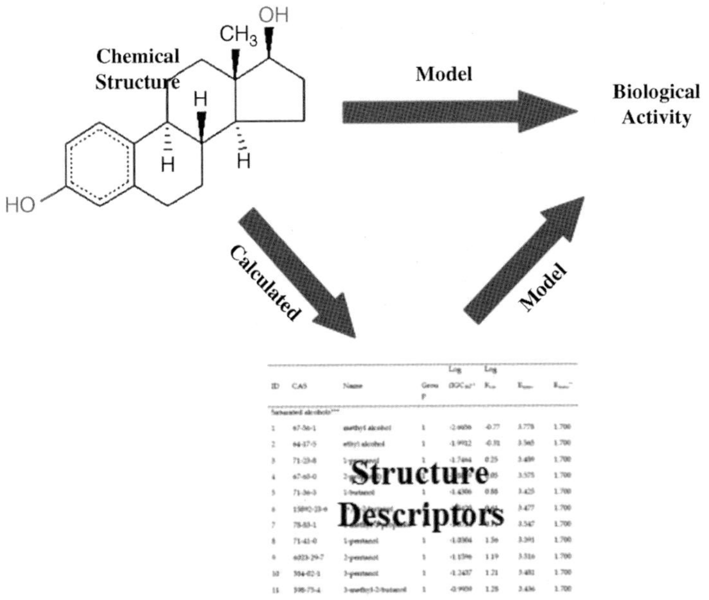


**Figure 3.1** Molecular descriptors bridge chemical structure to biological activity.

the molecular descriptors (or, in most cases, a subset of the descriptors) of the compounds (the blue path in Figure 3.1).

#### 3.1.3.1 Molecular Descriptors

An assumption of QSAR is that the molecular structure of a chemical must contain features responsible for its physical, chemical, and biological properties. As discussed above, when encoded as numerical values, these features are known as molecular descriptors. Those that are empirical in nature (such as substituent constants, partition coefficients, and various electronegativity-related parameters) are most frequently used for QSAR. Even today, electronegativity remains a popular and broadly employed molecular descriptor [31–37], while the substituent constants are usually neglected. With the increased computational power of the modern computers, the quantum-chemical, electronic, geometrical, constitutional, and topological descriptors become preferable. The quantum-chemical, electronic, and geometrical descriptors are usually derived from the results of empirical schemes or molecular orbital calculations, and they encode the molecule's ability to participate in polar or hydrogen bonding (donor, acceptor) interactions. The constitutional descriptors are fragment-additive and mostly reflect the general properties of compound structures. The topological descriptors are calculated using graph theory and may represent the structural flexibility and the degree of molecular branching or the information content.

Despite their different origin and mathematical formulation, most of the molecular descriptors are interconnected. For example, the topological changes in molecules can be related to the changes in their geometry and thus could act as a measure of their shape, symmetry, and degree of branching. However, the changes in symmetry or branching can affect the distribution of the electronic charge, which can cause changes in the chemical reactivity or polarity, respectively. Similarly, in some cases, a parameter depicting the distribution of the electronic density within the molecule might act as a better measure of branching, and therefore of symmetry and shape, than do the typical topological descriptors [38]. Thus, simply knowing the type of features measured by a certain descriptor could be insufficient in itself, especially when the final goal of the study is to interpret the QSAR model in terms of variation of certain structural characteristics.

Due to inherent covariance, descriptors involved in a particular QSAR model need to be tested for mutual orthogonality. It is recommended that an orthogonality threshold be imposed on all possible descriptor pairs, and that from each highly collinear pair, only the descriptor with a higher *t*-value be retained in the model. However, in some cases, it might be desirable to retain the descriptor with a lower *t*-value, especially if it has more intuitive physical/chemical or pharmacological interpretation.

#### 3.1.3.2 **Role of Molecular Descriptors**

The QSAR model is the final visible outcome of a lengthy process, whose starting point is the suitable and comprehensive description of the molecular structure. Due to increased interest toward cost-effective drug design, toxicology screening, and other related problems, hundreds of molecular descriptors were proposed during the past few decades. The early days of QSAR relied on simple schemes for the generation of molecular descriptors that were mostly empirical and constitutional. With the increased accessibility to high performance computing, more computationally demanding approaches for descriptor calculation became commonplace. Thus, the quantum mechanics that were considered to be the best description of the physical reality became preferred as the most appropriate method for calculating molecular descriptors. A trend developed that using quantum-chemical descriptors was a better approach for QSAR.

For a time, topological descriptors and chemical graph theory itself were met with suspicion and skepticism. To some extent this slowed acceptance of other innovative approaches (e.g., fingerprint descriptors based on the vectorial representation of the chemical structure). As time passed, it was demonstrated that some "innovative" molecular descriptors could yield more accurate models than the "classical" ones originating from quantum mechanics (QM). Moreover, topological descriptors derived from 2D molecular structures were shown to correlate well with some three-dimensional properties of compounds, such as collision cross-sections, dipole moments, solubility, and so on. Some descriptors (e.g., Kier flexibility index, Kier and Hall valence connectivity indices, etc.) were highly correlated (coefficients close to 0.8) with the partial positively or negatively charged surface areas and other quantum-chemical parameters. While such molecular descriptors tend to be difficult or even impossible to be physically interpreted, their contribution to forming accurate predictive models provides strong motivation for their use.

There is no consensus for a set of rules or criteria to guide new descriptor development for QSAR. However, some general criteria or goals to strive for in such an endeavor are listed as follows [39]:

1) A new descriptor should be able to discriminate between the molecules, preferably even between stereoisomers.
2) The numerical values of a descriptor should change gradually with gradual changes in molecular structure.
3) There should be at least to some degree a structural interpretation.
4) It must be proven useful; that is, it should correlate well with at least one property.
5) It should come from a simple computational scheme (if too complex, it likely correlates well with one or more existing descriptors (collinear descriptors)].
6) It should be translationally or rotationally invariant with respect to the chosen reference frame.

The field of molecular descriptors is interdisciplinary and involves the application of multiple theories: the graph theory, the information theory; the QM theory; the theory of organic reactivity and others. To use them efficiently, knowledge of chemoinformatics is necessary.

#### 3.1.3.3 **Types of Molecular Descriptors**

Historically, two major schemes for the classification of the molecular descriptors were established: one based on their origin, and another on their dimensionality.

Descriptors based on origin include topological (graph theory based), constitutional (functional group count), geometrical (distances, valence angles, surfaces, etc.), quantum-chemical (charge distribution related), and thermodynamic (heat of formation, entropy, etc.) descriptors.

Descriptors based on dimensionality include zero-dimensional descriptors (i.e., constitutional descriptors), 1D-descriptors (structural fragments, fingerprints), 2D-descriptors (topological indices), and 3D-descriptors [surface, volume, quantum-chemical descriptors, 3D-MoRSE (3D Molecule Representation of Structure based on Electron diffraction), etc.

#### 3.1.3.4 **Calculation of Molecular Descriptors (Software Packages)**

The ever increasing interest in activity/property prediction during the past decade resulted in a large number of software products brought to the market and the open source domain for scientists working in the field of QSAR. Table 3.1 lists some popular packages for calculating molecular descriptors. Additionally, some of them are complex products that also include modules for statistical analysis and data visualization.

## 3.2
## Mold$^2$ Molecular Descriptors

There is a long-standing debate regarding whether 3D molecular descriptors yield better predictive models than 2D descriptors. A preference for 3D has been argued

**Table 3.1** List of software packages for descriptors calculation.

| Name | Organization/Institution | Availability | Descriptors | Platform/Licence |
|---|---|---|---|---|
| ADAPT | Pennsylvania State University | http://research.chem.psu.edu/pcjgroup/adapt.html | >260 | Unix/Linux (freeware) |
| ADMET Predictor™ | Simulations Plus, Inc. | http://www.simulations-plus.com/ | 297 | Windows (commercial) |
| CODESSA | Semichem | http://www.semichem.com/codessa/default.php | >600 | Windows/Linux (commercial) |
| DRAGON | Talete SRL | http://www.talete.mi.it/products/dragon_description.htm | 3224 | Windows/Linux (commercial) |
| EPISUITE™ | EPA | http://www.epa.gov/opptintr/exposure/pubs/episuite.htm | ~20 | Windows (freeware) |
| MOE | Chemical Computing Group | http://www.chemcomp.com/software-moe2009.htm | >300 | Windows/Linux/SGI/MAC/Sun (freeware) |
| Molconn-Z™ | EduSoft | http://www.edusoft-lc.com/molconn/ | 327 | Windows/Unix/MAC (commercial) |
| MOLD$^2$ | NCTR/FDA | http://www.fda.gov/ScienceResearch/BioinformaticsTools/Mold2/default.htm | 777 | Windows (freeware) |
| MOLGEN | University of Bayreuth | http://www.molgen.de/?src=documents/molgenqspr.html | 707 | Windows (commercial) |
| PowerMV | National Institute of Statistical Sciences | http://nisla05.niss.org/PowerMV/?q=PowerMV/ | >1000 | Windows (freeware) |
| Sarchitect™ | Strand Life Sciences | http://www.strandls.com/sarchitect/index.html | 1084 | Windows/Linux (commercial) |
| SciQSAR™ | SciMatics | http://www.scimatics.com/jsp/qsar/QSARIS.jsp | >600 | Windows (commercial) |

from the standpoint of handling absolute stereochemistry and the advantages of using force fields to model complexities of ligand binding to a receptor. In contrast, some authors have shown in comparative studies [40–42] that 2D descriptors can perform as well as 3D descriptors in most applications. For many applications in QSAR and predictive toxicology, it appears that the simpler and less computationally expensive 2D descriptors are adequate and can be used to avoid the need to deal with estimating the bioactive conformations required for 3D methods. An optimal set of molecular descriptors in the absolute sense may well be indeterminable.

The Mold$^2$ software was developed to enable the rapid calculation of a large and diverse set of descriptors encoding both 1D and 2D chemical structure information [43]. The low computing cost for Mold$^2$ makes it suitable not only for small data sets, as is normal in QSAR applications, but especially for the large databases typical in virtual compound screening. Use of the Mold$^2$ descriptors does not require determination of 3D structures, and consequently, both the descriptors and models derived from them should exhibit high reliability and reproducibility.

The current version of Mold$^2$ calculates 777 1D and 2D molecular descriptors. The detail descriptions of the 777 molecular descriptors are given in Table 3.2. For simplicity, the 777 descriptors are organized and discussed in groups by their origin. Mold$^2$ is freely available to the public at: (http://www.fda.gov/nctr/science/centers/toxicoinformatics/index.htm).

## 3.2.1
## Description of Mold$^2$ Descriptors

### 3.2.1.1 Topological Descriptors

Most Mold$^2$ topological descriptors are developed from hydrogen-suppressed, undirected molecular graphs where the atoms are represented by vertices and the bonds by edges. Only the connectivity between the atoms constituting the molecule is required for their calculation. Since no quantum-chemical optimizations are needed, the calculation of hundreds of topological descriptors is expeditious even for huge data sets.

Examples of easily interpretable topological descriptors implemented in Mold$^2$ are as follows.

The *Wiener index* [44, 45] is defined as a sum of the number of edges in the shortest path between all pairs of non-hydrogen atoms in a molecule. It is the oldest topological index related to molecular branching and correlates well with molecular van der Waals surface area. The *Wiener index* was originally defined only for acyclic graphs. In Mold$^2$, *Wiener index* descriptors are derived from the Wiener matrix modified from the distance matrix of a molecule. Examples include the normalized *Wiener index*, *Wiener index degree*, and *resistance distance hyper-Wiener index*.

The *Kier index* [46–49] encodes the chemical structure flexibility, which depends on molecule size, branching, cycles, and heteroatom content. The flexibility index is made equal to the product of the kappa-one and kappa-two modified shape descriptors normalized to the number of atoms in the H-depleted molecular graph. The Mold$^2$ *Kier index* descriptors are determined from the Kier indices that are a set of topological shape indices defined in terms of the number of graph vertices and the

**Table 3.2** Description of Mold² molecular descriptors.

| Descriptor | Description |
|---|---|
| D001 | Number of 6-membered aromatic rings (only carbon atoms) |
| D002 | Number of 03-membered rings |
| D003 | Number of 04-membered rings |
| D004 | Number of 05-membered rings |
| D005 | Number of 06-membered rings |
| D006 | Number of 07-membered rings |
| D007 | Number of 08-membered rings |
| D008 | Number of 09-membered rings |
| D009 | Number of 10-membered rings |
| D010 | Number of 11-membered rings |
| D011 | Number of 12-membered rings |
| D012 | Number of multiple bonds |
| D013 | Number of circuits structure |
| D014 | Number of rotatable bonds |
| D015 | Rotatable bond fraction |
| D016 | Number of double bonds |
| D017 | Number of aromatic bonds |
| D018 | Sum of conventional bond orders (H-depleted) |
| D019 | Number of hydrogen |
| D020 | Number of helium |
| D021 | Number of lithium |
| D022 | Number of beryllium |
| D023 | Number of boron |
| D024 | Number of carbon |
| D025 | Number of nitrogen |
| D026 | Number of oxygen |
| D027 | Number of fluorine |
| D028 | Number of neon |
| D029 | Number of sodium |
| D030 | Number of magnesium |
| D031 | Number of aluminum |
| D032 | Number of silicon |
| D033 | Number of phosphorus |
| D034 | Number of sulfur |
| D035 | Number of chlorine |
| D036 | Number of argon |
| D037 | Number of potassium |
| D038 | Number of calcium |
| D039 | Number of scandium |
| D040 | Number of titanium |
| D041 | Number of vanadium |
| D042 | Number of chromium |
| D043 | Number of manganese |
| D044 | Number of iron |
| D045 | Number of cobalt |
| D046 | Number of nickel |

**Table 3.2** (*Continued*)

| Descriptor | Description |
|---|---|
| D047 | Number of copper |
| D048 | Number of zinc |
| D049 | Number of gallium |
| D050 | Number of germanium |
| D051 | Number of arsenic |
| D052 | Number of selenium |
| D053 | Number of bromine |
| D054 | Number of krypton |
| D055 | Number of rubidium |
| D056 | Number of strontium |
| D057 | Number of yttrium |
| D058 | Number of zirconium |
| D059 | Number of niobium |
| D060 | Number of molybdenum |
| D061 | Number of technetium |
| D062 | Number of ruthenium |
| D063 | Number of rhodium |
| D064 | Number of palladium |
| D065 | Number of silver |
| D066 | Number of cadmium |
| D067 | Number of indium |
| D068 | Number of tin |
| D069 | Number of antimony |
| D070 | Number of tellurium |
| D071 | Number of iodine |
| D072 | Number of xenon |
| D073 | Number of cesium |
| D074 | Number of barium |
| D075 | Number of lanthanum |
| D076 | Number of cerium |
| D077 | Number of praseodymium |
| D078 | Number of neodymium |
| D079 | Number of promethium |
| D080 | Number of samarium |
| D081 | Number of europium |
| D082 | Number of gadolinium |
| D083 | Number of terbium |
| D084 | Number of dysprosium |
| D085 | Number of holmium |
| D086 | Number of erbium |
| D087 | Number of thulium |
| D088 | Number of ytterbium |
| D089 | Number of lutetium |
| D090 | Number of hafnium |
| D091 | Number of tantalum |
| D092 | Number of tungsten |

(*Continued*)

**Table 3.2** (*Continued*)

| Descriptor | Description |
|---|---|
| D093 | Number of rhenium |
| D094 | Number of osmium |
| D095 | Number of iridium |
| D096 | Number of platinum |
| D097 | Number of gold |
| D098 | Number of mercury |
| D099 | Number of thallium |
| D100 | Number of lead |
| D101 | Number of bismuth |
| D102 | Number of polonium |
| D103 | Number of astatine |
| D104 | Number of radon |
| D105 | Number of francium |
| D106 | Number of radium |
| D107 | Number of actinium |
| D108 | Number of thorium |
| D109 | Number of protactinium |
| D110 | Number of uranium |
| D111 | Number of neptunium |
| D112 | Number of plutonium |
| D113 | Number of americium |
| D114 | Number of curium |
| D115 | Number of berkelium |
| D116 | Number of californium |
| D117 | Number of einsteinium |
| D118 | Number of fermium |
| D119 | Number of mendelevium |
| D120 | Number of nobelium |
| D121 | Number of lawrencium |
| D122 | Molecular weight |
| D123 | Average of molecular weight |
| D124 | Number of atoms in each molecule |
| D125 | Number of non-hydrogen atoms in each molecule |
| D126 | Number of bonds in each molecule |
| D127 | Number of non-hydrogen bonds in each molecule |
| D128 | Number of rings in each molecule |
| D129 | Number of triple bonds in each molecule |
| D130 | Number of halogen atoms in each molecule |
| D131 | Molecular size index |
| D132 | Atomic composition index |
| D133 | Mean value of atomic composition index |
| D134 | Branch index |
| D135 | Molecular structure connectivity index |
| D136 | Narumi-type topological index |
| D137 | Harmonic topological index |
| D138 | Geometric topological index |

**Table 3.2** (*Continued*)

| Descriptor | Description |
|---|---|
| D139 | Topological distance count order 3 |
| D140 | Log of vertex distance path count |
| D141 | Average of vertex distance path count |
| D142 | Balaban type of mean square vertex distance index |
| D143 | Sum of atomic van der Waals carbon-scale |
| D144 | Mean atomic van der Waals carbon-scale |
| D145 | Sum of atomic electronegativities Pauling-scale on carbon |
| D146 | Mean atomic electronegativities Pauling-scaled on carbon |
| D147 | Sum of atomic electronegativities Sanderson-scaled on carbon |
| D148 | Mean atomic electronegativity Sanderson-scaled on carbon |
| D149 | Sum of atomic electronegativity Allred-Rochow-scaled on carbon |
| D150 | Mean atomic electronegativity Allred-Rochow-scaled on carbon |
| D151 | Sum of atomic polarizabilities scaled on carbon-SP3 |
| D152 | Mean atomic polarizability scaled on carbon-SP3 |
| D153 | Zagreb order-1 index |
| D154 | Zagreb order-1 index with value of valence vertex degrees |
| D155 | Zagreb order-2 index |
| D156 | Vertex degree topological index |
| D157 | Second Zagreb order-2 index with a value of valence vertex degrees |
| D158 | Valence electrons of the principal quantum index |
| D159 | Schultz-type molecular topological index |
| D160 | Schultz-type molecular topological index of valence vertex degrees |
| D161 | Molecular topological distance index |
| D162 | Molecular topological distance index of valence vertex degrees |
| D163 | Molecular size and branching index |
| D164 | Index of terminal vertex matrix |
| D165 | Wiener index |
| D166 | Average path length in Wiener index |
| D167 | Reciprocal index of Wiener distance matrix |
| D168 | Harary index |
| D169 | Index of Laplacian matrix |
| D170 | First no-zero eigenvalue of Laplacian matrix |
| D171 | Wiener-path index |
| D172 | Reciprocal Wiener-path index |
| D173 | Mohar order-2 index |
| D174 | Maximum path index |
| D175 | Wiener type maximum path index |
| D176 | Reciprocal Wiener-type maximum path index |
| D177 | Minimum-path/maximum-path index |
| D178 | All-Path Wiener – sum of the edges in the shortest paths between all pairs of non-hydrogen atoms |
| D179 | Heteroatoms and multiple bonds weighted distance matrix |
| D180 | Mass weighted distance matrix |
| D181 | Index of van der Waals weighted distance matrix |
| D182 | Distance matrix of electronegativity weighted with electronegativities Pauling-scale |

(*Continued*)

**Table 3.2** (*Continued*)

| Descriptor | Description |
|---|---|
| D183 | Distance matrix of electronegativity weighted with Sanderson electronegativities |
| D184 | Distance matrix of electronegativity weighted with Allred-Rochow electronegativites |
| D185 | Polarizability weighted distance matrix |
| D186 | Average vertex distance connectivity index |
| D187 | Balaban heteroatoms bonds weighted index |
| D188 | Balaban mass weighted index |
| D189 | Balaban van der Waals weighted index |
| D190 | Balaban electronegativity weighted with the Pauling-scale index |
| D191 | Balaban electronegativity weighted with the Sanderson-scale index |
| D192 | Balaban electronegativity weighted with the Allred-Rochow-scale index |
| D193 | Balaban-type polarizability weighted index |
| D194 | Maximal valence vertex electrotopological negative variation |
| D195 | Maximal valence vertex electrotopological positive variation |
| D196 | Sum absolute electrotopological negative variation |
| D197 | Electrotopological index |
| D198 | Sum electrotopological states index |
| D199 | Mean electrotopological states index |
| D200 | Vertex connectivity order-0 index |
| D201 | Vertex connectivity order-1 index |
| D202 | Vertex connectivity order-2 index |
| D203 | Vertex connectivity order-3 index |
| D204 | Vertex connectivity order-4 index |
| D205 | Vertex connectivity order-5 index |
| D206 | Average vertex connectivity order-0 index |
| D207 | Average vertex connectivity order-1 index |
| D208 | Average vertex connectivity order-2 index |
| D209 | Average vertex connectivity order-3 index |
| D210 | Average vertex connectivity order-4 index |
| D211 | Average vertex connectivity order-5 index |
| D212 | Valence vertex connectivity order-0 index |
| D213 | Valence vertex connectivity order-1 index |
| D214 | Valence vertex connectivity order-2 index |
| D215 | Valence vertex connectivity order-3 index |
| D216 | Valence vertex connectivity order-4 index |
| D217 | Valence vertex connectivity order-5 index |
| D218 | Average valence vertex connectivity order-0 index |
| D219 | Average valence vertex connectivity order-1 index |
| D220 | Average valence vertex connectivity order-2 index |
| D221 | Average valence vertex connectivity order-3 index |
| D222 | Average valence vertex connectivity order-4 index |
| D223 | Average valence vertex connectivity order-5 index |
| D224 | Principal quantum vertex connectivity order-0 index |
| D225 | Principal quantum vertex connectivity order-1 index |
| D226 | Principal quantum vertex connectivity order-2 index |
| D227 | Principal quantum vertex connectivity order-3 index |

Table 3.2 (*Continued*)

| Descriptor | Description |
|---|---|
| D228 | Principal quantum vertex connectivity order-4 index |
| D229 | Principal quantum vertex connectivity order-5 index |
| D230 | Aromaticity valence vertex connectivity order-1 index |
| D231 | Sum of valence vertex connectivity order-1 index |
| D232 | Reciprocal distance order-1 sum product index |
| D233 | Squared reciprocal distance order-1 sum product index |
| D234 | Kier atom's 0-order path information index |
| D235 | Kier 1-path index |
| D236 | Kier 2-path index |
| D237 | Kier 3-path index |
| D238 | Molecular flexibility index |
| D239 | Atom's connectivity index in longest path |
| D240 | Sum of the longest path of the atom |
| D241 | Average longest path of the molecule |
| D242 | Average of deviation of average of longest path |
| D243 | Average of deviation of distance degree |
| D244 | Shortest path in the molecule |
| D245 | Shortest path centralization index |
| D246 | Maximum value of variation |
| D247 | EXP2 of Path-distance/Walk-distance over all atoms |
| D248 | EXP3 of Path-distance/Walk-distance over all atoms |
| D249 | EXP4 of Path-distance/Walk-distance over all atoms |
| D250 | EXP5 of Path-distance/Walk-distance over all atoms |
| D251 | Petitjean index |
| D252 | Structure centric index |
| D253 | Structure lopping centric group index |
| D254 | Radial centric index |
| D255 | Vertex distance count equality index |
| D256 | Vertex distance count magnitude index |
| D257 | Total vertex distance count equality index |
| D258 | Total vertex distance count magnitude index |
| D259 | Mean of distance degree equality index |
| D260 | Mean of distance degree magnitude index |
| D261 | Information of vertex degree equality index |
| D262 | Information of bonds index |
| D263 | Vertex distance path count index |
| D264 | Complexity vertex distance path count index |
| D265 | Vertex distance information index |
| D266 | Relative of vertex distance information index |
| D267 | Mean of vertex distance information index |
| D268 | Extended of vertex distance information index |
| D269 | Information content order-0 index |
| D270 | Information content order-1 index |
| D271 | Information content order-2 index |
| D272 | Information content order-3 index |
| D273 | Information content order-4 index |

(*Continued*)

**Table 3.2** (*Continued*)

| Descriptor | Description |
|---|---|
| D274 | Information content order-5 index |
| D275 | Total information content order-0 index |
| D276 | Total information content order-1 index |
| D277 | Total information content order-2 index |
| D278 | Total information content order-3 index |
| D279 | Total information content order-4 index |
| D280 | Total information content order-5 index |
| D281 | Structural information content order-0 index |
| D282 | Structural information content order-1 index |
| D283 | Structural information content order-2 index |
| D284 | Structural information content order-3 index |
| D285 | Structural information content order-4 index |
| D286 | Structural information content order-5 index |
| D287 | Complementary information content order-0 index |
| D288 | Complementary information content order-1 index |
| D289 | Complementary information content order-2 index |
| D290 | Complementary information content order-3 index |
| D291 | Complementary information content order-4 index |
| D292 | Complementary information content order-5 index |
| D293 | Bond information content order-0 index |
| D294 | Bond information content order-1 index |
| D295 | Bond information content order-2 index |
| D296 | Bond information content order-3 index |
| D297 | Bond information content order-4 index |
| D298 | Bond information content order-5 index |
| D299 | The largest eigenvalue |
| D300 | Spanning tree with log value |
| D301 | Maximum eigenvalue weighted by the heteroatoms and multiple bonds matrix |
| D302 | Maximum eigenvalue weighted by the mass distance matrix |
| D303 | Maximum eigenvalue weighted by the van der Waals distance matrix |
| D304 | Maximum eigenvalue weighted by the polarizability distance matrix |
| D305 | Maximum eigenvalue weighted by the electronegativity Pauling-scale distance matrix |
| D306 | Maximum eigenvalue weighted by the electronegativity Sanderson-scale weighted distance matrix |
| D307 | Maximum eigenvalue weighted by the electronegativity Allred-Rochow-scale distance matrix |
| D308 | Sum eigenvalue weighted by the heteroatoms and multiple bonds distance matrix |
| D309 | Sum eigenvalue weighted by the mass distance matrix |
| D310 | Sum eigenvalue weighted by the van der Waals distance matrix |
| D311 | Sum eigenvalue weighted by the polarizability distance matrix |
| D312 | Sum eigenvalue weighted by the electronegativity Pauling-scale distance matrix |
| D313 | Sum eigenvalue weighted by the electronegativity Sanderson-scale distance matrix |
| D314 | Sum eigenvalue weighted by the electronegativity Allred-Rochow-scale distance matrix |

Table 3.2 (*Continued*)

| Descriptor | Description |
|---|---|
| D315 | Sum absolute eigenvalue weighted by heteroatoms and multiple bonds distance matrix |
| D316 | Sum absolute eigenvalue weighted by the mass distance matrix |
| D317 | Sum absolute eigenvalue weighted by the van der Waals distance matrix |
| D318 | Sum absolute eigenvalue weighted by the polarizability distance matrix |
| D319 | Sum absolute eigenvalue weighted by the electronegativity Pauling-scale distance matrix |
| D320 | Sum absolute eigenvalue weighted by the electronegativity Sanderson-scale distance matrix |
| D321 | Sum absolute eigenvalue weighted by the electronegativity Allred-Rochow-scale distance matrix |
| D322 | Distance + detour path with ring index of order 3 |
| D323 | Distance + detour path with ring index of order 4 |
| D324 | Distance + detour path with ring index of order 5 |
| D325 | Distance + detour path with ring index of order 6 |
| D326 | Distance + detour path with ring index of order 7 |
| D327 | Distance + detour path with ring index of order 8 |
| D328 | Distance + detour path with ring index of order 9 |
| D329 | Distance + detour path with ring index of order 10 |
| D330 | Distance + detour path with ring index of order 11 |
| D331 | Distance + detour path with ring index of order 12 |
| D332 | Distance + detour path on ring index of order 3 (circuits) |
| D333 | Distance + detour path on ring index of order 4 (circuits) |
| D334 | Distance + detour path on ring index of order 5 (circuits) |
| D335 | Distance + detour path on ring index of order 6 (circuits) |
| D336 | Distance + detour path on ring index of order 7 (circuits) |
| D337 | Distance + detour path on ring index of order 8 (circuits) |
| D338 | Distance + detour path on ring index of order 9 (circuits) |
| D339 | Distance + detour path on ring index of order 10 (circuits) |
| D340 | Distance + detour path on ring index of order 11 (circuits) |
| D341 | Distance + detour path on ring index of order 12 (circuits) |
| D342 | Molecular topological path index of order 02 |
| D343 | Molecular topological path index of order 03 |
| D344 | Molecular topological path index of order 04 |
| D345 | Molecular topological path index of order 05 |
| D346 | Molecular topological path index of order 06 |
| D347 | Molecular topological path index of order 07 |
| D348 | Molecular topological path index of order 08 |
| D349 | Molecular topological path index of order 09 |
| D350 | Molecular topological path index of order 10 |
| D351 | Molecular topological multiple path index of order 03 |
| D352 | Molecular topological multiple path index of order 04 |
| D353 | Molecular topological multiple path index of order 05 |
| D354 | Molecular topological multiple path index of order 06 |
| D355 | Molecular topological multiple path index of order 07 |
| D356 | Molecular topological multiple path index of order 08 |

(*Continued*)

**Table 3.2** (*Continued*)

| Descriptor | Description |
|---|---|
| D357 | Molecular topological multiple path index of order 09 |
| D358 | Molecular topological multiple path index of order 10 |
| D359 | Molecular topological all path index |
| D360 | Conventional bond index |
| D361 | Ratio of convention bonds with total path counts |
| D362 | Ratio of difference of conventional bonds and total path counts |
| D363 | Randic index |
| D364 | Balaban All-Path index |
| D365 | Balaban Short-Path index |
| D366 | Sum of topological distance between the vertices N and N |
| D367 | Sum of topological distance between the vertices N and P |
| D368 | Sum of topological distance between the vertices N and O |
| D369 | Sum of topological distance between the vertices N and S |
| D370 | Sum of topological distance between the vertices N and F |
| D371 | Sum of topological distance between the vertices N and Cl |
| D372 | Sum of topological distance between the vertices N and Br |
| D373 | Sum of topological distance between the vertices N and I |
| D374 | Sum of topological distance between the vertices O and O |
| D375 | Sum of topological distance between the vertices O and S |
| D376 | Sum of topological distance between the vertices O and P |
| D377 | Sum of topological distance between the vertices O and F |
| D378 | Sum of topological distance between the vertices O and Cl |
| D379 | Sum of topological distance between the vertices O and Br |
| D380 | Sum of topological distance between the vertices O and I |
| D381 | Sum of topological distance between the vertices S and S |
| D382 | Sum of topological distance between the vertices S and P |
| D383 | Sum of topological distance between the vertices S and F |
| D384 | Sum of topological distance between the vertices S and Cl |
| D385 | Sum of topological distance between the vertices S and Br |
| D386 | Sum of topological distance between the vertices S and I |
| D387 | Sum of topological distance between the vertices P and P |
| D388 | Sum of topological distance between the vertices P and F |
| D389 | Sum of topological distance between the vertices P and Cl |
| D390 | Sum of topological distance between the vertices P and Br |
| D391 | Sum of topological distance between the vertices P and I |
| D392 | Sum of topological distance between the vertices F and F |
| D393 | Sum of topological distance between the vertices F and Cl |
| D394 | Sum of topological distance between the vertices F and Br |
| D395 | Sum of topological distance between the vertices F and I |
| D396 | Sum of topological distance between the vertices Cl and Cl |
| D397 | Sum of topological distance between the vertices Cl and Br |
| D398 | Sum of topological distance between the vertices Cl and I |
| D399 | Sum of topological distance between the vertices Br and Br |
| D400 | Sum of topological distance between the vertices Br and I |
| D401 | Sum of topological distance between the vertices I and I |
| D402 | Walk count order-01 |

**Table 3.2** (*Continued*)

| Descriptor | Description |
|---|---|
| D403 | Walk count order-02 |
| D404 | Walk count order-03 |
| D405 | Walk count order-04 |
| D406 | Walk count order-05 |
| D407 | Walk count order-06 |
| D408 | Walk count max-10 steps |
| D409 | Walk-returning count order 1 |
| D410 | Walk-returning count order 2 |
| D411 | Walk-returning count order 3 |
| D412 | Walk-returning count order 4 |
| D413 | Walk-returning count order 5 |
| D414 | Walk-returning count order 6 |
| D415 | Topological structure autocorrelation length-1 weighted by atomic masses |
| D416 | Topological structure autocorrelation length-2 weighted by atomic masses |
| D417 | Topological structure autocorrelation length-3 weighted by atomic masses |
| D418 | Topological structure autocorrelation length-4 weighted by atomic masses |
| D419 | Topological structure autocorrelation length-5 weighted by atomic masses |
| D420 | Topological structure autocorrelation length-6 weighted by atomic masses |
| D421 | Topological structure autocorrelation length-7 weighted by atomic masses |
| D422 | Topological structure autocorrelation length-8 weighted by atomic masses |
| D423 | Topological structure autocorrelation length-1 weighted by atomic van der Waals volumes |
| D424 | Topological structure autocorrelation length-2 weighted by atomic van der Waals volumes |
| D425 | Topological structure autocorrelation length-3 weighted by atomic van der Waals volumes |
| D426 | Topological structure autocorrelation length-4 weighted by atomic van der Waals volumes |
| D427 | Topological structure autocorrelation length-5 weighted by atomic van der Waals volumes |
| D428 | Topological structure autocorrelation length-6 weighted by atomic van der Waals volumes |
| D429 | Topological structure autocorrelation length-7 weighted by atomic van der Waals volumes |
| D430 | Topological structure autocorrelation length-8 weighted by atomic van der Waals volumes |
| D431 | Topological structure autocorrelation length-1 weighted by atomic Sanderson electronegativities |
| D432 | Topological structure autocorrelation length-2 weighted by atomic Sanderson electronegativities |
| D433 | Topological structure autocorrelation length-3 weighted by atomic Sanderson electronegativities |
| D434 | Topological structure autocorrelation length-4 weighted by atomic Sanderson electronegativities |
| D435 | Topological structure autocorrelation length-5 weighted by atomic Sanderson electronegativities |

(*Continued*)

Table 3.2 (*Continued*)

| Descriptor | Description |
|---|---|
| D436 | Topological structure autocorrelation length-6 weighted by atomic Sanderson electronegativities |
| D437 | Topological structure autocorrelation length-7 weighted by atomic Sanderson electronegativities |
| D438 | Topological structure autocorrelation length-8 weighted by atomic Sanderson electronegativities |
| D439 | Topological structure autocorrelation length-1 weighted by atomic polarizabilities |
| D440 | Topological structure autocorrelation length-2 weighted by atomic polarizabilities |
| D441 | Topological structure autocorrelation length-3 weighted by atomic polarizabilities |
| D442 | Topological structure autocorrelation length-4 weighted by atomic polarizabilities |
| D443 | Topological structure autocorrelation length-5 weighted by atomic polarizabilities |
| D444 | Topological structure autocorrelation length-6 weighted by atomic polarizabilities |
| D445 | Topological structure autocorrelation length-7 weighted by atomic polarizabilities |
| D446 | Topological structure autocorrelation length-8 weighted by atomic polarizabilities |
| D447 | Geary topological structure autocorrelation length-1 weighted by atomic masses |
| D448 | Geary topological structure autocorrelation length-2 weighted by atomic masses |
| D449 | Geary topological structure autocorrelation length-3 weighted by atomic masses |
| D450 | Geary topological structure autocorrelation length-4 weighted by atomic masses |
| D451 | Geary topological structure autocorrelation length-5 weighted by atomic masses |
| D452 | Geary topological structure autocorrelation length-6 weighted by atomic masses |
| D453 | Geary topological structure autocorrelation length-7 weighted by atomic masses |
| D454 | Geary topological structure autocorrelation length-8 weighted by atomic masses |
| D455 | Geary topological structure autocorrelation length-1 weighted by atomic van der Waals volumes |
| D456 | Geary topological structure autocorrelation length-2 weighted by atomic van der Waals volumes |
| D457 | Geary topological structure autocorrelation length-3 weighted by atomic van der Waals volumes |
| D458 | Geary topological structure autocorrelation length-4 weighted by atomic van der Waals volumes |
| D459 | Geary topological structure autocorrelation length-5 weighted by atomic van der Waals volumes |
| D460 | Geary topological structure autocorrelation length-6 weighted by atomic van der Waals volumes |
| D461 | Geary topological structure autocorrelation length-7 weighted by atomic van der Waals volumes |
| D462 | Geary topological structure autocorrelation length-8 weighted by atomic van der Waals volumes |
| D463 | Geary topological structure autocorrelation length-1 weighted by atomic Sanderson electronegativities |
| D464 | Geary topological structure autocorrelation length-2 weighted by atomic Sanderson electronegativities |
| D465 | Geary topological structure autocorrelation length-3 weighted by atomic Sanderson electronegativities |
| D466 | Geary topological structure autocorrelation length-4 weighted by atomic Sanderson electronegativities |

**Table 3.2** (*Continued*)

| Descriptor | Description |
|---|---|
| D467 | Geary topological structure autocorrelation length-5 weighted by atomic Sanderson electronegativities |
| D468 | Geary topological structure autocorrelation length-6 weighted by atomic Sanderson electronegativities |
| D469 | Geary topological structure autocorrelation length-7 weighted by atomic Sanderson electronegativities |
| D470 | Geary topological structure autocorrelation length-8 weighted by atomic Sanderson electronegativities |
| D471 | Geary topological structure autocorrelation length-1 weighted by atomic polarizabilities |
| D472 | Geary topological structure autocorrelation length-2 weighted by atomic polarizabilities |
| D473 | Geary topological structure autocorrelation length-3 weighted by atomic polarizabilities |
| D474 | Geary topological structure autocorrelation length-4 weighted by atomic polarizabilities |
| D475 | Geary topological structure autocorrelation length-5 weighted by atomic polarizabilities |
| D476 | Geary topological structure autocorrelation length-6 weighted by atomic polarizabilities |
| D477 | Geary topological structure autocorrelation length-7 weighted by atomic polarizabilities |
| D478 | Geary topological structure autocorrelation length-8 weighted by atomic polarizabilities |
| D479 | Moran topological structure autocorrelation length-1 weighted by atomic masses |
| D480 | Moran topological structure autocorrelation length-2 weighted by atomic masses |
| D481 | Moran topological structure autocorrelation length-3 weighted by atomic masses |
| D482 | Moran topological structure autocorrelation length-4 weighted by atomic masses |
| D483 | Moran topological structure autocorrelation length-5 weighted by atomic masses |
| D484 | Moran topological structure autocorrelation length-6 weighted by atomic masses |
| D485 | Moran topological structure autocorrelation length-7 weighted by atomic masses |
| D486 | Moran topological structure autocorrelation length-8 weighted by atomic masses |
| D487 | Moran topological structure autocorrelation length-1 weighted by atomic van der Waals volumes |
| D488 | Moran topological structure autocorrelation length-2 weighted by atomic van der Waals volumes |
| D489 | Moran topological structure autocorrelation length-3 weighted by atomic van der Waals volumes |
| D490 | Moran topological structure autocorrelation length-4 weighted by atomic van der Waals volumes |
| D491 | Moran topological structure autocorrelation length-5 weighted by atomic van der Waals volumes |
| D492 | Moran topological structure autocorrelation length-6 weighted by atomic van der Waals volumes |
| D493 | Moran topological structure autocorrelation length-7 weighted by atomic van der Waals volumes |

(*Continued*)

**Table 3.2** (*Continued*)

| Descriptor | Description |
|---|---|
| D494 | Moran topological structure autocorrelation length-8 weighted by atomic van der Waals volumes |
| D495 | Moran topological structure autocorrelation length-1 weighted by atomic Sanderson electronegativities |
| D496 | Moran topological structure autocorrelation length-2 weighted by atomic Sanderson electronegativities |
| D497 | Moran topological structure autocorrelation length-3 weighted by atomic Sanderson electronegativities |
| D498 | Moran topological structure autocorrelation length-4 weighted by atomic Sanderson electronegativities |
| D499 | Moran topological structure autocorrelation length-5 weighted by atomic Sanderson electronegativities |
| D500 | Moran topological structure autocorrelation length-6 weighted by atomic Sanderson electronegativities |
| D501 | Moran topological structure autocorrelation length-7 weighted by atomic Sanderson electronegativities |
| D502 | Moran topological structure autocorrelation length-8 weighted by atomic Sanderson electronegativities |
| D503 | Moran topological structure autocorrelation length-1 weighted by atomic polarizabilities |
| D504 | Moran topological structure autocorrelation length-2 weighted by atomic polarizabilities |
| D505 | Moran topological structure autocorrelation length-3 weighted by atomic polarizabilities |
| D506 | Moran topological structure autocorrelation length-4 weighted by atomic polarizabilities |
| D507 | Moran topological structure autocorrelation length-5 weighted by atomic polarizabilities |
| D508 | Moran topological structure autocorrelation length-6 weighted by atomic polarizabilities |
| D509 | Moran topological structure autocorrelation length-7 weighted by atomic polarizabilities |
| D510 | Moran topological structure autocorrelation length-8 weighted by atomic polarizabilities |
| D511 | Molecular topological order-1 charge index |
| D512 | Molecular topological order-2 charge index |
| D513 | Molecular topological order-3 charge index |
| D514 | Molecular topological order-4 charge index |
| D515 | Molecular topological order-5 charge index |
| D516 | Molecular topological order-6 charge index |
| D517 | Molecular topological order-7 charge index |
| D518 | Molecular topological order-8 charge index |
| D519 | Molecular topological order-9 charge index |
| D520 | Molecular topological order-10 charge index |
| D521 | Mean molecular topological order-1 charge index |
| D522 | Mean molecular topological order-2 charge index |

Table 3.2 (Continued)

| Descriptor | Description |
|---|---|
| D523 | Mean molecular topological order-3 charge index |
| D524 | Mean molecular topological order-4 charge index |
| D525 | Mean molecular topological order-5 charge index |
| D526 | Mean molecular topological order-6 charge index |
| D527 | Mean molecular topological order-7 charge index |
| D528 | Mean molecular topological order-8 charge index |
| D529 | Mean molecular topological order-9 charge index |
| D530 | Mean molecular topological order-10 charge index |
| D531 | Sum of molecular topological mean charge index |
| D532 | Lowest eigenvalue from Burden matrix weighted by masses order 1 |
| D533 | Lowest eigenvalue from Burden matrix weighted by masses order 2 |
| D534 | Lowest eigenvalue from Burden matrix weighted by masses order 3 |
| D535 | Lowest eigenvalue from Burden matrix weighted by masses order 4 |
| D536 | Lowest eigenvalue from Burden matrix weighted by masses order 5 |
| D537 | Lowest eigenvalue from Burden matrix weighted by masses order 6 |
| D538 | Lowest eigenvalue from Burden matrix weighted by masses order 7 |
| D539 | Lowest eigenvalue from Burden matrix weighted by masses order 8 |
| D540 | Lowest eigenvalue from Burden matrix weighted by van der Waals order 1 |
| D541 | Lowest eigenvalue from Burden matrix weighted by van der Waals order 2 |
| D542 | Lowest eigenvalue from Burden matrix weighted by van der Waals order 3 |
| D543 | Lowest eigenvalue from Burden matrix weighted by van der Waals order 4 |
| D544 | Lowest eigenvalue from Burden matrix weighted by van der Waals order 5 |
| D545 | Lowest eigenvalue from Burden matrix weighted by van der Waals order 6 |
| D546 | Lowest eigenvalue from Burden matrix weighted by van der Waals order 7 |
| D547 | Lowest eigenvalue from Burden matrix weighted by van der Waals order 8 |
| D548 | Lowest eigenvalue from Burden matrix weighted by electronegativities Sanderson-scale order 1 |
| D549 | Lowest eigenvalue from Burden matrix weighted by electronegativities Sanderson-scale order 2 |
| D550 | Lowest eigenvalue from Burden matrix weighted by electronegativities Sanderson-scale order 3 |
| D551 | Lowest eigenvalue from Burden matrix weighted by electronegativities Sanderson-scale order 4 |
| D552 | Lowest eigenvalue from Burden matrix weighted by electronegativities Sanderson-scale order 5 |
| D553 | Lowest eigenvalue from Burden matrix weighted by electronegativities Sanderson-scale order 6 |
| D554 | Lowest eigenvalue from Burden matrix weighted by electronegativities Sanderson-scale order 7 |
| D555 | Lowest eigenvalue from Burden matrix weighted by electronegativities Sanderson-scale order 8 |
| D556 | Lowest eigenvalue from Burden matrix weighted by polarizabilities order 1 |
| D557 | Lowest eigenvalue from Burden matrix weighted by polarizabilities order 2 |
| D558 | Lowest eigenvalue from Burden matrix weighted by polarizabilities order 3 |
| D559 | Lowest eigenvalue from Burden matrix weighted by polarizabilities order 4 |
| D560 | Lowest eigenvalue from Burden matrix weighted by polarizabilities order 5 |

(Continued)

**Table 3.2** (*Continued*)

| Descriptor | Description |
|---|---|
| D561 | Lowest eigenvalue from Burden matrix weighted by polarizabilities order 6 |
| D562 | Lowest eigenvalue from Burden matrix weighted by polarizabilities order 7 |
| D563 | Lowest eigenvalue from Burden matrix weighted by polarizabilities order 8 |
| D564 | Highest eigenvalue from Burden matrix weighted by masses order 1 |
| D565 | Highest eigenvalue from Burden matrix weighted by masses order 2 |
| D566 | Highest eigenvalue from Burden matrix weighted by masses order 3 |
| D567 | Highest eigenvalue from Burden matrix weighted by masses order 4 |
| D568 | Highest eigenvalue from Burden matrix weighted by masses order 5 |
| D569 | Highest eigenvalue from Burden matrix weighted by masses order 6 |
| D570 | Highest eigenvalue from Burden matrix weighted by masses order 7 |
| D571 | Highest eigenvalue from Burden matrix weighted by masses order 8 |
| D572 | Highest eigenvalue from Burden matrix weighted by van der Waals order 1 |
| D573 | Highest eigenvalue from Burden matrix weighted by van der Waals order 2 |
| D574 | Highest eigenvalue from Burden matrix weighted by van der Waals order 3 |
| D575 | Highest eigenvalue from Burden matrix weighted by van der Waals order 4 |
| D576 | Highest eigenvalue from Burden matrix weighted by van der Waals order 5 |
| D577 | Highest eigenvalue from Burden matrix weighted by van der Waals order 6 |
| D578 | Highest eigenvalue from Burden matrix weighted by van der Waals order 7 |
| D579 | Highest eigenvalue from Burden matrix weighted by van der Waals order 8 |
| D580 | Highest eigenvalue from Burden matrix weighted by electronegativities Sanderson-scale order 1 |
| D581 | Highest eigenvalue from Burden matrix weighted by electronegativities Sanderson-scale order 2 |
| D582 | Highest eigenvalue from Burden matrix weighted by electronegativities Sanderson-scale order 3 |
| D583 | Highest eigenvalue from Burden matrix weighted by electronegativities Sanderson-scale order 4 |
| D584 | Highest eigenvalue from Burden matrix weighted by electronegativities Sanderson-scale order 5 |
| D585 | Highest eigenvalue from Burden matrix weighted by electronegativities Sanderson-scale order 6 |
| D586 | Highest eigenvalue from Burden matrix weighted by electronegativities Sanderson-scale order 7 |
| D587 | Highest eigenvalue from Burden matrix weighted by electronegativities Sanderson-scale order 8 |
| D588 | Highest eigenvalue from Burden matrix weighted by polarizabilities order 1 |
| D589 | Highest eigenvalue from Burden matrix weighted by polarizabilities order 2 |
| D590 | Highest eigenvalue from Burden matrix weighted by polarizabilities order 3 |
| D591 | Highest eigenvalue from Burden matrix weighted by polarizabilities order 4 |
| D592 | Highest eigenvalue from Burden matrix weighted by polarizabilities order 5 |
| D593 | Highest eigenvalue from Burden matrix weighted by polarizabilities order 6 |
| D594 | Highest eigenvalue from Burden matrix weighted by polarizabilities order 7 |
| D595 | Highest eigenvalue from Burden matrix weighted by polarizabilities order 8 |
| D596 | Number of total primary C-sp3 |
| D597 | Number of total secondary C-sp3 |
| D598 | Number of total tertiary C-sp3 |

**Table 3.2** (*Continued*)

| Descriptor | Description |
|---|---|
| D599 | Number of total quaternary C-sp3 |
| D600 | Number of ring secondary C-sp3 |
| D601 | Number of ring tertiary C-sp3 |
| D602 | Number of ring quaternary C-sp3 |
| D603 | Number of unsubstituted aromatic C-sp2 |
| D604 | Number of substituted aromatic C-sp2 |
| D605 | Number of primary C-sp2 |
| D606 | Number of secondary C-sp2 |
| D607 | Number of tertiary C-sp2 |
| D608 | Number of group allenes |
| D609 | Number of terminal C-sp |
| D610 | Number of nonterminal C-sp |
| D611 | Number of group cyanates (aliphatic) |
| D612 | Number of group cyanates (aromatic) |
| D613 | Number of group isocyanates (aliphatic) |
| D614 | Number of group isocyanates (aromatic) |
| D615 | Number of group thiocyanates (aliphatic) |
| D616 | Number of group thiocyanates (aromatic) |
| D617 | Number of group isothiocyanates (aliphatic) |
| D618 | Number of group isothiocyanates (aromatic) |
| D619 | Number of group carboxylic acids (aliphatic) |
| D620 | Number of group carboxylic acids (aromatic) |
| D621 | Number of group esters (aliphatic) |
| D622 | Number of group esters (aromatic) |
| D623 | Number of group primary amides (aliphatic) |
| D624 | Number of group primary amides (aromatic) |
| D625 | Number of group secondary amides (aliphatic) |
| D626 | Number of group secondary amides (aromatic) |
| D627 | Number of group tertiary amides (aliphatic) |
| D628 | Number of group tertiary amides (aromatic) |
| D629 | Number of group carbamates (aliphatic) |
| D630 | Number of group carbamates (aromatic) |
| D631 | Number of group acyl halogenides (aliphatic) |
| D632 | Number of group acyl halogenides (aromatic) |
| D633 | Number of group thioacids (aliphatic) |
| D634 | Number of group thioacids (aromatic) |
| D635 | Number of group ditioacids (aliphatic) |
| D636 | Number of group ditioacids (aromatic) |
| D637 | Number of group thioesters (aliphatic) |
| D638 | Number of group thioesters (aromatic) |
| D639 | Number of group dithioesters (aliphatic) |
| D640 | Number of group dithioesters (aromatic) |
| D641 | Number of group aldehydes (aliphatic) |
| D642 | Number of group aldehydes (aromatic) |
| D643 | Number of group ketones (aliphatic) |
| D644 | Number of group ketones (aromatic) |

(*Continued*)

**Table 3.2** (*Continued*)

| Descriptor | Description |
|---|---|
| D645 | Number of group urea derivatives |
| D646 | Number of group urea derivatives (aromatic) |
| D647 | Number of group primary amines (aliphatic) |
| D648 | Number of group primary amines (aromatic) |
| D649 | Number of group secondary amines (aliphatic) |
| D650 | Number of group secondary amines (aromatic) |
| D651 | Number of group tertiary amines (aliphatic) |
| D652 | Number of group tertiary amines (aromatic) |
| D653 | Number of group N-hydrazines (aliphatic) |
| D654 | Number of group N-hydrazines (aromatic) |
| D655 | Number of group N-azo (aliphatic) |
| D656 | Number of group N-azo (aromatic) |
| D657 | Number of group nitriles (aliphatic) |
| D658 | Number of group nitriles (aromatic) |
| D659 | Number of group imines (aliphatic) |
| D660 | Number of group imines (aromatic) |
| D661 | Number of group ammonia groups (aliphatic) |
| D662 | Number of group ammonia groups (aromatic) |
| D663 | Number of group hydroxylamines (aliphatic) |
| D664 | Number of group hydroxylamines (aromatic) |
| D665 | Number of group oximes (aliphatic) |
| D666 | Number of group oximes (aromatic) |
| D667 | Number of group N-nitroso (aliphatic) |
| D668 | Number of group N-nitroso (aromatic) |
| D669 | Number of group nitroso (aliphatic) |
| D670 | Number of group nitroso (aromatic) |
| D671 | Number of group nitro (aliphatic) |
| D672 | Number of group nitro (aromatic) |
| D673 | Number of group imides |
| D674 | Number of group total hydroxyl groups |
| D675 | Number of group phenols |
| D676 | Number of group primary alcohols (aliphatic) |
| D677 | Number of group secondary alcohols (aliphatic) |
| D678 | Number of group tertiary alcohols (aliphatic) |
| D679 | Number of group ethers (aliphatic) |
| D680 | Number of group ethers (aromatic) |
| D681 | Number of group hypohalogenydes (aliphatic) |
| D682 | Number of group hypohalogenydes (aromatic) |
| D683 | Number of group water molecules |
| D684 | Number of group sulfoxides |
| D685 | Number of group sulfones |
| D686 | Number of group sulfates |
| D687 | Number of group thioles |
| D688 | Number of group thioketones |
| D689 | Number of group sulfides |
| D690 | Number of group disulfides |

Table 3.2 (Continued)

| Descriptor | Description |
|---|---|
| D691 | Number of group sulfonic acids |
| D692 | Number of group sulfonamides |
| D693 | Number of group phosphites |
| D694 | Number of group phosphates |
| D695 | Number of group phosphothionates |
| D696 | Number of group phosphodithionates |
| D697 | Number of group phosphothioates |
| D698 | Number of group CH2X |
| D699 | Number of group CR2HX |
| D700 | Number of group CR3X |
| D701 | Number of group R=CHX |
| D702 | Number of group R=CRX |
| D703 | Number of group R#CX |
| D704 | Number of group CHRX2 |
| D705 | Number of group CR2X2 |
| D706 | Number of group R=CX2 |
| D707 | Number of group RCX3 |
| D708 | Number of group X–C on aromatic ring |
| D709 | Number of group X–C– on ring |
| D710 | Number of group X–C= on ring |
| D711 | Number of group X–C on conjugated C |
| D712 | Number of group donor atoms for H-bonds (with N and O) |
| D713 | Number of group acceptor atoms for H-bonds (N O F) |
| D714 | Number of group $CH_3R$ and $CH_4$ |
| D715 | Number of group $CH_2R_2$ |
| D716 | Number of group $CHR_3$ |
| D717 | Number of group $CR_4$ |
| D718 | Number of group $CH_3X$ |
| D719 | Number of group $CH_2RX$ |
| D720 | Number of group $CH_2X_2$ |
| D721 | Number of group $CHR_2X$ |
| D722 | Number of group $CHRX_2$ |
| D723 | Number of group $CHX_3$ |
| D724 | Number of group $CR_3X$ |
| D725 | Number of group $CR_2X_2$ |
| D726 | Number of group $CRX_3$ |
| D727 | Number of group $CX_4$ |
| D728 | Number of group $=CH_2$ |
| D729 | Number of group =CHR |
| D730 | Number of group $=CR_2$ |
| D731 | Number of group =CHX |
| D732 | Number of group =CRX |
| D733 | Number of group $=CX_2$ |
| D734 | Number of group #CH |
| D735 | Number of group #CR or R=C=R |
| D736 | Number of group #CX |

(Continued)

**Table 3.2** (*Continued*)

| Descriptor | Description |
|---|---|
| D737 | Number of group R~CH~R |
| D738 | Number of group R~CR~R |
| D739 | Number of group R~CX~R |
| D740 | Number of group Al−CH=X |
| D741 | Number of group Ar−CH=X |
| D742 | Number of group Al−C(=X)−Al |
| D743 | Number of group Ar−C(=X)−R |
| D744 | Number of group R−C(=X)−X/R−C#X |
| D745 | Number of group X−C(=X)−X |
| D746 | Number of group H attached to C0(sp3) no X attached to next C |
| D747 | Number of group H attached to heteroatom |
| D748 | Number of group H attached to C0(sp3) with 1X attached to next C |
| D749 | Number of group H attached to C0(sp3) with 2X attached to next C |
| D750 | Number of group H attached to C0(sp3) with 3X attached to next C |
| D751 | Number of group H attached to C0(sp3) with 4X attached to next C |
| D752 | Number of group alcohol |
| D753 | Number of group phenol or enol or carboxyl OH |
| D754 | Number of group O= |
| D755 | Number of group Al−O−Al |
| D756 | Number of group Al−O−Ar or Ar−O−Ar or R−O−C=X |
| D757 | Number of group $Al-NH_2$ |
| D758 | Number of group $Al_2-NH$ |
| D759 | Number of group $Al_3-N$ |
| D760 | Number of group $Ar-NH_2$ or $X-NH_2$ |
| D761 | Number of group Ar−NH−Al |
| D762 | Number of group $Ar-NAl_2$ |
| D763 | Number of group $RCO-N<$ or $>N-X=X$ |
| D764 | Number of group $Ar_2NH$ or $Ar_3N$ or $Ar_2N-Al$ |
| D765 | Number of group R#N or R=N− |
| D766 | Number of group $Ar-NO_2$ or $RO-NO_2$ |
| D767 | Number of group $Al-NO_2$ |
| D768 | Number of group Ar−N=X or X−N=X |
| D769 | Number of group R−SH |
| D770 | Number of group $R_2S$ or RS−SR |
| D771 | Number of group R=S |
| D772 | Number of group R−SO−R |
| D773 | Number of group $R-SO_2-R$ |
| D774 | Unsaturation index weighted by conventional bonds order |
| D775 | Hydrophilic factor index |
| D776 | Aromatic bonds ratio |
| D777 | Molecular regresson coefficients surface Log$P$ index |

number of paths with fixed length. The descriptors include *Kier shape descriptors, Kier steric descriptors, Kier flexibility descriptors,* and so on.

The Mold$^2$ *Balaban index* descriptors [50, 51] are obtained from the Balaban distance connectivity index $J$ that is calculated using the following formula:

$$J = \frac{B}{C+1}\sum_{k=1}^{B}\sqrt{(v_i v_j)_k}$$

where $v_i$ and $v_j$ are the vertex distance degrees of two atoms connected by bond $k$, $B$ is the number of bonds of the molecule, and $C$ is the cyclomatic number.

The *Randić index* [50–56], also called the *connectivity index,* is a measure of branching of the carbon-atom skeleton. It is related to boiling point, Kovats constants, and a calculated surface. The Mold$^2$ *Randić index* descriptors are calculated from the topological structure of a molecule by using a formula similar to the index $J$. The *total connectivity index, local connectivity index,* and *Randić connectivity index* are examples of such connectivity index descriptors.

The 2D autocorrelation descriptors [58, 59], $A(d)$, are calculated from the 2D structure based on the autocorrelation function:

$$A(d) = \sum_{j=1}^{a}\sum_{i=1}^{a}\sigma(d_{ij}-d)p_i p_j$$

$$\sigma = \begin{cases} 1(d_{ij}=d) \\ 0(d_{ij}\neq d) \end{cases}$$

where $d$ is a topological distance that can be any number between 1 and the maximum of distance in a molecule, $\sigma$ is a function of the variable $d_{ij}$ (the topological distance between atoms $i$ and $j$), $a$ is the number of atoms in the molecule, and $p_i$ and $p_j$ are the properties of atoms $i$ and $j$, respectively. The *ATS* (autocorrelation of topological structures) and *Moran coefficient* (general spatial autocorrelation index) are the members of this subclass of descriptors.

Descriptors such as the *cyclicity index* and *average cyclicity index* are part of the detour *index* [55, 60] based descriptors that are calculated from the detour distance matrix of a molecule. The detour *matrix* (or the maximum path matrix) $[\Delta]_{ij}$ is a square symmetric matrix of the form

$$[\Delta]_{ij} = \begin{cases} \Delta_{ij} = {}^{\max}p_{ij} & (i \neq j) \\ 0 & (i = j) \end{cases}$$

where the value of the $i$th row element at column $j$ (${}^{\max}p_{ij}$) is the maximum number of edges between nodes $i$ and $j$ and is zero from a node to itself.

The *topological distance index* [50, 52, 57, 58] descriptors are calculated from the distance matrix $D$ (or the vertex distance matrix) of a molecule:

$$[D]_{ij} = \begin{cases} d_{ij} = {}^{\min}p_{ij} & (i \neq j) \\ 0 & (i = j) \end{cases}$$

which is a square symmetric matrix summing up the topological distance information between all pairs of atoms. Its elements represent the shortest paths in the number of edges from an atom to another. By convention, the distance from an atom to itself is zero. The *average atom connectivity index* and *Rouvray index* (total connectivity index) are examples of this type of descriptors.

Mold$^2$ descriptors extracted from the eigenvalues [61] of the adjacent matrix and other matrices of a molecule are grouped together as eigenvalue-based descriptors, including the *Lovasz–Pelikan index, folding degree index, characteristic root index,* and others.

Descriptors related to molecular walk counts [54, 62] are calculated based on the graph walks. They are extracted from the adjacency matrix of a molecule. *Total walk count, weighted walk degrees,* and *walk connectivity indices* are examples.

The *Schultz index SI* [63] descriptors are calculated based on the formula

$$SI = \sum_{i=1}^{a} [(M + D) \bullet v]_i$$

where $a$ is the number of atoms or bonds in a molecule, $M$ is the adjacency matrix (nodes or edges), $D$ is the distance matrix (nodes or edges), and $v$ is the vertex or edge degree vector.

Descriptors related to the *topological charge index* [64, 65] are derived from the adjacency matrix and distance matrix of a molecule, which estimate the charge transfer between pairs of atoms, and therefore the global charge transfer in the molecule, including *total topological charge, mean topological charge,* and so on.

The *Zagreb index* [58] descriptors are derived from the vertex degree of atoms of a molecule, including the *quadratic index* and *binormalized quadratic index.*

#### 3.2.1.2 **Constitutional Descriptors**

The constitutional descriptors are the simplest in nature and depend only on the structural constitution of the molecule. They include the number of specific atoms or functional groups, the number of chiral centers, the number of rotatable bonds, and so on. Descriptors such as the *number of aromatic rings, number of double bonds, numbers of function groups,* and so on are examples of the constitutional descriptors calculated in Mold$^2$.

#### 3.2.1.3 **Information Content-based Descriptors**

The information content [66] based descriptors are calculated from the information content of a molecule ($I_c$). $I_c$ is used to measure the degree of diversity of the atoms or bonds in a molecule and defined by the formula

$$I = \sum_{c=1}^{C} n_c \log_2 n_c$$

where $C$ is the number of different types of atoms or bonds and $n_c$ is the number of atoms or bonds of the *c*th type. *Mean information content, mean information content on edge equality,* and *redundancy index* are some examples of information content-based descriptors.

### 3.2.2
### Calculation of Mold$^2$ Descriptors

Molecular descriptors are calculated from 2D molecular structures of chemicals in Mold$^2$. A flowchart of calculations is given in Figure 3.2. It accepts an SDfile of the

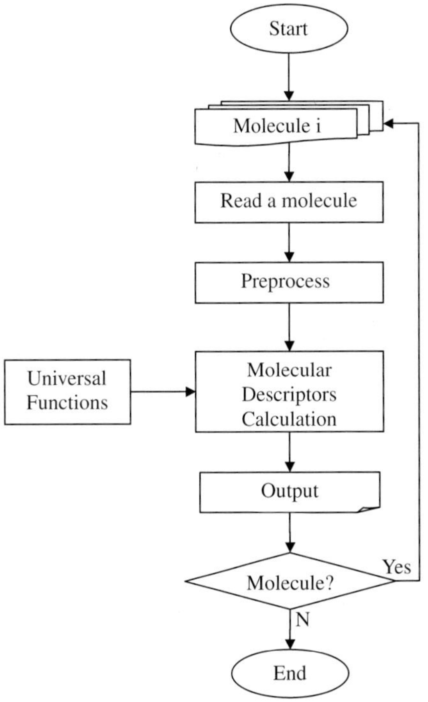


**Figure 3.2** Flowchart of calculating descriptors in Mold$^2$.

molecules for which descriptors are to be calculated. Other formats for representing the molecular structure have to be converted to an SDfile format for use by Mold$^2$. The chemical structures in an SDfile are sequentially processed. Mold$^2$ preprocesses the input structure prior to descriptor calculation to check whether the structure is as expected and correct. The preprocessing is a tedious process of standardizing the structures, which includes counter ion removal, adding hydrogen atoms to heavy atoms, correcting errors of an input structure, and transforming the initial Molfile to the connection table that is operated on by Mold$^2$. The main part of Mold$^2$ is a module for computing the 777 descriptors. A set of generalized functions are used to speed up calculations. A module for the perception of the smallest set of smallest rings (SSSR) is among the most important that is adopted from a previously reported algorithm [67]; the related perception of the aromatic ring system was implemented based on the same algorithm. The functions important for generating walk, distance, and path matrices were developed by modifying the publicly available algorithm [67] while adopting the canonical representation of a structure [68]. Descriptors are calculated in a set of subroutines, where related descriptors are simultaneously calculated in the same subroutine to minimize redundant calculations. Once all 777 molecular descriptors for a molecule are calculated, the results are outputted to a file (not kept in memory), and molecules are serially processed until all molecules in the SDfile are processed. The processing status and errors are recorded in a log file.

Mold$^2$ was developed in C++ on the Windows XP system. It is operated by using the command line. The descriptors of all molecules in a single SDfile (exported from any database) can be calculated in a single run. An API (application program interface) for Mold$^2$ is in development. The executable file for Mold$^2$ is free for use and can be downloaded from http://www.fda.gov/ScienceResearch/BioinformaticsTools/Mold2/default.htm.

### 3.2.3 Evaluation of Mold$^2$ Descriptors

Selection of proper molecular descriptors is an essential step for QSAR. Accordingly, evaluation of the performance of molecular descriptors is integral to guiding their proper use. There are two ways of evaluating the utility of descriptors. The first is to assess information that is presented in a data set represented by a set of descriptors. Here the variance of a descriptor among molecules in a data set is a measure of the information represented by the descriptor. Higher variance in the descriptors portends higher probability of valid model development. The second way is minimizing redundancy of information encoded in descriptors by minimizing descriptor covariance. The Mold$^2$ descriptors were evaluated using both approaches. Specifically, Mold$^2$ descriptors were compared with descriptors from three commercial software packages using information entropy analysis and analysis of correlations between descriptors for several reported data sets. The results showed that Mold$^2$ descriptors hold up to current scrutiny on the basis of information and descriptor correlation, along with the compared sets of descriptors.

#### 3.2.3.1 Information Content by Shannon Entropy Analysis

The concept of Shannon entropy [66, 69], also called information entropy, has played a central role in information theory. It can be used as a measure of uncertainty. Recently, Dehmer and Mowshowitz [70] had a comprehensive review on methods for measuring entropy of graphs and the applications in a variety of research fields such as chemoinformatics and bioinformatics. In the approach, the entropy of a random variable is associated with its probability distribution, which is formulated as

$$H_n(p_1, p_2, \ldots, p_n) = -\sum_{i=1}^{n} p_i \log_2 p_i$$

where $p_i$ is the probability of outcome $i$. In information theory, the entropy is conceptually the actual amount of (information theoretic) information in a data set. It is reasonable to use the entropy of descriptors as an estimation or measure of the structural information inherently encoded in the descriptors for a data set. Golden *et al.* [71] used Shannon entropy to analyze different descriptors by ranking them according to variance, where high variance descriptors are considered more discriminatory among compounds.

Shannon entropy analysis was conducted for comparing Mold$^2$ descriptors with three sets of descriptors from commercial software packages (i.e., Molconn-Z,

**Table 3.3** Shannon entropy analysis results.

| Data set | | NCTR_AR | NCTR_ER | ER_Comb | EPA |
|---|---|---|---|---|---|
| Size (compounds) | | 202 | 232 | 1086 | 57 453 |
| Cerius$^2$ | Descriptors | 205 | 197 | 228 | a) |
| | Mean entropy | 1.962 | 1.995 | 1.5983 | |
| Dragon | Descriptors | 616 | 604 | 671 | a) |
| | Mean entropy | 2.453 | 2.431 | 1.996 | |
| Molconn_Z | Descriptors | 331 | 312 | 378 | 450 |
| | Entropy | 1.792 | 1.798 | 1.021 | 0.801 |
| Mold$^2$ | Descriptors | 590 | 578 | 626 | 742 |
| | Entropy | 2.408 | 2.380 | 2.058 | 1.34 |

a) The licenses of Cerius$^2$ and Dragon are not available in our labs now; thus, the descriptors for EPA data set were not analyzed.

Dragon, and Cerius$^2$) in four published data sets, listed in Table 3.3 as NCTR_ER [72], NCTR_AR [73], ER_comb [74], and EPA [75]. First, descriptors for which values could not be calculated across all the chemicals in a data set were discarded. For example, the descriptor for the number of phosphorus atoms was discarded if no compounds in a data set had the atom. Next, descriptors were binned for each descriptor in the data sets and the probability distributions calculated using 20 even bins that spread from the minimum to the maximum values of the descriptor. The binning approach allows the comparison of different units and values, a necessity because the number and type of descriptors from different software packages vary substantially. Then, the comparison of descriptor sets was done assuming that mean Shannon entropy was proportional to the average information encoded in different sets of descriptors for the same data set. As shown in Table 3.3, for all of the four data sets, the mean Shannon entropy of descriptors from Mold$^2$ is comparable to that from Dragon, and slightly higher than that from Cerius$^2$ and Molconn-Z. This analysis demonstrates that the Mold$^2$ descriptors are equal to or more informative than those from the compared commercial software packages.

The mean Shannon entropy only estimates the average information presented in the descriptors. The Shannon entropy distribution among descriptors is another important property since those with higher entropy are expected to yield more informative predictive models. Figure 3.3 plots the Shannon entropy distribution for all descriptors of each of the four data sets as the reverse cumulative probability distributions against Shannon entropy. A point $(x, y)$ in a curve indicates that the probability (percentage) of descriptors ($y$) has Shannon entropy greater than or equal to $x$. For example, for the NCTR_ER data set (Figure 3.3b), the probability of Mold$^2$ descriptors with Shannon entropy greater than or equal to 2 is 0.721, compared to 0.551 for Cerius$^2$, 0.723 for Dragon, and 0.505 for Molconn-Z. The slightly larger probability value in the high entropy region for Mold$^2$ across all of the four data sets indicates that Mold$^2$ descriptors encode sufficient information and are thus equal to or better than those from the compared commercial software. The entropies of top

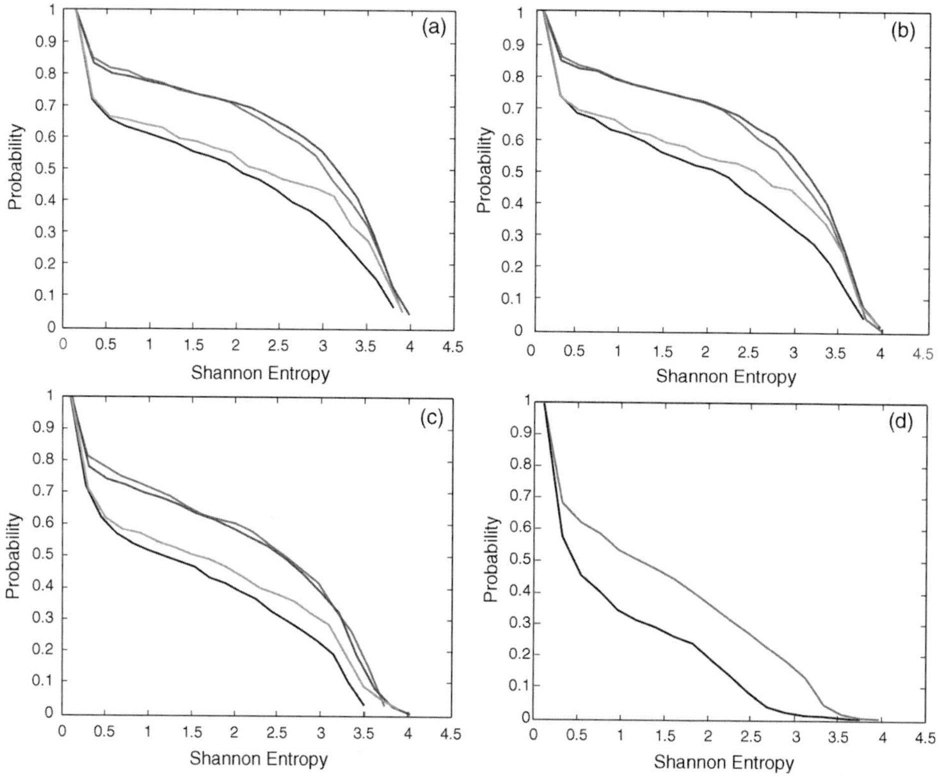


**Figure 3.3** Reverse cumulative probability versus the Shannon entropy of descriptors for data sets NCTR_AR (a), NCTR_ER (b), ER_Comb (c), and EPA (d). The *x*-axis represents the Shannon entropies of descriptors, while the *y*-axis is the reverse cumulative probability of descriptors having Shannon entropy greater than or equal to the *x* value. The curves of Mold² descriptors are colored in red; Cerius² in green; Dragon in blue; and Molconn-Z in black [43].

20 descriptors listed in Table 3.4 indicate that the most informative descriptors in the compared sets are equivalent.

#### 3.2.3.2 **Correlations between Descriptors**

It is expected that few redundant or highly correlated descriptors will exist in a set of descriptors, and that none will be completely redundant. Nonetheless, even more informative models can be expected when correlations between descriptors is minimized. Correlation coefficients ($r$) between the descriptors in each of the compared descriptors sets were calculated for the four data sets. The frequencies of correlation coefficients ($r$) are plotted in Figure 3.4. For all four of the data sets, uncorrelated descriptors comprise the majority of those compared. The low correlations ($0.25 < r^2 < 0.8$) are less than 15% for all the compared descriptor sets in all of the data sets. The high correlations ($r^2 > 0.8$) are less than 1%. Furthermore, the high correlations were from descriptors with many missing values replaced with zeros. The correlation

**Table 3.4** Shannon entropy of top 20 descriptors.

| NCTR_AR | | | | NCTR_ER | | | | ER_Comb | | | | EPA | |
|---|---|---|---|---|---|---|---|---|---|---|---|---|---|
| M2 | MZ | C2 | DG | M2 | MZ | C2 | DG | M2 | MZ | C2 | DG | M2 | MZ |
| 4.02 | 3.92 | 4.03 | 4.11 | 4.12 | 3.89 | 4.05 | 4.08 | 3.82 | 3.58 | 3.98 | 4.13 | 4.30 | 4.30 |
| 4.02 | 3.90 | 3.98 | 4.05 | 3.93 | 3.87 | 4.03 | 4.07 | 3.82 | 3.58 | 3.92 | 3.99 | 4.30 | 3.79 |
| 3.97 | 3.90 | 3.93 | 4.03 | 3.89 | 3.84 | 3.91 | 3.99 | 3.81 | 3.56 | 3.89 | 3.97 | 4.17 | 3.79 |
| 3.94 | 3.90 | 3.92 | 4.02 | 3.86 | 3.79 | 3.91 | 3.98 | 3.80 | 3.53 | 3.86 | 3.95 | 3.68 | 3.41 |
| 3.93 | 3.86 | 3.91 | 4.02 | 3.85 | 3.79 | 3.87 | 3.97 | 3.79 | 3.47 | 3.78 | 3.91 | 3.65 | 3.22 |
| 3.93 | 3.85 | 3.90 | 4.00 | 3.82 | 3.79 | 3.85 | 3.97 | 3.78 | 3.46 | 3.78 | 3.91 | 3.60 | 3.09 |
| 3.93 | 3.85 | 3.87 | 3.99 | 3.77 | 3.78 | 3.84 | 3.92 | 3.78 | 3.44 | 3.71 | 3.90 | 3.58 | 2.95 |
| 3.92 | 3.83 | 3.86 | 3.99 | 3.75 | 3.77 | 3.83 | 3.92 | 3.76 | 3.41 | 3.71 | 3.89 | 3.57 | 2.95 |
| 3.91 | 3.82 | 3.86 | 3.98 | 3.75 | 3.76 | 3.81 | 3.89 | 3.72 | 3.41 | 3.71 | 3.86 | 3.54 | 2.92 |
| 3.90 | 3.80 | 3.84 | 3.97 | 3.75 | 3.75 | 3.75 | 3.89 | 3.72 | 3.41 | 3.70 | 3.83 | 3.53 | 2.81 |
| 3.90 | 3.80 | 3.83 | 3.97 | 3.74 | 3.75 | 3.74 | 3.89 | 3.70 | 3.41 | 3.65 | 3.81 | 3.48 | 2.80 |
| 3.90 | 3.78 | 3.81 | 3.96 | 3.74 | 3.70 | 3.74 | 3.85 | 3.70 | 3.41 | 3.63 | 3.80 | 3.47 | 2.78 |
| 3.89 | 3.77 | 3.81 | 3.95 | 3.74 | 3.70 | 3.73 | 3.85 | 3.69 | 3.40 | 3.55 | 3.78 | 3.47 | 2.76 |
| 3.88 | 3.77 | 3.80 | 3.94 | 3.74 | 3.70 | 3.70 | 3.83 | 3.68 | 3.38 | 3.51 | 3.77 | 3.45 | 2.69 |
| 3.87 | 3.77 | 3.80 | 3.93 | 3.73 | 3.67 | 3.70 | 3.83 | 3.67 | 3.38 | 3.44 | 3.75 | 3.42 | 2.69 |
| 3.87 | 3.77 | 3.79 | 3.93 | 3.73 | 3.67 | 3.67 | 3.82 | 3.67 | 3.38 | 3.41 | 3.74 | 3.42 | 2.66 |
| 3.87 | 3.77 | 3.78 | 3.93 | 3.72 | 3.65 | 3.67 | 3.82 | 3.66 | 3.37 | 3.41 | 3.74 | 3.41 | 2.58 |
| 3.86 | 3.75 | 3.78 | 3.93 | 3.72 | 3.65 | 3.64 | 3.81 | 3.65 | 3.36 | 3.41 | 3.72 | 3.40 | 2.58 |
| 3.86 | 3.73 | 3.77 | 3.93 | 3.72 | 3.63 | 3.64 | 3.79 | 3.63 | 3.36 | 3.41 | 3.71 | 3.40 | 2.57 |
| 3.86 | 3.73 | 3.75 | 3.93 | 3.72 | 3.61 | 3.64 | 3.79 | 3.62 | 3.36 | 3.40 | 3.71 | 3.38 | 2.57 |

analysis demonstrated that Mold² descriptors hold up to current scrutiny on the basis of descriptor correlation along with the compared sets of descriptors.

## 3.3 QSAR Using Mold² Descriptors

Shannon entropy offers a way of assessing information content and distribution encoded in descriptors. However, the Shannon entropy compares variance of encoded structural information, without offering insight regarding the correlation of structural variance with the biological activities or physicochemical properties of interest. In other words, a better set of descriptors not only carries sufficient structural information, but also relevant biological and physicochemical information. In a sense, the quality of a QSAR model is solely dependent on the correlation of chemical structure-independent variables with biologically or physicochemically related dependent variables.

To demonstrate the usefulness of Mold² descriptors for QSAR, examples of qualitative (classification) and quantitative (regression) models developed based on Mold² descriptors are given in this section. Furthermore, similar QSAR models generated from the same data sets by using the same modeling algorithms based on other sets of molecular descriptors are compared with those based on Mold².

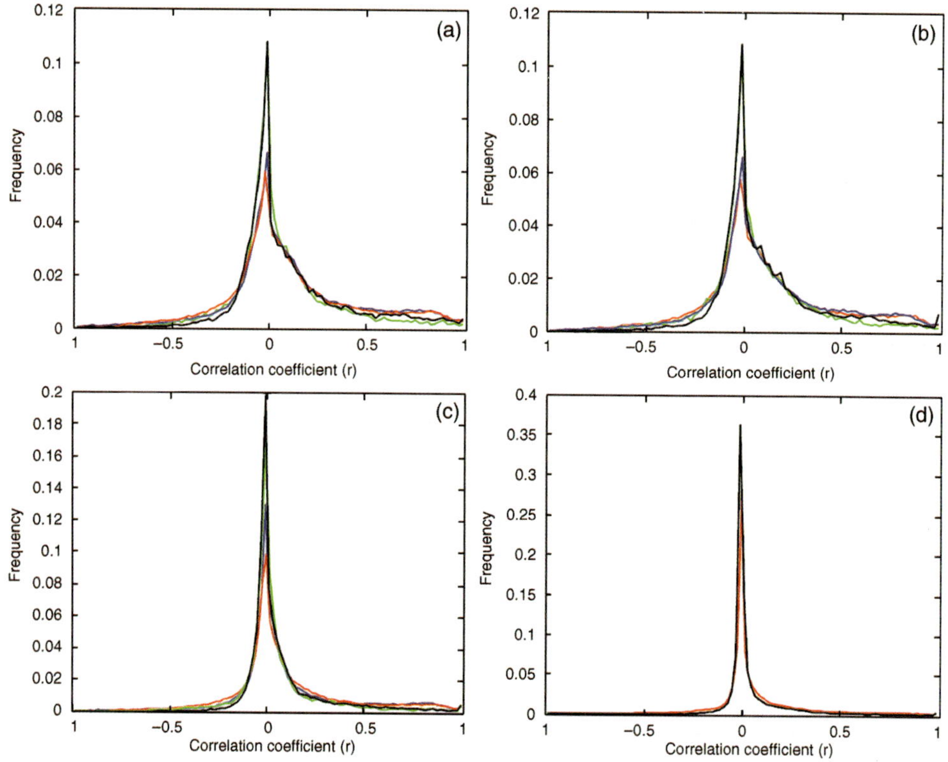


**Figure 3.4** Distribution of correlation coefficients (*r*) among descriptors for data sets NCTR_AR (a), NCTR_ER (b), ER_Comb (c), and EPA (d). The *x*-axis represents the correlation coefficients (*r*) among descriptors, while the *y*-axis is the frequency of *r*. The curves of Mold$^2$ descriptors are colored in red; Cerius$^2$ in green; Dragon in blue; and Molconn-Z in black [43].

### 3.3.1 Classification Models based on Mold$^2$ Descriptors

Decision Forest (DF) [17–19] is a classification algorithm developed in our laboratories. In principle, DF combines several accurate decision tree models that use distinct sets of descriptors. The flowchart of the DF algorithm is given in Figure 3.5. Briefly, each individual decision tree model is developed using a distinct set of descriptors that is explicitly excluded from all other models, thus ensuring each individual model's unique contribution to prediction. DF was applied to three previously published data sets (NCTR_ER [72], NCTR_AR [73], and ER_comb [74] as given in Table 3.3). Molecular descriptors were separately calculated from Cerius$^2$, Dragon, Molconn-Z, and Mold$^2$ for the chemicals in these three data sets. Thereafter, QSAR was conducted separately for each of the three data sets based on each of the four sets of molecular descriptors. Model performance estimates were made using 10-fold cross-validation, in which the data sets were randomly divided into 10 equal portions; each portion was successively excluded from the training set and predicted by the model developed from

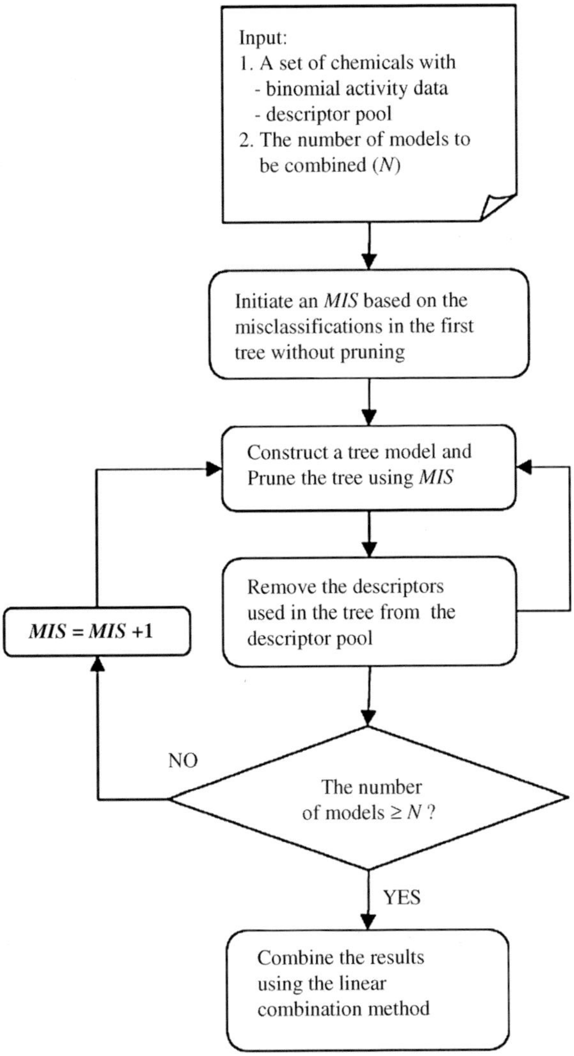


**Figure 3.5** Flowchart of the Decision Forest algorithm. The parameter *MIS* determines the number of misclassified chemicals allowed in pruning [19].

the remaining nine portions. The prediction accuracy of the cross-validation was taken as the average accuracy of the 10 models. Each random division of the data set into 10 portions leads to 10 specific pairs of training and test sets that could be biased in terms of prediction accuracy. Therefore, the 10-fold cross-validation was repeated 100 times to achieve a statistically unbiased estimation of predictive accuracy, sensitivity, and specificity. The performances (average prediction accuracy, sensitivity, and specificity of the 100 runs of 10-fold cross-validations as well as the corresponding standard deviations) of DF models from different data sets and descriptors are listed in Table 3.5. Mold$^2$ descriptors were found to yield slightly more accurate predictions (at the same

**Table 3.5** Prediction results of 100 runs of 10-fold cross-validations (%)[a)].

| Statistics | Data set | Cerius$^2$ | Dragon | Molconn_Z | Mold$^2$ |
|---|---|---|---|---|---|
| Accuracy | NCTR_AR | 79.09(±1.53) | 77.69(±1.73) | 78.18(±1.68) | 80.11(±1.75) |
| | NCTR_ER | 80.69(±0.65) | 80.98(±0.72) | 80.27(±0.66) | 80.90(±0.68) |
| | ER_Comb | 80.49(±1.72) | 80.20(±1.80) | 81.55(±1.51) | 81.69(±1.64) |
| Sensitivity | NCTR_AR | 91.01(±1.59) | 88.83(±1.62) | 90.01(±1.65) | 90.66(±1.66) |
| | NCTR_ER | 59.19(±1.36) | 60.42(±1.57) | 60.46(±1.36) | 61.26(±1.67) |
| | ER_Comb | 85.72(±2.28) | 85.19(±1.91) | 87.79(±1.70) | 86.95(±1.75) |
| Specificity | NCTR_AR | 48.02(±4.48) | 48.66(±4.35) | 47.36(±4.14) | 52.63(±4.54) |
| | NCTR_ER | 90.83(±0.79) | 90.75(±0.73) | 89.70(±0.72) | 90.28(±0.73) |
| | ER_Comb | 73.71(±2.86) | 73.73(±3.12) | 73.45(±2.59) | 74.88(±2.87) |

*a)* The data are the average values of 100 runs. The numbers in the parentheses are the standard deviations.

level of quality) than those using the descriptors from the compared commercial software. The results demonstrate the utility of Mold$^2$ descriptors for qualitative QSAR.

## 3.3.2
## Regression Models based on Mold$^2$ Descriptors

Generally speaking, all currently available descriptor software packages share a firm foundation in theory and practice spanning many decades. Likewise, there is a rich literature and proven track record of contributions to chemistry, medicinal chemistry, and a myriad of other QSAR applications. Different software packages may emphasize specific descriptor space domains corresponding to distinct aspects of chemical structure space, or may offer various statistical or graphical functionalities. It can be expected that good results will be obtained from quality data by the experienced and careful practitioner.

While not significantly expanding the descriptor pool, some products emphasize different aspects of the description of the molecular structure. For example, while Mold$^2$ and DRAGON primarily calculate constitutional and topological parameters, CODESSA (COmprehensive DEscriptors for Structural and Statistical Analysis) emphasizes descriptors obtained from quantum mechanical calculations.

To illustrate the mathematical consistency of results (statistical phenomena) we compared the performance of several products employed to correlate $C_{max}$ (the maximum or "peak" concentration of a drug observed after its administration) to the compound's structure expressed in terms of various molecular descriptor types. The following strategy was applied: (1) all of the 777 Mold$^2$ descriptors were scaled within the range of 0 to 1, and a forward stepwise multiple linear regression method was used to construct an optimal regression model and (2) all of the CODESSA descriptors (scaled automatically) were further processed by the integrated best multiple linear regression (BMLR) algorithm. For comparative purposes, a model involving the same number of parameters as the one generated using Mold$^2$ descriptors was selected.

The BMLR method implemented in CODESSA is a modification of the forward stepwise multiple linear regression algorithm [76] that utilizes the following strategy: (1) in a given descriptor space BMLR searches and selects all orthogonal (within some user defined limit) descriptor pairs; (2) using the restricted descriptor pool from step 1, BMLR generates two-parameter regression equations and ranks them according to their values of $R^2$ with the property; and (3) using the best two-parameter models saved in step 2, the algorithm adds a third noncollinear descriptor to obtain three-parameter equations. Then the Fisher criterion ($F$) and the cross-validated squared correlation coefficient ($R^2_{cv}$) are assessed. If there is no improvement over the best two-parameter regressions derived in step 2, the procedure is halted. Otherwise, BMLR continues with the generation of models of higher rank and stores them in a descending order in accordance with $R^2$.

The descriptor space generated by Mold$^2$ was processed in a similar way using a stepwise multiple linear regression algorithm implemented within the framework of the R language [77].

The two alternative regression models reported in Tables 3.6 and 3.7 are based on a dataset of 410 compounds with $C_{max}$ that were collected in the liver toxicity knowledge base (LTKB) project in our laboratory. The initial set was randomly split into training (273 compounds) and test (137 compounds) subsets. The modeling results from using Mold$^2$ and CODESSA descriptors are plotted in Figure 3.6a and b.

**Table 3.6** The best five-parameter regression model obtained using CODESSA descriptors [a].

| $X$ | $\Delta X$ | t | Descriptor |
|---|---|---|---|
| −31.28 | 11.27 | −2.775 | Intercept |
| −0.7139 | 0.1430 | −4.994 | Average Information content (order 2) |
| 7.935 | 2.127 | 3.731 | FHACA Fractional HACA (HACA/TMSA) (MOPAC PC) |
| −0.06850 | 0.02003 | −3.419 | Number of C atoms |
| 35.90 | 11.79 | 3.044 | Max SIGMA-SIGMA bond order |
| −0.0008475 | 0.0002969 | −2.855 | WNSA-2 Weighted PNSA (PNSA2*TMSA/1000) (MOPAC PC) |

a) $R^2 = 0.319$; $R^2_{test} = 0.229$; $F = 24.9$; Standard error of estimate = 1.10.

**Table 3.7** The best five-parameter regression model obtained using Mold$^2$ descriptors [a].

| $X$ | $\Delta X$ | $T$ | Descriptor |
|---|---|---|---|
| −0.1678 | 0.5709 | −0.2939 | Intercept |
| 3.335 | 0.7269 | 4.589 | Mean of vertex distance information index |
| −1.744 | 0.5480 | −3.183 | Information content order-3 index |
| −3.416 | 1.097 | −3.114 | Lowest eigenvalue from Burden matrixweighted by mass order 1 |
| 0.9708 | 0.3904 | 2.486 | Maximal valence vertex electrotopological negative variation |
| 1.008 | 0.6775 | 1.487 | Mean electrotopological states index |

a) $R^2 = 0.302$; $R^2_{test} = 0.212$; $F = 23.1$; Standard error of estimate = 1.12.

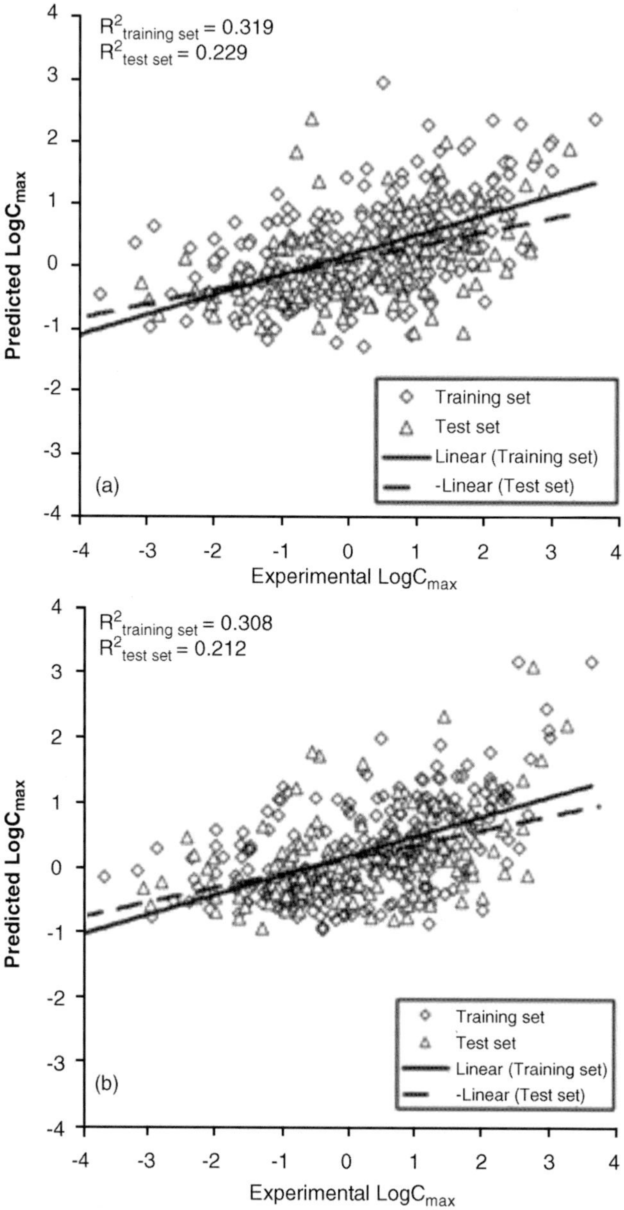


**Figure 3.6** Regression models based on CODESSA (a) and Mold² (b) molecular descriptors. Experimental log ($C_{max}$) values were plotted against the predicted log ($C_{max}$) values. The training results were depicted by the diamond points while the testing results were given by the triangle markers.

As shown in Figure 3.6, both sets of molecular descriptors produced regression models of almost identical quality. This not only emphasizes the equivalence between the different descriptor spaces, but also demonstrates that Mold$^2$ can be advantageously used for the purpose of quantitative QSAR.

## 3.4
## Conclusion Remarks

Molecular descriptors are used to represent chemical structures and have continually contributed to the many disciplines employing QSAR and chemoinformatics. Diverse applications include virtual library generation and screening, similarity and diversity analysis, and predictive toxicology. Descriptors can be either computationally or empirically derived. The 777 Mold$^2$ descriptors presented here are calculated from both 1D and 2D chemical structures. The molecular descriptors have been demonstrated to generate qualitative and quantitative models of similar quality as do molecular descriptors from other popular software. It is free of charge to the scientific community for efficiently obtaining a large set of informative descriptors with wide applicability to QSAR.

## References


1 Brown, C. and Frazer, T. (1868) On the connection between chemical constitution and physiological action: Part I. On the physiological action of the salts of the ammonia bases derived from Strycnia, Brucia, Thebaia, Codeia, Morphia, and Nicotinia. *Transactions of the Royal Society of Edinburgh-Earth Sciences*, **25**, 151–203.

2 Richardson, B. (1869) Physiological research on alcohols. *Medical Times and Gazette*, **2**, 703–706.

3 Portier, P. and Richet, C. (1902) De l'action anaphylactique de certain venins. *Comptes Rendus Des Seances De La Societe De Biologie Et De Ses Filiales*, **54**, 170–172.

4 Meyer, H. (1899) Zur Theorie der Alkoholnarkose. Erste Mittheilung. Welche Eigenschaft der Anästhetica bedingt ihre narkotische Wirkung? *Archiv für experimentelle Pathologie und Pharmakologie*, **42**, 109–118.

5 Overton, E. (1897) Ueber die osmotischen eigenschaften der zelle In ihrer bedcutung fur die toxikologie und pharmakologie. *Zeitschrift für Physikalische Chemie*, **22**, 189–209.

6 Bell, R.P. (1978) *Correlation Analysis in Chemistry: Recent Advances* (eds N.B. Chapman and J. Shorter), Plenum Press, New York, pp. 55–84.

7 Charton, M. (1996) *Advances in Quantitative Structure-Property Relationships* (ed. M. Charton), JAI Press, Inc, Greenwich, pp. 171–219.

8 Hansch, C., Maloney, P.P., Fujita, T., and Muir, R.M. (1962) Correlation of biological activity of phenoxyacetic acids with Hammett substituent constants and partition coefficients. *Nature*, **194**, 178–180.

9 Hansch, C., Muir, R.M., Fujita, T., Maloney, P.P., Geiger, F., and Streich, M. (1963) The correlation of biological activity of plant growth regulators and chloromycetin derivatives with Hammett constants and partition coefficients. *Journal of the American Chemical Society*, **85**, 2817–2824.

10 Taft, R.W. and Lewis, I.C. (1959) Evaluation of resonance effects on

reactivity by application of the linear inductive energy relationship: V. Concerning a $\sigma_R$ scale of resonance effects. *Journal of the American Chemical Society*, **81**, 5343–5352.

11 Cramer, R.D. and Milne, M. (1979) The lattice model: a general paradigm for shape-related structure/activity correlation. *Abstracts of Papers of the American Chemical Society M*, Computer Chemistry Section, no. 44.

12 Cramer, R.D., Patterson, D.E., and Bunce, J.D. (1988) Comparative molecular field analysis (CoMFA): Effect of shape on binding of steroids to carrier proteins. *Journal of the American Chemical Society*, **110**, 5959–5967.

13 SYBYL/QSAR (1699) Molecular Modelling Software, Tripos Inc., St. Louis, MO 63944, USA.

14 Strynadka, N.C., Eisenstein, M., Katchalski-Katzir, E., Shoichet, B.K., Kuntz, I.D., Abagyan, R., Totrov, M., Janin, J., Cherfils, J., Zimmerman, F., Olson, A., Duncan, B., Rao, M., Jackson, R., Sternberg, M., and James, M.N. (1996) Molecular docking programs successfully predict the binding of a beta-lactamase inhibitory protein to TEM-1 beta-lactamase. *Nature Structural Biology*, **3**, 233–239.

15 Kubinyi, H. (1993) QSAR: Hansch analysis and related approaches, in *Methods and Principles in Medicinal Chemistry*, vol. **1** (eds R. Mannhold, P. Kroogsgard-Larsen, and H. Timmerman), VCH, Weinheim.

16 Kubinyi, H. (1995) *Burger's Medicinal Chemistry*, 5th edn, vol. **I** (ed. M.E. Wolff), John Wiley & Sons Inc, New York, pp. 497–571.

17 Hong, H., Tong, W., Xie, Q., Fang, H., and Perkins, R. (2005) An *in silico* ensemble method for lead discovery: Decision forest. *SAR and QSAR in Environmental Research*, **16**, 339–347.

18 Hong, H., Tong, W., Perkins, R., Fang, H., Xie, Q., and Shi, L. (2004) Multiclass decision forest – a novel pattern recognition method for multiclass classification in microarray data analysis. *DNA and Cell Biology*, **23**, 685–694.

19 Tong, W., Hong, H., Fang, H., Xie, Q., and Perkins, R. (2003) Decision forest: Combining the predictions of multiple independent decision tree models. *Journal of Chemical Information and Computer Sciences*, **43**, 525–531.

20 Su, Z., Hong, H., Perkins, R., Shao, X., Cai, W., and Tong, W. (2007) Consensus analysis of multiple classifiers using non-repetitive variables: Diagnostic application to microarray gene expression data. *Computational Biology and Chemistry*, **31**, 48–56.

21 Efroymson, M.A. (1960) Multiple regression analysis, in *Mathematical Methods for Digital Computers* (eds A. Ralston and H.S. Wilf), Wiley, New York, pp. 191–203.

22 Jolliffe, I.T. (1982) A note on the use of principal components in regression. *Journal of the Royal Statistical Society, Series C (Applied Statistics)*, **31**, 300–303.

23 Dayal, B.S. and MacGregor, J.F. (1997) Improved PLS algorithms. *Journal of Chemometrics*, **11**, 73–85.

24 Rogers, D. and Hopfinger, A.J. (1994) Application of genetic function approximation to quantitative structure–activity relationships and quantitative structure–property relationships. *Journal of Chemical Information and Computer Sciences*, **34**, 854–866.

25 Bishop, C.M. (1995) *Neural Networks for Pattern Recognition*, Oxford University Press, Oxford.

26 Cortes, C. and Vapnik, V. (1995) Support-vector networks. *Machine Learning*, **20**, 273–297.

27 Breiman, L. (2001) Random forests. *Machine Learning*, **45**, 5–32.

28 Picard, R. and Cook, D. (1984) Cross-validation of regression models. *Journal of the American Statistical Association*, **79**, 575–583.

29 Efron, B. (1987) Better bootstrap confidence intervals. *Journal of the American Statistical Association*, **82**, 171–185.

30 Eriksson, L., Johansson, E., Müller, M., and Wold, S. (2000) On the selection of the training set in environmental QSAR analysis when compounds are clustered. *Journal of Chemometrics*, **14**, 599–616.

**31** Yang, Z.Z. and Wang, C.S. (2003) Atom-bond electronegativity equalization method and its applications based on density functional theory. *Journal of Theoretical and Computational Chemistry*, **2**, 273–299.

**32** Arulmozhiraja, S. and Morita, M. (2004) Structure–activity relationships for the toxicity of polychlorinated dibenzofurans: Approach through density functional theory-based descriptors. *Chemical Research in Toxicology*, **17**, 348–356.

**33** Liu, S.S., Cui, S.H., Yin, D.Q., Shi, Y.Y., and Wang, L.S. (2003) QSAR studies on the COX-2 inhibition by 3,4-diarylcycloxazolones based on MEDV descriptor. *Chinese Journal of Chemistry*, **21**, 1510–1516.

**34** Chiu, T.-L. and So, S.-S. (2004) Development of neural network QSPR models for Hansch substituent constants: 1. Method and validations. *Journal of Chemical Information and Computer Sciences*, **44**, 147–153.

**35** Chin, T.-L. and So, S.-S. (2004) Development of neural network QSPR models for Hansch substituent constants: 2. Applications in QSAR studies of HIV-1 reverse transcriptase and dihydrofolate reductase inhibitors. *Journal of Chemical Information and Computer Sciences*, **44**, 154–160.

**36** Lin, Z.H., Wu, Y.Z., Quan, X.J., Zhou, Y.G., Ni, B., and Wan, Y. (2002) Use of a novel electrotopological descriptor for the prediction of biological activity of peptide analogues. *Letters in Peptide Science*, **9**, 273–281.

**37** Agrawal, V.K., Mishra, K., and Khadikar, P.V. (2003) Multivariate analysis for modelling some antibacterial agents. *Oxidation Communications*, **26**, 14–21.

**38** Stanton, D.T. and Jurs, P.C. (1992) Computer-assisted study of the relationship between molecular structure and surface tension of organic compounds. *Journal of Chemical Information and Computer Sciences*, **32**, 109–115.

**39** Randic, M. (1996) Molecular bonding profiles. *Journal of Mathematical Chemistry*, **19**, 375–392.

**40** McGregor, M.J. and Pallai, P.V. (1997) Clustering large databases of compounds: using the MDL 'keys' as structural descriptors. *Journal of Chemical Information and Computer Sciences*, **37**, 443–448.

**41** Brown, R.D. and Martin, Y.C. (1997) The information content of 2D and 3D structural descriptors relevant to ligand–receptor binding. *Journal of Chemical Information and Computer Sciences*, **37**, 1–9.

**42** Matter, H. and Potter, T. (1999) Comparing 3D pharmacophore triplets and 2D fingerprints for selecting diverse compound subsets. *Journal of Chemical Information and Computer Sciences*, **39**, 1211–1225.

**43** Hong, H., Xie, Q., Ge, W., Qian, F., Fang, F., Shi, L., Su, Z., Perkins, R., and Tong, W. (2008) Mold$^2$, molecular descriptors from 2D structures for chemoinformatics and toxicoinformatics. *Journal of Chemical Information and Modeling*, **48**, 1337–1344.

**44** Wiener, H. (1947) Structural determination of paraffin boiling points. *Journal of the American Chemical Society*, **69**, 17–20.

**45** Diudea, M.V. (1996) Wiener and hyper-Wiener numbers in a single matrix. *Journal of Chemical Information and Computer Sciences*, **36**, 833–836.

**46** Kier, L.B. (1989) An index of molecular flexibility from kappa shape attributes. *Quantitative Structure–Activity Relationships*, **8**, 221–224.

**47** Hall, L.H. and Kier, L.B. (1977) Structure–activity studies using valence molecular connectivity. *Journal of Pharmaceutical Sciences*, **66**, 642–644.

**48** Hall, L.H. and Kier, L.B. (1991) The molecular connectivity Chi indexes and Kappa shape indexes in structure-property modeling, in *Reviews in Computational Chemistry*, vol. **2** (eds K.B. Lipkowitz and D.B. Boyd), VCH Publishers, New York, pp. 367–422.

**49** Kier, L.B. and Hall, L.H. (1999) *Molecular Structure Description. The Electrotopological State*, Academic Press, London.

**50** Balaban, A.T. (1983) Topological indices based on topological distances in molecular graphs. *Pure and Applied Chemistry*, **55**, 199–206.

51 Balaban, A.T. (1998) Topological and stereochemical molecular descriptors for databases useful in QSAR, similarity/dissimilarity and drug design. *SAR and QSAR in Environmental Research*, **8**, 1–21.

52 Randić, M. (1974) On the recognition of identical graphs representing molecular topology. *Journal of Chemical Physics*, **60**, 3920–3928.

53 Randić, M. (1975) On characterization of molecular branching. *Journal of the American Chemical Society*, **97**, 6609–6615.

54 Randić, M. and Wilkins, C.L. (1979) Graph theoretical ordering of structures as a basis for systematic search for regularities in molecular data. *Journal of Chemical Physics*, **83**, 1525–1540.

55 Randić, M. (1991) Generalized molecular descriptors. *Journal of Mathematical Chemistry*, **7**, 155–168.

56 Randić, M. and Basak, S.C. (1999) Optimal molecular descriptors based on weighted path numbers. *Journal of Chemical Information and Computer Sciences*, **39**, 261–266.

57 Randić, M., Basak, N., and Plavsic, D. (2004) Novel graphical index and distance-based molecular descriptors. *Croatica Chemica Acta*, **77**, 251–257.

58 Ivanciuc, O. and Ivanciuc, T. (2000) Matrices and structural descriptors computed from molecular graphs distances, in *Topological Indices and Related Descriptors in QSAR and QSPR* (eds J. Devillers and A.T. Balaban), Gordon & Breach, Amsterdam, The Netherlands, pp. 221–277.

59 Krenkel, G., Castro, E.A., and Toropov, A.A. (2001) Improved molecular descriptors based on the optimization of correlation weights of local graph invariants. *International Journal of Molecular Sciences*, **2**, 57–65.

60 Buckley, F. and Harary, F. (1990) *Distance Matrix in Graphs*, Addison-Wesley, Redwood City, CA.

61 Burden, F.R. (1997) A chemically intuitive molecular index based on the eigenvalues of a modified adjacency matrix. *Quantitative Structure–Activity Relationships*, **16**, 309–314.

62 Rücker, G. and Rücker, C. (1993) Counts of all walks as atomic and molecular descriptors. *Journal of Chemical Information and Computer Sciences*, **33**, 683–695.

63 Schultz, H.P. (1989) Topological organic chemistry: 1. Graph theory and topological indices of alkanes. *Journal of Chemical Information and Computer Sciences*, **29**, 227–228.

64 Gasteiger, J., Li, X., Rudolph, C.J., Sadowski, J., and Zupan, J. (1994) Representation of molecular electrostatic potentials by topological feature maps. *Journal of the American Chemical Society*, **116**, 4608–4620.

65 Galvez, J., Garcia, R., Salabert, M.T., and Soler, R. (1989) Charge indexes. New topological descriptors. *Journal of Chemical Information and Computer Sciences*, **29**, 520–525.

66 Shannon, C.E. (1948) A mathematical theory of communication. *Bell System Technical Journal*, **27**, 379–423.

67 Hong, H. and Xin, X. (1992) ESSESA, an expert system for structure elucidation from spectral analysis: Part II. Novel algorithm of perception of the linear independent smallest set of smallest rings. *Analytica Chimica Acta*, **262**, 179–191.

68 Hong, H. and Xin, X. (1994) ESSESA, an expert system for structure elucidation from spectra: 4. Canonical representation of structures. *Journal of Chemical Information and Computer Sciences*, **34**, 730–734.

69 Bonchev, D. (1983) *Information Theoretic Indices for Characterization of Chemical Structures*, Research Studies Press, Chichester.

70 Dehmer, M. and Mowshowitz, A. (2011) A history of graph entropy measures. *Information Sciences*. **181**, 57–78.

71 Godden, J.W., Stahura, F.L., and Bajorath, J. (2000) Variability of molecular descriptors in compound databases revealed by Shannon entropy calculations. *Journal of Chemical Information and Computer Sciences*, **40**, 796–800.

72 Blair, R., Fang, H., Branham, W.S., Hass, B., Dial, S.L., Moland, C.L., Tong, W., Shi, L., Perkins, R., and Sheehan, D.M. (2000) Estrogen receptor relative binding affinities of 188 natural and

xenocompounds: structure diversity of ligands. *Toxicological Sciences*, **54**, 138–153.

73 Fang, H., Tong, W., Branham, W., Moland, C.L., Dial, S.L., Hong, H., Xie, Q., Perkins, R., Owens, W., and Sheehan, D.M. (2003) Study of 202 natural, synthetic and environmental chemicals for binding to the androgen receptor. *Chemical Research in Toxicology*, **16**, 1338–1358.

74 Tong, W., Xie, Q., Hong, H., Fang, H., Shi, L., and Perkins, R. (2004) Assessment of prediction confidence and domain extrapolation of two structure-activity relationship models for predicting estrogen receptor binding activity. *EHP Toxicogenomics*, **112**, 1249–1254.

75 Hong, H., Tong, W., Fang, H., Shi, L., Xie, Q., Wu, J., Perkins, R., Walker, J.D., Branham, W., and Sheehan, D.M. (2002) Prediction of estrogen receptor binding for 58,000 compounds using an integrated system of a tree-based model with structural alerts. *Environ Health Perspect*, **110**, 29–36.

76 Drapper, N.R. and Smith, H. (1981) *Applied Regression Analysis*, Wiley, New York.

77 The R Project for Statistical Computing (http://www.r-project.org/).

# 4
# Multivariate Analysis of Molecular Descriptors

*Viviana Consonni and Roberto Todeschini*

## 4.1
## Introduction

In the last few decades, much interest has been addressed to studying how to encompass and convert the information encoded in the molecular structure into one or more numbers used to establish quantitative relationships between structures and properties, biological activities, or other experimental properties. Molecular descriptors are numerical indices encoding some information related to the molecular structure. They can be both experimental physicochemical properties of molecules and theoretical indices calculated by mathematical formulas or computational algorithms [1].

Matrices are the most common mathematical tool to encode structural information of molecules. They usually are the starting point for the calculation of many molecular descriptors; moreover, they constitute the mathematical form used as the molecule input in the majority of software packages for calculation of molecular descriptors.

Important matrices are the molecular matrix, which collects atom spatial coordinates, and all the matrices related to molecular geometry, such as the geometry matrix, whose 3D molecular descriptors are derived from and computational chemistry approaches are based on.

Other important and very popular matrices are the *graph-theoretical matrices*, a huge number of which were proposed in the last few decades in order to derive topological indices and describe molecules from a topological point of view. Graph-theoretical matrices are matrices derived from a molecular graph G, often from an H-depleted molecular graph. However, in a less restrictive sense, graph-theoretical matrices are all the matrices derived from a molecular graph, even if they encode additional contributions from the molecular geometry or other nontopological quantities. A comprehensive collection of graph-theoretical matrices is reported by Janežič *et al.* [2] and extended overviews in [3, 4].

Graph-theoretical matrices can be either vertex matrices if both rows and columns refer to graph vertices (atoms) and matrix elements encode some property of pairs of

*Statistical Modelling of Molecular Descriptors in QSAR/QSPR*. First Edition.
Edited by M. Dehmer, K. Varmuza, and D. Bonchev

vertices, or edge matrices, if both rows and columns refer to graph edges (bonds) and matrix elements encode some property of pairs of edges. Vertex matrices are square matrices of dimension $A \times A$, $A$ being the number of graph vertices, while edge matrices are square matrices of dimension $B \times B$, $B$ being the number of graph edges.

Vertex matrices are undoubtedly the graph-theoretical matrices most frequently used for characterizing a molecular graph. The matrix entries encode some information about pairs of vertices such as their connectivities, topological distances, sums of the weights of the vertices along the connecting paths; the diagonal entries can encode chemical information about the vertices. The most important vertex matrices are the *adjacency matrix* **A** that encodes information about vertex connectivities and the *distance matrix* **D** that also encodes information about relative locations of graph vertices.

The weighted graph-theoretical matrices are derived from vertex- and/or edge-weighted molecular graphs representing molecules containing heteroatoms and/or multiple bonds. A vertex- and edge-weighted molecular graph is obtained by defining a vertex weight set and an edge weight set according to a specific weighting scheme $w$. Usually, in a vertex- and edge-weighted graph, hydrogen atoms are not considered, thus resulting an H-depleted molecular graph, and the weight of a vertex corresponding to a carbon atom is zero, while the weight of an edge corresponding to a carbon–carbon single bond is 1 [5]. Typical vertex weights are physicochemical atomic properties such as atomic numbers, atomic mass, and atomic electronegativity, and local vertex invariants such as the vertex degree, which is the number of adjacent vertices. Edge weights usually are bond parameters such as conventional bond order, force constant, ionic character, dipole moment, and bond distance. Edge weights can also be derived from some combination of the weights of the two vertices incident with the edge.

*Topological indices* (TIs) or *molecular topological indices* (MTIs) are numerical quantifiers of molecular topology which are mathematically derived in a direct and unambiguous manner from the structural graph of a molecule; they are, therefore, graph invariants (Scheme 4.1). They can be sensitive to one or more structural features of the molecule such as size, shape, symmetry, branching and cyclicity, and can also encode chemical information concerning atom type and bond multiplicity. In fact, topological indices were divided into two categories: *topostructural* and *topochemical indices* [6, 7]. Topostructural indices encode only information about the adjacency and distances between atoms in the molecular structure; topochemical indices quantify information about topology but also specific chemical properties of atoms such as their chemical identity and hybridization state. Topological indices are mainly based on distances between atoms calculated by the number of intervening bonds and are thus considered through-bond indices; they differ from topographic indices and geometrical descriptors which are, instead, considered through-space indices because they are based on interatomic geometric distances [8, 9].

In general, TIs do not uniquely characterize molecular topology; different structures may have some of the same TIs. A consequence of nonuniqueness of topological indices is that they do not, in general, allow reconstructing molecule. Therefore, instead of single TIs suitably defined ordered sequences of TIs can be used to characterize molecules with higher discrimination.

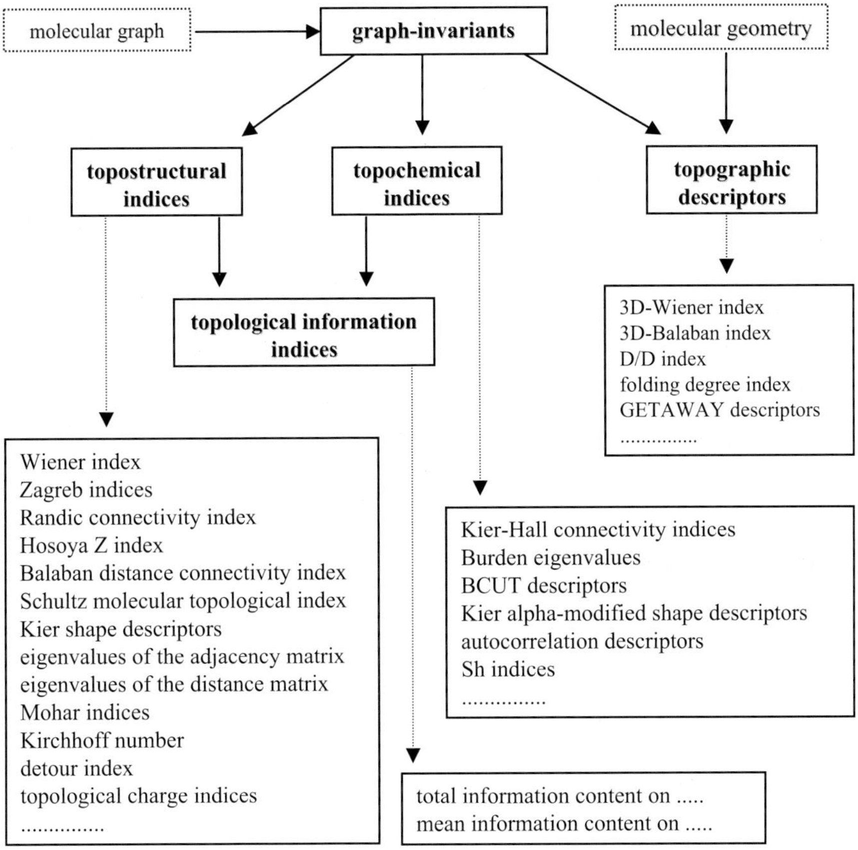


**Scheme 4.1** Different classes of graph invariants.

There are several ways to calculate topological indices. Simple topological indices consist in the counting of some specific graph elements; examples are the *Hosoya Z index* [10], *path counts, walk counts, self-returning walk counts, Kier shape descriptors* [11, 12], and *path/walk shape indices* [13]. However, the most common TIs are single numerical indices obtained by the application of algebraic operators to *graph-theoretical matrices* and whose values are independent of vertex numbering or labeling [8, 14–38]. Among them there are the *Wiener index* [39], *Randić connectivity index* [40], *Balaban distance connectivity index* [41], *Schultz molecular topological index* [42], *hyper-Wiener index* [3], *spectral indices* [43], *determinant-based descriptors* [1], *Harary indices* [1].

The most common functions to derive graph invariants from graph-theoretical matrices are listed in Table 4.1. Note that, in functions $D_1$ and $D_2$, the most common parameter values are $\alpha = 1/2$ and $\lambda = 1$. Function $D_3$ is used to generate descriptors derived from the matrix determinant and function $D_4$ descriptors that are linear combinations of the coefficients of the characteristic polynomial of a graph-

**Table 4.1** Algebraic operators for the calculation of molecular descriptors from graph-theoretical matrices.

| Operator | Operator |
|---|---|
| | $D_5(\mathbf{M}) = f(\Lambda(\mathbf{M}))$ |
| $D_1(\mathbf{M}; \alpha, \lambda) = \alpha \cdot \sum_{i=1}^{n}\sum_{j=1}^{n} [\mathbf{M}]_{ij}^{\lambda}$ | $D_6(\mathbf{M}; \alpha, \lambda) = \alpha \cdot \sum_{k=1}^{K} \left( \prod_{i=1}^{n_k} VS_i(\mathbf{M}) \right)_k^{\lambda}$ |
| $D_3(\mathbf{M}; \alpha) = \alpha \cdot \det(\mathbf{M})$ | $D_7(\mathbf{M}; \alpha, \lambda) = \alpha \cdot \sum_{i=1}^{n}\sum_{j=1}^{n} \left( [\mathbf{M}]_{ij}^{\lambda} + [\mathbf{M}]_{ij} \right)$ |
| $D_4(\mathbf{M}; \alpha, \lambda) = \alpha \cdot \sum_{i=0}^{n} \left\| c(Ch(\mathbf{M}; x))_i^{\lambda} \right\|$ | $D_8(\mathbf{M}; \alpha, \lambda) = \alpha \cdot \max_{ij} \left( [\mathbf{M}]_{ij}^{\lambda} \right)$ |

**M** is a graph-theoretical matrix, $n$ is the matrix dimension, $c(Ch(\mathbf{M};x))_i$ is the $i$th coefficient of the characteristic polynomial of **M**, $\Lambda(\mathbf{M})$ indicates the graph spectrum (i.e., the set of eigenvalues of **M**), $\alpha$ and $\lambda$ are real parameters. In function $D_6$, $VS_i(\mathbf{M})$ is the $i$th matrix row sum, $K$ is the total number of selected graph fragments, and $n_k$ is the number of vertices in the $k$th fragment. $a_{ij}$ indicates the elements of the adjacency matrix which are equal to one for pairs of adjacent vertices, and zero otherwise.

theoretical matrix, such as the *Hosoya-like indices* [44, 45]. Function $D_5$ is based on the eigenvalues calculated from graph-theoretical matrices and the related molecular descriptors are the so-called *spectral indices* [43]. Function $D_6$ makes use of the matrix row sums $VS_i$ as the local vertex invariants and, then, adds up the contributions from different graph fragments (e.g., edges), each weighted by the product of the local invariants of all the vertices contained in the fragment; *Randić connectivity index* [40] and *Balaban-like indices* [3] are calculated by means of this function. Function $D_7$ for $\alpha = 1/2$ and $\lambda = 2$ generates the *hyper-Wiener-like indices* [3].

Other topological indices can be obtained by using suitable functions applied to *local vertex invariants*; the most common functions are atom and/or bond additive, resulting in descriptors which correlate well physicochemical properties that are atom and/or bond additive themselves. *Zagreb indices* [46] and *ID numbers* [47–49] are derived according to this approach.

Another way to derive topological indices is by generalizing the existing indices and/or graph-theoretical matrices. *Kier–Hall connectivity indices* [31, 50], *higher order Wiener numbers* [51], *generalized Wiener indices* [1], and *variable Zagreb indices* [52, 53] are all examples of the generalization of the existing indices, while the *generalized distance matrix* [1], *expanded distance matrices* [2], and *graphical matrices* [54] are examples of generalized matrices.

## 4.2
## 2D Matrix-Based Descriptors

In the present study the 2D matrix-based descriptors calculated by the software DRAGON 6.0 were considered [55]. These are topological indices calculated by applying a set of basic algebraic operators to different graph-theoretical matrices representing an H-depleted molecular graph of molecules.

In the following, the algebraic operators used to calculate 2D matrix-based descriptors are listed. In all the equations, **M** will indicate any graph-theoretical matrix, *w* the vertex weighting scheme, and nSK the total number of non-H atoms in the molecule.

*Wiener-like indices* are indicated by the general symbol Wi_M(w) and calculated according to the following:

$$\text{Wi\_M(w)} = \sum_{i=1}^{\text{nSK}} [\mathbf{M}(w)]_{ii} + \sum_{i=1}^{\text{nSK}-1} \sum_{j=i+1}^{\text{nSK}} [\mathbf{M}(w)]_{ij} \tag{4.1}$$

where the summation of diagonal matrix elements is added to the summation of half-matrix nondiagonal elements. When no vertex weights are applied, Equation 4.1 reduces to the classical formula for calculation of the Wiener index [39].

*Average Wiener-like indices,* indicated by WiA_M(w), are calculated as

$$\text{WiA\_M(w)} = \begin{cases} \dfrac{2 \cdot \text{Wi\_M(w)}}{\text{nSK} \cdot (\text{nSK}-1)} & \text{if} [\text{M}(w)]_{ii} = 0 \\ \dfrac{2 \cdot \text{Wi\_M(w)}}{\text{nSK} \cdot (\text{nSK}+1)} & \text{if} [\text{M}(w)]_{ii} \neq 0 \end{cases} \tag{4.2}$$

where Wi_M(w) is a Wiener-like index and two different equations are applied depending on the presence of vertex weights on the main matrix diagonal.

*Average vertex sum indices,* AVS_M(w), are calculated as

$$\text{AVS\_M(w)} = \frac{\sum_{i=1}^{\text{nSK}} VS_i(\mathbf{M}; w)}{\text{nSK}} = \frac{\sum_{i=1}^{\text{nSK}} \sum_{j=1}^{\text{nSK}} [\mathbf{M}(w)]_{ij}}{\text{nSK}} \tag{4.3}$$

where $VS_i(\mathbf{M};w)$ is the *i*th matrix row sum and represents the simplest local invariant derived from a graph-theoretical matrix.

*Harary-like indices,* H_M(w), are defined as the half-sum of matrix elements like the Wiener indices; however, they are calculated from the reciprocal matrix as

$$\text{H\_M(w)} = \sum_{i=1}^{\text{nSK}} [\mathbf{M}(w)]_{ii} + \sum_{i=1}^{\text{nSK}-1} \sum_{j=i+1}^{\text{nSK}} \frac{1}{[\mathbf{M}(w)]_{ij}} \tag{4.4}$$

Note that the inverse function is only applied to nondiagonal elements different from zero.

*Randić-like indices,* Chi_M(w), are calculated by a generalization of the classical formula of the Randic connectivity index [40] based on the vertex degrees:

$$\text{Chi\_M(w)} = \sum_{i=1}^{\text{nSK}-1} \sum_{j=i+1}^{\text{nSK}} a_{ij} \cdot \left[VS_i(\mathbf{M}; w) \cdot VS_j(\mathbf{M}; w)\right]^{-1/2} \tag{4.5}$$

where the vertex degrees are replaced by the matrix row sums $VS(\mathbf{M};w)$ and elements $a_{ij}$ of the adjacency matrix are introduced to account only for contributions from bonded atom pairs.

*Average Randić-like indices*, ChiA_M(w), are calculated as

$$\mathrm{ChiA\_M(w)} = \frac{\mathrm{Chi\_M(w)}}{\mathrm{nBO}} \tag{4.6}$$

where nBO is the total number of edges in the H-depleted molecular graph.

*Balaban-like indices*, J_M(w), are similar to *Randić-like indices*, but calculated with a normalization factor that makes them independent of the molecule size and ciclicity degree:

$$\mathrm{J\_M(w)} = \frac{\mathrm{nBO}}{\mathrm{nCIC}+1} \cdot \sum_{i=1}^{\mathrm{nSK}-1} \sum_{j=i+1}^{\mathrm{nSK}} a_{ij} \cdot \left[VS_i(\mathbf{M};w) \cdot VS_j(\mathbf{M};w)\right]^{-1/2} \tag{4.7}$$

where nBO is the number of graph edges and nCIC the number of independent rings in the molecule.

*Hyper-Wiener-like indices*, HyWi_M(w), are a generalization of the hyper-Wiener index [3], calculated as

$$\mathrm{HyWi\_M(w)} = \ln\left\{1 + \frac{1}{2} \cdot \sum_{i=1}^{\mathrm{nSK}} \sum_{j=i}^{\mathrm{nSK}} \left([\mathbf{M}(w)]_{ij}^2 + [\mathbf{M}(w)]_{ij}\right)\right\} \tag{4.8}$$

Note that, unlike the original formula, Dragon software applies the natural logarithm to prevent too large values that are not suitable for QSAR modeling.

*Graph energy indices*, SpAbs_M(w), are calculated by applying to the matrix spectrum (i.e., the set of matrix eigenvalues) the same formula as the graph energy proposed by Gutman [56, 57] and based on the eigenvalues of the adjacency matrix:

$$\mathrm{SpAbs\_M(w)} = \sum_{i=1}^{\mathrm{nSk}} |\lambda_i| \tag{4.9}$$

where $\lambda$ represent the matrix eigenvalues.

*Spectral positive sum indices*, SpPos_M(w), are the function of the positive matrix eigenvalues:

$$\mathrm{SpPos\_M(w)} = \sum_{i=1}^{n^+} (\lambda_i^+) \tag{4.10}$$

where $n^+$ indicates that the summation only runs on the positive eigenvalues of matrix **M**.

*Normalized spectral positive sum indices*, SpPosA_M(w), are derived from Equation 4.10 by applying a normalization factor to reduce molecule size influence [58]:

$$\mathrm{SpPosA\_M(w)} = \frac{1}{\mathrm{nSK}} \cdot \sum_{i=1}^{n^+} (\lambda_i^+) \tag{4.11}$$

where nSK is the total number of graph vertices.

*Logarithmic spectral positive sum indices,* SpPosLog_M(w), are obtained by a logarithmic transformation of Equation 4.10 [58]:

$$\mathrm{SpPosLog\_M(w)} = \frac{\mathrm{nSK}}{10} \cdot \log\left(\sum_{i=1}^{n^+} \lambda_i^+\right) \quad (4.12)$$

*Leading eigenvalues,* SpMax_M(w), are defined as the largest eigenvalue of the matrix spectrum:

$$\mathrm{SpMax\_M(w)} = \max_i\{\lambda_i\} \quad (4.13)$$

When the matrix **M** is the adjacency matrix **A**, then the Lovasz–Pelikan index [59] is obtained.

The *normalized leading eigenvalues,* SpMaxA_M(w), are obtained by division of the leading eigenvalue by the number of graph vertices:

$$\mathrm{SpMaxA\_M(w)} = \frac{\max_i\{\lambda_i\}}{\mathrm{nSK}} \quad (4.14)$$

The *spectral diameters,* SpDiam_M(w), are calculated as the difference between the largest and smallest matrix eigenvalue:

$$\mathrm{SpDiam\_M(w)} = \max_i\{\lambda_i\} - \min_i\{\lambda_i\} \quad (4.15)$$

*Spectral absolute deviations,* SpAD_M(w), are a generalization of the graph energy indices accounting for those cases for which the average of eigenvalues is not zero [43]:

$$\mathrm{SpAD\_M(w)} = \sum_{i=1}^{\mathrm{nSK}} |\lambda_i - \bar{\lambda}| \quad (4.16)$$

where $\bar{\lambda}$ indicates the average of the matrix eigenvalues.

*Spectral mean absolute deviations,* SpMAD_M(w), are calculated by normalizing Equation 4.16 as [43]

$$\mathrm{SpMAD\_M(w)} = \left(\sum_{i=1}^{\mathrm{nSK}} |\lambda_i - \bar{\lambda}|\right) / \mathrm{nSK} \quad (4.17)$$

where nSK is the total number of matrix eigenvalues.

*Hosoya-like indices,* Ho_M(w), are based on the characteristic polynomial of the matrix $\mathbf{M}(w)$ as

$$\mathrm{Ho\_M(w)} = \ln\left(1 + \sum_{i=0}^{\mathrm{nSK}} |c_i|\right) \quad (4.18)$$

where $c_i$ the $i$th coefficient of the characteristic polynomial of **M**. Note that Dragon software applies the logarithmic transformation to prevent too large values that these indices tend to have with big molecules.

*Estrada-like indices*, EE_M(w), are calculated as generalization of the Estrada index, defined as average subgraph centrality and based on the eigenvalues of the adjacency matrix [60]. They are calculated as

$$\mathrm{EE\_M(w)} = \ln\left(1 + \sum_{i=1}^{\mathrm{nSK}} e^{\lambda_i}\right) \tag{4.19}$$

where the exponential function gives the advantage to take contemporarily into account both positive and negative eigenvalues, without compensation effects, and the logarithmic transformation was introduced to lower descriptor values.

The *spectral moments* of a matrix, indicated by the general symbol SM$k$_M(w) where $k$ is the order of the spectral moment, are calculated as summation of the $k$th power of the matrix eigenvalues as

$$\mathrm{SM}k\mathrm{\_M(w)} = \mathrm{sgn}\left(\sum_{i=1}^{\mathrm{nSK}} \lambda_i^k\right) \cdot \ln\left(1 + \left|\sum_{i=1}^{\mathrm{nSK}} \lambda_i^k\right|\right) \tag{4.20}$$

where the logarithmic transformation is applied to avoid too large values and the absolute value of the summation is calculated to avoid that the argument of the logarithm was negative. Then, the sign function was introduced to preserve the correct sign of the original spectral moment. Spectral moments from order 1 to 6 were calculated by Dragon.

The *coefficient sum of the last eigenvector*, VE1_M(w), is calculated as

$$\mathrm{VE1\_M(w)} = \sum_{i=1}^{\mathrm{nSK}} |l_i| \tag{4.21}$$

where $l_i$ indicates the $i$th coefficient of the last eigenvector of $\mathbf{M}(w)$ and the summation runs over all the eigenvector coefficients that equal the number of graph vertices. The last eigenvector is that associated with the largest negative eigenvalue of the matrix **M**; these indices were originally proposed as derived from the adjacency matrix [58].

The *average coefficient of the last eigenvector*, VE2_M(w), is obtained by normalization of VE1_M(w) as

$$\mathrm{VE2\_M(w)} = \frac{1}{\mathrm{nSK}} \cdot \sum_{i=1}^{\mathrm{nSK}} |l_i| \tag{4.22}$$

while the *logarithmic coefficient sum of the last eigenvector*, VE3_M(w), by a logarithmic transformation as

$$\mathrm{VE3\_M(w)} = \frac{\mathrm{nSK}}{10} \cdot \log(\mathrm{VE1\_M(w)}) \tag{4.23}$$

*Randic-like eigenvector based-indices*, VR1_M(w), are calculated by a Randic-type formula using the coefficients of the last eigenvector in place of the vertex degrees [58]:

$$\mathrm{VR1\_M(w)} = \sum_{i=1}^{\mathrm{nSK}-1} \sum_{j=i+1}^{\mathrm{nSK}} a_{ij} \cdot \left[1 + \left(|l_i| \cdot |l_j|\right)\right]^{-1/2} \tag{4.24}$$

where $a_{ij}$ are the elements of the adjacency matrix, which are equal to 1 for pairs of adjacent vertices, and zero otherwise. Note that the Randic-type formula was modified in order to avoid too large values for big and symmetryc molecules for which some eigenvector coefficients tend toward zero.

*Normalized Randic-like eigenvector-based indices*, VR2_M(w), are calculated as

$$\mathrm{VR2\_M(w)} = \frac{\mathrm{VR1\_M(w)}}{\mathrm{nSK}} \tag{4.25}$$

and *logarithmic Randic-like eigenvector-based indices*, VR3_M(w), as

$$\mathrm{VR3\_M(w)} = \frac{\mathrm{nSK}}{10} \cdot \log(\mathrm{VR1\_M(w)}) \tag{4.26}$$

where VR1_M(w) is the Randic-like eigenvector-based index as defined in Equation 4.24.

Four additional descriptors were considered, only defined for and calculated from the Laplace matrix **L**, as described below. These are the quasi-Wiener index, first and second Mohar indices, and the spanning tree number.

The *quasi-Wiener index* (QW_L), also called the *Kirchhoff number*, is calculated as the product of the number of non-H atoms (nSK) and the sum of the reciprocal nonzero eigenvalues of the Laplacian matrix [56]; [61]:

$$\mathrm{QW\_L} = \mathrm{nSK} \cdot \sum_{i=1}^{\mathrm{nSK}-1} \frac{1}{\lambda_i} \tag{4.27}$$

The *first Mohar index* (TI1_L) and the *second Mohar index* (TI2_L) are calculated from the eigenvalues of the Laplace matrix as follows:

$$\mathrm{TI1\_L} = 2 \cdot \log\left(\frac{\mathrm{nBO}}{\mathrm{nSK}}\right) \cdot \mathrm{QW\_L} \tag{4.28}$$

$$\mathrm{TI2\_L} = \frac{4}{\mathrm{nSK} \cdot \lambda_{\mathrm{nSK}-1}} \tag{4.29}$$

where QW_L is the quasi-Wiener index, nBO and nSK are the number of non-H bonds and non-H atoms, respectively; and $\lambda_{\mathrm{nSK}-1}$ is the smallest nonzero eigenvalue of the Laplace matrix [61].

The *spanning tree number* (STN_L) is the count of all the spanning trees of a molecular graph G, the spanning tree being a connected acyclic subgraph containing all the vertices of G [61]. It is calculated as the product of the positive nSK–1 eigenvalues of the Laplacian matrix divided by the number of non-H atoms (nSK):

$$\mathrm{STN\_L} = \frac{1}{\mathrm{nSK}} \cdot \prod_{i=1}^{\mathrm{nSK}-1} \lambda_i \tag{4.30}$$

The number of spanning trees of a graph is used as a measure of molecular complexity for polycyclic graphs; it increases with the complexity of the molecular structure.

## 4.3 Graph-Theoretical Matrices

The mathematical definitions of the graph-theoretical matrices [1, 2] considered in this study are collected below. The following notation is used: $i$ and $j$ indicate two vertices in the molecular graph and $\mathcal{E}(\mathcal{G})$ is the set of the graph edges.

The *adjacency matrix* **A** is a square binary matrix, whose elements equal 1 if they correspond to pairs of adjacent vertices:

$$[\mathbf{A}]_{ij} = \begin{cases} 1 & \text{if}\,(i,j) \in \mathcal{E}(\mathcal{G}) \\ 0 & \text{otherwise} \end{cases} \tag{4.31}$$

The *topological distance matrix* **D** collects topological distances between pairs of vertices:

$$[\mathbf{D}]_{ij} = \begin{cases} d_{ij} = |{}^{\min}p_{ij}| & \text{if } i \neq j \\ 0 & \text{if } i = j \end{cases} \tag{4.32}$$

where $d_{ij}$ is called the topological distance and is defined as the length (i.e., number of edges) of the shortest path ${}^{\min}p_{ij}$ between two vertices $i$ and $j$.

The *Laplace matrix* **L** is obtained by the difference between a diagonal vertex degree matrix and the adjacency matrix **A**:

$$[\mathbf{L}]_{ij} = \begin{cases} \delta_i & \text{if } i = j \\ -1 & \text{if}\,(i,j) \in \mathrm{E(G)} \\ 0 & \text{if}\,(i,j) \notin \mathrm{E(G)} \end{cases} \tag{4.33}$$

where $\delta_i$ is the $i$th vertex degree, that is, the number of vertices adjacent to vertex $i$.

The *Chi matrix* **X** is a modified adjacency matrix, obtained by weighting each edge between pairs of vertices by the edge connectivity as

$$[\mathbf{X}]_{ij} = \begin{cases} 1/\sqrt{\delta_i \cdot \delta_j} & \text{if}\,(i,j) \in \mathrm{E(G)} \\ 0 & \text{otherwise} \end{cases} \tag{4.34}$$

where $\delta$ indicates vertex degrees.

The *reciprocal squared distance matrix* **H2** collects reciprocal squared topological distances between any pair of non-H atoms; it is defined as

$$[\mathbf{H2}]_{ij} = \begin{cases} 1/d_{ij}^2 & \text{if } i \neq j \\ 0 & \text{if } i = j \end{cases} \tag{4.35}$$

where $d_{ij}$ is the topological distance between vertices $i$ and $j$.

The *detour matrix* **Dt** is a square symmetric matrix, whose entry *i–j* the detour distance $\Delta_{ij}$ , i.e., is the length of the longest path from vertex $v_i$ to vertex $v_j$; it is defined as

$$[\mathbf{Dt}]_{ij} = \begin{cases} \Delta_{ij} = |^{\max}p_{ij}| & \text{if } i \neq j \\ 0 & \text{if } i = j \end{cases} \tag{4.36}$$

where $^{\max}p_{ij}$ refers to the longest path between the two vertices $i$ and $j$.

The *distance/detour matrix* **D/Dt**, derived from distance and detour matrices, is a square symmetric matrix, whose off-diagonal entries are the ratios of the lengths of the shortest over the longest path between any pair of vertices:

$$[\mathbf{D}/\mathbf{Dt}]_{ij} = \begin{cases} \dfrac{d_{ij}}{\Delta_{ij}} & \text{if } i \neq j \\ 0 & \text{if } i = j \end{cases} \tag{4.37}$$

where $d_{ij}$ and $\Delta_{ij}$ are the topological and detour distances between vertices $v_i$ and $v_j$, respectively.

*Barysz matrices* **Dz(w)** are weighted distance matrices accounting contemporarily for the presence of heteroatoms and multiple bonds in the molecule. They were defined on the basis of a generalization of the Barysz weighting scheme in terms of conventional bond orders $\pi^*$ and any atomic property [5]:

$$[\mathbf{Dz}(\mathbf{w})]_{ij} = \begin{cases} d_{ij}(w,\pi^*) & \text{if } i \neq j \\ 1 - \dfrac{w_C}{w_i} & \text{if } i = j \end{cases} \qquad d_{ij}(w,\pi^*) = \sum_{b=1}^{d_{ij}} \left( \frac{1}{\pi_b^*} \cdot \frac{w_C^2}{w_{b(1)} \cdot w_{b(2)}} \right) \tag{4.38}$$

where $w_C$ is any atomic property of the carbon atom and $w_i$ the corresponding value of the $i$th atom; $\pi^*$ is the conventional bond order, equal to 1, 2, 3, and 1.5 for single, double, triple, and aromatic bonds, respectively; $d_{ij}(w,\pi^*)$ is a weighted topological distance calculated by summing the edge weights over all $d_{ij}$ bonds involved in the shortest path between vertices $v_i$ and $v_j$, $d_{ij}$ being the topological distance, and the subscripts $b(1)$ and $b(2)$ represent the two vertices incident to the considered $b$ edge. Note that diagonal elements are weights for vertices based on relative atomic properties and the quantity in parenthesis is an edge weight based on conventional bond order and atomic properties of the vertices incident to the considered edge.

*Burden matrices* **B(w)** are augmented adjacency matrices derived from an H-depleted molecular graph defined to account for heteroatoms and bond multiplicity as follows:

$$[\mathbf{B}]_{ij} = \begin{cases} \sqrt{\pi_{ij}^*} & \text{if } (i,j) \in E(G) \\ \dfrac{w_i}{w_C} & \text{if } i = j \\ 0.001 & \text{if } (i,j) \notin E(G) \end{cases} \tag{4.39}$$

The diagonal elements are atomic carbon-scaled properties ($w_i/w_C$); the off-diagonal elements corresponding to pairs of bonded atoms are the square roots of conventional bond orders; entries corresponding to terminal bonds are augmented by 0.1; all other matrix elements are set at 0.001.

### 4.3.1
### Vertex Weighting Schemes

Dragon software [55] uses the following atomic properties as the atomic weightings for descriptor calculation: *atomic mass* (*m*), *van der Waals volume* (*v*), *atom electronegativity* (*e*), *atom polarizability* (*p*), *ionization potential* (*i*), and *intrinsic state* (s).

The intrinsic state of the *i*th atom is a local vertex invariant calculated from the molecular graph as follows [62, 63]:

$$I_i = \frac{(2/L_i)^2 \cdot \delta_i^v + 1}{\delta_i} \tag{4.40}$$

where $L$ is the principal quantum number, $\delta^v$ is the number of valence electrons (valence vertex degree), and $\delta$ is the number of sigma electrons (simple vertex degree) of the *i*th atom in the H-depleted molecular structure.

Carbon-scaled values were used for calculation of all the weightings except for the intrinsic state. All the Dragon weighting schemes were settled for calculation of Barisz and Burden matrices. The original Barisz matrix based on the atomic numbers (*Z*) was also calculated, whereas the Barisz matrix based on the intrinsic states (s) was not provided by Dragon and then not considered in this study.

## 4.4
## Multivariate Similarity Analysis of Chemical Spaces

Since molecules can be described in different ways depending on the selected molecular representation and algorithms for generating molecular descriptors, methods for comparison of the different viewpoints are required. In QSAR and drug design field, different results can derive if the same collection of molecules is analyzed in different chemical spaces. Thus, information about differences between chemical spaces should be acquired before attempting analysis of molecule relationships in different chemical spaces.

The *chemical space* is defined as the *p*-dimensional space constituted by a set of *p* molecular descriptors selected to represent the studied compounds; chemical space design is generally recognized as a crucial step for the successful application of QSPR/QSAR methods [64–67].

In different chemical spaces, the same molecules can behave differently with the consequence that different relationships may arise. Then, comparison among chemical spaces can enhance knowledge about a pool of molecules and selection of chemical spaces that are really diverse can help to find the most diverse molecules when the objective is molecule library design or to select the most similar molecules to a lead compound when the objective is to find out new drug candidates.

Let A and B be two different chemical spaces, that is, two different sets of molecular descriptors calculated for the same $n$ molecules. The simplest way to measure the diversity between the two chemical spaces disregards the actual descriptor values and simply consists in computing the number of diverse descriptors defining the two spaces, that is, the squared Hamming distance:

$$d_{\mathrm{H}}^2 = b + c \tag{4.41}$$

where $b$ is the number of molecular descriptors defining the space A but not the space B, and $c$ the number of molecular descriptors defining the space B but not the space A. Hamming distance usually has an upward bias since it overestimates the actual distance between two chemical spaces, due to the fact that variable correlation is not accounted for.

In order to overcome this drawback, a distance measure between two sets of variables was proposed with the name *Canonical Measure of Distance*, or simply *CMD index* [68–70]; it is defined as

$$CMD_{\mathrm{AB}} = p_{\mathrm{A}} + p_{\mathrm{B}} - 2 \cdot \sum_{j=1}^{M} \sqrt{\lambda_j} \quad 0 \leq CMD_{\mathrm{AB}} \leq (p_{\mathrm{A}} + p_{\mathrm{B}}) \tag{4.42}$$

where A and B are the two sets of molecular descriptors being compared, $p_{\mathrm{A}}$ and $p_{\mathrm{B}}$ are the number of descriptors in sets A and B, respectively; $\lambda$ are the eigenvalues of the symmetrical cross-correlation matrix and $M$ is the number of nonvanishing eigenvalues.

The cross-correlation matrix contains the pairwise correlation coefficients between descriptors of the two sets; it is an unsymmetrical matrix $\mathbf{C_{AB}}$ of size $(p_{\mathrm{A}} \times p_{\mathrm{B}})$ or $\mathbf{C_{BA}}$ of size $(p_{\mathrm{B}} \times p_{\mathrm{A}})$. The symmetrical cross-correlation matrix is derived by the following inner product:

$$\mathbf{Q}_{\mathrm{A}} = \mathbf{C}_{\mathrm{AB}} \times \mathbf{C}_{\mathrm{BA}} \quad \text{or} \quad \mathbf{Q}_{\mathrm{B}} = \mathbf{C}_{\mathrm{BA}} \times \mathbf{C}_{\mathrm{AB}} \tag{4.43}$$

where $\mathbf{Q}_{\mathrm{A}}$ and $\mathbf{Q}_{\mathrm{B}}$ are two different square symmetrical matrices, one of size $p_{\mathrm{A}} \times p_{\mathrm{A}}$ and the other of size $p_{\mathrm{B}} \times p_{\mathrm{B}}$. Although these symmetrical matrices are different, their $M$ nonzero eigenvalues coincide, $M$ being the minimum rank between $\mathbf{Q}_{\mathrm{A}}$ and $\mathbf{Q}_{\mathrm{B}}$.

The *Canonical Measure of Correlation* or the *CMC index* was also derived from the nonvanishing eigenvalues $\lambda$ of the symmetrical cross-correlation matrices as follows:

$$CMC_{\mathrm{AB}} = \frac{\sum_{j=1}^{M} \sqrt{\lambda_j}}{\sqrt{p_{\mathrm{A}} \cdot p_{\mathrm{B}}}} \quad 0 \leq CMC_{\mathrm{AB}} \leq 1 \tag{4.44}$$

where the numerator measures the interset common variance and the denominator is its theoretical maximum value. This index is related to the multidimensional correlational structure between two sets of variables and thus is a suitable index for measuring the degree of similarity between two chemical spaces. If no correlation exists between any pair of molecular descriptors from the two spaces, then $CMC = 0$ and the *CMD* index reduces to the squared Hamming distance.

## 4.5
## Analysis of Chemical Information of Descriptors from Graph-Theoretical Matrices

This study aimed at comparing the chemical information encoded by molecular descriptors derived from the different types of graph-theoretical matrices to find out the types that are the most diverse and hence provide the most diverse information. The total number of considered molecular descriptors is $31 \times 9 + 4 = 283$ descriptors, and thus their investigation requires statistical multivariate analysis techniques. The common techniques working on all the variables together are not much useful for analysis of these data, because information is mixed and interpretation of the roles of single descriptors is not an easy task. Then, a stepwise analysis was followed. First, the similarity/diversity of chemical spaces defined by descriptor vectors derived from the different types of matrices was globally analyzed, focusing on the role of graph matrices as sources of chemical information. A second analysis was carried out considering separately the chemical spaces derived from the different matrices in order to deeply investigate the role of single matrix operators in defining each chemical space. The final analysis focused on the different types of matrix operators, trying to find out how the same operator can generate different information depending on the matrix it is applied to.

Since the final results of analysis depend on the specific data set used for comparison, in order to draw more general conclusion this study was extended to a number of benchmark data sets.

### 4.5.1
### Data Sets

In order to perform an extensive comparison among molecular descriptors calculated from different graph-theoretical matrices, several data sets with different characteristics were used (Table 4.2).

The data set *Dragon* is a set of 42 small diverse organic molecules distributed along with the software Dragon; the first molecule is methane and was excluded because some descriptors are missing for this molecule. The data set *C10* is comprised of

**Table 4.2** Basic characteristics of the data sets in analysis.

| Data set | No. of compounds | Type | Heteroatoms | Rings |
|---|---|---|---|---|
| Dragon | 41 | Diverse – small molecules | O, N, S, F, Cl, Br, I | Yes |
| C10 | 75 | Constitutional isomers | No | No |
| Haloalkanes | 198 | Chemical class | F, Cl, Br, I | No |
| Phenylureas | 55 | Congeneric class | N, O, S, F, Cl, Br | Yes |
| PCDD | 75 | Congeneric class | O, Cl | Yes |
| Flavonoids | 82 | Congeneric class | O, N, F, Cl, Br, I | Yes |
| Pesticides | 153 | Diverse | O, N, S, P, F, Cl, Br | Yes |
| ED | 911 | Diverse | O, N, S, P, F, Cl, Br, I | Yes |

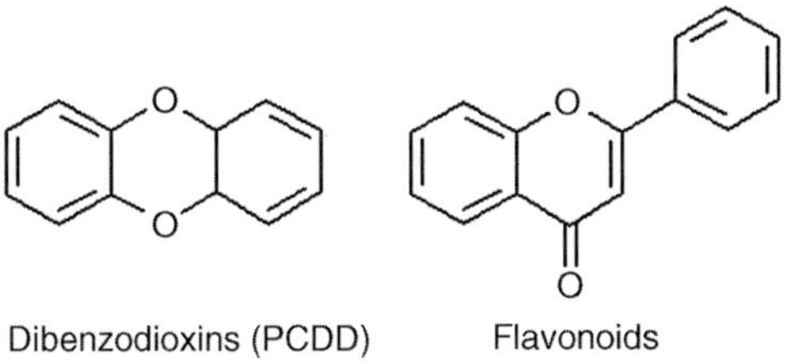


**Figure 4.1** Common scaffold of *PCDD* and *Flavonoids* data sets.

75 constitutional isomers of all the acyclic alkanes with 10 carbon atoms. The data set *Haloalkanes* is comprised of 198 methane-, ethane-, and propane-substituted alkanes with different number and combinations of the four halogen atoms. The data set *PCDD* includes the 75 polychlorodibenzodioxins, that is, a set of cyclic congeneric molecules differring only for the number and position of chlorine atoms. The data set *Flavonoids* is comprised of 82 flavonoids, containing oxygen, nitrogen and halogens as heteroatoms. The set *Pesticides* includes 153 diverse molecules belonging to four chemical classes: phenyleureas, triazines, carbamates, and organophospor-ous compounds. The set *ED* is a large data set of 911 molecules sharing different activities as endocrine disruptors and with very different chemical structures. In Figure 4.1, the common scaffolds of PCDD and flavonoids are shown.

### 4.5.2 Comparison of Graph-Theoretical Matrices

For all the considered data sets, a first analysis was carried out on the 2D matrix-based descriptors calculated from the following nine graph-theoretical matrices: adjacency matrix (**A**), topological distance matrix (**D**), Laplace matrix (**L**), Chi matrix (**X**), reciprocal squared distance matrix (**H2**), detour matrix (**Dt**), distance/detour matrix (**D/Dt**), original Barysz matrix (**Dz**(**Z**)), and mass-weighted Burden matrix (**B**(**m**)). Molecular descriptors were calculated by the software DRAGON 6.0 [55]. Dragon provides about 5000 descriptors ranging from the simplest counts of structural features to more complex geometrical descriptors. All the molecular structures, used as the DRAGON input, were designed by means of the software HyperChem 8.0 (2007); geometry optimization was not performed since only topological indices were considered in this study.

The list of 2D matrix-based descriptors calculated by Dragon from each of the considered matrices is given in Table 4.3. Note that Randic-like and average Randic-like index from the adjacency matrix (**A**) are not directly provided by Dragon under the block of 2D matrix-based descriptors; they, in effect, are Randic connectivity index and its average counterpart, which are included in the block of connectivities indices. Also the spectral moments (SMs) of adjacency matrix are the self-returning walks of order up to 6.

Depending on the starting matrix, some operators are not applied when giving trivial or redundant molecular descriptors. For instance, the Wiener-like index of the

**Table 4.3** 2D matrix-based descriptors calculated by Dragon software.

| No. | Descriptor | A | D | L | X | H2 | Dt | D/Dt | Dz(Z) | B(m) |
|---|---|---|---|---|---|---|---|---|---|---|
| 1 | Wi | | X | | | X | X | X | X | X |
| 2 | WiA | | X | | | X | X | X | X | X |
| 3 | AVS | | X | | X | X | X | X | X | X |
| 4 | H | | X | | X | | X | X | X | |
| 5 | Chi | X | X | | X | X | X | X | X | X |
| 6 | ChiA | X | X | | X | X | X | X | X | X |
| 7 | J | X | X | | X | X | X | X | X | X |
| 8 | HyWi | | X | | X | X | X | X | X | X |
| 9 | SpAbs | | | | | | | | X | X |
| 10 | SpPos | X | X | X | X | X | X | X | X | X |
| 11 | SpPosA | X | X | X | X | X | X | X | X | X |
| 12 | SpPosLog | X | X | X | X | X | X | X | X | X |
| 13 | SpMax | X | X | X | X | X | X | X | X | X |
| 14 | SpMaxA | X | X | X | X | X | X | X | X | X |
| 15 | SpDiam | X | X | X | X | X | X | X | X | X |
| 16 | SpAD | X | X | X | X | X | X | X | X | X |
| 17 | SpMAD | X | X | X | X | X | X | X | X | X |
| 18 | Ho | X | X | X | X | X | X | X | X | X |
| 19 | EE | X | X | X | X | X | X | X | X | X |
| 20 | SM1 | | | | | | | | X | X |
| 21 | SM2 | X | X | X | X | X | X | X | X | X |
| 22 | SM3 | X | X | X | X | X | X | X | X | X |
| 23 | SM4 | X | X | X | X | X | X | X | X | X |
| 24 | SM5 | X | X | X | X | X | X | X | X | X |
| 25 | SM6 | X | X | X | X | X | X | X | X | X |
| 26 | VE1 | X | X | X | X | X | X | X | X | X |
| 27 | VE2 | X | X | X | X | X | X | X | X | X |
| 28 | VE3 | X | X | X | X | X | X | X | X | X |
| 29 | VR1 | X | X | X | X | X | X | X | X | X |
| 30 | VR2 | X | X | X | X | X | X | X | X | X |
| 31 | VR3 | X | X | X | X | X | X | X | X | X |
| 32 | QW | | | X | | | | | | |
| 33 | T1 | | | X | | | | | | |
| 34 | T2 | | | X | | | | | | |
| 35 | STN | | | X | | | | | | |
| | Total | 17 | 29 | 25 | 27 | 28 | 29 | 29 | 31 | 30 |

adjacency matrix **A** coincides with the number of non-H bonds and, thus, this index and its related ones are not calculated. The Harary-like index (H) of the reciprocal squared distance matrix (**H2**) is not provided since it would be the half-sum of squared topological distances and then would give too large values. Graph-energy indices (SpAbs) are calculated only for Barisz and Burden matrices since for all the other matrices they coincide with spectral absolute deviations (SpAD). Also the spectral moment of first order (SM1) is calculated only for weighted matrices since in all the other cases it equals zero for any molecule. The Wiener-like index of

**Table 4.4** Number of descriptors calculated for each graph-theoretical matrix and for each data set.

| Data set | A | D | L | X | H2 | Dt | D/Dt | Dz(Z) | B(m) |
|---|---|---|---|---|---|---|---|---|---|
| Dragon | 17 | 29 | 25 | 25 | 28 | 29 | 29 | 31 | 30 |
| C10 | 17 | 29 | 15 | 22 | 28 | — | — | — | — |
| Haloalkanes | 17 | 29 | 24 | 23 | 28 | — | — | 31 | 30 |
| Phenylureas | 17 | 29 | 25 | 25 | 28 | 29 | 29 | 31 | 30 |
| PCDD | 17 | 29 | 24 | 23 | 28 | 29 | 29 | 31 | 30 |
| Flavonoids | 17 | 29 | 25 | 23 | 28 | 29 | 29 | 31 | 30 |
| Pesticides | 17 | 29 | 25 | 26 | 28 | 29 | 29 | 31 | 30 |
| ED | 17 | 29 | 25 | 26 | 28 | 28 | 29 | 31 | 30 |

the Chi matrix (**X**) coincides with the Randic-like index from the adjacency matrix (**A**). Harary-like indices of the adjacency (**A**), Laplace (**L**), and Burden (**B**) matrices, which all codify information about atom connectivities, are not calculated since inverting connectivities, that are commonly expressed by one, is meaningless.

In Table 4.4, the total number of descriptors calculated from each matrix and for each data set is reported. Constant descriptors were excluded from analysis together with descriptors with some missing values. Moreover, the number of molecular descriptors derived from each matrix differs because, for a given graph-theoretical matrix, some descriptors have no meaning or the algorithm for their calculation is not applicable.

For acyclic molecules, the detour matrix (**Dt**) coincides with the distance topological matrix (**D**) and the distance/detour matrix (**D/DT**) is a trivial unity matrix. Then, for the data sets *C10* and *Haloalkanes*, detour and distance/detour matrices were excluded from analysis. Moreover, Burden (**B**) and Barysz (**Dz**) matrices, which are defined for vertex- and/or edge-weighted molecular graphs, were not calculated for the data set *C10*, it being comprised of molecules without heteroatoms nor multiple bonds. For this kind of molecules, the Barysz matrix (**Dz**) coincides with the topological distance matrix (**D**) and, accordingly, all the derived molecular descriptors.

For all the considered data sets, the descriptor SpMax from the Chi matrix (**X**) resulted constant and therefore was excluded. The Estrada-like index from the detour matrix (EE_Dt) was not considered for the data set *ED* because it has two missing values. Likewise, when considering the other data sets, some descriptors were not calculated or excluded from analysis and this explains why the total number of descriptors from each matrix for each data set sometimes differs from the default total numbers of Dragon (Table 4.3).

After pretreatment analysis, to quantitatively measure the extent to which the chemical spaces defined by the considered descriptors differ, the Canonical Measure of Distance (*CMD*) and Canonical Measure of Correlation (*CMC*) were calculated for each pair of graph-theoretical matrices. While *CMD* measures the dissimilarity between two chemical spaces, *CMC* measures their similarity. The complement to 1 of *CMC*, that is $CMD^* = 1 - CMC$, was also calculated because, unlike *CMD*, this is a dissimilarity measure independent of the number of descriptors defining the chemical space. Then, the multidimensional scaling (MDS) was applied on the final square (9 × 9) dissimilarity matrix comprised of the $CMD^*$ values calculated for all the possible pairs of graph-theoretical matrices.

The MDS is a multivariate analysis technique that enables the identification of a subspace of the original $p$-dimensional space into which the points can be projected and in which the interobject dissimilarities are approximated as well as possible by the corresponding interpoint distances [71]. The final result of such a technique is a geometrical model of the objects in analysis, which allows a visual investigation of the relationships between the objects.

In the MDS scatter plot of the first two dimensions each vector of descriptors calculated from the same matrix is represented by a point; near points indicate similar types of descriptor vectors (i.e., graph matrices).

To have an additional way of comparing the different graph-theoretical matrices, their average distance and Gini's concentration index were also calculated. The former is a measure of average matrix dissimilarity, low values indicating that matrices are not much diverse on average, while the latter focuses more on the presence of some matrices that can significantly differ from the others. Gini's concentration index was calculated as

$$C = \frac{\sum_{i=1}^{n}\sum_{j=1}^{n}\sum_{s=1}^{n}\sum_{t=1}^{n}\left|CMD_{ij}^{*}-CMD_{st}^{*}\right|}{n\cdot(n-1)} \quad i \neq j \text{ and } s \neq t \tag{4.45}$$

where $n$ is the number of considered graph-theoretical matrices and $CMD^*$ is the standardized distance between the chemical spaces defined by two matrices; smaller Gini's index larger the uniformity of the matrix distribution. Values of this index for each analyzed data set are collected in Table 4.5.

In Figure 4.2, the MDS scatter plots of the first two dimensions are reported for each data set. One point in these plots represents the chemical space defined by the descriptor vector derived from a graph matrix. From these MDS plots some considerations can be drawn. First of all, the matrices are in general spread in the MDS

**Table 4.5** Overview information on the dissimilarity analysis of graph-theoretical matrices.

| Data set | No. of compounds | No. of matrices | No. of descriptors | Average distance | Concentration index |
|---|---|---|---|---|---|
| Dragon | 41 | 9 | 243 | 0.132 | 0.085 |
| C10 | 75 | 5 | 111 | 0.132 | 0.085 |
| Haloalkanes | 198 | 7 | 182 | 0.167 | 0.179 |
| Phenylureas | 55 | 9 | 243 | 0.113 | 0.088 |
| PCDD | 75 | 9 | 240 | 0.028 | 0.019 |
| Flavonoids | 82 | 9 | 241 | 0.099 | 0.105 |
| Pesticides | 153 | 9 | 244 | 0.151 | 0.085 |
| ED | 911 | 9 | 243 | 0.148 | 0.092 |

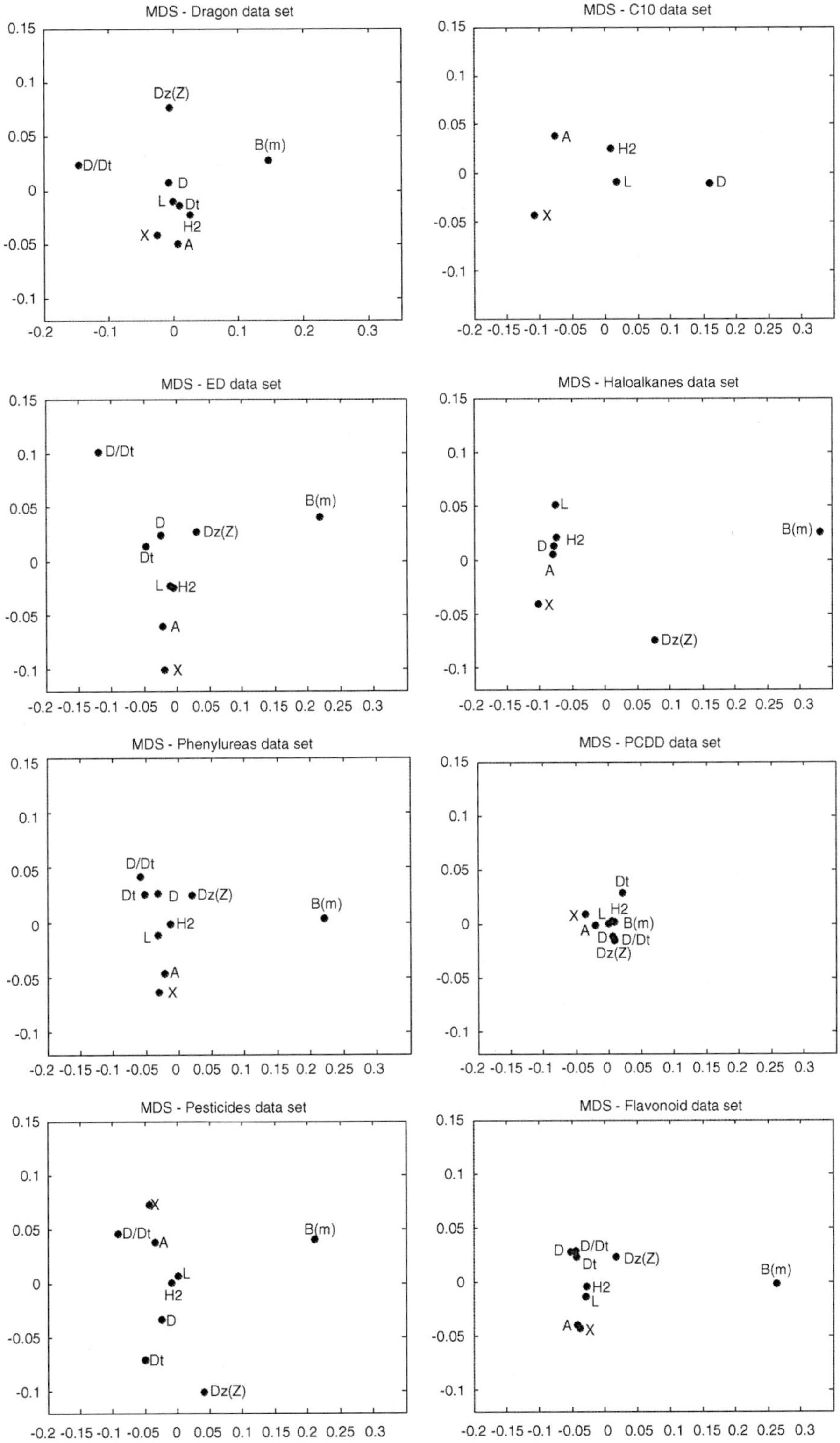


**Figure 4.2** Multidimensional scaling plots of the first two dimensions calculated by using the dissimilarity matrix of nine different graph-theoretical matrices for all the data sets in analysis.

plots meaning that they define different chemical spaces except for the data set *PCDD* for which dissimilarity among all the matrices is very small.

For all the studied cases, the descriptors derived from the adjacency (**A**) and Chi (**X**) matrices give almost the same information as well as the Laplace (**L**) and reciprocal square distance matrix (**H2**). Some larger differences among the adjacency (**A**) and Chi (**X**) matrices were detected for the data sets *C10* and *Haloalkanes*, they being comprised of small acyclic molecules, where branching could be particularly relevant. For the data sets including heterogeneous compounds (*ED*, *Pesticides* and *Dragon*), the descriptor vectors derived from the adjacency (**A**), Laplace (**L**), and reciprocal square distance (**H2**) matrices give the same information, while for congeneric compounds (*Phenylureas*, *PCDD*, and *Flavonoids*) the adjacency matrix behaves in a quite different way. The Laplace (**L**) and reciprocal squared distance (**H2**) matrices partially differ only in the case of molecules not containing rings, such as *C10* and *Haloalkanes*. The descriptors derived from the Chi matrix (**X**) appear in almost all the cases very similar to those obtained from the adjacency matrix (**A**), with the exceptions for the sets of acyclic compounds (*C10* and *Haloalkanes*) and the *ED* data set.

It is noteworthy that for data sets being comprised of heterogeneous molecules ranging from a small to very large dimension, information encoded by descriptors calculated from the reciprocal squared distance matrix (**H2**) is very similar to that from the adjacency matrix (**A**). In effect, when there are a lot of pairs of atoms located far apart in the molecule, their reciprocal squared distance is very near zero and thus the matrix **H2** tends to have a structure very similar to that of **A**, with the only nonvanishing elements being those corresponding to bonded atoms and pairs of atoms very near each other.

In general, the descriptors derived from the Burden matrix (**B(m)**, mass weighted) appear to give specific information different from that obtained by the descriptors derived from the other matrices.

The descriptor vectors derived from the distance (**D**) and detour (**Dt**) matrices seem to give the same information for all the data sets with a remarkable exception for the case of *PCDD*, whose molecules are characterized by a common cyclic scaffold and the substituents (chlorine atoms) differ only for number and position, thus enhancing the relative role of the detour distance. The descriptors calculated from the distance/detour matrix (**D/Dt**) contain some specific structural information only for diverse data sets (*ED*, *Pesticides* and *Dragon*). Finally, the Barysz matrix (**Dz(Z)**, weighted by the atomic number Z) seems to generate descriptors that give specific information especially for small molecules with a number of different heteroatoms, such as for data sets *Haloalkanes*, *Dragon*, *Pesticides* and, partially, in the cases of *Phenylureas* and *Flavonoids*.

#### 4.5.2.1 **Comparison of Weighted Graph-Theoretical Matrices**

In order to analyze the role of the weighting schemes in defining different chemical spaces, Barysz (**Dz**) and Burden (**B**) graph-theoretical matrices were calculated, using the following atomic properties: atomic mass ($m$), atomic van der Waals volume ($v$), atomic polarizability ($p$), Sanderson's electronegativity ($e$), and atomic ionization potential ($i$); moreover, the Barysz matrix was also calculated as originally proposed in

**Table 4.6** Statistics of the multidimensional scaling calculation from $CMD^*$ distances for weighted matrices.

| Data set | No. of compounds | No. of matrices | No. of descriptors | Average distance | Concentration index |
|---|---|---|---|---|---|
| Dragon | 41 | 12 | 366 | 0.103 | 0.072 |
| PCDD | 75 | 12 | 366 | 0.011 | 0.009 |
| Flavonoids | 82 | 12 | 366 | 0.103 | 0.096 |
| Phenylureas | 55 | 12 | 366 | 0.106 | 0.089 |
| Haloalkanes | 198 | 12 | 366 | 0.178 | 0.103 |
| Pesticides | 153 | 12 | 366 | 0.152 | 0.090 |
| ED | 911 | 12 | 366 | 0.119 | 0.083 |

the literature, that is, weighted by the atomic number $Z$, while the Burden matrix was also weighted by the intrinsic state (s) that is a local vertex invariant calculated from the molecular graph.

For all the data sets, the total number of analyzed matrices was then 12 and the number of descriptors 366 (Table 4.6). The data set *C10* was excluded from calculation, it being comprised of molecules with only carbon atoms and single bonds.

The MDS plot relative to the *PCDD* data set was not presented, because all the matrices gave the same information, that is, all their interdistances were very close to zero and, thus, they were concentrated in a single point of the MDS plot, as it can be seen from the data of Table 4.6. This means that for this kind of congeneric molecules using different weighting schemes is useless since they provide the same chemical information.

In Figure 4.3, the MDS plots of the first two dimensions for weighted matrices are shown. In general, the Burden (**B**) and Barysz (**Dz**) matrices give different information, regardless of the weighting scheme used. This difference is quite apparent in the MDS plot of the *Dragon* data set, where the two types of matrices give rise to two different well separated groups. Moreover, there is not much variability among chemical spaces defined by the different Barysz matrices, except for the *Haloalkanes* data set; this means that applying different weighting schemes does not give much different descriptors from those calculated from the original $Z$-weighted Barysz matrix.

Then, for all the studied data sets, the descriptors derived from atomic polarizability ($p$) and ionization potential ($i$) for both Barysz and Burden matrices are very similar, as well as the descriptors derived from atomic number ($Z$) and atomic mass ($m$) for the Barysz matrix. Moreover, in all the cases with the exception of the *Haloalkanes* data set, the weighting schemes based on electronegativity ($e$) and van der Waals volume ($v$) give similar information to that of atomic polarizability ($p$) and ionization potential ($i$) for both Barysz and Burden matrices.

The most diverse weighting schemes are the atomic mass and topological intrinsic state applied to the Burden matrix (**B(m)** and **B(s)**). Unlike the atomic mass ($m$) that is widely used to distinguisch heteroatoms, the use of the intrinsic state as the

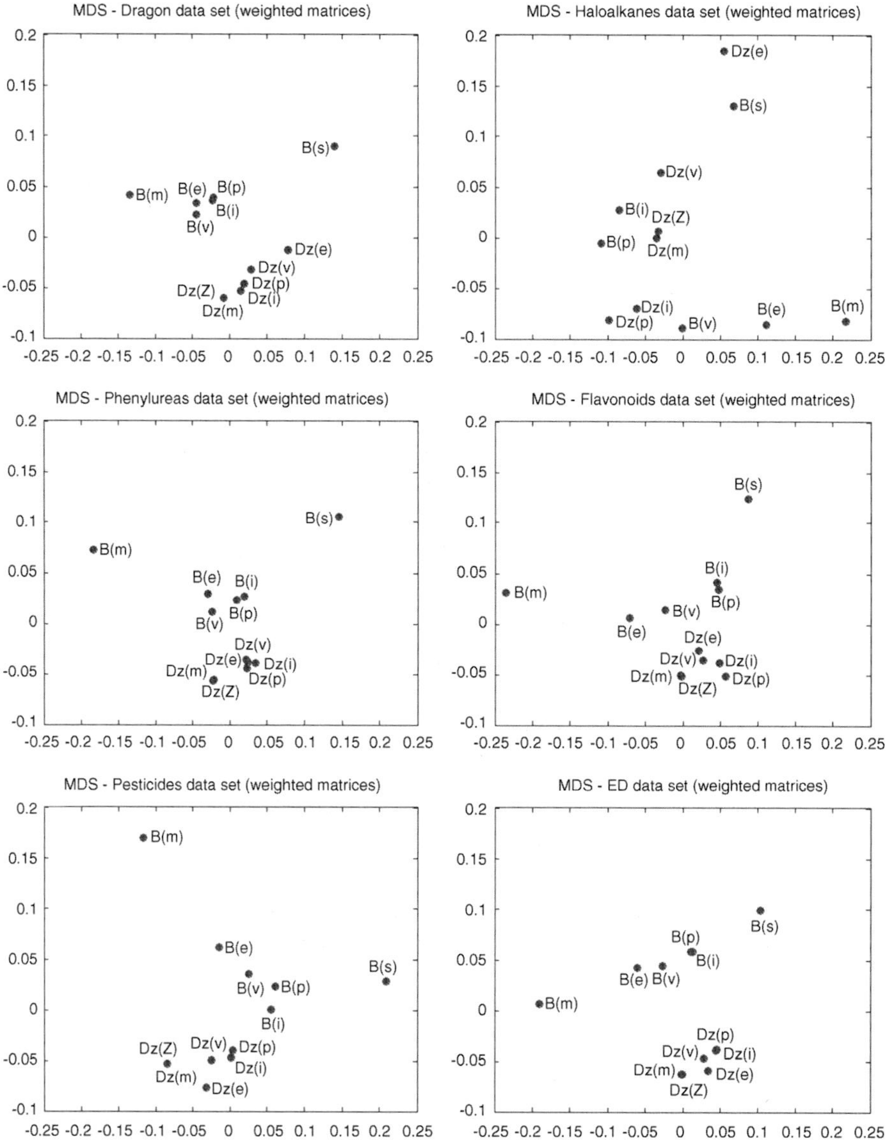


**Figure 4.3** Multidimensional scaling plots of the first two dimensions calculated by using the dissimilarity matrix of 12 different weighted graph-theoretical matrices for all the data sets in analysis.

weighting scheme is not so common; nevertheless, it seems to be able to encompass specific information not only related to single atoms but also to their neighborhood.

Finally, for the *Haloalkanes* data set, also the electronegativity-weighted Barysz matrix (**Dz(e)**) gives some specific information because the four halogens strongly differ for their electronegativity, their differences resulting relevant for small molecules.

### 4.5.3
### Comparison of Matrix Operators

Diversity between chemical spaces was further investigated by the aid of principal component analysis (PCA). PCA is a multivariate statistical technique widely used for approximating relevant information in data matrices [71–73]. The score plots of the first four principal components (PCs) allowed visual inspection of the chemical spaces defined by the different descriptor vectors of graph-theoretical matrices; the corresponding loading plots allowed analysis of the role of chemical descriptors in defining the largest variance directions (i.e., the principal components).

PCA was carried out on descriptors derived from single graph-theoretical matrices (see Table 4.3). In order to draw considerations as general as possible, PCA was carried out by using the data set *ED* that is comprised of 911 heterogeneous compounds. The results of the nine independent PCAs are summarized in the following. In Table 4.7, for the first four components, explained variance (E.V. in percentage) and cumulative explained variance of each component are collected, together with the total number of descriptors used for PCA. For each analysis, only the loading plots of the first four components are reported (Figures 4.4–4.12).

As was expected from the early MDS analysis on the *ED* data set (Figure 4.2), distance (**D**), detour (**Dt**), and Barysz (**Dz(Z)**) matrices give descriptors that define quite similar chemical spaces and, therefore, their first two PCs approximately give the same information. Looking at Figures 4.5, 4.9 and 4.11, it is apparent that the correlation structure of descriptors defining the three chemical spaces is quite the same; this especially holds for the distance (**D**) and Barysz matrix (**Dz**), the only relevant difference being the descriptor SM1_Dz(Z) that is not calculated for the distance matrix and mostly influences the second and third PCs for the Barysz matrix. In Figure 4.9, unlike the loading plots of the distance matrix, the role of descriptor H_Dt in defining the second PC of the chemical space of detour matrix is noteworthy.

Also adjacency (**A**) and reciprocal squared distance (**H2**) matrices define chemical spaces with similar correlation structure among descriptors (Figure 4.4 and 4.8,

**Table 4.7** Summary of explained variances of the first four PCs for PCAs on single matrix descriptors.

| Matrix | No. of descriptors | E.V. 1 | E.V. 2 | E.V. 3 | E.V. 4 | Sum (1–4) |
|---|---|---|---|---|---|---|
| A | 17 | 73.1 | 11.1 | 6.1 | 4.3 | 94.7 |
| D | 29 | 86.8 | 6.2 | 4.1 | 1.9 | 99.0 |
| L | 25 | 78.4 | 11.4 | 4.4 | 3.1 | 97.3 |
| X | 27 | 66.8 | 11.1 | 7.6 | 4.8 | 90.3 |
| H2 | 28 | 83.0 | 9.2 | 4.5 | 1.7 | 98.4 |
| Dt | 29 | 82.7 | 8.5 | 5.2 | 1.6 | 98.0 |
| D/Dt | 29 | 64.7 | 28.4 | 2.7 | 1.6 | 97.4 |
| Dz(Z) | 31 | 83.1 | 6.6 | 4.9 | 3.0 | 97.6 |
| B(m) | 30 | 55.3 | 32.5 | 4.9 | 2.8 | 95.5 |

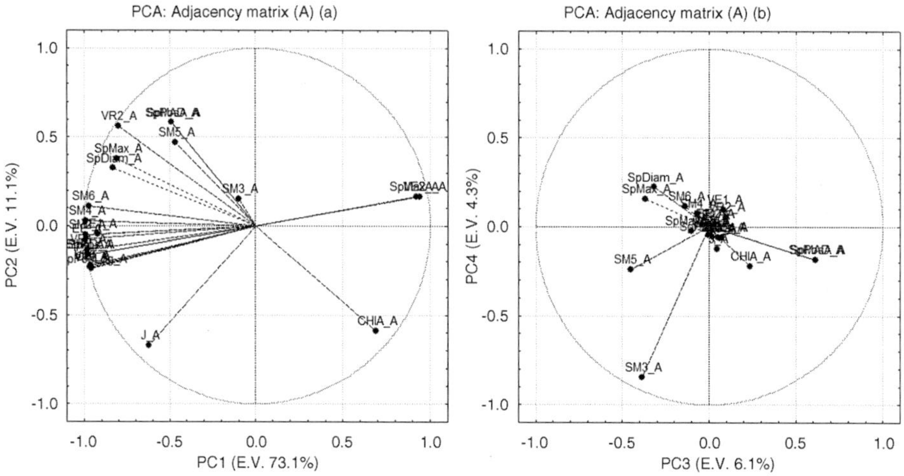


Figure 4.4 Loading plots of PC2 vs. PC1 (a) and PC4 vs. PC3 (b) for the adjacency matrix.

respectively), the only relevant difference being the descriptor SM3_A that is largely influent only in the chemical space defined by the adjacency matrix.

Only for the Burden matrix (**B(m)**), descriptors seem to be less correlated among themselves, explaining different information in the first PCs (Figure 4.12). In particular, only for this matrix the ordered sequence of SMs is apparent from left to right in the loading plot of the first two PCs (Figure 4.12a). These descriptors are all correlated along with the first PC, whereas they differ along the second component where the higher order spectral moments SM6, SM5, and SM4 seem to be more relevant than SM1 and SM2. All these descriptors are based on the matrix eigenvalues and calculated by the same formula but using different powers of eigenvalues.

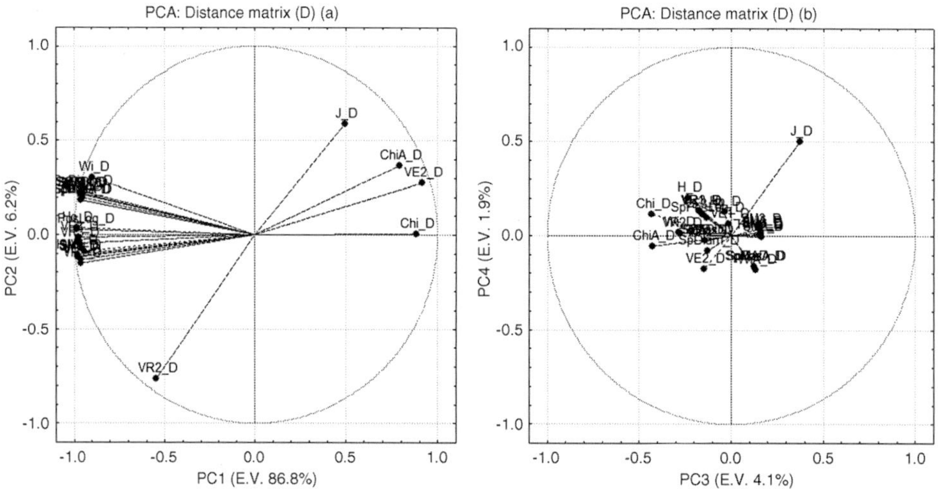


Figure 4.5 Loading plots of PC2 vs. PC1 (a) and PC4 vs. PC3 (b) for the distance matrix.

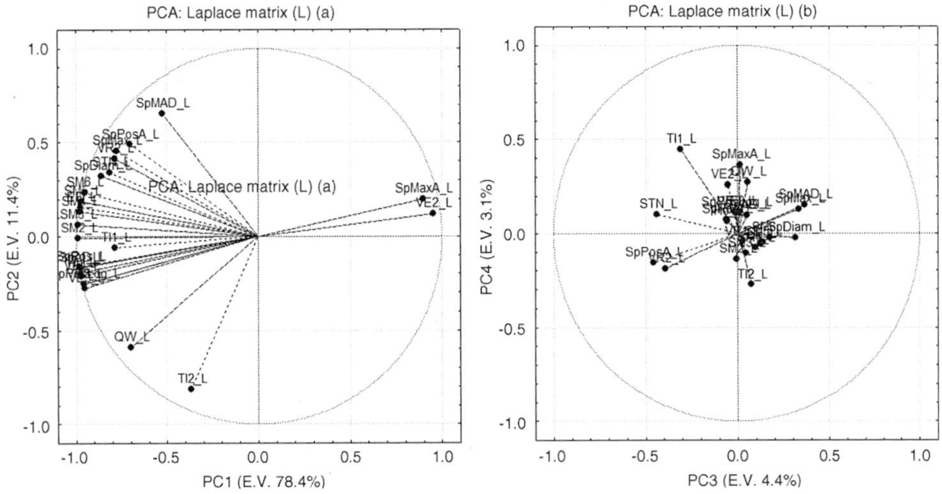


**Figure 4.6** Loading plots of PC2 vs. PC1 (a) and PC4 vs. PC3 (b) for the Laplacian matrix.

Moreover, for all the matrices, the first PC encodes information about the molecule size; most of the descriptors encode this information, being concentrated around the maximum loading absolute values. The second PC seems to be related to cyclicity, that is, high number of cycles (low values of the Balaban index (*J*)) as opposite to molecules having linear shape and/or long chains (high values of the Balaban index). This behavior can be easily noted, for example, at the bottom of the plot of the adjacency (**A**, Figure 4.4a), reciprocal squared distance (**H2**, Figure 4.8a), detour (**Dt**, Figure 4.9a) and Barysz (**Dz(Z)**, Figure 4.11a) matrix and at the top of the plot of the distance matrix (**D**, Figure 4.5a).

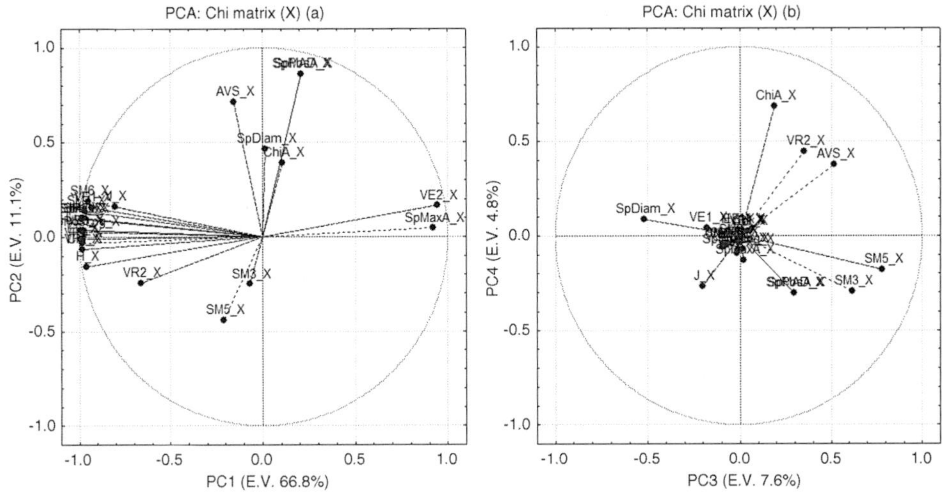


**Figure 4.7** Loading plots of PC2 vs. PC1 (a) and PC4 vs. PC3 (b) for the Chi matrix.

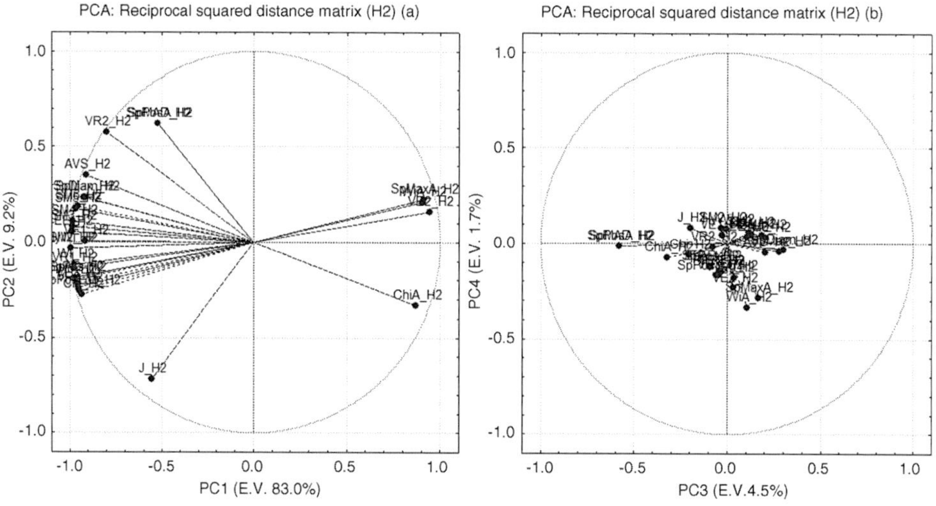


**Figure 4.8** Loading plots of PC2 vs. PC1 (a) and PC4 vs. PC3 (b) for the reciprocal squared distance matrix.

It is interesting to note the relationships between the Balaban index (*J*) and the average connectivity index (CHIA) for the different matrices along the first two PCs. In fact, these two descriptors seem to encode specific and similar information along the second PC, but they have opposite behavior along the first PC for the adjacency (**A**, Figure 4.4a) and the reciprocal squared distance matrix (**H2**, Figure 4.8a): they encode the same information but the Balaban index is multiplied by the number of bonds, thus increasing with the molecular size, while the average connectivity index decreases with the molecular size.

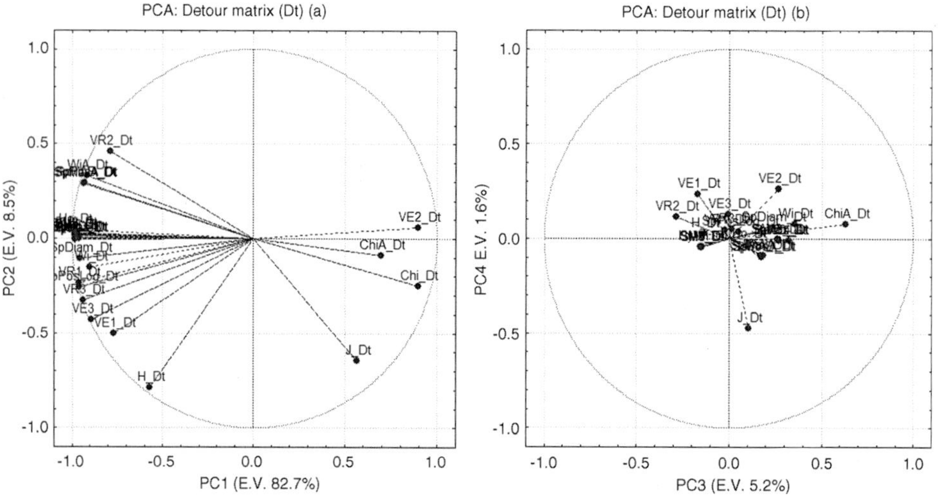


**Figure 4.9** Loading plots of PC2 vs. PC1 (a) and PC4 vs. PC3 (b) for the detour matrix.

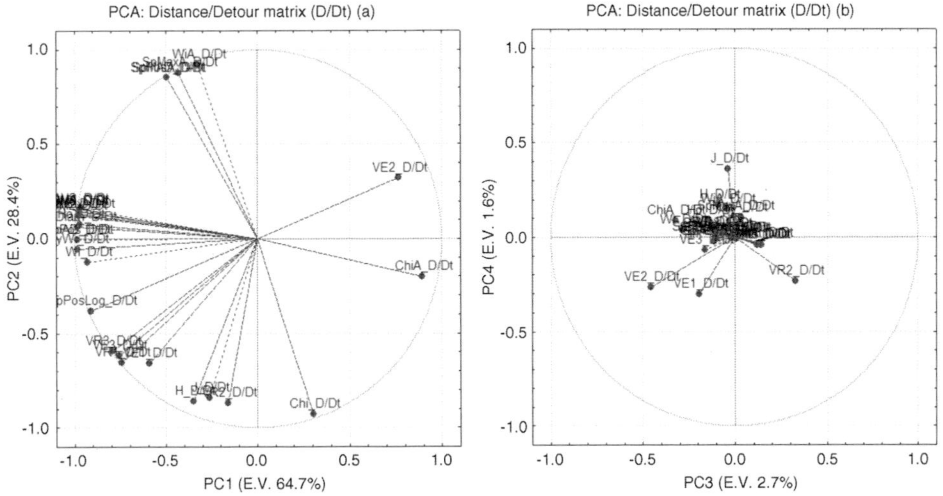


**Figure 4.10** Loading plots of PC2 vs. PC1 (a) and PC4 vs. PC3 (b) for the distance/detour matrix.

## 4.5.4
## Comparison of Single Operators from Different Graph-Theoretical Matrices

The third analysis focussed on comparison of the same type of descriptors calculated from different graph matrices. In particular, the following types of descriptors were analyzed: Wiener-like (Wi), Randic-like (Chi), Balaban-like (J), Hosoya-like (Ho), Estrada-like (EE), spectral absolute deviations (SpAD), spectral mean absolute deviations (SpMAD).

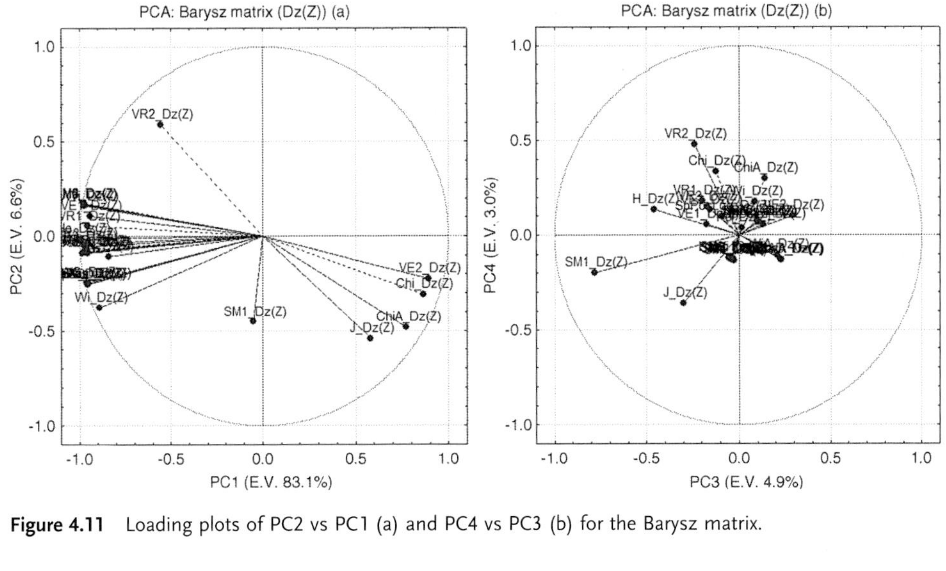


**Figure 4.11** Loading plots of PC2 vs PC1 (a) and PC4 vs PC3 (b) for the Barysz matrix.

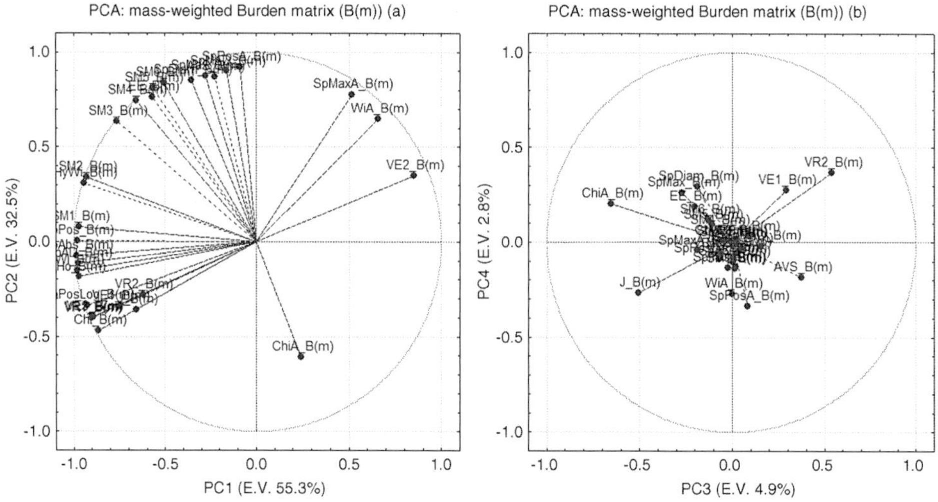


**Figure 4.12** Loading plots of PC2 vs. PC1 (a) and PC4 vs. PC3 (b) for the mass-weighted Burden matrix.

The comparison was performed by PCA, considering the first four PCs, calculated on the *ED* data set, constituted by 911 diverse molecules.

In Table 4.8, the explained variances (E.V. in percentage) of the first four components together with the total explained variance of these components are shown. The loading plots of the seven PCAs are shown in Figures 4.13–4.19.

It can be easily observed that the information encoded in the descriptors is distributed in different ways. The first component of the Wiener-like, Hosoya-like, and spectral absolute deviations (SpAD) explains the largest amount of information and is related to the quantitative trend of each descriptor set. For these descriptors, different chemical information from that related to quantitative trend seems to be given only by the second PC.

In Figures 4.13a and b, the loading plots of the Wiener-like indices are shown. The first two principal components, explaining almost 95% of the total variance,

**Table 4.8** Summary of explained variances of the first four PCs for PCAs on single operators.

| Descriptor | No. of descriptors | E.V. 1 | E.V. 2 | E.V. 3 | E.V. 4 | Sum (1–4) |
|---|---|---|---|---|---|---|
| Wi-like | 6 | 87.9 | 6.9 | 4.4 | 0.6 | 99.8 |
| Chi-like | 8 | 76.8 | 16.2 | 4.7 | 1.6 | 99.3 |
| J-like | 7 | 53.9 | 36.4 | 7.3 | 1.6 | 99.2 |
| Ho-like | 9 | 91.2 | 7.1 | 1.3 | 0.2 | 99.8 |
| EE-like | 9 | 74.4 | 10.9 | 9.8 | 3.8 | 99.0 |
| SpAD-like | 9 | 90.0 | 6.1 | 2.6 | 0.7 | 99.4 |
| SpMAD-like | 9 | 37.8 | 22.8 | 21.9 | 9.9 | 92.4 |
| All | 58 | 69.1 | 14.0 | 4.7 | 3.8 | 91.6 |

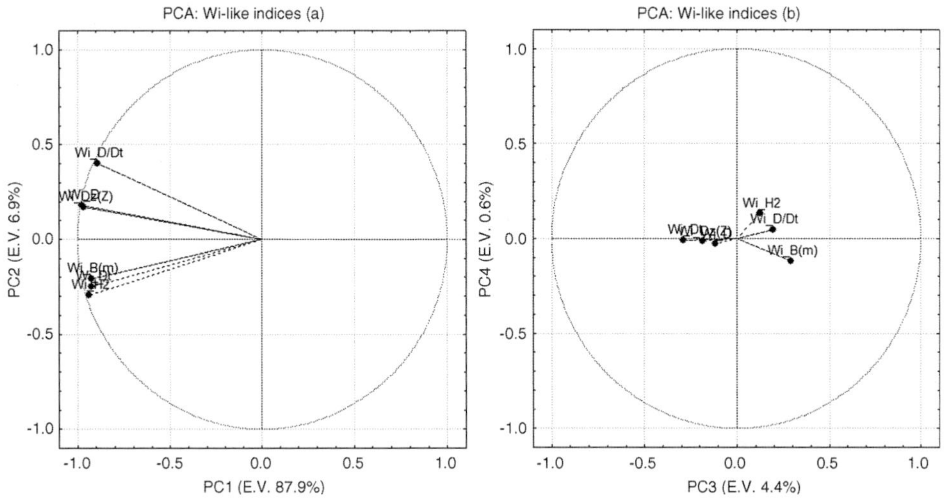


**Figure 4.13** Loading plots of PC2 vs. PC1 (a) and PC4 vs. PC3 (b) for Wiener-like indices.

encode all the chemical information. The first PC (E.V. 87.9%) is related to molecular size, high values of the descriptors (on the left) representing big molecules; the second PC (E.V. 6.9%) is characterized by a contrast between Wi_D, and Wi_B(m), Wi_D/Dt, and Wi_H2 and distinguishes molecules with highly chained small rings (at the top) versus condensed rings (at the bottom). The residual information in the third PC (4.4%) is mainly related to Wi(B(m)): high values characterize molecules with a high fraction of heteroatoms.

In Figures 4.14a and b, the loading plots of Randic-like indices are shown. The first PC (E.V. 87.9%) is still related to molecular size: large molecules are located on the

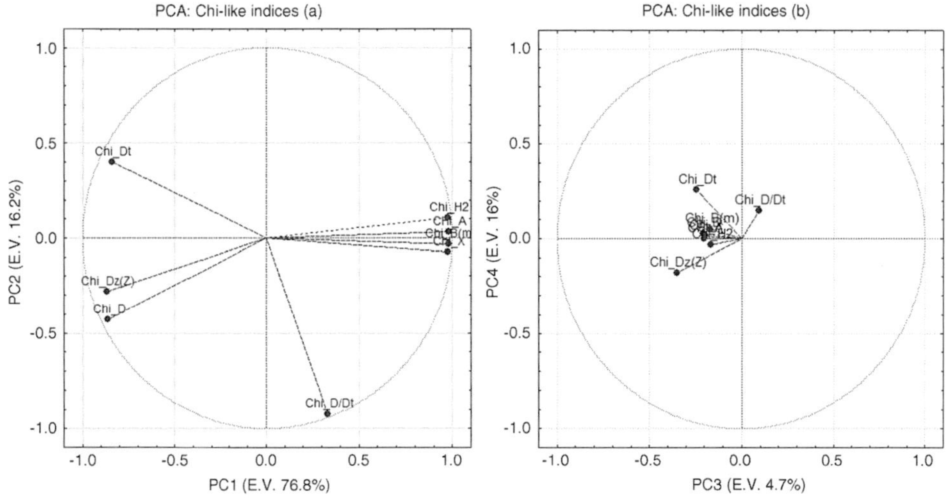


**Figure 4.14** Loading plots of PC2 vs. PC1 (a) and PC4 vs. PC3 (b) for Randic-like indices.

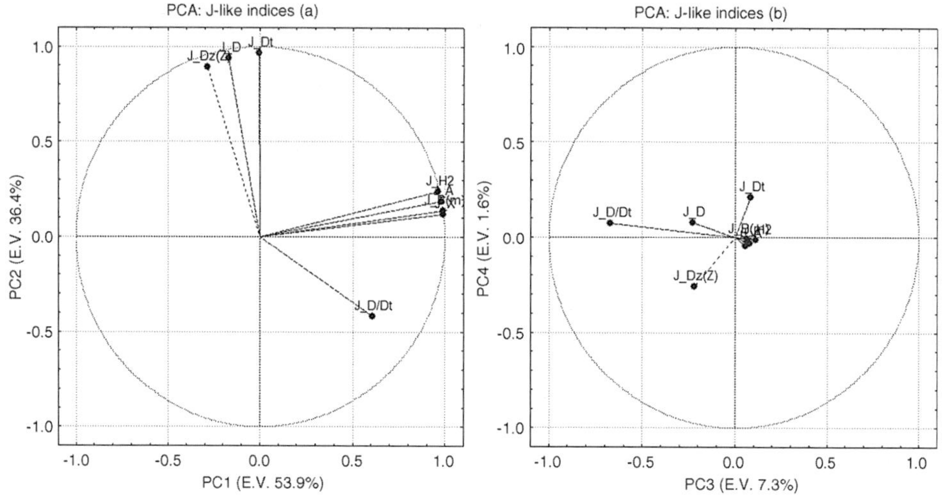


**Figure 4.15** Loading plots of PC2 vs. PC1 (a) and PC4 vs. PC3 (b) for Balaban-like indices.

right of the loading plot, characterized by high values of most of the Randic-like indices except for indices derived from distance (**D**), detour (**Dt**), and Barysz (**Dz(Z)**) matrices which have small values for big molecules and high values for small molecules (including also heteroatoms). At the bottom of the second PC are located molecules having several condensed rings, which are characterized by high values of the index calculated from the distance/detour matrix and low values of the index calculated from the detour matrix (at the top). Low values of the third PC (on the left) are expected for small cyclic molecules with high fraction of heteroatoms (high values of Chi_Dz(Z) and Chi_Dt), whereas high values of this PC mainly characterize linear

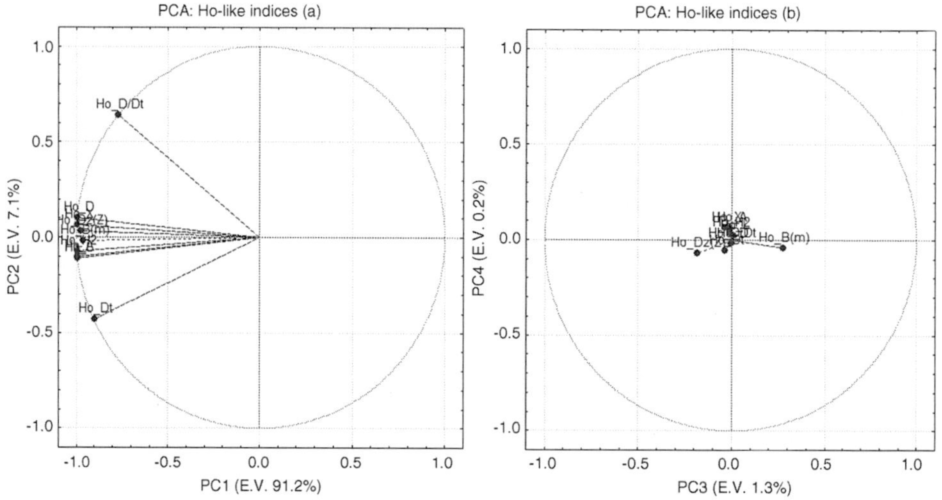


**Figure 4.16** Loading plots of PC2 vs. PC1 (a) and PC4 vs. PC3 (b) for Hosoya-like indices.

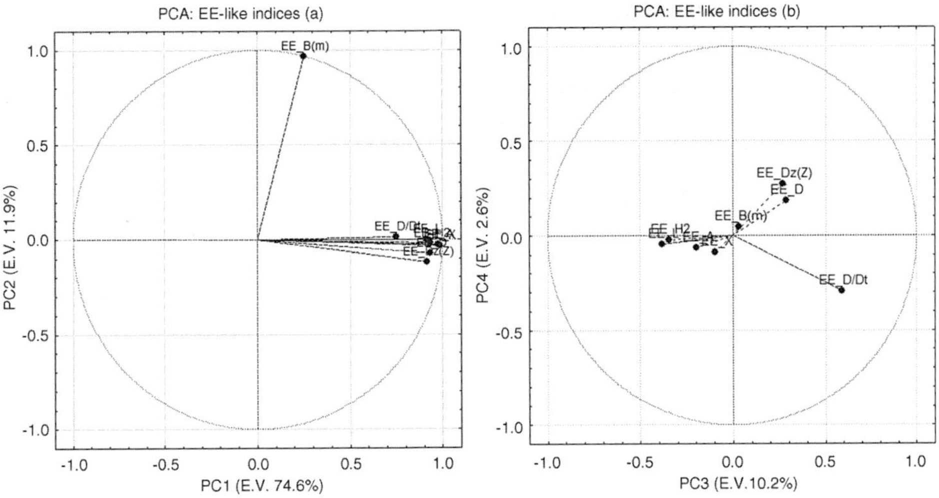


**Figure 4.17** Loading plots of PC2 vs. PC1 (a) and PC4 vs. PC3 (b) for Estrada-like indices.

molecules. High values of the fourth PC refer to small acyclic molecules (high values of Chi_Dt and Chi_D/Dt).

In Figures 4.15a and b, the loading plots of the Balaban-like indices are shown. High values of J_A, J_X, J_H2, and J_B(m) lead to high values of the first PC and identify highly chained molecules (including also heteroatoms); on the other hand, J_D, J_Dt, and J_Dz(Z) strongly influence the second PC and mainly identify star-graph or highly branched molecules with short chains, while highly condensed rings are well characterized by low values of the aforementioned three indices and high values of J_D/Dt.

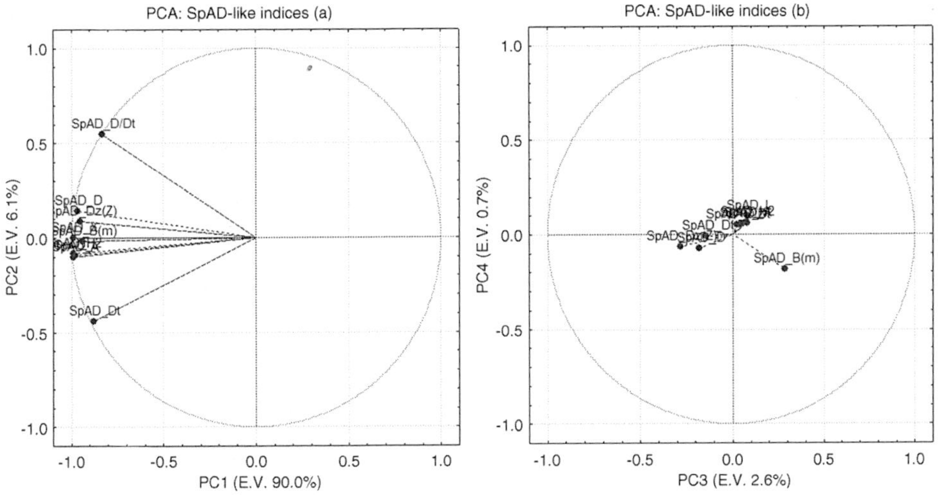


**Figure 4.18** Loading plots of PC2 vs. PC1 (a) and PC4 vs. PC3 (b) for spectral absolute deviations.

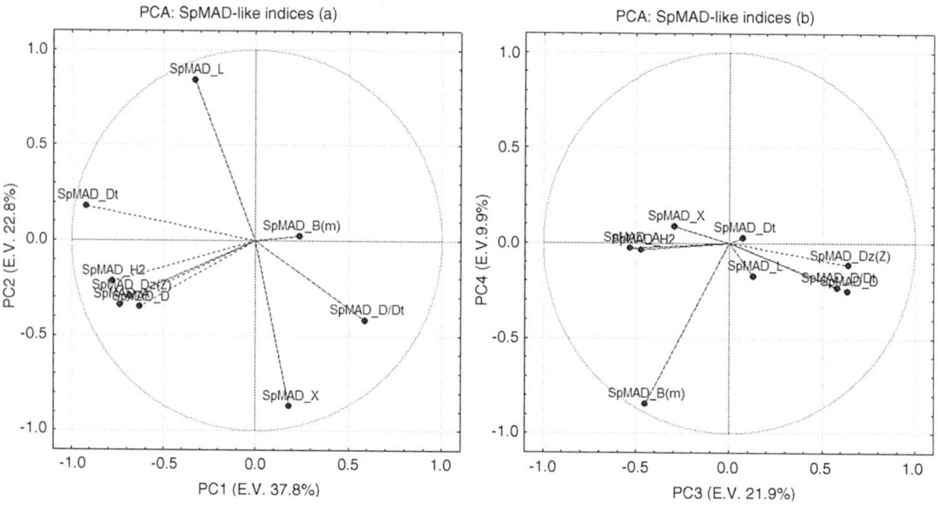


**Figure 4.19** Loading plots of PC2 vs. PC1 (a) and PC4 vs. PC3 (b) for spectral mean absolute deviations.

High values of J_D/Dt and J_Dz(Z) along the third PC (on the left) are related to small molecules having highly condensed rings with heteroatoms; high values of J_Dz(Z) along the fourth PC (at the bottom) are more related to monocyclic molecules with heteroatoms as opposite to acyclic molecules (at the top).

As for the Wiener-like indices, the first PC of Hosoya-like (E.V. 91.2%, Figure 4.16a), Estrada-like indices (E.V. 74.6%, Figure 4.17a) and spectral absolute deviations (E.V. 90.0%, Figure 4.18a) is quantitative (i.e., all the descriptors having similar loadings in PC1) and related to molecule size.

Hosoya-like indices and spectral absolute deviations give almost the same correlation structure, all their PCs being very similar to each other as can be seen from Figures 4.16 and 4.18. Moreover, all their studied PCs are quite correlated with the PCs of Wiener-like indices and, accordingly, have the same interpretation.

For Estrada-like indices, the second PC (Figure 4.17a) explains only the index EE_B(m) that seems to be important in distinguishing molecules with a few not-condensed rings with heteroatoms; the third PC (Figure 4.17b) is mainly influenced by EE_D/Dt with minor contributions of EE_D and EE_Dz(Z), characterizing molecules with rings and heteroatoms.

A very different correlation structure is observed for the spectral mean absolute deviations (SpMADs), for which different chemical information is distributed on all the four PCs; moreover, the fifth and the sixth components (not reported) further explain 4.1% and 2.7% of the total variance, respectively. In Figures 4.19a and b, the loading plots of spectral mean absolute deviation indices are shown. Low values of the first PC are mainly related to big molecules with many condensed rings and heteroatoms which have high values of SpMAD descriptors derived from the **A**, **D**, **Dt**, **H2**, and **Dz(Z)** matrices; on the opposite side, there are small acyclic molecules, with high values of SpMAD_D/Dt. The second PC is characterized by high values of

SpMAD_X (at the bottom) which are related to linear (or less branched) molecules with low values of SpMAD_L. The third PC is mainly influenced by SpMAD descriptors derived from the **D**, **Dz(Z)**, and **D/Dt** matrices and identifies on the right molecules having long chains; high values of SpMAD_B(m) along the fourth PC are related to small molecules having rings and heteroatoms.

## 4.6 Conclusions

From all the multivariate analyses carried out on topological indices derived from different graph-theoretical matrices some general considerations can be drawn.

From the MDS analysis of matrix dissimilarity (Figures 4.2 and 4.3) it was demonstrated that, for the data sets in analysis, the Laplace (**L**) and reciprocal squared distance (**H2**) matrices seem to give the same sets of descriptors, defining very similar chemical spaces. In almost all the cases, the adjacency (**A**), Chi (**X**), and Laplace (**L**) matrices also appear to give very similar chemical information. The most diverse chemical information is then provided by the adjacency **A**, mass-weighted Burden **B(m)**, intrinsic state-weighted Burden **B(s)**, Barysz **Dz(Z)**, and quotient distance/detour **D/Dt** matrices.

Regarding the weighting schemes, those used for Barysz matrices (**Dz(w)**) seem to give the same information, except for small heteroatom substituted molecules such as those in the *Haloalkanes* data set for which Sanderson electronegativity gives diverse information.

Most of the studied 2D matrix-based descriptors are largely correlated, explaining the same information regardless of the matrix they derive from: this especially holds for Wiener-like (Wi), hyper-Wiener-like (HyWi), graph energy-like (SpAbs), spectral positive sum (SpPos), spectral absolute deviation (SpAD), Hosoya-like (Ho), Estrada-like (EE), the second (SM2) and fourth (SM4) spectral moment, which mainly encode information about molecule size for all the studied data sets.

Finally, Balaban-like indices (*J*) confirm to give information quite different from all the others due to their independence of molecule size. Its relationship with the normalized Randic-like eigenvector-based indices (VR2) should be further investigated.

## References

1 Todeschini, R. and Consonni, V. (2009) *Molecular Descriptors for Chemoinformatics (2 volumes)*, Wiley-VCH Verlag GmbH, Weinheim, Germany, p. 1257.

2 Janežic, D., Milicevic, A., Nikolic, S., and Trinajstic, N. (2007) *Graph Theoretical Matrices in Chemistry*, University of Kragujevac, Kragujevac, Serbia, p. 205.

3 Ivanciuc, O., Ivanciuc, T., and Diudea, M.V. (1997) Molecular graph matrices and derived structural descriptors. *SAR and QSAR in Environmental Research*, **7**, 63–87.

4 Ivanciuc, O. and Ivanciuc, T. (1999) Matrices and structural descriptors computed from molecular graphs

distances, in *Topological Indices and Related Descriptors in QSAR and QSPR* (eds J. Devillers and A.T. Balaban), Gordon & Breach Science Publishers, Amsterdam, The Netherlands, pp. 221–277.

5 Ivanciuc, O. (2000) QSAR Comparative study of Wiener descriptors for weighted molecular graphs. *Journal of Chemical Information and Computer Sciences*, **40**, 1412–1422.

6 Basak, S.C., Gute, B.D., and Grunwald, G.D. (1997) Use of topostructural, topochemical, and geometric parameters in the prediction of vapor pressure: A hierarchical QSAR approach. *Journal of Chemical Information and Computer Sciences*, **37**, 651–655.

7 Gute, B.D., Grunwald, G.D., and Basak, S.C. (1999) Prediction of the dermal penetration of polycyclic aromatic hydrocarbons (PAHs): A hierarchical QSAR approach. *SAR and QSAR in Environmental Research*, **10**, 1–15.

8 Balaban, A.T. (1997) From chemical graphs to 3D molecular modeling, in *From Chemical Topology to Three-Dimensional Geometry* (ed. A.T. Balaban), Plenum Press, New York, NY, pp. 1–24.

9 Diudea, M.V., Horvath, D., and Graovac, A. (1995) Molecular topology: 15. 3D distance matrices and related topological indices. *Journal of Chemical Information and Computer Sciences*, **35**, 129–135.

10 Hosoya, H. (1971) Topological index. A newly proposed quantity characterizing the topological nature of structural isomers of saturated hydrocarbons. *Bulletin of the Chemical Society of Japan*, **44**, 2332–2339.

11 Kier, L.B. (1985) A shape index from molecular graphs. *Quantitative Structure–Activity Relationships*, **4**, 109–116.

12 Kier, L.B. (1986) Shape indexes of orders one and three from molecular graphs. *Quantitative Structure–Activity Relationships*, **5**, 1–7.

13 Randic, M. (2001) Novel shape descriptors for molecular graphs. *Journal of Chemical Information and Computer Sciences*, **41**, 607–613.

14 Balaban, A.T. (1998) Topological and stereochemical molecular descriptors for databases useful in QSAR, similarity/dissimilarity and drug design. *SAR and QSAR in Environmental Research*, **8**, 1–21.

15 Balaban, A.T. and Ivanciuc, O. (1999) Historical development of topological indices, in *Topological Indices and Related Descriptors in QSAR and QSPR* (eds J. Devillers and A.T. Balaban), Gordon and Breach Science Publishers, Amsterdam, The Netherlands, pp. 21–57.

16 Balaban, A.T., Motoc, I., Bonchev, D., and Mekenyan, O. (1983) Topological indices for structure-activity correlations, in *Steric Effects in Drug Design (Topics in Current Chemistry)*, vol. **114** (eds M. Charton and I. Motoc), Springer, Berlin, Germany, pp. 21–55.

17 Basak, S.C., Grunwald, G.D., and Niemi, G.J. (1997) Use of graph-theoretic and geometrical molecular descriptors in structure–activity relationships, in *From Chemical Topology to Three-Dimensional Geometry* (ed. A.T. Balaban), Plenum Press, New York, NY, pp. 73–116.

18 Basak, S.C., Magnuson, V.R., and Veith, G.D. (1987) Topological indices: Their nature, mutual relatedness, and applications, in *Mathematical Modelling in Science and Technology* (eds X.J.R. Avula, G., Leitmann, C.D., MoteJr., and E.Y. Rodin), Pergamon Press, Oxford, UK, pp. 300–305.

19 Basak, S.C., Niemi, G.J., and Veith, G.D. (1990) Recent developments in the characterization of chemical structure using graph-theoretic indices, in *Computational Chemical Graph Theory* (ed. D.H. Rouvray), Nova Science Publishers, New York, NY, pp. 235–277.

20 Bonchev, D. (2003) On the complexity of directed biological networks. *SAR and QSAR in Environmental Research*, **14**, 199–214.

21 Bonchev, D., Mekenyan, O., von Knop, J., and Trinajstic, N. (1979) On characterization of monocyclic structures. *Croatica Chemica Acta*, **52**, 361–367.

22 Bonchev, D. and Rouvray, D.H. (eds) (2000) *Topology in Chemistry. Applications*, Gordon and Breach, Reading, UK, p. 351.

**23** Bonchev, D. and Trinajstic, N. (1977) Information theory, distance matrix, and molecular branching. *Journal of Chemical Physics*, **67**, 4517–4533.

**24** Devillers, J. and Balaban, A.T. (eds) (1999) *Topological Indices and Related Descriptors in QSAR and QSPR*, Gordon and Breach Science Publishers, Amsterdam, The Netherlands, p. 824.

**25** Diudea, M.V. and Gutman, I. (1998) Wiener-type topological indices. *Croatica Chemica Acta*, **71**, 21–51.

**26** Hansen, P.J. and Jurs, P.C. (1988) Chemical applications of graph theory. Part I. Fundamentals and topological indices. *Journal of Chemical Education*, **65**, 574–580.

**27** Ivanciuc, O. (2003) Graph theory in chemistry, in *Handbook of Chemoinformatics*, vol. **1** (ed. J. Gasteiger), Wiley-VCH, Weinheim, Germany, pp. 103–138.

**28** Ivanciuc, O. and Balaban, A.T. (1999) The graph description of chemical structures, in *Topological Indices and Related Descriptors in QSAR and QSPR* (eds J. Devillers and A.T. Balaban), Gordon & Breach Science Publishers, Amsterdam, The Netherlands, pp. 59–167.

**29** Kerber, A., Laue, R., Meringer, M., and Rücker, C. (2004) Molecules *in silico*: The generation of structural formulae and its applications. *Journal of Combinatorial Chemistry Japan*, **3**, 85–96.

**30** Kier, L.B. and Hall, L.H. (1976) *Molecular Connectivity in Chemistry and Drug Research*, vol. **14**, Academic Press, New York, NY, p. 257.

**31** Kier, L.B. and Hall, L.H. (1986) *Molecular Connectivity in Structure-Activity Analysis*, Research Studies Press–Wiley, Chichester, UK, p. 262.

**32** Randic, M. (1993) Comparative regression analysis. regressions based on a single descriptor. *Croatica Chemica Acta*, **66**, 289–312.

**33** Randic, M. (1998) Topological indices, in *Encyclopedia of Computational Chemistry* (ed. P.R. von Schleyer), John Wiley & Sons Ltd, London, UK, pp. 3018–3032.

**34** Randic, M. (2003) Chemical graph theory – Facts and fiction. *Indian Journal of Chemistry*, **42**, 1207–1218.

**35** Rouvray, D.H. (1983) Should we have designs on topological indices? in *Chemical Applications of Topology and Graph Theory. Studies in Physical and Theoretical Chemistry* (ed. R.B. King), Elsevier, Amsterdam, The Netherlands, pp. 159–177.

**36** Rouvray, D.H. (1989) The limits of applicability of topological indices. *Journal of Molecular Structure (Theochem)*, **185**, 187–201.

**37** Rouvray, D.H. (1995) A rationale for the topological approach to chemistry. *Journal of Molecular Structure (Theochem)*, **336**, 101–114.

**38** Trinajstic, N. (1992) *Chemical Graph Theory*, CRC Press, Boca Raton, FL, p. 322.

**39** Wiener, H. (1947) Structural determination of paraffin boiling points. *Journal of the American Chemical Society*, **69**, 17–20.

**40** Randic, M. (1975) On characterization of molecular branching. *Journal of the American Chemical Society*, **97**, 6609–6615.

**41** Balaban, A.T. (1982) Highly discriminating distance-based topological index. *Chemical Physics Letters*, **89**, 399–404.

**42** Schultz, H.P. (1989) Topological organic chemistry: 1. Graph theory and topological indices of alkanes. *Journal of Chemical Information and Computer Sciences*, **29**, 227–228.

**43** Consonni, V. and Todeschini, R. (2008) New spectral indices for molecule description. *MATCH – Communications in Mathematical and in Computer Chemistry*, **60**, 3–14.

**44** Ivanciuc, O. (1999) Design of topological indices: Part 11. Distance-valency matrices and derived molecular graph descriptors. *Revue Roumaine De Chimie*, **44**, 519–528.

**45** Ivanciuc, O. (2001) Design of topological indices: Part 19. Computation of vertex and molecular graph structural descriptors with operators. *Revue Roumaine De Chimie*, **46**, 243–253.

**46** Gutman, I. and Trinajstic, N. (1972) Graph theory and molecular orbitals. Total $\pi$-electron energy of alternant hydrocarbons. *Chemical Physics Letters*, **17**, 535–538.

47 Carter, S., Trinajstic, N., and Nikolic, S. (1987) A note on the use of ID numbers in QSAR studies. *Acta Pharmaceutica Jugoslovenska*, **37**, 37–42.
48 Randic, M. (1984) On molecular identification numbers. *Journal of Chemical Information and Computer Sciences*, **24**, 164–175.
49 Szymanski, K., Müller, W.R., von Knop, J., and Trinajstic, N. (1985) On Randic's molecular identification numbers. *Journal of Chemical Information and Computer Sciences*, **25**, 413–415.
50 Kier, L.B. and Hall, L.H. (1977) The nature of structure–activity relationships and their relation to molecular connectivity. *European Journal of Medicinal Chemistry*, **12**, 307–312.
51 Randic, M., Guo, X., Oxley, T., and Krishnapriyan, H. (1993) Wiener matrix: Source of novel graph invariants. *Journal of Chemical Information and Computer Sciences*, **33**, 709–716.
52 Li, X. and Zhao, H. (2004) Trees with the first three smallest and largest generalized topological indices. *MATCH – Communications in Mathematical and in Computer Chemistry*, **50**, 57–62.
53 Milicevic, A. and Nikolic, S. (2004) On variable Zagreb indices. *Croatica Chemica Acta*, **77**, 97–101.
54 Randic, M., Plavšic, D., and Razinger, M. (1997) Double invariants. *MATCH – Communications in Mathematical and in Computer Chemistry*, **35**, 243–259.
55 DRAGON (Software for molecular descriptor calculations). Ver. 6. Talete s.r.l., via V.Pisani, 13-20124 Milano, Italy, http://www.talete.mi.it/dragon.htm.
56 Gutman, I. (1978) The energy of a graph. *Berichte der Mathematisch-Statistischen Sektion im Forschungszentrum (Graz, German)*, **103**, 1–22.
57 Gutman, I. (2001) The energy of a graph: Old and new results, in *Algebraic Combinatorics and Applications* (eds A. Betten, A. Kohnert, R. Laue, and A. Wassermann), Springer, Berlin, Germany, pp. 196–211.
58 Balaban, A.T., Ciubotariu, D., and Medeleanu, M. (1991) Topological indices and real vertex invariants based on graph eigenvalues or eigenvectors. *Journal of Chemical Information and Computer Sciences*, **31**, 517–523.
59 Lovasz, L. and Pelikan, J. (1973) On the eigenvalue of trees. *Periodica Mathematica Hungarica*, **3**, 175–182.
60 Estrada, E. and Hatano, N. (2007) Statistical-mechanical approach to subgraph centrality in complex networks. *Chemical Physics Letters*, **439**, 247–251.
61 Trinajstic, N., Babic, D., Nikolic, S., Plavšic, D., Amic, D., and Mihalic, Z. (1994) The Laplacian matrix in chemistry. *Journal of Chemical Information and Computer Sciences*, **34**, 368–376.
62 Ivanciuc, O. (2008) Electrotopological state indices, in *Molecular Drug Properties*, vol. **37** (ed. R. Mannhold), Wiley-VCH Verlag GmbH, Weinheim, Germany, pp. 85–109.
63 Kier, L.B. and Hall, L.H. (1990) An electrotopological-state index for atoms in molecules. *Pharmaceutical Research*, **7**, 801–807.
64 Dutta, D., Guha, R., Jurs, P.C., and Chen, T. (2006) Scalable partitioning and exploration of chemical spaces using geometric hashing. *Journal of Chemical Information and Modeling*, **46**, 321–333.
65 Eckert, H., Vogt, I., and Bajorath, J. (2006) Mapping algorithms for molecular similarity analysis and ligand-based virtual screening: Design of DynaMAD and comparison with MAD and DMC. *Journal of Chemical Information and Modeling*, **46**, 1623–1634.
66 Landon, M.R. and Schaus, S.E. (2006) JEDA: Joint entropy diversity analysis. An information-theoretic method for choosing diverse and representative subsets from combinatorial libraries. *Molecular Diversity*, **10**, 333–339.
67 Oprea, T.I., Zamora, I., and Ungell, A.-L. (2002) Pharmacokinetically based mapping device for chemical space navigation. *Journal of Combinatorial Chemistry*, **4**, 258–266.
68 Ballabio, D., Consonni, V., Mauri, A., and Todeschini, R. (2010) Canonical measure of correlation (*CMC*) and canonical measure of distance (*CMD*) between sets of data: Part 3. Variable selection in classification. *Analytica Chimica Acta*, **657**, 116–122.

**69** Consonni, V., Ballabio, D., Manganaro, A., Mauri, A., and Todeschini, R. (2009) Canonical measure of correlation (*CMC*) and canonical measure of distance (*CMD*) between sets of data: Part 2. Variable reduction. *Analytica Chimica Acta*, **648**, 52–59.

**70** Todeschini, R., Ballabio, D., Consonni, V., Manganaro, A., and Mauri, A. (2009) Distance and correlation measures between sets of data: Part 1. Theory and simple chemometric applications. *Analytica Chimica Acta*, **648**, 45–51.

**71** Krzanowski, W.J. (1988) *Principles of Multivariate Analysis*, Oxford University Press, New York, NY, p. 564.

**72** Jolliffe, I.T. (1986) *Principal Component Analysis*, Springer, New York, NY, p. 272.

**73** Varmuza, K. and Filzmoser, P. (2009) *Introduction to Multivariate Statistical Analysis in Chemometrics*, Taylor & Francis, Boca Raton, FL.

# 5
# Partial-Order Ranking and Linear Modeling: Their Use in Predictive QSAR/QSPR Studies

*Andrew G. Mercader and Eduardo A. Castro*

## 5.1
## Introduction

Most data analysis attempt to fit a proposed quantitative model to a set of experimental measurements or observations. There are numerous reasons to perform such a task. For instance, the model can be purely empirical and employed to perform predictions to assist new experiments. On the other hand, the model can be based on a type of law or theory; thus, its evaluation could be employed for gaining some insight of underlying processes. In many circumstances, the ability of fitting a model to a particular set of data may offer new elements to formulate innovative hypotheses. The type of model that can be employed in a certain dataset depends not only on its nature, but also on the use that the model would be given. Mathematical techniques serve as valuable tools to estimate a given specific data with respect to their particular properties [1].

Whenever the main objective is to understand the behavior of macroscopic properties exhibited by physicochemical and biological systems, which depend on the molecular structure and chemical reactivity, approaches based on quantum chemical mechanics need to be adopted in order to accurately represent the involved phenomena. This translates into the necessity of considering all possible interactions present in the physical system of particles. Actual quantum mechanical calculations can be solved to a good approximation only when they involve molecules with few atoms free of molecular interactions. Although various methods with severe approximations can be postulated for solving the problem of interrelated particles, their fuzzy nature sometimes makes them unsatisfactorily explain the quality of the results. Hence, a different approach will be preferable to accurately predict the properties exhibited by a collection of interacting molecules [2].

Quantitative structure–property/activity relationships (QSPR/QSAR) are generally applied to overcome the lack of experimental data in complex chemical phenomena [3]. Therefore, there exists a permanently renewed interest focused on the development of such kind of predictive techniques [4, 5]. The ultimate role of QSPR/QSAR theory is to suggest mathematical models capable of estimating relevant properties of interest. Such studies rely on the basic assumption that the structure of a

*Statistical Modelling of Molecular Descriptors in QSAR/QSPR.* First Edition.
Edited by M. Dehmer, K. Varmuza, and D. Bonchev

compound determines entirely its properties, which can therefore be translated into so-called molecular descriptors. These parameters are calculated through mathematical formulae obtained from several theories, such as chemical graph theory, information theory, quantum mechanics, and so on [6, 7]. Several types of descriptors can be employed as numerical variables to reflect topological, geometrical, or electronic aspects of the molecule. The developed model enables us to estimate the property of substances that have not been tested yet for different reasons, such as being unstable, toxic, or simply because their measurement requires too much time. There are several motivations to establish and develop QSPR/QSAR studies; in terms of economical aspects, the design of QSPR/QSAR models gives chances for a rational use of the available resources present in the laboratory or even in a plant, avoiding the performance of expensive and unnecessary experimental determinations. In terms of moral aspects, the QSPR/QSAR models applied to toxicology have reached a great importance in the virtual screening of potentially toxic compounds before their synthesis, and thus representing an effective alternative that reduces animal testing in biological essays. Finally, from theoretically point of view, the model can enlighten the mechanisms of physicochemical properties or biological activities of compounds.

Linear modeling algorithms are considered to be most convenient for analyzing QSPR/QSAR datasets for numerous reasons, the main ones are (i) they exhibit a higher predictive capability and perform more efficiently on external test sets not considered during the model calibration; (ii) when few experimental observations are available, it is necessary to employ the lowest number of optimized parameters during the model development, a condition that linear models fulfill [8]; (iii) the nature of the achieved models allows the interpretation of the influence of each descriptor in the property or activity.

Ranking methods [9] are considered a very useful mathematical approach that has been continuously developed during the past few decades to perform data exploration, investigate the interrelationships of objects and/or variables, and set priorities [10–12]. Furthermore, they constitute a useful tool also for modeling purposes.

An important focus of the partial-order ranking (POR) theory, also known in the literature as Hasse diagram technique (HDT), relies on the prediction of endpoints of interest. The POR methodology, although being a discipline of statistics, provides an interesting alternative and simplified approach as it does not depend on statistical distribution assumptions, and in this sense, it is a parameter-free technique. Consequently, an advantage of this fact is the potential to surmount the lack of availability of experimental data, since the calculation of reliable optimized parameters for a mathematical model normally requires many known data.

## 5.2
## Linear QSAR Methodology, ERM, RM and GA

At the present time, thousands of descriptors encoding different aspects of the molecular structure have been developed and are available in the literature [13]. As in any QSAR/QSPR study, it must be determined how to select those that characterize the property/activity under consideration in the most efficient way. First, the optimal number of parameters to include in the model has to be defined. Afterwards the mathematical problem of selecting a subset $\mathbf{d}$ of $d$ descriptors from a much larger set $\mathbf{D}$ of $D \gg d$ ones has to be addressed.

The search for the optimal set of descriptors may be monitored by the minimization or maximization of a chosen function; for example, one option is to look for a model that makes the standard deviation ($S$) as small as possible. In other words, the global minimum of $S(\mathbf{d})$ is sought, where $\mathbf{d}$ is a point in a space of $D!/(D-d)!d!$ ones. A full search (FS) of the optimal variables is impractical since it requires $D!/[d!(D\text{-}d)!]$ linear regressions and $D$ normally is higher than 1000; an example is presented in Table 5.1.

For that reason, some time ago the replacement method (RM) was proposed [14–16], and later the enhanced replacement method (ERM) [17], which produces linear regression QSPR/QSAR models that are quite close the FS ones with much less computational work. These alternative techniques approache the minimum of $S$ by judiciously taking into account the relative errors of the coefficients of the least-squares model given by the subset of $d$ descriptors. The RM gives models with better statistical parameters than the forward stepwise regression (FSR) procedure [18] and variants of the more elaborated genetic algorithms (GAs) [19]. The ERM leads to even better statistical parameters with slightly more computational work [17].

The RM is a rapidly convergent iterative algorithm that produced linear regression models with small $S$ in a remarkably little computer time [20, 21]. However, in some cases, the RM gets trapped in a local minimum of $S$ not been able to continue without some kind of constraint. Although such local minima provided acceptable models, as shown in all earlier applications of the RM [20–23], still there was room for improvement and the ERM was developed.

The ERM followed the same RM philosophy but exhibited fewer propensities for remaining in local minima and at the same time was less dependent on the initial set of descriptors. It has a resemblance with simulated annealing that is an adaptation of the Metropolis–Hastings algorithm, a Monte Carlo method [24], to generate sample states of a thermodynamic system. The name and inspiration came from annealing

**Table 5.1** Number of necessary linear regressions and calculation time (using an AMD Athlon 64 2800+ processor) to carry out a FS with D = 1187.

| *D* | 1 | 2 | 3 | 4 | 5 | 6 | 7 |
|---|---|---|---|---|---|---|---|
| Regression Number | 1187 | 703891 | 2.8E + 08 | 8.2E + 10 | 1.9E + 13 | 3.8E + 15 | 6.5E + 17 |
| Minutes | **0.003** | **1.95** | 7.7E + 02 | 2.3E + 05 | 5.4E + 07 | 1.1E + 10 | 1.8E + 12 |
| Hours | 5.5E − 05 | 3.2E − 02 | **12.8** | 3791.6 | 9.0E + 05 | 1.8E + 08 | 3.0E + 10 |
| Days | 2.3E − 06 | 1.4E − 03 | 5.3E − 01 | **158.0** | 3.7E + 04 | 7.4E + 06 | 1.2E + 09 |
| Years | 6.2E − 09 | 3.7E − 06 | 1.5E − 03 | 4.3E − 01 | **1.0E + 02** | **2.0E + 04** | **3.4E + 06** |

in metallurgy, a technique involving heating, and controlled cooling of a material to increase the size of its crystals and reduce their defects. The heat causes the atoms to become unstuck from their initial positions (a local minimum of the internal energy) and lets them wander randomly through states of higher energy; the slow cooling gives them more chances of finding configurations with lower internal energy than the initial one [25].

A recent work presents an extensive and reliable comparison of the ERM and RM against the GA. The results for 10 different experimental databases suggested that the ERM is clearly preferable to GA, supporting earlier results based on comparisons made one case at a time [26–28]. In addition to the quality of the results, it was reflected the fact that the ERM is much simpler than GA. It was also shown that although the GA is slightly better than the RM, the simplicity and lower computational cost make the latter still attractive. In that work, ten different experimental datasets were used to test and contrast the performance of ERM, RM, and GA. These datasets were: fluorophilicity dataset (FLUOR), consisting of 116 organic compounds characterized by 1268 theoretical descriptors [21]; growth inhibition (GI) dataset, with growth inhibition values to the ciliated protozoan *Tetrahymena pyriformis* by 200 mechanistically diverse phenolic compounds and 1338 structural descriptors [23]; a GABA receptor dataset, containing 78 inhibition data for flavone derivatives and 1187 molecular descriptors [22]; 100 $pED_{50}$ antiepileptic activities of enaminones with 1306 descriptors (MES) [29]; 166 aqueous solubility of drug like compounds with 1497 descriptors (SOL) [30]; 470 $pIGC_{50}$ aqueous toxicities of heterogeneous aliphatic compounds with 1505 descriptors (TOX1) [31]; 392 $pIGC_{50}$ aqueous toxicities of benzene derivatives with 1497 descriptors (TOX2) [32]; 17 acetyl cholinesterase (ACET) inhibitor activities of substituted indanone and benzylpiperidine analogs using 300 descriptors [33]; 35 glass transition temperatures (GTTs) of structurally diverse polymers and 442 descriptors (GTT) [33]; and finally 30 melt transition temperatures (MTTs) of structurally diverse polymers and 368 descriptors [33].

In the first seven datasets the structures of the compounds were firstly preoptimized with the molecular mechanics force field (MM+) procedure included in HyperChem version 6.03, and the resulting geometries were further refined by means of the semiempirical molecular orbital theory PM3 method (parametric method-3) using the Polak–Ribiere algorithm and a gradient norm limit of 0.01 kcal $Å^{-1}$. For each database, more than a thousand molecular descriptors were calculated using the software Dragon [34], including parameters of all types such as constitutional, topological, geometrical, charge, GETAWAY (geometry, topology and atoms-weighted assembly), WHIM (weighted holistic invariant molecular descriptors), 3D-MoRSE (3D-molecular representation of structure based on electron diffraction), molecular walk counts, BCUT descriptors, 2D-autocorrelations, aromaticity indices, Randić molecular profiles, radial distribution functions, functional groups, atom-centered fragments, empirical, and properties. All the algorithms were programmed in the computer system MATLAB [35]. The calculation of de descriptors in the ACET, GTT, and MTT datasets differs from the rest [33].

The numerical tests were carried out by taking values of *d* from 3 to 7, the last one as an example of a computationally demanding search with a reasonable number of descriptors for a potential model in common QSPR/QSAR studies. The number of variables $d = 1$ and $d = 2$ were excluded because in such cases one can carry out a FS in relatively short times as shown in Table 5.1; where it can also be seen that FS is impractical for greater $d$ even for a small database (GABA, $D = 1187$). Using a more powerful computer the required computational time will still be impractical for large number of $d$. In Table 5.2 it can be seen that ERM outperforms or equals GA, for all cases except when $d = 3$ for the TOX2 dataset. This particular case appears to be fortuitous since ERM is preferable to GA in all the other cases. It should to be kept in mind that since the GA is a nondeterministic methodology then its results may change for different runs using exactly the same initial conditions. The computational demand used on both algorithms was similar.

The following subsections briefly describe the theory of RM, ERM, and GA as variable subset selection methods.

### 5.2.1
### Replacement Method

An optimal subset $\mathbf{d}_m = \{X_{m1}, X_{m2}, \ldots, X_{md}\}$ of $d \ll D$ is chosen, from a large set $\mathbf{D} = \{X_1, X_2, \ldots, X_D\}$ of $D$ descriptors provided by some available commercial program, with minimum standard deviation $S$:

$$S = \sqrt{\frac{1}{(N-d-1)} \sum_{i=1}^{N} \mathrm{res}_i^2} \tag{5.1}$$

**Table 5.2** Standard deviation ($S$) and number of linear regressions for ERM and GA[a].

| | d | 3 | 4 | 5 | 6 | 7 |
|---|---|---|---|---|---|---|
| MES | ERM | **0.3507** | **0.3341** | **0.3230** | **0.3030** | **0.2896** |
| | GA | **0.3507** | **0.3341** | 0.3234 | 0.3084 | 0.2973 |
| GI | ERM | **0.5124** | **0.4949** | **0.4781** | **0.4550** | **0.4367** |
| | GA | **0.5124** | **0.4949** | 0.4788 | 0.4608 | 0.4421 |
| GABA | ERM | **0.6075** | **0.5531** | **0.4896** | **0.4440** | **0.3961** |
| | GA | **0.6075** | **0.5531** | 0.5129 | **0.4440** | **0.3961** |
| FLUOR | ERM | **0.6729** | **0.6021** | **0.5376** | **0.4994** | **0.4328** |
| | GA | **0.6729** | **0.6021** | 0.5625 | 0.5003 | 0.4603 |
| SOL | ERM | **0.9207** | **0.8613** | **0.8277** | **0.8057** | **0.7788** |
| | GA | **0.9207** | 0.9051 | 0.8459 | 0.8166 | 0.8028 |
| TOX1 | ERM | **0.4740** | **0.4476** | **0.4272** | **0.4081** | **0.3982** |
| | GA | **0.4740** | **0.4476** | **0.4272** | 0.4096 | 0.3982 |
| TOX2 | ERM | 0.3676 | **0.3305** | **0.3113** | **0.3026** | **0.2891** |
| | GA | **0.3623** | **0.3305** | 0.3197 | 0.3077 | 0.2992 |
| Regression number | ERM | 962031 | 1765060 | 2586950 | 4251480 | 5014618 |
| | GA | 1167781 | 1881094 | 2681219 | 4310438 | 5045313 |

[a] Using seven different datasets and $d = 3$, 4, 5, 6, and 7 (the best results appear in boldface numbers).

where $N$ is the number of molecules in the training set, and $res_i$ the residual for molecule $i$ (difference between the experimental and predicted property). The fact that $S(\mathbf{d}_n)$ is a distribution on a discrete space of $D!/d!(D-d)!$ disordered points $\mathbf{d}_n$ should be noticed. The FS that consists of calculating $S(\mathbf{d}_n)$ on all those points always allows us to arrive at the global minimum, but as mentioned is computationally prohibitive if $D$ is sufficiently large (refer to Table 5.1). The RM consists of the following steps:

- An initial set of descriptors $\mathbf{d}_k$ is chosen at random; one of the descriptors is replaced, denoted as $X_{ki}$, with all the remaining $D-d$ descriptors, one by one, and the set with the smallest value of $S$ is kept. That is what is defined as a "step."
- From this resulting set, the descriptor with the greatest standard deviation in its coefficient is chosen (the one changed previously is not considered) and substituted with all the remaining $D-d$ descriptors, one by one. This procedure is repeated until the set remains unmodified. In each cycle, the descriptor optimized in the previous one is not modified. Thus, the candidate $\mathbf{d}_m^{(i)}$ that comes from the so-constructed path $i$ is obtained.
- It should be noticed that if the replacement of the descriptor with the largest error by those in the pool does not decrease the value of $S$, then that descriptor is not changed.
- The above process is carried on for all the possible paths $i = 1, 2, \ldots, d$ and the point $\mathbf{d}_m$ with the smallest standard deviation: $\min_i S(\mathbf{d}_m^{(i)})$ is kept.

### 5.2.2 Enhanced Replacement Method

The ERM is a three-step combination of two algorithms: first the RM already described above, then the modified RM (MRM), and finally the RM is used again. The MRM follows the same strategy as RM except that in each step the descriptor with the largest error is substituted even if that substitution is not accompanied by a smaller value of $S$ (the next smallest value of $S$is chosen). The main difference in the MRM is that it adds some sort of noise that prevents the selected model to stay in a local minimum of $S$ [17].

### 5.2.3 Genetic Algorithm

The GA is a search technique based on natural evolution where variables play the role of genes (in this case the descriptors) in an individual (in this case a set of $d$ descriptors) of a species. An initial group of random individuals (population) evolves according to a fitness function (in this case the standard deviation) that determines the survival of the individuals. The algorithm searches for those individuals who lead to better values of the fitness function through selection, mutation, and crossover genetic operations. The selection operators guarantee the propagation of individuals with better fitness in future populations. The GAs explore the solution space combining genes from two individuals (parents) using the crossover operator to form two new individuals (children) and also by randomly mutating individuals using the mutation operator. The GA offers a combination of the hill-climbing ability (natural selection) and a stochastic method (crossover and mutation), and explores many solutions in parallel, processing information in a very efficient manner. The practical application of GAs requires the tuning of some parameters such as population size, generation gap, crossover probability, and mutation probability. These parameters typically interact among themselves nonlinearly and cannot be optimized one at a time. There is considerable discussion about parameter settings and approaches to parameter adaptation in the evolutionary computation literature; however, there does not seem to be conclusive results on which may be the best [36]. Consequently, they have to be set for each individual application in a very laborious task.

With the purpose of providing a graphical visualization of the behavior of ERM and RM algorithms, Figures 5.1 and 5.2 show $S$ as a function of the number of steps on them, as an example the optimization of a seven-parameter model using the FLUOR dataset [21] was used. Figure 5.1 reveals that the ERM exhibits three sections: first one due to RM, second one that simulates a higher temperature or "a higher noise" than the RM, although maintaining the overall decreasing tendency of the $S$ function (MRM), and finally a third section where a second RM makes $S$ to further decrease. This apparent thermal agitation makes the ERM less likely to get trapped by a local minimum at the cost of a slower convergence and a slightly greater computer time [17].

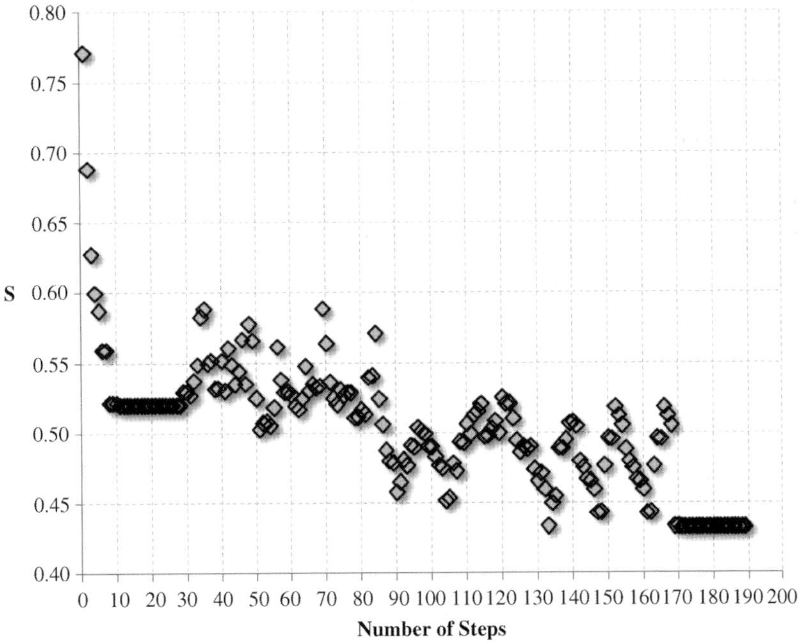


**Figure 5.1** Standard deviation vs. number of steps for the ERM.

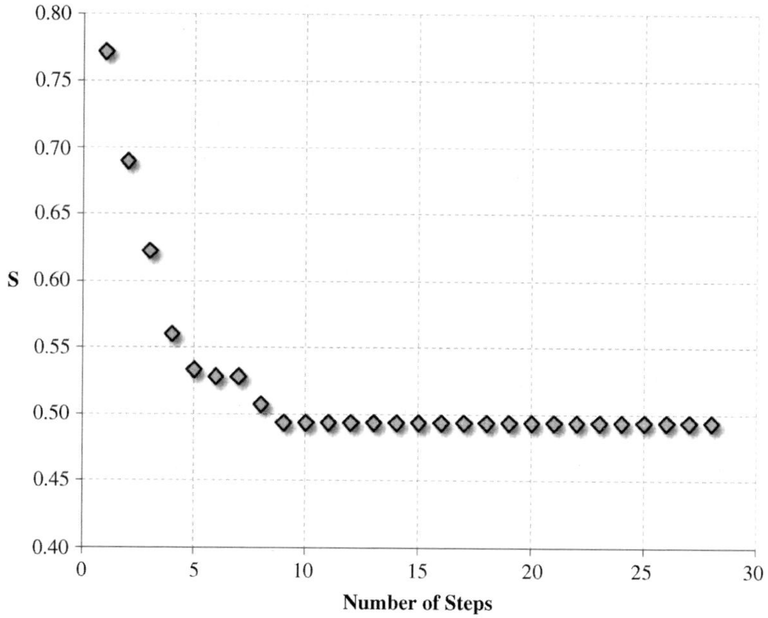


**Figure 5.2** Standard deviation vs. number of steps of the RM.

Figure 5.2 displays the high convergence rate of RM as the number of steps is increased.

Finally, in order to illustrate the behavior of the GA with variations in its parameters, Figures 5.3–5.5 show $S$ as a function of the population number for three cases where the number of individuals was 20, 5, and 100, respectively. For all those cases, the generation gap was 0.9, single point crossover probability 0.6, and mutation probability of $0.7/d$.

### 5.2.4
### Main Differences between MRM and RM

In order to clarify the difference between MRM and R, M their application to one of the datasets is presented as an example. The fluorophilicity dataset (FLUOR) that consists of 116 organic compounds characterized by 1268 theoretical descriptors was used. The optimal model with $d = 7$ topological descriptors out of the pool of $D = 1268$ ones will be obtained.

First it was arbitrarily selected an initial set of descriptors $\mathbf{d} = \{X_1, X_2, X_3, X_4, X_5, X_6, X_7\}$ which yielded $S_{(0)} = 0.771$; following path 1 the behavior is similar to Figure 5.2.

Table 5.3 displays a summary of the procedure showing the relative error of the regression coefficients for the descriptors, the regression constant (C) and how $S$ decreases and R increases in each step.

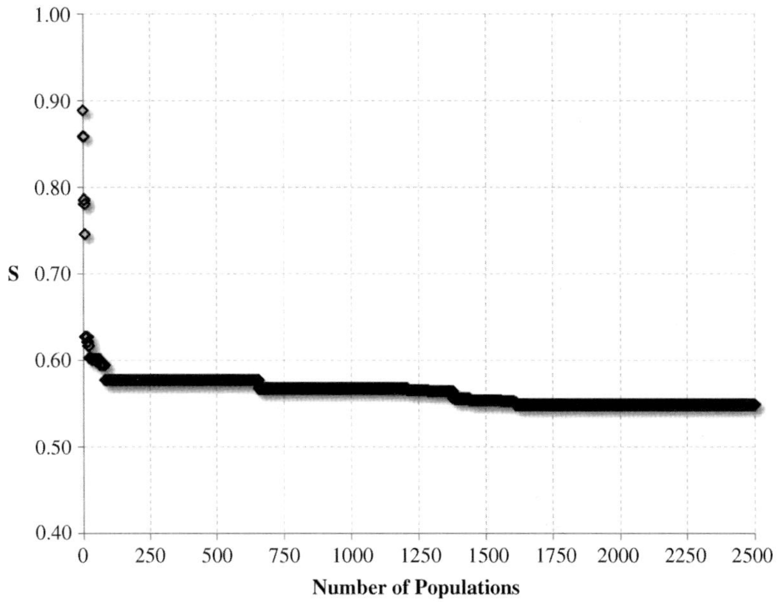


**Figure 5.3** Standard deviation vs. population number for GA with parameters: number of individuals = 20; generation gap = 0.9; single-point crossover probability = 0.6; mutation probability = 0.7/*d*.

Following path 1 the first replaced descriptor was $X_1$; each change was indicated by the notation ($X_{old}$, $X_{new}$).

Of all the 1261 ($D-d$) variables, the substitution that minimized $S$ was ($X_1$,$X_{1068}$) yielding $S(1) = 0.689$.

The variable with the greatest relative error $X_6$ was then replaced by all the rest of 1261 descriptors ($X_{1068}$ was left out of the descriptor pool and $X_1$ was back in it) and it was found that the substitution ($X_6$, $X_{40}$) yielded the smallest standard deviation $S(2) = 0.634$.

Then the variable with greatest relative error was $X_3$. After its replacement by all the 1261 descriptors, it was concluded that the substitution ($X_3$, $X_{411}$) yielded the minimal value $S(3) = 0.602$.

In the following step the variable with greatest relative error was $X_7$ and after its replacement by all the 1261 descriptors, the best alternative was ($X_7$, $X_{697}$) and $S(4) = 0.574$.

Of all the variables not yet replaced, $X_2$ was the one with the largest relative error. Its replacement by all the 1261 descriptors did not lead to a model with lower $S$.

Up to this point, MRM and RM had exactly the same behavior. Their difference becomes visible afterwards.

First, it will be described how RM would have continued. Since the replacement of $X_2$ did not lead to a model with lower $S$, $X_2$ remains in its position and is not replaced. Exactly the same situation occurred with the next descriptors $X_5$ and $X_4$. Restarting the

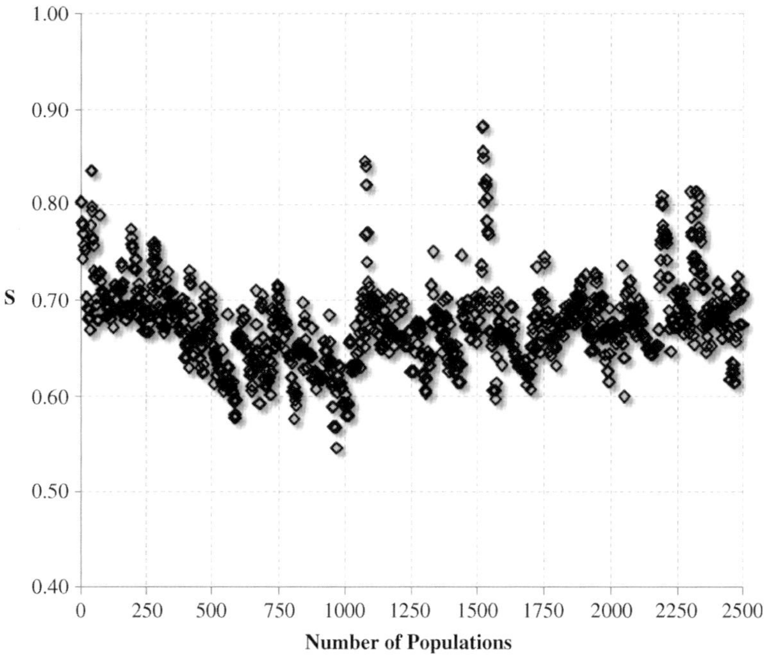


**Figure 5.4** Standard deviation vs. population number for GA with parameters: number of individuals = 5; generation gap = 0.9; single-point crossover probability = 0.6; mutation probability = 0.7/*d*.

process once again did not lead to a model with lower $S$, consequently the best model found in this case yielded $S(4) = 0.574$.

MRM continued as follows, even when the replacement of $X_2$ did not lead to a model with lower $S$, the descriptor was replaced anyway by the descriptor that leaded to the lowest $S$ from the 1261 remaining descriptors; thus, the following replacement was $(X_2, X_{1110})$ with $S(5) = 0.580$. Note that in this step $S$ has increased slightly; this is far from being a problem since an even lower S was found later on, showing that the increase in $S$ was necessary to get out of a local S minimum.

In the next step once again it was found that the replacement of the descriptor $X_5$ with higher error in the coefficient that was not previously replaced by all the 1261 descriptors leaded to a substitution $(X_5, X_{394})$ that yielded an even higher standard deviation $S(6) = 0.593$.

Nevertheless in the following step the replacement of $X_4$ (the descriptor with higher error in the coefficient that remained untouched) by all the 1261 descriptors led to the substitution $(X_4, X_{1050})$ that yielded $S(7) = 0.545$ that is even lower than the local minimum found in step four: $S(4) = 0.574$.

As the procedure continued, $S$ continued the decreasing tendency, as can be seen in Figure 5.1, in this case arriving to the lowest value after 222 steps. The best model found yielded $S(222) = 0.4572$ and $R = 0.9835$, having the form

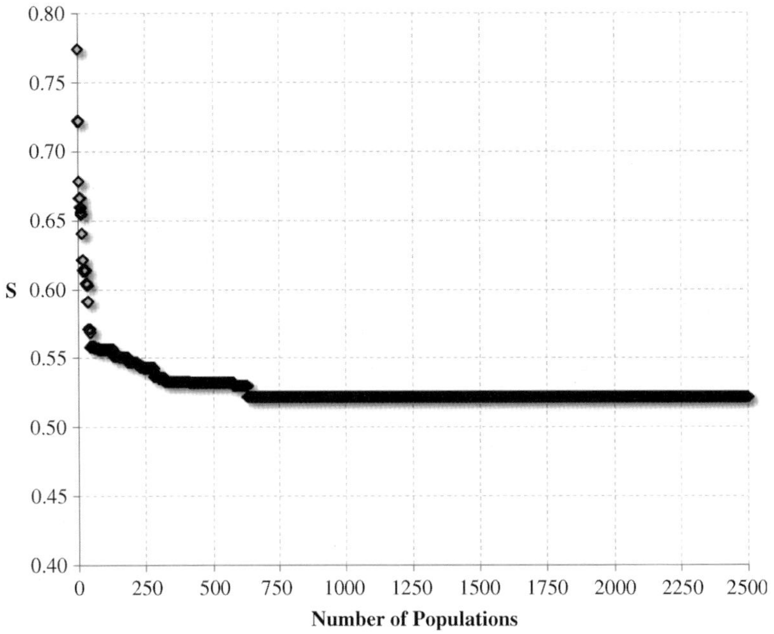


**Figure 5.5** Standard deviation versus population number for GA with parameters: number of individuals = 100; generation gap = 0.9; single-point crossover probability = 0.6; mutation probability = 0.7/*d*.

$$\ln \mathrm{P} = 0.065(\pm 0.3) - 3.9029(\pm 0.6)X_{425} - 0.0544(\pm 0.005)X_{240} - 0.063(\pm 0.004)X_{40} - 0.3749(\pm 0.01)X_{200} + 1.7051(\pm 0.2)X_{480} - 23.6913(\pm 2.7)X_{1095} + 0.008(\pm 0.001)X_{256} \quad (5.2)$$

The molecular descriptors appearing in the equation combine several two- and three-dimensional aspects of the molecular structure, and can be classified as a 2D autocorrelations, four topological descriptors, an aromaticity index, and a GETWAY descriptor [34]. The names of this descriptors and their meanings a can be found in Table 5.4.

## 5.3 Principles of Ranking Methods

The methodology consisting on partial order enables a suitable way of making comparisons among a set of objects (N), that is, molecules exhibiting investigated experimental values, by means of comparing their associated numerical attributes, that is, a set of molecular descriptors, usually called an information basis (IB) [37]. If a molecule A is characterized with a set of attributes $d_i$(A), $i = 1, \ldots, a$ and a molecule B with the set $d_i$(B), then two molecules A and B can be compared (ranked) through

**Table 5.3** Evolution of the MRM[a].

| Step N° | Descriptor number/relative errors in the regression coefficients | | | | | | | | S | R |
|---|---|---|---|---|---|---|---|---|---|---|
| | *C* | 1 | 2 | 3 | 4 | 5 | 6 | 7 | 0.771 | 0.952 |
| 0 | 28.29 | 90.12 | 38.95 | 59.59 | 20.36 | 194.94 | 84.91 | 50.21 | | |
| | *C* | 1068 | 2 | 3 | 4 | 5 | 6 | 7 | 0.689 | 0.962 |
| 1 | 21.34 | 18.58 | 41.89 | 67.67 | 15.74 | 66.44 | 796.66 | 35.89 | | |
| 2 | *C* | 1068 | 2 | 3 | 4 | 5 | 40 | 7 | 0.634 | 0.968 |
| | 15.62 | 16.69 | 31.43 | 43.14 | 10.58 | 35.56 | 22.60 | 27.24 | | |
| 3 | *C* | 1068 | 2 | 411 | 4 | 5 | 40 | 7 | 0.602 | 0.971 |
| | 16.16 | 15.48 | 19.93 | 23.67 | 6.45 | 9.06 | 20.34 | 82.96 | | |
| 4 | *C* | 1068 | 2 | 411 | 4 | 5 | 40 | 697 | 0.574 | 0.974 |
| | 8.76 | 17.92 | 12.77 | 18.99 | 5.29 | 7.07 | 18.49 | 28.74 | | |
| 5 | *C* | 1068 | 1110 | 411 | 4 | 5 | 40 | 697 | 0.580 | 0.973 |
| | 9.24 | 23.11 | 13.12 | 18.97 | 6.49 | 6.91 | 21.53 | 23.96 | | |
| 6 | *C* | 1068 | 1110 | 411 | 4 | 394 | 40 | 697 | 0.593 | 0.972 |
| | 6.50 | 26.58 | 10.69 | 16.79 | 7.41 | 7.14 | 23.55 | 15.56 | | |
| 7 | *C* | 1068 | 204 | 411 | 1050 | 394 | 40 | 697 | 0.545 | 0.974 |
| | 45.13 | 27.90 | 25.83 | 21.09 | 4.84 | 5.85 | 13.76 | 12.49 | | |
| 222 | *C* | 425 | 240 | 40 | 200 | 480 | 1095 | 256 | 0.457 | 0.984 |
| | 414.91 | 16.10 | 8.35 | 6.05 | 3.37 | 10.05 | 11.34 | 12.30 | | |

[a] Number of the descriptors in the model with the corresponding relative errors in the regression coefficients, *S* and *R* for each step of the algorithm. *C* stands for regression constant.

**Table 5.4** Information about the descriptors of the best model found in the example shown in Equation 5.2.

| Descriptor | | | |
|---|---|---|---|
| **Number** | **Name** | **Type** | **Meaning** |
| $X_{425}$ | MATS1p | 2D autocorrelations | Moran autocorrelation – lag 1/weighted by atomic polarizabilities |
| $X_{240}$ | piPC03 | Topological | Molecular multiple path count of order 03 |
| $X_{40}$ | IAC | Topological | Total information index of atomic composition |
| $X_{200}$ | SEigv | Topological | Eigenvalue sum from van der Waals weighted distance matrix |
| $X_{480}$ | AROM | Aromaticity indices | Aromaticity (trial) |
| $X_{1095}$ | R3u+ | GETAWAY | R maximal autocorrelation of lag 3/unweighted |
| $X_{256}$ | D/Dr10 | Topological | Distance/detour ring index of order 10 |

comparison (ranking) of their single descriptor values, resorting to the binary relation "$\leq$":

$$\mathrm{B} \leq \mathrm{A} \leftrightarrow d_i(\mathrm{B}) \leq d_i(\mathrm{A}) \text{for all } i = 1, \ldots, a. \tag{5.3}$$

The requirement "for all $i$" to set up the order relation is called the generality principle, and this condition transforms partial ordering into a vectorial approach. Each molecule is characterized with a vector whose elements are its attribute values [38]. Also, the following axioms have to be satisfied:

$$\begin{array}{lll} \mathrm{A} \leq \mathrm{A} & \mathrm{A} \leq \mathrm{B} \text{ and } \mathrm{B} \leq \mathrm{A} \Rightarrow \mathrm{A} = \mathrm{B} & \mathrm{A} \leq \mathrm{B} \text{ and } \mathrm{B} \leq \mathrm{C} \Rightarrow \mathrm{A} \leq \mathrm{C} \\ \text{reflexivity} & \text{antisymmetry} & \text{transitivity} \end{array} \tag{5.4}$$

When the inequality of Equation 5.3 is true, it is said that compound A is ranked higher than compound B (A dominates B), and that at least one descriptor for A is higher than the corresponding descriptor for B, and no descriptor for A is lower than the respective descriptor for B. If the rank of a molecule X is denoted by $\mathrm{Rk_X}$, the previously formulated inequality can be expressed as $\mathrm{Rk_B} \leq \mathrm{Rk_A}$. If all the descriptors for A are equal to the corresponding descriptors for B, the two compounds will have identical order (rank) and will be considered as "equivalent" rather than "identical" belonging to the same equivalence class. In consequence, the binary relation "$\leq$" is called a quasi order [39]. If Equation 5.3 is false, then both A and B are incomparable (denoted as $||$) and cannot be assigned a mutual order.

For the special case where Equation 5.3 is valid for all the $N$ compounds of the set under study, a complete comparability among the objects holds, and a total-order ranking can be developed. On the contrary, sets equipped with an order relation like Equation 5.3 for some of the $N$ molecules are called partially ordered sets (posets), denoted by taking into account the number of elements $N$ and the IB involved as ($N$, IB).

Note that Equation 5.3 *a priori* includes "$\leq$" as the only mathematical relation, and thus constitutes a parameter-free method that assigns equal preference and designation to the descriptors employed in the analysis. One possible disadvantage of the POR technique in this form is that, as the attributes are given equal preferences, it does not directly result in a unique decision. As no compensation among indicators is carried out, conflicting evaluations of these indicators cannot be methodologically removed, and consequently, multiple favorable options can be identified as incomparable winner options. This is the reason why the POR technique may be considered as an evaluation tool rather than a support system [40].

It is possible to aggregate a series of attributes for complex systems in a very simple and compressed picture by resorting to the HDT, a graphical visualization of the existing partial orderings in a set of objects considered [12, 38, 41] and that is able to establish a ranking net of comparisons among the objects. This kind of representation has been typically used to rank chemicals according to environmental hazards [42]. The order relations included in a HD are expressed in the plane, thus constituting an oriented acyclic graph; instead of drawing arrows, indicating that object A is "greater" than object B, object A is located above B in the plane. All objects are drawn with small circles or small rectangles, with comparable elements being connected with lines, while lines due to transitivity are omitted to simplify the scheme. Those elements that result incomparable or contradictory appear as unconnected, since there is no sequence of lines with the same orientation. Different classification levels are established in the ranking net, with the higher ones corresponding to objects that have a higher rank, and with lower levels for lower ranked objects. Objects that are located in the same levels in the diagram are not necessarily equivalents, as a result of their possible different attribute values. Figure 5.6a presents a HD for a total order of four contaminants. From this diagram, it is clear that compound A is less hazardous than B, B less than C, and C less than D, which is the most hazardous of the four. On the contrary, if a HD for three contaminants corresponds to the poset appearing in Figure 5.6b, where contradictions between objects arise, it is not immediately clear which chemical should be chosen as the safest; both A and B are better than D, but they are incomparable with each other.

The uncertainties due to the experimentally measured property or owing to the variation of the single descriptors values may influence the ranking of the substances, especially in the case of environmental data [43–47]. This influence of the data uncertainty on the POR has been previously analyzed by generating random artificial uncertainties [43]. Since the number of comparisons per compound is a crucial parameter in this context, the uncertainty of ranking increases significantly if the average number of comparisons ($E$) is less than 4–5 per compound. In cases where the average number of comparisons per compound is 5, or a higher number, the uncertainty of the ranking apparently is virtually constant and thus robust. It is possible to propose general guidelines for the critical number of molecules ($e_{\mathrm{cr}}$) needed to ensure a ranking that is robust in relation to uncertainties in the data:

$$e_{\mathrm{cr}} = E_{cr} 0.5^{-a} + 1 \tag{5.5}$$

This equation predicts that the necessary number of molecules doubles when one new molecular descriptor is added in the HD, which makes the number of attributes ($a$) extremely critical. Therefore, when establishing a model order, two main conclusions arise: (i) the model must be able to rank the compounds correctly according

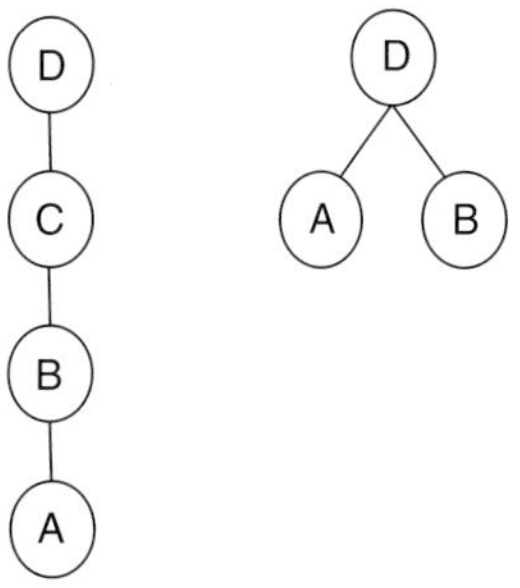


**Figure 5.6** (a) HD for a total order. (b) HD for a partially ordered set (poset).

to the experimental data, and (ii) the number of compounds used in the study should be large enough to secure a suitable fine-meshed net taking the number of descriptors into account, that is to say, a ranking net that is capable of performing comparisons among the set of objects adequately and that is also able to rank these objects according to the selected criteria. In those cases where the number of descriptors exceeds that corresponding to the number of compounds under investigation, the data reduction technique should be applied preliminary.

Hasse diagrams are able not only to present information on the ranking but, most importantly, also show whether the criteria, characterizing two investigated objects, lead to ambiguities in the ranking. These objects are not ordered simply because their data are "contradictory," and this ambiguity is hidden when an index for ranking is employed. A necessary previous step before ranking a series of substances is to ensure that all the descriptors involved exhibit identical "high-low" orientation in their numerical values. Both the ranking of the compounds and the HD can be obtained applying the Whasse software [39, 48].

The number of incomparable elements in the partial ordering constitutes a limitation in the attempt to rank, for example, a series of chemical substances based on their potential environmental or human health hazard. To some extent this problem can be remedied through the application of the so-called linear extensions of the POR [49, 50]. A linear extension is a total order, where all comparable elements of the partial order are reproduced [9, 12]. Due to the incomparable elements in the POR, a number of possible linear extensions correspond to one partial order. If all possible linear extensions are found, a ranking probability can be calculated, that is, based on the linear extensions the probability that a certain element has a certain absolute rank can be derived. If all possible linear extensions are found, it is possible to calculate the averaged ranks of the single elements in a partially ordered set [51, 52]. In every linear extension, an unknown compound will be directly comparable to others, while that is not necessarily the case in the partial order. Unfortunately, although this method is expected to lead to a probabilistic and more specific prediction of the property of interest, it presents some drawbacks in the case of partial-order sets including more than 20–25 compounds, as is not practically possible to identify the total set of linear extensions. This is basically a consequence of the extensive combinatorial demand. However, a newly developed method to analyze partial orders that are too large to be analyzed using the total set of linear extensions is available as a tool for QSAR modeling [53, 54]. The basic principle of this technique is to apply a randomly chosen fraction (subset) of all possible linear extensions.

## 5.4 Selection of the Molecular Descriptors for Ranking

The relative ranking of two compounds A and B will depend on the choice of the adopted molecular descriptors, which can lead to several PORs of substances by using different kinds of combinations of variables. The situation is reminiscent of the situation characterizing consideration of molecular similarity [55, 56]; the same set of molecular structures will yield distinct similarity/dissimilarity table, depending on the set of molecular descriptors employed.

Owing to the fact that complex systems are described by several structural variables, a major goal in system analysis is the extraction of relevant information, together with the exclusion of redundant and noisy structural information. Furthermore, the decision rules applying to the single descriptors should be explicable in chemical/physicochemical terms. It should also be mentioned that the HD is able to provide the possibility of estimating the quantitative importance of the attributes involved in its design. By removing the descriptors one at a time, the changes in the HD reflects the influence from that particular attribute [12]. Thus, if the removal of a descriptor induces a large number of changes, it means that it is important for the structure of the HD.

An alternative systematically way to select the best descriptors for ranking consists on an exhaustive examination of all the possible $a$ variables models (the model size) obtained from a set containing $D$ variables. However, when many variables are available, an exhaustive combinatorial FS [57, 58] of all possible models is not feasible since as mentioned it requires extensive computational resources and is time consuming. In such cases, a different variable selection technique is needed.

There are many approximate numerical algorithms reported in the literature that replace the FS method [59–62]. The above presented ERM emerges as one optimal alternative, since it gives better results than the more complicated genetic algorithms. For an even simper alternative, the RM is convenient since it presents similar results compared to GA and better results than the commonly used "step by step" procedure FSR (Forward Stepwise Regression) [63]. The FSR process selection begins without any independent variable in the model and in each step is introduced a variable that presents the higher correlation with the property. This method is stopped when there are not independent variables outside the model that satisfy the selection criterion.

The simple mathematics outlined above makes the POR technique a method sensitive to noise, since even minor fluctuations in the attribute values may lead to noncomparability or reversed ordering of the objects. The remedy to this problem is to generate "noise deficient" descriptors for use in the ranking, resorting to QSAR models [44–47]. The descriptors obtained in this way are not hampered by random fluctuations, as they are forced to obey a first-order equation. In example, it is employed the following general formula for the descriptor $d_i$ instead of using the original descriptor $d'_i$:

$$d_i = a_i d'_i + b_i \tag{5.6}$$

with $a_i$ and $b_i$ being constants. In noise deficient QSAR models, the natural variation on both the experimental and primary models data has been suppressed. Thus, the estimated property values increase or decrease monotonously over the entire range of values analyzed, eliminating circumstantial fluctuations during the eventual mutual ranking/prioritization of the substances [2].

## 5.5 QSAR Based on Hasse Diagrams

The application of the POR technique for predicting properties of interest requires ordering twice the compounds investigated. First, these are sorted according to a given experimental endpoint, based on the numerical values of the selected descriptors (model ranking). This order can then be compared to an order based on the experimental endpoints (experimental ranking). If the model ranking is able to reproduce the experimental ranking, then other compounds that do not poses an experimental value can be assigned a position in the ranking and then their experimental properties can be predicted.

A type of similarity index necessary to define for QSAR based on POR is the modified Tanimoto index $T$ (0,0), reflecting the percentage of rankings in the model that can be found in the experimental data [64]. It is defined as

$$T(0,0) = \frac{N_a}{N_a - RN_d} \tag{5.7}$$

with $N_a$ being the number of agreements for the mutual rank of two compounds in the model and in the experimental data, and $RN_d$ is the number of disagreements. With the help of two keys numbers, that is, Above and Below, the numbers of comparable elements lying above (higher rank) and below (lower rank) an element under investigation, respectively, it is possible to perform predictions. The predicted value for a given compound X (value $X$) belonging to the training series can be obtained through arithmetic means between the lowest value of the comparable compounds ranked above X (min Above) and the highest values of the comparable compounds ranked below X (max Below):

$$\text{Value } X = \frac{\text{min Above} + \text{max Below}}{2}. \tag{5.8}$$

Thus, the predictions obtained via Equation 5.8 are compared to the corresponding experimental values. For the special case when X is not enclosed in any of the intervals of known compounds, only one lower or upper bound can be given as the predicted value for X. When $T$ (0,0) shows a close agreement (value close to unity), the uncertainty of the value $X$ is equally distributed in the interval $\pm 0.5$(min Above-max Below). If $T$ (0,0) is far from unity, then suitable confidence intervals are to be considered. The predicted value for a compound X (value $X$) belonging to the test series can also be obtained through arithmetic means resorting to the Equation 5.8. Figure 5.7 includes an example for predicting the property value for compound X, located in the HD between compounds A and E.

## 5.6 Discussion

In POR – in contrast to standard multidimensional-statistical analysis – neither any assumptions about linearity nor any assumptions about distribution properties are

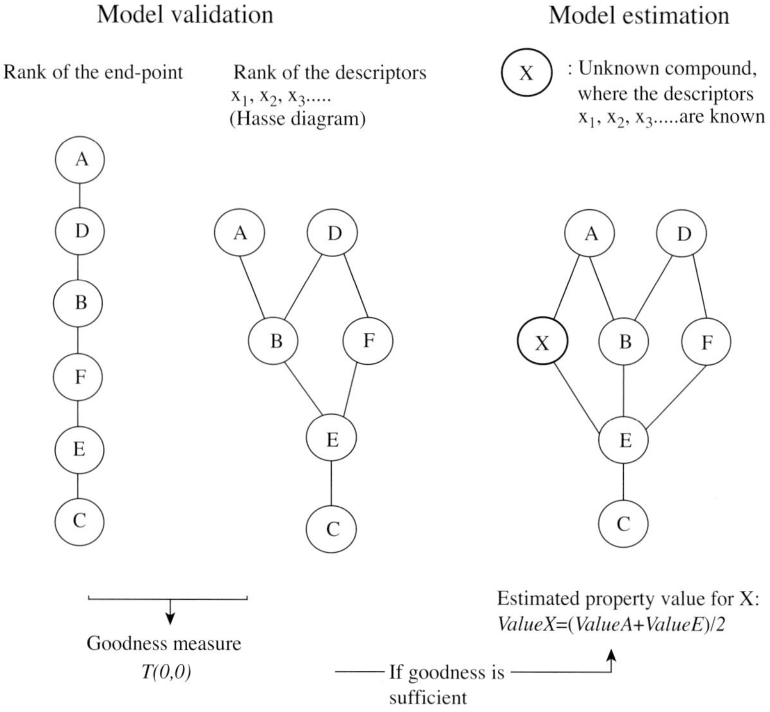


**Figure 5.7** Principles of model validation and model estimation.

made. In this way, the POR can be considered a nonparametric method. Therefore, the method may be more general compared to many standard statistical techniques [65]. Thus, there is no preference among the descriptors. However, due to its simple mathematics, it must be emphasized that as mentioned the method *a priori* is rather sensitive to noise, since even minor fluctuations in the descriptor values may lead to noncomparability or reversed ordering [66].

Hence, in principle predictions using POR could be preferable over QSAR based on multilinear regression analysis for those cases were the distribution of the input experimental values do not present a normal distribution. Nevertheless, in practice there are many cases that the available data does not have a normal distribution and predictive models can still be derived using linear QSAR methods, indicating that the technique is robust to departures from normality [67].

The number of available experimental observations is another main issue to take into account in the use of POR and linear QSAR. In its application, linear QSAR needs a minimum amount of experimental data to carry sufficient structural information by the selected descriptors to the model. The number of necessary experimental values depends on each case, nevertheless a guideline is the rule of thumb that at least five data points should be present for each fitting parameter [68]. POR on the other hand in principle needs only two observations to predict the value of a given molecule (min Above and max Below) as shown in Equation 5.8; however, as stated above a critical number of molecules is needed to ensure a robust ranking in relation to uncertainties as the general guidelines presented in Equation 5.5. An analysis of Equation 5.8 allows making noticeable that any POR prediction will have a great dependence on the quality of the two observations used to predict its values in addition to the gap between them.

The number of successful predictive studies using linear QSAR is enormous and their enumeration escapes this text. As an example, a predictive QSPR model of the luminescence lifetime ($\tau$) of $Eu(fod)_3$ in 23 different solvents using four molecular descriptors that take into account 2D- and 3D-aspects of the molecular structure was presented some time ago [28]. The ERM and RM were used to select the optimal set of four descriptors from a pool of 1057 parameters. The nature of the presented model (similar to that in Equation 5.2) allows its analysis, which suggested that $\tau$ depends significantly on the polarizability and electronegativity of the solvent. In this example, a small databases with a departure from normality was employed, nevertheless the robustness of the linear QSAR methodology employed achieved a model with good predictive power.

In practice, POR presents some complications; one main point is that all descriptors may not have identical orientations, that is, "high" and "low." As a consequence of this, it may be necessary to multiply some descriptors by –1 in order to achieve identical directions. As an example, bioaccumulation and toxicity can be mentioned. In the case of bioaccumulation, the higher the number the higher a chemical substance tends to bioaccumulate and thus the more problematic the substance, whereas in the case of toxicity, the lower the figure the more toxic the substance. Thus, in order to secure identical directions of the two descriptors, one of them, for example, the toxicity numbers has to be multiplied by −1. Consequently, both in the case of bioaccumulation and in the case of toxicity higher figures will now correspond to more problematic sites [66]. In addition, as previously mentioned a limitation arises from the number of incomparable elements in the partial ordering; however, this problem to some extent can be remedied through the application of the so-called linear extensions of the POR [49, 50].

Despite the above-mentioned complications in the practical use of POR, applications have been published in the specific literature and some examples will be presented aiming to provide a general picture.

In a QSPR developed by Carlsen *et al.* [11] a dataset of octanol–water partition coefficients (log $K_{ow}$) composed of the 46 organic compounds was successfully analyzed using the HDT. A comparison of the model order with the experimental order for octanol–water partitioning gives a "goodness of fit" $T(0,0) = 0.998$. This means that out of a total of 408 comparisons 407 were found in agreement and 1 in disagreement.

A QSAR study [65] employed HD for predicting ecotoxicological data (ME) in a sample set (G) constituted by 12 chlorinated alkanes. Four estimated properties were studied: (i) *Daphnia* effect concentration (D); (ii) algae lethal concentration (A); (iii) oxygen consumption, measured as effect concentration (O); and (iv) photo bacteria effect concentration (P). Three molecular descriptors families were used: constitutionals (MS), topologicals (MG), and quantum chemicals (MC). The estimated values achieved using MC and MS descriptor families showed that many chemicals got the same predicted values of the ecotoxicological data, due to the specific character of the HD: the chemicals under investigation had nearly the same upper and lower neighbor(s).

Another application of considerable interest in QSAR based on POR was proposed by Randić [69] and relies on the search of pharmacophores; "a molecular framework that carries (phoros) the essential features responsible for a drug's (pharmacon's) biological activity" [70]. In such investigations, a group of atoms in a biologically active compound are believed to be responsible for its activity; the atoms not necessarily need to be connected forming a fragment. In summary, various molecular fragments are considered and analyzed to determine which fragment descriptors best reproduce the experimentally observed ordering of structures in terms of a given activity. A possible way to distinguish among the structures is to select a common (relevant) fragment present in all of these, instead of considering the whole sequence, and to compare mutual similarities among the molecules with respect to the selected fragment.

Such identifications were illustrated in publications on nitrosamines [71], nitroarenes [72], the antiviral activity of substituted benzimidazoles [73], and the ordering of folded structures [69].

In a 2004 study [74], a total ranking model of solubility ($-\log S_w$), aqueous activity coefficient ($-\log I_w$), and octanol–water partitioning coefficient ($-\log K_{ow}$) of 64 polychlorinated biphenyls (PCBs) was illustrated, with the aim of providing a list of priority PCB according to their environmental impact. The chemical structures of the PCB were described by the use of more than 1500 constitutional, topological, geometrical, and electronic molecular descriptors obtained from the Dragon software [34], with the purpose of carrying as much structural information as possible. The geometries were optimized by means of the HyperChem package [75] (PM3 semiempirical method). Variable subset selection was performed with the GA technique [76, 77] starting with a population of 100 random GA models with a number of variables equal to or less than three, employing the Spearman's rank coefficient (rexp-mod) [78] as optimization parameter (fitness function of GA) to quantify the agreement between the experimental and model rankings. Very good results were provided by the simple model made of two variables, the average connectivity index (X1A), and a Burden descriptor (*BELm2*). The first is a topological descriptor calculated from the vertex degree of the atoms in the H-depleted molecular graph and the second is the second lowest eigenvalue of the Burden matrix weighted by atomic masses. Both descriptors were inversely correlated with the environmental impact of the PCB, because their values decreased with the increasing size of the PCB. The correlation between the experimental and model ranking was as high as 97.93. As the quality of the simple two descriptors model leads to satisfactory results, this methodology enabled a priority list of PCB to be defined according to their environmental impact, taking into account their solubility, aqueous activity coefficient, and hydrophobicity simultaneously. Moreover, to enable comparison of ranking modeling with a conventional modeling method, MLR was performed on the environmental impact score of the 64 PCB starting from the same molecular descriptors used for the ranking model. The two selected descriptors were the second lowest eigenvalue of the Burden matrix weighted by atomic masses (*BELm2*) and a 3D-MoRSE descriptor (*Mor29u*). The selection of *BELm2* descriptor confirms the agreement between the regression and the ranking model. The reason why the descriptors solution using ordering and regression approach differed is that if a descriptor functions well in a regression, it does not guarantee that it would also work similarly well for ranking. The opposite situation is also valid: If a descriptor orders itself just like the property do, then it does not mean that a good correlation between the descriptor and the property is expected [79].

In the later example, linear QSAR and POR are used as complementary tools; the results show that their combined use has a high potential for future studies. This is in agreement of what was mentioned by Carlsen in 2009; whereas the POR methodologies lead to a prioritization of the studied chemicals simultaneous taking a multitude of parameters into account, the formal concept analysis leads to valuable information on possible links between the studied chemicals and the associated parameters. As such, the combination QSAR/QSPR and POR constitutes a highly effective decision support tool [66].

In two resent studies, the use of POR as a nonpredictive tool for ordering chemical compounds was employed. In the first one, the possible association between human maternal exposure to organochlorine compounds used as pesticides and cryptorchidism among male children in Finland and Denmark was evaluated. In the HDT, the ordering of the chemicals, their different positions in the Hasse diagrams, and the quantification using partial-order theoretical method were new and important amendments in the findings of chemicals influencing the state of health of human beings [80]. In the second case the possible impact of the rocket fuel 1,1-dimethyl hydrazine and its transformation products on human health was ranked applying POR methodologies according to the predicted human health impact, the eventual ranking disclosed the seven compounds that on an cumulative basis should receive the major attention [81].

## 5.7 Conclusions

Partial-order ranking (POR) was presented and its potential to be applied in predictive studies was explored. In addition, recently developed linear QSAR methodology, ERM and RM were described. The ERM has previously shown to give QSAR models of better quality than the GAs) with the bonus that ERM is a much simpler methodology, making this new technique very promising for its use in future QSAR studies.

The characteristics of linear QSAR methodologies suggest that they are optimal for their simplicity in practical application and high quality of the predictive results; in addition, they allow indentifying and analyzing those descriptors that influence the studied experimental property or activity. Accordingly, they have been widely used and many applications are described in the specialized literature.

In practice, POR shows some complications, one main point is that all descriptors may not have identical orientations and as a consequence it might be necessary to multiply some descriptors by −1 in order to achieve identical directions. In addition, a limitation arises from the number of incomparable elements in the partial ordering; however, this problem to some extent can be remedied through the application of the so-called linear extensions of the POR. Applications have been described and published in the specific literature as seen in the presented examples.

The use of both POR and linear QSAR methodologies combined as complementary tools is a very attractive alternative with high potential in future studies. In addition, POR appears as a good back up method in those cases where linear QSAR does not present acceptable predictive ability as a result of some sort of difficulty in the dataset, that is, severe departures from normality or shortage of experimental data.

Uses of POR in nonpredictive studies for the ordering of chemical compounds was shown; two recent examples were presented showing its application in the ordering of organochlorine compounds used as pesticides in terms of the association between human maternal exposure and cryptorchidism among male children and on the ordering of rocket fuel 1,1-dimethyl hydrazine and its transformation products in terms of human health impact.

## References


1 Duchowicz, P.R. and Castro, E.A. (2008) Partial order theory applied to QSPR-QSAR studies. *Combinatorial Chemistry and High Throughput Screening*, **11**, 783–793.

2 Duchowicz, P.R. and Castro, E.A. (2008) *The Order Theory in QSPR-QSAR Studies. Mathematical Chemistry Monographs*, University of Kragujevac, Kragujevac, Serbia.

3 Hansch, C. and Leo, A. (1995) *Exploring QSAR: Fundamentals and Applications in Chemistry and Biology*, American Chemical Society, Washington, D.C.

4 Puzyn, T., Leszczynski, J., and Cronin, M.T.D. (2009) *Recent Advances in QSAR Studies: Methods and Applications*, 1st edn, Springer, New York.

5 Kubinyi, H. (2008) *QSAR: Hansch Analysis and Related Approaches*, Wiley-Interscience., New York.

6 Katritzky, A.R., Lobanov, V.S., and Karelson, M. (1995) QSPR: The correlation and quantitative prediction of chemical and physical properties from structure. *Chemical Society Reviews*, **24**, 279–287.

7 Trinajstic, N. (1992) *Chemical Graph Theory*, CRC Press., Boca Raton, FL.

8 Mercader, A.G., Duchowicz, P.R., Fernández, F.M., and Castro, E.A. (2010) Replacement method and enhanced replacement method versus the genetic algorithm approach for the selection of molecular descriptors in QSPR/QSAR theories. *Journal of Chemical Information and Modeling*, **50** (9), 1542–1548.

9 Davey, B.A. and Priestley, H.A. (1990) *Introduction to Lattices and Order*, Cambridge University Press, Cambridge, UK.

10 Carlsen, L., Sorensen, P.B., and Thomsen, M. (2001) Partial order ranking-based QSAR's: Estimation of solubilities and octanol–water partitioning. *Chemosphere*, **43** (3), 295–302.

11 Carlsen, L., Sørensen, P.B., Thomsen, M., and Brüggemann, R. (2002) QSARs based on partial order ranking. *SAR and QSAR in Environmental Research*, **13** (1), 153–165.

12 Brüggemann, R., Halfen, E., Welzl, G., Voigt, K., and Steinberg, C.E.W. (2001) Applying the concept of partially ordered

sets on the ranking of near-shore sediments by a battery of tests. *Journal of Chemical Information and Computer Sciences*, **41** (4), 918–925.

13 Todeschini, R. and Consonni, V. (2000) *Handbook of Molecular Descriptors*, Wiley VCH., Weinheim, Germany.

14 Duchowicz, P.R., Castro, E.A., Fernández, F.M., and González, M.P. (2005) A new search algorithm of QSPR/QSAR theories: Normal boiling points of some organic molecules. *Chemical Physics Letters*, **412**, 376–380.

15 Duchowicz, P.R., Castro, E.A., and Fernández, F.M. (2006) Alternative algorithm for the search of an optimal set of descriptors in QSAR-QSPR studies. *MATCH-Communications in Mathematical and in Computer Chemistry*, **55**, 179–192.

16 Duchowicz, P.R., Fernández, M., Caballero, J., Castro, E.A., and Fernández, F.M. (2006) QSAR of non-nucleoside inhibitors of HIV-1 reverse transcriptase. *Bioorganic and Medicinal Chemistry*, **14**, 5876–5889.

17 Mercader, A.G., Duchowicz, P.R., Fernandez, F.M., and Castro, E.A. (2008) Modified and enhanced replacement method for the selection of molecular descriptors in QSAR and QSPR theories. *Chemometrics and Intelligent Laboratory Systems*, **92**, 138–144.

18 Draper, N.R. and Smith, H. (1981) *Applied Regression Analysis*, John Wiley & Sons, Inc, New York.

19 So, S.-S. and Karplus, M. (1996) Evolutionary optimization in quantitative structure–activity relationship: An application of genetic neural networks. *Journal of Medicinal Chemistry*, **39**, 1521–1530.

20 Helguera, A.M., Duchowicz, P.R., Pérez, M.A.C., Castro, E.A., Cordeiro, M.N.D.S., and González, M.P. (2006) Application of the replacement method as novel variable selection strategy in QSAR: 1. Carcinogenic potential. *Chemometrics and Intelligent Laboratory*, **81**, 180–187.

21 Mercader, A.G., Duchowicz, P.R., Sanservino, M.A., Fernandez, F.M., and Castro, E.A. (2007) QSPR analysis of fluorophilicity for organic compounds. *Journal of Fluorine Chemistry*, **128** (5), 484–492.

22 Duchowicz, P.R., González, M.P., Helguera, A.M., Cordeiro, M.N.D.S., and Castro, E.A. (2007) Application of the replacement method as novel variable selection in QSPR. 2. Soil sorption coefficients. *Chemometrics and Intelligent Laboratory Systems*, **88**, 197–203.

23 Duchowicz, P.R., Mercader, A.G., Fernández, F.M., and Castro, E.A. (2008) Prediction of aqueous toxicity for heterogeneous phenol derivatives by QSAR. *Chemometrics and Intelligent Laboratory*, **90**, 97–107.

24 Metropolis, N., Rosenbluth, A.W., Rosenbluth, M.N., Teller, A.H., and Teller, E. (1953) Equation of state calculations by fast computing machines. *Journal of Chemical Physics*, **21** (6), 1087–1092.

25 Kirkpatrick, S., Gelatt, C.D. Jr., and Vecchi, M.P. (1983) Optimization by simulated annealing. *Science*, **220** (4598), 671–680.

26 Mercader, A.G., Duchowicz, P.R., Fernández, F.M., Castro, E.A., Bennardi, D.O., Autino, J.C., and Romanelli, G.P. (2008) QSAR prediction of inhibition of aldose reductase for flavonoids. *Bioorganic and Medicinal Chemistry*, **16**, 7470–7476.

27 Mercader, A.G., Duchowicz, P.R., Fernández, F.M., Castro, E.A., Cabrerizo, F.M., and Thomas, A.H. (2009) Predictive modeling of the total deactivation rate constant of singlet oxygen by heterocyclic compounds. *Journal of Molecular Graphics & Modelling*, **28**, 12–19. j.jmgm.2009.03.002

28 Mercader, A.G., Duchowicz, P.R., Fernández, F.M., Castro, E.A., and Wolcan, E. (2008) QSPR Study of solvent quenching of the $^5D_0 \rightarrow {}^7F_2$ emission of Eu(6,6,7,7,8,8,8-heptafluoro-2,2-dimethyl-3,5-octanedionate)$_3$. *Chemical Physics Letters*, **462**, 352–357.

29 Garro-Martínez, J.C., Duchowicz, P.R., Estrada, M.R., Zamarbide, G.N., and Castro, E.A. (2009) Anticonvulsant activity of ringed enaminones: A QSAR study. *The QSAR and Combinatorial Science*, **28**, 1376–1385.

30 Duchowicz, P.R., Talevi, A., Bruno-Blanch, L.E., and Castro, E.A. (2008) New QSPR study for the prediction of aqueous

solubility of drug-like compounds. *Bioorganic and Medicinal Chemistry*, **16**, 7944–7955.
31 Duchowicz, P.R. and Ocsachoque, M.A. (2009) Quantitative structure-toxicity models for heterogeneous aliphatic compounds. *The QSAR & Combinatorial Science*, **28**, 281–295.
32 Castillo-Garit, J.A., Marrero-Ponce, Y., Escobar, J., Torrens, F., and Rotondo, R. (2008) A novel approach to predict aquatic toxicity from molecular structure. *Chemosphere*, **73**, 415–427.
33 Rogers, D. and Hopfinger, A.J. (1994) Application of genetic function approximation to quantitative structure–activity relationships and quantitative structure–property relationships. *Journal of Chemical Information and Computer Sciences*, **34**, 854.
34 DRAGON, release 5.0 Evaluation Version http://www.disat.unimib.it/chm.
35 Matlab The MathWorks, Inc http://www.mathworks.com/.
36 Melanie, M. (1998) *An Introduction to Genetic Algorithms*, A Bradford Book The MIT Press, Cambridge, Massachusetts, London, UK, pp. 3–9, 130–131.
37 Brüggemann, R., Schwaiger, J., and Negele, R.D. (1995) Applying Hasse diagram technique for the evaluation of toxicological fish tests. *Chemosphere*, **30** (9), 1767–1780.
38 Halfon, E. and Reggiani, M.G. (1986) On ranking chemicals for environmental hazard. *Environmental Science & Technology*, **20** (11), 1173–1179.
39 Brüggemann, R., Bücherl, C., Pudenz, S., and Steinberg, C.E.W. (1999) Application of the concept of partial order on comparative evaluation of environmental chemicals. *Acta Hydrochimica et Hydrobiologica*, **27** (3), 170–178.
40 Wiegleb, G. (1997) Beziehungen zwischen naturschutzfachlichen Bewertungsverfahren und Leitbildentwicklung. *NNA-Berichte*, **3**, 40–47.
41 Hasse, H. (1952) *Über die Klassenzahl Abelscher Zahlkörper*, Akademie Verlag., Berlin.
42 Brüggemann, R. (1998) *Proceedings of the Workshop on Order Theoretical Tools in Environmental Sciences*, Berichte des IGB, Institut für Gewässerökologie and Binnenfischerei., Berlin, pp. 1–9.
43 Sørensen, P.B., Mogensen, B.B., Carlsen, L., and Thomsen, M. (2000) The influence on partial order ranking from input parameter uncertainty. Definition of a robustness parameter. *Chemosphere*, **41** (4), 595–601.
44 Carlsen, L. (2006) A combined QSAR and partial order ranking approach to risk assessment. *SAR and QSAR in Environmental Research*, **17** (2), 133–146.
45 Carlsen, L. (2005) A QSAr approach to physico-chemical data for organophosphates with special focus on known and potential nerve agents. *Internet Electronic Journal of Molecular Design*, **4**, 355–366.
46 Carlsen, L. (2005) Partial order ranking of organophosphates with special emphasis on nerve agents. *Match*, **54** (3), 519–534.
47 Carlsen, L. (2004) Giving molecules an identity. On the interplay between QSARs and partial order ranking. *Molecules*, **9** (12), 1010–1018.
48 Brüggemann, R., Halfon, E., and Bücherl, C. (1995) Theoretical base of the program "Hasse", GSF Forschungzentrum für Umwelt und Gesundheit, Neuherberg, Germany.
49 Fishburn, P.C. (1974) On the family of linear extensions of a partial order. *Journal of Combinatorial Theory Series B*, **17** (3), 240–243.
50 Graham, R.L. (1982) Linear extensions of partial orders and the FKG inequality, in *Ordered Sets* (ed. I. Rival), Reidel Publishing Company, Dordrecht, The Netherlands, pp. 213–236.
51 Winkler, P.M. (1982) Average height in a partially ordered set. *Discrete Mathematics*, **39**, 337–341.
52 Winkler, P.M. (1983) Correlation among partial orders. *SIAM Journal on Algorithm Discrete Mathematics.*, **4**, 1–7.
53 Sørensen, P.B., Lerche, D.B., Carlsen, L., and Brüggemann, R. (2001) *In Order Theoretical Tools in Environmental Science and Decision Systems* (ed. R. Brüggemann), Berichte des IGB, Berlin, pp. 87–97.
54 Carlsen, L., Lerche, D.B., and Sørensen, P.B. (2002) Improving the predicting

power of partial order based QSARs through linear extensions. *Journal of Chemical Information and Computer Sciences*, **42** (4), 806–811.

55 Randic, M. (1996) *Mathematical Methods in Contemporary Chemistry* (ed. I.S. Kuchanov), Gordon & Breach Science Publ., Amsterdam.

56 Randic, M. (1990) Design of molecules with desired properties. Molecular similarity approach to property optimization, in *Concepts and Applications of Molecular Similarity* (eds M.A. Johnson and G. Maggiora), John Wiley & Sons, Inc, New York, pp. 77–145.

57 Miller, A.J. (1990) *Subset Selection in Regression*, Chapman & Hall, London. *Subset Selection in Regression.*

58 Hocking, R.R. (1976) The analysis and selection of variables in linear regression. *Biometrics*, **32** (1), 1–49.

59 McFarland, J.W. and Gans, D.J. (1994) On identifying likely determinants of biological activity in high dimensional QSAR problems. *Quantitative Structure–Activity Relationships*, **13** (1), 11–17.

60 Livingstone, D.J. and Rahr, E. (1989) CORCHOP – An interactive routine for the dimension reduction of large QSAR data sets. *Quantitative Structure-Activity Relationships*, **8** (2), 103–108.

61 Héberger, K. and Rajkó, R. (2002) Generalization of pair correlation method (PCM) for non-parametric variable selection. *Journal of Chemometrics*, **16** (8–10), 436–443.

62 Héberger, K. and Rajkó, R. (2002) Variable selection using pair-correlation method. Environmental applications. *SAR and QSAR in Environmental Research*, **13** (5), 541–554.

63 Draper, N.R. and Smith, H. (1981) *Applied Regression Analysis*, Wiley, New York.

64 Sørensen, P.B., Brüggemann, R., Carlsen, L., Mogensen, B.B., Kreuger, J., and Pudenz, S. (2003) Analysis of monitoring data of pesticide residues in surface waters using partial order ranking theory. *Environmental Toxicology and Chemistry*, **22** (3), 661–670.

65 Brüggemann, R., Pudenz, S., Carlsen, L., Sørensen, P.B., Thomsen, M., and Mishra, R.K. (2001) The use of Hasse diagrams as a potential approach for inverse QSAR. *SAR and QSAR in Environmental Research*, **11** (5–6), 473–487.

66 Carlsen, L. (2009) The interplay between QSAR/QSPR studiesand partial order ranking and formal concept analyses. *International Journal of Molecular Sciences*, **10** (4), 1628–1657.

67 Livingstone, D.J. (1995) *Data analysis for chemists: applications to QSAR and chemical product design*, Oxford University Press, New York.

68 Hansch, C. (1990) *Comprehensive Drug Design*, Vol. 6, Pergamon Press., New York.

69 Randić, M., Vračko, M., Novič, M., and Basak, S.C. (2000) On ordering of folded structures. *Match*, **42**, 181–231.

70 Ehrlich, P. (1909) Über den jetzigen stand der chemotherapie. *Deutschen Chemischen Gesellschaft*, **42**, 17–47.

71 Randić, M., Jerman-Blažič, B., Rouvray, D.H., Seybold, P.G., and Grossman, S.C. (1987) The search for active substructure in structure–activity studies. *International Journal of Quantum Chemistry. Quantum Biology Symposium*, **14**, 245–260.

72 Randić, M., Grossman, S.C., Jerman-Blažič, B., Rouvray, D.H., and El-Basil, S. (1988) An approach to modeling the mutagenicity of nitroarenes. *Mathematical and Computer Modelling*, **11C**, 837–842.

73 Randić, M. and Jurs, P.C. (1989) On a fragment approach to structure–activity correlations. *Quantitative Structure–Activity Relationships*, **8** (1), 39–48.

74 Pavan, M., Mauri, A., and Todeschini, R. (2004) Total ranking models by the genetic algorithm variable subset selection (GA-VSS) approach for environmental priority settings. *Analytical and Bioanalytical Chemistry*, **380** (3 SPECISS), 430–444.

75 HYPERCHEM, 6.03 (Hypercube) http://www.hyper.com.

76 Leardi, R. (1994) Application of a genetic algorithm to feature selection under full validation conditions and to outlier detection. *Journal of Chemometrics*, **8**, 65–79.

77 Leardi, R. (1996) Genetic algorithms in feature selection, in Genetic algorithms in molecular modeling, (ed. J. Devillers), Academic Press, London, pp. 67–86.

78 Kendall, M.G. (1948) *Rank Correlation Methods*, **195**, 202–204.

**79** Castro, E.A., Fernández, F.M., and Duchowicz, P.R. (2005) QSPR modeling of the enthalpy of formation based on partial order ranking. *Journal of Mathematical Chemistry*, **37** (4), 433–441.

**80** Voigt, K., Brueggemann, R., Scherb, H., Shen, H., and Schramm, K.-W. (2010) Evaluating the relationship between chemical exposure and cryptorchidism. *Environmental Modelling & Software*, **25** (12), 1801–1812.

**81** Carlsen, L., Kenessov, B.N., and Batyrbekova, S.Y. (2009) A QSAR/QSTR study on the human health impact of the rocket fuel 1,1-dimethyl hydrazine and its transformation products: Multicriteria hazard ranking based on partial order methodologies. *Environmental Toxicology and Pharmacology*, **27** (3), 415–423.

# 6
# Graph-Theoretical Descriptors for Branched Polymers

*Koh-hei Nitta*

## 6.1
## Introduction

A polymer is a chain-like molecule that comprises a huge number of repeating structural units or atoms connected by chemical bonds. Branching formation of such chain molecules has been known from the beginning of polymer chemistry, and modern polymerization techniques make it possible to produce many types of chain branching. These branched polymers possess specific properties that are of practical importance to processability and rheological features [1–6], which cannot be attained by linear polymers.

The specification of a linear polymer requires only one parameter, namely the degree of polymerization or molecular weight; however, additional parameters are necessary to specify various branched polymers. One such quantitative factor is the average size of isolated single polymers determined by measuring the properties of the polymer in a dilute solution. The smaller spatial expansion of isolated branched polymers compared with isolated linear polymers of the same molecular weight is the basis of most fundamental methods for estimating branching. Zimm and Stockmayer [7] have succeeded in calculating the radius of gyration of chain molecules with various types of branching such as star-shaped, comb-shaped, and randomly branched polymers.

For the purpose of estimating branching, therefore, considerable effort has been devoted for obtaining chain dimensions expressed by, for example, the mean square radius of gyration $\langle s^2 \rangle$, estimated by the light scattering (LS) measurements, and/or the mean hydrodynamic (Stokes) radius $r_H$, estimated from the intrinsic viscosity $[\eta]$ data. In addition, gel permeation chromatograph [8–10] is one of the most popular methods used for the fractionation of a polymer according to the volume dimension $[\eta]M$, where $M$ is the molecular weight. Many efforts to relate $[\eta]$ to chain dimensions, $\langle s^2 \rangle$ or $r_H$, have been made using a combination of gel permeation chromatography and LS or $[\eta]$ techniques to obtain experimentally the relationship between the chain dimensions and the molecular weight. However, these approaches cannot satisfactorily provide the structural details of branched polymers such as the number

*Statistical Modelling of Molecular Descriptors in QSAR/QSPR*. First Edition.
Edited by M. Dehmer, K. Varmuza, and D. Bonchev

of branches, branch length, and their position on the backbone. It is necessary to develop graph-theoretical methods for analyzing the branching architecture of polymers and to find topological descriptors and quantitative measures that reflect molecular branching.

Alkane molecules are the most basic organic compounds, comprising a set of carbon and hydrogen atoms connected by covalent bonds. The topological structure of their isomers can be expressed by a tree-like drawing in which the carbon atoms are depicted as points, their hydrogen atoms are omitted, and the bonds linking the carbons are depicted as straight lines. Then, there exists a one-to-one correspondence between these isomers and the tree-like drawings. This type of drawings whose points have maximum functionality of four is called the "chemical graph" [11]. If the chemical graph is allowed to have a branch point with functionality greater than four, the extended chemical graph may be referred to as the "molecular graph". Enumeration of chemical isomers, in particular the constructional isomers of alkanes, has been the subject matter of a mathematical discipline known as graph theory since the pioneering work of Cayley [12] and Sylvester [13].

Ideal polymer molecules have been modeled as points jointed by bonds of fixed length from the beginning of polymer science. This is called the random flight or freely jointed chain [14] that corresponds to the molecular graph in graph theory. Random flight statistics can characterize the statics of linear as well as branched polymers.

An important achievement of polymer dynamics is representation of a flexible polymer (in the ideal, isolated state) as a sequence of Hookean springs connecting point-like beads, the whole being termed a Gaussian chain. Rouse [15], Bueche [16], and Zimm [17] demonstrated that the dynamics of a dilute solution of linear polymers can be analyzed by considering a Gaussian chain suspended in a flowing viscous liquid. This model was subsequently applied to any branched molecule by Ham [18] and Zimm and Kilb [19]. Interestingly, the Gaussian chain description can be transformed into the random-flight statistics by means of the graph theoretical treatments.

Application of graph theory concepts to polymer statics and dynamics is most attractive for providing topological and physical insights into the branching in polymers [20–24]. This chapter shows that the problems of the dynamics and statistics of any branched homopolymer composed of the same repeat units without loops or rings are completely reduced to the problem of the eigen-polynomial of the algebraic graph. The associated significance of this approach lies in the identification of physical meaning of the graph-theoretical descriptors for quantitative structure–property and structure–activity relationships in branched tree-like polymers.

## 6.2
## Algebraic Graph Theory

In chemical or molecular graphs, points and beads are generally referred to as vertices, bonds are referred to as edges, and the functionality of the atoms is called the

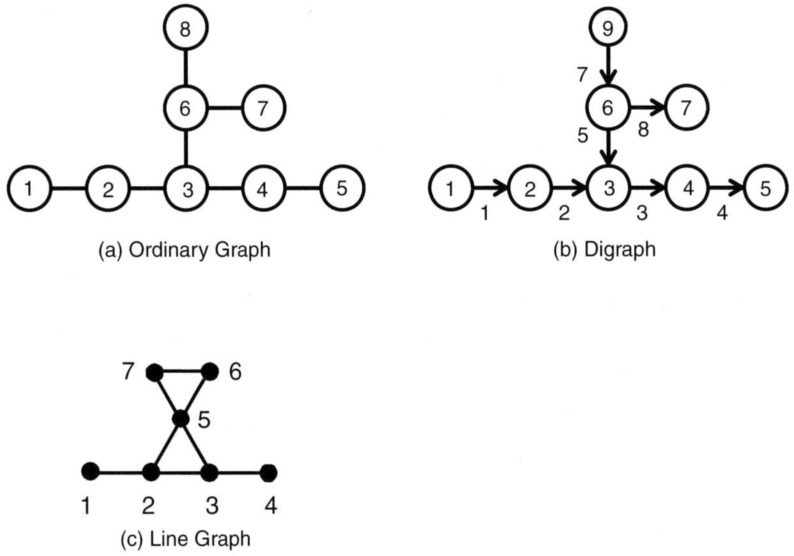


**Figure 6.1** Represetation of a branched molecule: (a) the ordinary graph, (b) the digraph, and (c) the line graph.

vertex degree. Chain molecules can be modeled as beads jointed by bonds of fixed length and are represented as a tree graph $G$ in which each vertex and each edge are arbitrarily numbered, and a digraph $D$ in which each edge of graph $G$ is arbitrarily directed, as exemplified in Figure 6.1. In algebraic graph theory, several matrices reflecting the connectivity in labeled graph $G$ [25] are introduced for determining the topological feature of graphs, and the algebraic properties of the characteristic polynomials have been extensively examined. For graph $G$, the adjacency matrix $\mathbf{A} = (a_{ij})$ is the most fundamental matrix for the graph representation [25], and is defined as a square matrix with the following elements:

$$a_{ij} = \begin{cases} 1 & \text{if the vertices } i \text{ and } j \text{ are adjacent} \\ 0 & \text{otherwise} \end{cases} \tag{6.1}$$

The order of $\mathbf{A}$ is identical with the total number of vertices in $G$. The adjacency matrix $\mathbf{A}$ of $G$ is useful for characterizing and encoding the skeletal structure of the corresponding molecules.

The characteristic polynomial of a molecular graph represents an important, though not unique, molecular invariant defined as

$$\Phi(\mathbf{A};\lambda) = \mathrm{Det}\,|\mathbf{A}-\lambda\mathbf{I}| \tag{6.2}$$

where $\mathbf{I}$ is a unit matrix of the same order as $\mathbf{A}$. The list of eigenvalues of a matrix calculated from the characteristic polynomial is called the spectrum of the matrix, and includes much quantitative information on the topological nature of the molecular graph.

Another important invariant of a molecular graph is its distance matrix **D** [26] whose element $d_{ij}$ is the number of edges for the shortest path between $i$th and $j$th vertices. The entries in the distance matrix **D** are related to the entries in powers of the adjacency matrix **A**.

The connectivity between edge and vertex is represented by the incidence matrix [26] that determines the homology of the graph. The incidence matrix $\mathbf{B} = (b_{ij})$ of a digraph $D$ is defined by

$$b_{ij} = \begin{cases} 1 & \text{if the edge } j \text{ starts from the vertex } i \\ -1 & \text{if the edge } j \text{ terminates in the vertex } i \\ 0 & \text{otherwise} \end{cases} \tag{6.3}$$

The incidence matrix **B** is constructed for any graph having $n$ vertices and $m$ edges by setting up an $n \times m$ matrix array; the rows and columns of the matrix then correspond to the vertices and the edges, respectively, of the digraph. In addition, the signless incidence matrix $\mathbf{C} = (c_{ij})$ of an ordinary graph $G$ can be defined and its elements are the absolute values of those of **B**. Thus, the numeral 1 is inserted in the $(i, j)$th position of the array if the $j$th edge is coincident with the $i$th vertex and all other entries in the array are zeros.

The line graph $L(G)$ of $G$ is formed by replacing the edges of $G$ by vertices such that the vertices in $L(G)$ are connected whenever the corresponding edges in $G$ are adjacent. Kirchhoff [27] introduced a connectivity matrix defined by $\mathbf{K} = \mathbf{C}^{\mathrm{T}}\mathbf{C}$ while solving problems involving the calculation of currents in electrical networks. He showed that the matrix **K** can be represented by the adjacency matrix $\mathbf{A}_L$ of $L(G)$ for tree graphs $G$:

$$\mathbf{K} = \mathbf{A}_L + 2\mathbf{I} \tag{6.4}$$

An example of a line graph is shown in Figure 6.1c. Combination of (6.4) and (6.2) gives

$$\Phi(\mathbf{K}; \lambda) = \Phi(\mathbf{A}_L; \lambda - 2) \tag{6.5}$$

Consequently, the eigenvalues of **K** of a tree graph can be calculated from the eigenvalues of $\mathbf{A}_L$ of its line graph.

The topological analysis of graphs has been widely performed using Laplacian matrices related to both the adjacency and incidence matrices, and they are defined by

$$\mathbf{L} = \mathbf{V} - \mathbf{A} \quad \text{and} \quad \mathbf{L}^{+} = \mathbf{V} + \mathbf{A} \tag{6.6}$$

where **V** is a diagonal matrix whose entries are the vertex degrees, and $\mathbf{L}^{+}$ is called the signless Laplacian matrix [28, 29], the entries of which are the absolute values of the entries of **L**. Both Laplacian matrices can be represented using the incidence matrices [30, 31]

$$\mathbf{L} = \mathbf{B}\mathbf{B}^{\mathrm{T}} \quad \text{and} \quad \mathbf{L}^{+} = \mathbf{C}\mathbf{C}^{\mathrm{T}} \tag{6.7}$$

where the superscript T indicates the transpose of a matrix including that of a vector.

The characteristic polynomials of both matrices are the same:

$$\Phi(\mathbf{L};\lambda) = \Phi(\mathbf{L}^+;\lambda) \tag{6.8}$$

indicating that the eigenvalues of the signless matrix $\mathbf{L}^+$ are the same with those of the Laplacian matrix $\mathbf{L}$. Furthermore, comparing the characteristic polynomials of $\mathbf{L}^+ = \mathbf{C}\mathbf{C}^\mathrm{T}$ and $\mathbf{K} = \mathbf{C}^\mathrm{T}\mathbf{C}$, we can find the relation

$$\Phi(\mathbf{L}^+;\lambda) = \lambda\Phi(\mathbf{K};\lambda) \tag{6.9}$$

The eigenvalues of $\mathbf{L}^+$ contain one zero eigenvalue, and the nonzero eigenvalues of $\mathbf{L}^+$ are identical with those of $\mathbf{K}$. It should be noted here that the elements of these square matrices are dependent on the assignment and sense of the edges and vertices, but these eigenvalues are independent of those.

One of the earliest graph invariants, or topological descriptors, is known to mathematicians as the vertex number and to polymer scientists as the total number of monomeric units (polymerization degree or molecular weight). The polymerization degree is an appropriate index only for linear polymer molecules, however, and is not well suited to branched polymers, which may have a number of different skeletal structures even if they have the same polymerization degree. Therefore, the polymerization degree may be considered to be a descriptor with low discriminating power. Hence, it is necessary to develop other descriptors that can effectively distinguish nonlinear polymers with various types of branched structures.

The first topological descriptor capable of characterizing the branchedness of alkanes was proposed by Wiener [32, 33] to predict the boiling points of isomeric alkanes. The Wiener index $W$ is defined as the sum of the distances between any two carbon atoms in an alkane molecule:

$$W = \sum_{i=1}^{N-1} ik_i \tag{6.10}$$

where $N$ is the total number of carbon and $k_i$ stands for the total number of pairs of carbons whose separation is $i$. Platt attempted to interpret this index as $W^{1/3}$ corresponding to mean molecular diameter [34]. The validity of this interpretation is clarified in Section 6.3. Bonchev *et al.* [35] have succeeded in formulating structural features such as the branching point and the branch length from the Wiener index of any branched graph.

The Wiener index of an alkane molecule, like the carbon number, becomes generally larger for molecules with higher molecular weights; however, $W$ also provides a measure of the branched structure of a molecule and has a possibility to be a useful device for ordering the sequence of structural isomers. Several researchers have found that the Wiener index correlates surprisingly well with properties such as heat capacity, viscosity, surface tension, refractive index, and electron energy for certain types of hydrocarbon molecules as well as conjugated polymers [36–41].

Hosoya [42] found that $W$ turns out to be the half sum of the off-diagonal elements of a distance matrix **D**. This not only offers an alternative method to determine $W$ but also allows extension of $W$ to cyclic structures. Thus, $W$ can be rewritten as

$$W = \frac{1}{2}\sum_{i=1}^{N}\sum_{j=1}^{N} d_{ij} \tag{6.11}$$

The Laplacian matrix **L** is a real symmetric matrix. The diagonalization of **L** for a graph $G$ with $N$ vertices produces $N$ real eigenvalues, $\lambda_1 \geq \lambda_2 \geq \cdots > \lambda_N = 0$, where the smallest eigenvalue is always zero. Let a graph $G$ be a tree; then, the Wiener index of the tree can be calculated from the inverse of its Laplacian nonzero eigenvalues as follows [43]:

$$W = N\sum_{i=1}^{N-1} \frac{1}{\lambda_i} \tag{6.12}$$

In a number of papers, the problem of molecular branching has been related to the properties of the characteristic polynomial. Hosoya *et al.* [44] introduced a topological index which is equal to the sum of the absolute values of the polynomial coefficients of the adjacency matrix in the case of acyclic molecules, and which was used to describe the boiling points of various molecules. Lovász and Pelikán [45] have found that the maximal eigenvalue of a tree graph is a fairly reliable measure of branching.

As well known, the mathematical characterization of molecules has led to a large number of molecular descriptors, and their number continues to grow. These descriptors are expected to play an important role in structure–property and structure–activity studies. The important task for the purpose is to find the physical meanings of these topological descriptors.

## 6.3 Ideal Chain Models

In this chapter, we deal only with the ideal state of polymer chains in the unperturbed state where only short-range interactions are considered. Flory proposed that polymer molecules in the molten state are unperturbed [46]. Small-angle neutron scattering data [47, 48] support the Flory theorem. An elegant explanation was given by de Gennes [49]. Thus, the concept of the ideal chain model plays a central role in analyzing the fundamental topological features of branched polymers.

The chemical bonds of a flexible polymer in the unperturbed state are fairly rigid with respect to stretching and to bending of the valence angles between adjacent bonds. However, a single polymer molecule has many internal degrees of rotational freedom about each chemical bond, resulting in the ability to adopt many different conformations, thereby necessitating the use of statistical mechanics. Each conformation exists for only a very short time and the conformations are temporal

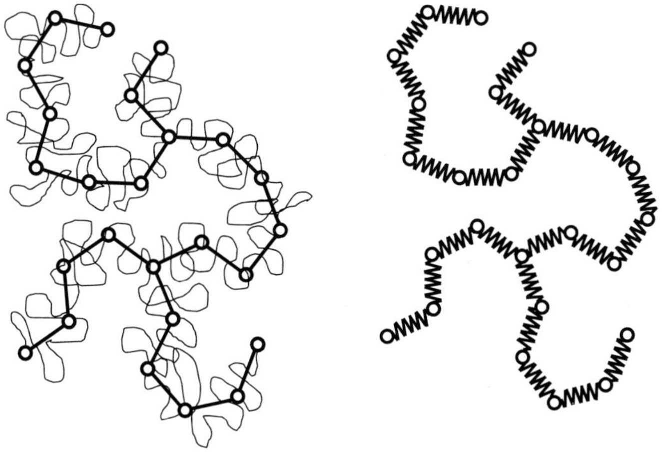


**Figure 6.2** Representation of a branched polymer: (a) a real chain and its equivalent chain and (b) its Gaussian chain model.

averages over all molecules, resulting in the polymer chain forming a random coil. Consequently, any actual flexible polymer can be treated as a freely jointed chain consisting of effective bonds with fixed length where several actual bonds are considered as one fictive bond with a fixed length. The details of the microconformations of real chains can be smeared out and such a coarse-grained polymer chain is often referred to as the "equivalent chain" (Figure 6.2).

The random-flight or freely jointed chain is regarded as a molecular graph composed of $N-1$ statistical edges of a length $b$ joining $N$ vertices of a unit mass. The quantity $b$ is referred to as the effective bond length, and it depends on the conformational characteristics. The square of the radius of gyration is the means of all bead position from the center of mass vector. The mean square radius of gyration for the unperturbed chains can be derived from the Lagrange theorem [50–52]:

$$\langle s^2 \rangle_0 = \frac{1}{2N^2} \sum_{i=1}^{N} \sum_{j=1}^{N} \langle \mathbf{r}_{ij}^2 \rangle \tag{6.13}$$

The hydrodynamic radius $r_{\mathrm{H}}$ of the random coil is obtained from the mean reciprocal distance:

$$\langle r_{\mathrm{H}}^{-1} \rangle_0 = \frac{1}{2N^2} \sum_{i=1}^{N} \sum_{j=1}^{N} \langle |\mathbf{r}_{ij}|^{-1} \rangle \tag{6.14}$$

where $\mathbf{r}_{ij}$ is the vector from beads $i$ to $j$.

According to the central limit theorem, the random-flight statistics of a flexible polymer can be described by a Gaussian statistics [52]. The equilibrium distribution function for the freely jointed chains is considered to be identical to the random walk based on Markov process [53]. This is because the random walk distribution is inconsistent with the equilibrium statistical distribution but the difference bacomes inconsequential for large $N$.

It is well known that a polymer chain can take many different conformations at a temperature above the molecular glass temperature [49]. This corresponds to the micro-Brownian motion and chains are instantly converted into other conformations. The instantaneous shape obtained by time-averaging over many conformations can be described by a Gaussian distribution for large $N$. The distribution function $W(\mathbf{b}_i)$ of the bond vector $\mathbf{b}_i$ between adjacent units $i$ and $i+1$ can be defined either as the time-averaged incidence of $\mathbf{b}_i$ within the specified range for a given molecule or as the average incidence for an ensemble of many identical units subject to identical conditions:

$$W(\mathbf{b}_i) = \left(\frac{3}{2\pi\langle \mathbf{b}^2\rangle}\right)^{3/2} \exp\left(-\frac{3\mathbf{b}_i^{\mathrm{T}}\cdot \mathbf{b}_i}{2\langle \mathbf{b}^2\rangle}\right) \tag{6.15}$$

where $\langle \mathbf{b}^2\rangle$ is the time-averaged bond length. The distribution of the effective bond vector $\mathbf{b}_i$ is given by the Gaussian distribution function (6.15), and hence the probability distribution of such a chain is given by $\prod_{i=1}^{N-1} W(\mathbf{b}_i)$. Then, the equilibrium state of this chain can be represented by a distribution function $\exp(-V/k_{\mathrm{B}}T)$, where $V$ is the potential energy, $k_{\mathrm{B}}$ is the Boltzmann constant, and $T$ is the absolute temperature. Therefore, if we choose

$$V = \frac{1}{2}\kappa \sum_{i=1}^{N-1} \mathbf{b}_i^{\mathrm{T}}\cdot \mathbf{b}_i \tag{6.16}$$

where $\kappa = 3k_{\mathrm{B}}T/\langle \mathbf{b}^2\rangle$, the equilibrium distribution of the chain can be modeled as a chain of beads connected by Hookean springs with spring constant $\kappa$ as shown in Figure 6.2b. If the value of $\langle \mathbf{b}^2\rangle^{1/2}$ is identical with the effective bond length $b$, that is, $\langle \mathbf{b}^2\rangle^{1/2} = b$, the random-flight model of a flexible polymer can be treated as a Gaussian chain model (see Figure 6.2). Then, the quantity $b$ is called segment length that is determined by the conformational characteristics, depending on the polymer species.

## 6.4 Graph-Theoretical Approach to Chain Dynamics and Statistics

### 6.4.1 Radius of Gyration

One of the quantitative measures that reflect molecular branching is the average size of isolated single-branched polymers. This is because the expansion in the space of isolated branched polymers with a fixed molecular weight is smaller as increasing branching.

When the branched chains obey random-flight statistics, the average of the scalar product extending over all sets of bond vectors $\mathbf{b}_i$ is given by $\langle \mathbf{b}_i^T \cdot \mathbf{b}_j \rangle = b^2 \delta_{ij}$, where $\delta_{ij}$ is the Kronecker delta function. Considering that the number of bonds between $i$th and $j$th bead is identical to the graph-theoretical distance $d_{ij}$, the mean square of $\mathbf{r}_{ij}$ becomes

$$\langle \mathbf{r}_{ij}^2 \rangle = \sum_{i,j} b^2 = d_{ij} b^2 \tag{6.17}$$

Combination of (6.11) and (6.13) with (6.17) yields

$$\langle s^2 \rangle_0 = \left( \frac{b}{N} \right)^2 W \tag{6.18}$$

This is a graph-theoretical expression for the mean square radius of gyration for any unperturbed random-flight chain [24, 54]. The mean radius of chain molecules becomes proportional to $W^{1/2} N^{-1}$.

As a measure of branching, Zimm–Stockmayer proposed a parameter g [7] that is defined by the ratio of $\langle s^2 \rangle_0$ for a given branched chain to that for the linear chain with the same molecular weight. Consequently, the g-factor is equal to the Wiener index normalized by the Wiener index of the linear chain, as proposed by Bonchev *et al.* [55, 56]:

$$g = \frac{\langle s^2 \rangle_0}{\langle s^2 \rangle_{0,\mathrm{L}}} = \frac{W}{W_\mathrm{L}} \tag{6.19}$$

where subscript L indicates the linear chain. The Wiener index of linear chains $W_\mathrm{L}$ can be easily obtained by substituting the relation $k_i = N - i$ into (6.10):

$$W_\mathrm{L} = \frac{1}{6} N(N^2 - 1) \tag{6.20}$$

Substituting (6.20) into (6.18) yields the equation for $\langle s^2 \rangle_0$ of linear chains, which was in accordance with the first deviation by Kramers [14, 57, 58]. Substituting (6.20) into (6.19) shows that the g-factor of any branched polymer can be represented by the Wiener index $W$ and the number of beads $N$.

$$g = 6 \frac{W}{N(N^2 - 1)} \cong 6 N^{-3} W \tag{6.21}$$

The Wiener indices for various types of branched molecular graphs are presented by Bonchev and Trinajstić [35].

Since real polymers have distribution in chain lengths (or molecular weight) and in number and position of branches, they can be considered to be mixture of such isomers and homologs. For simplicity, we consider a randomly branched polymer in which all possible arrangements of subchains of varying lengths are assumed to occur with equal frequency [7]. The term subchain refers to a portion of the molecule between two adjacent branched beads or between adjacent end and branched beads. An example of a randomly branched chain is shown in Figure 6.3.

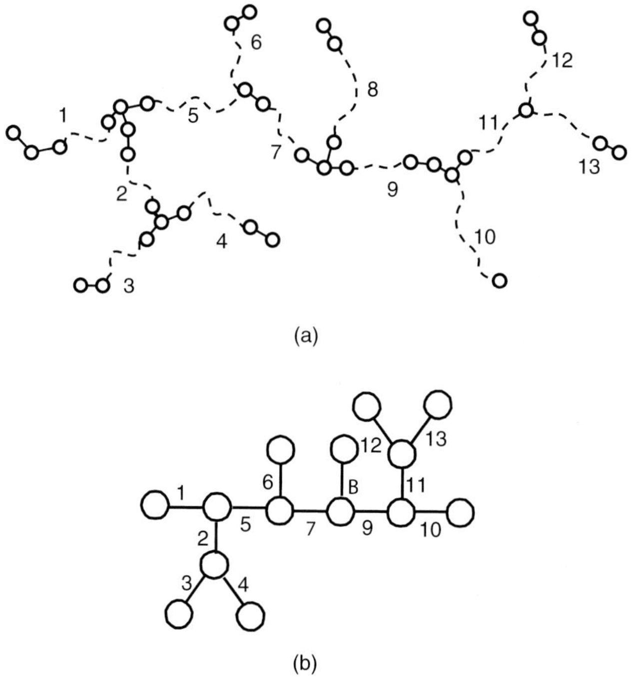


**Figure 6.3** (a) The molecular graph of a randomly branched polymer chain and (b) its proper graph where branched and end beads are transformed to uniform vetices and subchains are transformed to uniform edges.

Let $n_k$ be the number of the beads in the $k$th subchian, then the total number of beads is given by $N = n_1 + n_2 + \cdots + n_p$, where $p$ is the total number of subchains per molecule. Here we consider the g-factor of randomly branched chains having the random distribution of $n_k$ which is defined such that the number of beads in each subchain varies with an equal frequency of occurrence of a set of $(n_1, n_2, \ldots, n_p)$ under the restriction that $N$ and $p$ are fixed. Then the g-factor given as a function of $(n_1, n_2, \ldots, n_p)$ can be averaged by integrating the g value over all arrangements of $n_k$:

$$\bar{g} = Av[g] = \frac{\int_0^N dn_{p-1} \int_0^{N-n_{p-1}} dn_{p-2} \cdots \int_0^{N-\sum_{k=1}^{p-1} n_k} dn_1\, g(n_1, n_2, \ldots, n_p)}{\int_0^N dn_{p-1} \int_0^{N-n_{p-1}} dn_{p-2} \cdots \int_0^{N-\sum_{k=1}^{p-1} n_k} dn_1} \tag{6.22}$$

The integrations in (6.22) were performed by Saito and Kataoka [59], giving

$$\bar{g} = \frac{6}{p(p+1)(p+2)} \left( p^2 + \sum_{(\alpha,\beta)} \nu_{\alpha\beta} \right) \tag{6.23}$$

where $\nu_{\alpha\beta}$ represents the number of subchains lying between the $\alpha$th and $\beta$th subchains, and $\sum_{(\alpha,\beta)}$ is the sum extending over all distinguishable pairs of subchains.

Nitta [24] showed that the g-factor of any randomly branched chain can be estimated from the Wiener index $\omega$ of its reduced graph where all subchains are transferred to single edges, as shown in Figure 6.3b. Thus,

$$\bar{g} = \mathrm{Av}[g] = \mathrm{Av}\left[\frac{W}{W_\mathrm{L}}\right] = \frac{\omega}{\omega_\mathrm{L}} \tag{6.24}$$

where $\omega_\mathrm{L}$ is the Wiener index of a linear graph with the same number of edge that can be readily evaluated from (6.20) by putting $N = p+1$. Thus,

$$\omega_\mathrm{L} = \frac{p(p+1)(p+2)}{6} \tag{6.25}$$

The reduced graph corresponds to the proper graph, which contains only two types of vertices-terminal and branched ones, introduced by Bonchev *et al.* [56]. Moreover, since $W_\mathrm{L}$ may be considered to be constant for the averaging process in (6.22), the average Wiener index can be obtained as follows:

$$\bar{W} = \mathrm{Av}[W] = \mathrm{Av}[g]\,W_\mathrm{L} = \frac{N(N+1)(N+2)}{p(p+1)(p+2)}\omega \tag{6.26}$$

Consequently, the average Wiener index of any randomly branched chain can be determined by the total number of subchain $p$, the total number of beads $N$, and the Wiener index $\omega$ of the corresponding proper graph. If the system consists of polymer homologs differing in molecular weight, the mean Wiener index is given using the mean-cube molecular weight $\langle N^3 \rangle$ for large $N$.

$$\langle \bar{W} \rangle = \langle N^3 \rangle \mathrm{Av}[g] \tag{6.27}$$

It should be noted here that the mean Wiener index of actual polymers with distribution of molecular weight and position of branches is expressed by the product of the term of molecular weight distribution and the term of the g factor of the proper graph. The mean Wiener index may be useful for analyzing the statistics as well as dynamics of actual branched polymers.

### 6.4.2
### Rouse Dynamics

For the purpose of considering the polymer dynamics, we consider a Gaussian chain composed of $N$ beads connected by $N-1$ Hookean springs with spring constant of $\kappa = 3k_\mathrm{B}T/b^2$. The laws that govern the behavior of a linear flexible chain are assumed to hold for branched flexible polymers [18]. For a Gaussian chain suspended in a flowing liquid, the equation of motion is given by [60]

$$-\varsigma_0 \dot{\mathbf{r}} = \kappa \mathbf{Z} \cdot \mathbf{r} \tag{6.28}$$

where $\mathbf{r}$ is a $3 \times N$ matrix whose rows contain the dimensional component of the $N$ position vectors of beads, $\dot{\mathbf{r}}$ is the time derivative of $\mathbf{r}$, $\varsigma_0$ is the friction constant of each bead, and $\mathbf{Z}$ is the $N \times N$ connectivity matrix called the Zimm matrix. According to the Rouse theory [15], (6.28) can be rewritten in terms of bond vectors $\mathbf{b}(= \mathbf{r} \cdot \mathbf{B})$ as

$$-\varsigma_0 \dot{\mathbf{b}} = \kappa \mathbf{R} \cdot \mathbf{b} \tag{6.29}$$

where $\mathbf{b}$ is a $3 \times (N-1)$ matrix whose rows contain the dimensional component of the $N-1$ bond vectors. The matrix $\mathbf{R}$ is the $(N-1) \times (N-1)$ connectivity matrix called the Rouse matrix [15]. Comparing the characteristic polynomials of $\mathbf{Z}$ and $\mathbf{R}$, we can find

$$\Phi(\mathbf{Z}; \lambda) = \lambda \Phi(\mathbf{R}; \lambda) \tag{6.30}$$

The eigenvalues of $\mathbf{Z}$ contain one zero eigenvalue and hence, $\mathbf{Z}$ does not possess an ordinary inverse. The zero eigenvalue of $\mathbf{Z}$ represents the mode of chain translation [21]. Equation (6.30) shows that the nonzero eigenvalues of $\mathbf{Z}$ are identical with those of $\mathbf{R}$. It should be noted here that the springs and beads are assigned in any arbitrary fashion but the eigenvalues of $\mathbf{Z}$ and $\mathbf{R}$ are independent of how their elements are numbered [21, 61, 62]. Relation (6.30) in polymer physics corresponds to the relation (6.9) in graph theory.

Each eigenvalue $\lambda_i (i = 1, 2, \ldots, N-1)$ of $\mathbf{R}$ or nonzero eigenvalue of $\mathbf{Z}$ is associated with the relaxation time $\tau_i$ of the $i$th mode for dynamic molecular motions [22, 61]:

$$\tau_i = \frac{\varsigma_0 b^2}{6 k_{\mathrm{B}} T} \lambda_i^{-1} \tag{6.31}$$

According to the theory of linear viscoelasticity [63], the relaxation time spectrum $H(\tau)$ is defined by

$$H(\tau) = \frac{c k_{\mathrm{B}} T}{N} \sum_{i=1}^{N-1} \delta(\ln \tau - \ln \tau_i) \tag{6.32}$$

where $c$ is the concentration of beads per unit volume and $\delta(x)$ is a Dirac delta function. Consequently, various rheological functions describing flow and deformation of polymers can be completely determined through the spectrum of the matrix $\mathbf{R}$ or $\mathbf{Z}$.

The Zimm matrix $\mathbf{Z}$ or the Rouse matrix $\mathbf{R}$ for branched tree-like chains can also be constructed in a different manner by making use of the incidence matrix $\mathbf{B}$ of a digraph $D$. Forsman [61] showed that $\mathbf{Z}$ and $\mathbf{R}$ are given by

$$\mathbf{Z} = \mathbf{B}\mathbf{B}^{\mathrm{T}} \quad \text{and} \quad \mathbf{R} = \mathbf{B}^{\mathrm{T}}\mathbf{B} \tag{6.33}$$

Interestingly, the Zimm matrix $\mathbf{Z}$ in the polymer physics is identical to the Laplacian matrix $\mathbf{L}(= \mathbf{V} - \mathbf{A})$ in graph theory.

$$\mathbf{Z} = \mathbf{L} \tag{6.34}$$

Furthermore, it is apparent that all elements of a connectivity matrix $\mathbf{K}(=\mathbf{C}^{T}\mathbf{C})$ are the absolute values of those of $\mathbf{R}$. Combination of the relations (6.8,6.9,6.30) and (6.34) gives

$$\Phi(\mathbf{R};\lambda)=\Phi(\mathbf{K};\lambda) \tag{6.35}$$

From Kirchhoff's relation (6.4), we have

$$\Phi(\mathbf{R};\lambda)=\Phi(\mathbf{A}_{\mathrm{L}};\lambda-2) \tag{6.36}$$

Consequently, the eigenvalues $\lambda_i$ of $\mathbf{R}$ can be calculated from the eigenvalues $\mu_i$ of $\mathbf{A}_L$ as follows:

$$\lambda_i=\mu_i+2 \tag{6.37}$$

Remembering (6.31), we have

$$\tau_i=\frac{\varsigma_0 b^2}{6k_{\mathrm{B}}T}(\mu_i+2)^{-1} \tag{6.38}$$

It is interesting to note here that the relaxation spectrum of a chain molecule can be determined entirely by the set of eigenvalues of the adjacency matrix $\mathbf{A}_{\mathrm{L}}$ of its line graph.

Forsman [61] showed that the square of the radius of gyration can be related to the reciprocal of Rouse matrix $\mathbf{R}$ and is given by $s^2=N^{-1}\mathbf{b}^{\mathrm{T}}\cdot\mathbf{R}^{-1}\cdot\mathbf{b}$. When the chains obey random-flight statistics, we have

$$\langle s^2\rangle_0=\frac{b^2}{N}\mathrm{Tr}\left[\mathbf{R}^{-1}\right]=\frac{b^2}{N}\sum_{i=1}^{N-1}\frac{1}{\lambda_i}=\frac{b^2}{N}\sum_{i=1}^{N-1}\frac{1}{\mu_i+2} \tag{6.39}$$

where Tr denotes the trace of a matrix. The mean radius of gyration is related to the sum of the reciprocal eigenvalues of the Rouse matrix $\mathbf{R}$, so that it largely depends on the minimum eigenvalue of $\mathbf{R}$ or the second smallest eigenvalue of $\mathbf{L}$. This corresponds to Mohar's suggestion that the second smallest eigenvalue of $\mathbf{L}$ is related to the diameter and mean distance in a graph [62].

Comparison of (6.39) and (6.18) gives

$$W=N\mathrm{Tr}\left[\mathbf{R}^{-1}\right]=N\sum_{i=1}^{N-1}\frac{1}{\lambda_i}=N\sum_{i=1}^{N-1}\frac{1}{\mu_i+2} \tag{6.40}$$

Moreover, a combination of (6.31) and (6.40) gives a relation between the Wiener index and relaxation times:

$$W=\frac{6Nk_{\mathrm{B}}T}{\varsigma_0 b^2}\sum_{i=1}^{N-1}\tau_i \tag{6.41}$$

This may be a physical meaning of the Wiener index.

### 6.4.3
### Intrinsic Viscosity

The steady-state viscosity in a dilute solution can be obtained from the relaxation time spectrum [63]:

$$\eta = \eta_s + c\frac{N_A}{M}k_B T\sum_{i=1}^{N-1}\tau_i \tag{6.42}$$

where $N_A$ is Avogadro's number. Consequently, the intrinsic viscosity in the freely draining condition can be written as

$$[\eta]_{FD} = \frac{N_A}{M\eta_s}k_B T\sum_{i=1}^{N-1}\tau_i \tag{6.43}$$

where $\eta_s$ is the viscosity of pure solvent. Since the total sum of the relaxation times is proportional to the mean square of radius of gyration, we have

$$[\eta]_{FD} = \frac{N_A\varsigma_0}{6M_0\eta_s}\langle s^2\rangle = \frac{N_A\varsigma_0 b^2}{6M_0\eta_s N^2}W \tag{6.44}$$

where $M_0(= M/N)$ is the molar mass of segment. Equation (6.44) shows that the freely draining intrinsic viscosity $[\eta]_{FD}$ is proportional to $WN^{-2}$ for branched chains and to the first power of $N$ for linear chains. This equation may be referred to as Staudinger's relation. This suggests that the Wiener index of tree-like polymer molecules can be experimentally evaluated from the steady-flow viscosity of dilute solusition in the freely draining condition. Sheridan *et al.* [64] computationally examined the relation between the intrinsic viscosity and the Wiener index for hyperbranched polymers. Their computer simulations gave the following relation $[\eta] \sim W^a N^b$, where $a = 1.0$ and $b = -2.2$ which is similar to (6.44) in the freely draining condition.

The nondraining intrinsic viscosity $[\eta]_{ND}$ of suspended spheres with hydrodynamic radius $r_H$ was given by Einstein as

$$[\eta]_{ND} = \frac{5}{2}\left(\frac{N_A}{M}\right)\frac{4}{3}\pi r_H^3 \tag{6.45}$$

According to (6.14), the mean reciprocal hydrodynamic radius $r_H^{-1}$ can be calculated from the mean reciprocal of $\mathbf{r}_{ij}$ which is given by [52, 65]

$$\langle|\mathbf{r}_{ij}|^{-1}\rangle = \left(\frac{6}{\pi}\right)^{1/2} d_{ij}^{-1/2} b^{-1} \tag{6.46}$$

Thus, the hydrodynamic radius can be obtained using the modified Harary index [25] as follows:

$$\langle\mathbf{r}_H^{-1}\rangle = \frac{1}{bN^2}\left(\frac{6}{\pi}\right)^{1/2} H_{1/2} \tag{6.47}$$

where the modified Harary index $H_{1/2}$ is defined by [66]

$$H_{1/2} = \frac{1}{2}\sum_{i=1}^{N}\sum_{j=1}^{N} d_{ij}^{-0.5} \tag{6.48}$$

As a result, we can obtain the hydrodynamic radius using the distance matrix. Assuming that the radius $r_{\mathrm{H}}$ is proportional to the square root of the mean-square radius of gyration with proportional constant $\chi$

$$r_{\mathrm{H}} = \chi\left(6\langle s^2\rangle_0\right)^{1/2} = \chi\left(\frac{b}{N}\right)(6W)^{1/2} \tag{6.49}$$

then the intrinsic viscosity in the nondraining condition [67] is also related to the Wiener index. Consequently, we have

$$[\eta]_{\mathrm{ND}} = \frac{5}{2}\Phi\frac{\left(6\langle s^2\rangle_0\right)^{3/2}}{N} = \frac{5}{2}\Phi\frac{b^3}{N^4}(6W)^{3/2} \tag{6.50}$$

where $\Phi = (N_{\mathrm{A}}/M_0)(4/3)\pi\chi^3$. The non-free-draining intrinsic viscosity $[\eta]_{\mathrm{ND}}$ is proportional to $W^{3/2}N^{-4}$ as shown in (6.50).

The intrinsic viscosity of flexible chain molecules is found empirically to be represented by $KN^{\varepsilon}$ with exponent $\varepsilon$ varying from 0.5 to 1, depending on the nature of the solvent and on temperature. This is because the inertial beads of the chains perturb the flow around the beads in such a manner that they are partially shielded from intrachain hydrodynamic interaction with the external viscous fluid. According to the theoretical treatments of Brinkman [68] and Debye and Bueche [69], the hydrodynamic interactions are approximated by considering the hindered flow of the external fluid through its interior to be proportional to the average beads density $\nu = N/\left(\frac{4}{3}\pi r_{\mathrm{H}}^3\right)$. Then the intrinsic viscosity is given by

$$[\eta] = \left(\frac{N_{\mathrm{A}}}{M}\right)\frac{4}{3}\pi r_{\mathrm{H}}^3\psi(\mu) \tag{6.51}$$

with

$$\psi(\mu) = \frac{5}{2}\frac{1+3\mu-3\mu^{1/2}\coth\mu^{-1/2}}{1+10\mu\left(1+3\mu-3\mu^{1/2}\coth\mu^{-1/2}\right)} \tag{6.52}$$

where $\mu = L^2/r_{\mathrm{H}}^2$ and $L = \sqrt{\eta_{\mathrm{s}}/(\nu\varsigma_0)}$. The parameter $\mu$ represents the dimensionless permeability of the penetrable sphere, which is the reciprocal second power of the shielding ratio by Debye and Bueche [69], and $L$ represents the shielding length. In the limit $\mu\to 0$, $\psi(\mu)$ becomes 5/2 and (6.51) is identical with Einstein's relation (6.45). On the other hand, the limiting values of $\psi(\mu)$ as $\mu\to\infty$ is $(10\mu)^{-1}$ so that we have

$$\lim_{\mu\to\infty}[\eta] = \frac{1}{10}\left(\frac{N_{\mathrm{A}}}{M_0}\right)\frac{\varsigma_0}{\eta_{\mathrm{s}}}r_{\mathrm{H}}^2 \tag{6.53}$$

Let the proportionality constant $\chi$ be $\sqrt{10}/6$. Then (6.53) is identical with the Staudingers relation (6.44). This indicates that (6.51) can be a general formula for the intrinsic viscosity in any draining condition if and only if $\chi = \sqrt{10}/6$.

### 6.4.4 Scattering Function

The size of a polymer molecule is measured by various scattering experiments such as LS, small-angle X-ray scattering, and neutron scattering. When a light beam is scattered from single polymers in dilute solution, the beam intensity depends on the location of scattering centers in a scattering volume. Here we assume that the scattering centers are the beads of an assembly of random flight chains and the centers of mass are uncorrelated in the scattering volume. The intensity of scattered light depends on the size of the scattering molecules and the magnitude of the scattering vector $\mathbf{Q}$:

$$Q = |\mathbf{Q}| = \frac{4\pi}{\lambda}\sin\frac{\theta}{2} \tag{6.54}$$

where $\lambda$ is the wavelength in the medium and $\theta$ is the scattering angle. The incident wave is exactly in phase but the scattered waves arriving at the detector are out-of phase since the scattered wave and the incidence wave travel different distances. The phase difference at the detector of the two beams scattered by the $i$th bead and the $j$th bead is $\mathbf{r}_{ij} \cdot \mathbf{Q}$. The phase difference causes the beams to interfere, and the resultant intensity at the detector is proportional to

$$I(\theta) \propto \sum_i \sum_j \exp i(\mathbf{r}_{ij} \cdot \mathbf{Q}) \tag{6.55}$$

Considering that the distribution of $\mathbf{r}_{ij}$ is Gaussian, the Debye scattering function $P(Q)$ is given by the average of $I(\theta)/I(0)$ that can be written as

$$P(Q) = \frac{1}{N(N-1)} \sum_{i=1}^{N} \sum_{j=1}^{N} \exp\left(-\frac{b^2 Q^2}{6} d_{ij}\right) \tag{6.56}$$

Here we can introduce a new topological index as an exponential type of Wiener index:

$$W_{\exp}(\alpha) = \frac{1}{2} \sum_{i=1}^{N} \sum_{j=1}^{N} \exp(-\alpha d_{ij}) \tag{6.57}$$

where $\alpha$ is the control parameter. This may be a descriptor for scattering intensity and the structural factor.

### 6.4.5
### High Moments of Relaxation Time and Radius of Gyration

The potential energy of the chain molecule given by (6.16) can be rewritten as

$$V = \frac{1}{2}\kappa \mathrm{Tr}\left[\mathbf{b}\cdot\mathbf{b}^{\mathrm{T}}\right] = \frac{1}{2}\kappa \mathrm{Tr}\left[\mathbf{r}\cdot\mathbf{Z}\cdot\mathbf{r}^{\mathrm{T}}\right] \tag{6.58}$$

According to Fixman [70], the distribution function of the square of radius of gyration $s^2$ is given by

$$P(s^2)\mathrm{d}s^2 = \frac{\mathrm{d}s^2}{Z}\int \delta(s^2 - N^{-1}\mathrm{Tr}\left[\mathbf{r}\cdot\mathbf{r}^{\mathrm{T}}\right])e^{-V/k_{\mathrm{B}}T}\mathbf{dr} \tag{6.59}$$

where $Z$ is the configuration partition function. Using the Fourier representation of the delta function, we have

$$P(s^2)\mathrm{d}s^2 = \frac{\mathrm{d}s^2}{2\pi}\int_{-\infty}^{\infty} \varrho(ik)e^{iks^2}\mathbf{dr} \tag{6.60}$$

and

$$\varrho(ik) = \frac{1}{Z}\int \exp\left[-(3/2b^2)\mathrm{Tr}\left[\mathbf{r}\cdot(\mathbf{Z}+ik\gamma\mathbf{I})\cdot\mathbf{r}^{\mathrm{T}}\right]\right]\mathbf{dr} \tag{6.61}$$

where $\gamma = 2b^2/(3N)$. $\varrho(ik)$ becomes the Fourier transform of $P(s^2)$, that is, the characteristic function. Thus,

$$P(s^2) = F^{-1}[\varrho(ik)] \tag{6.62}$$

Employing a transformation to normal coordinates, the integration in (6.61) can be solved by making use of (6.30) as follows:

$$\varrho(ik) = \mathrm{Det}|\mathbf{I} + ik\gamma\mathbf{R}^{-1}|^{-3/2} \equiv \varphi(ik)^{-3/2} \tag{6.63}$$

Using the change of variable $z = ik$, $\varrho(z)$ can be transformed to the Laplace transform of the characteristic function: $\varrho(z) = \mathcal{L}[P(s^2)]$. Then we obtain

$$\mathcal{L}[P(s^2)] = \int_0^{\infty} e^{-zs^2}P(s^2)\mathbf{d}s^2 = \varphi(z)^{-3/2} \tag{6.64}$$

and

$$\varphi(z) = \mathrm{Det}|\mathbf{I} + \gamma z\mathbf{R}^{-1}| \tag{6.65}$$

The Laplace transform of $P(s^2)$, that is, the generating function of $P(s^2)$, provides the average $\langle \mathbf{e}^{-zs^2}\rangle$ and hence the average of powers of $s^2$ can be computed by making use of the expansion in the form of a power series in $z$. Thus, we have

$$\langle s^{2n}\rangle_0 = (-1)^n \frac{\partial^n}{\partial z^n}\varphi(z)^{-3/2}|_{z\to 0} \tag{6.66}$$

Since Det|**R**| = $N$ for any tree graph, (6.65) can be rewritten as

$$\varphi(z) = \frac{1}{N}\mathrm{Det}|\mathbf{R}+\gamma z\mathbf{I}| = \frac{1}{N}\Phi(\mathbf{R}; -\gamma z) \tag{6.67}$$

Using (6.36), we have

$$\varphi(z) = \frac{1}{N}\mathrm{Det}\,|\,\mathbf{A}_{\mathrm{L}} + (\gamma z + 2)\mathbf{I}\,| = \frac{1}{N}\Phi(\mathbf{A}_{\mathrm{L}}; -\gamma z - 2) \tag{6.68}$$

It was shown that the characteristic polynomial of the line graph gives the radius of gyration of a Gaussian chain with any type of branching [71].

Employing Eichinger's mathematical treatments [21], we can derive the relation between eigenvalues $\lambda_i$ of the Rouse matrix **R** and the higher moments of the radius of gyration:

$$\langle s^4\rangle - \langle s^2\rangle = \frac{2b^4}{3N^2}\mathrm{Tr}[\mathbf{R}^{-1}]^2 = \frac{2b^4}{3N^2}\sum_{i=1}^{N-1}\frac{1}{\lambda_i^2} \tag{6.69}$$

$$\langle s^6\rangle - 3\langle s^4\rangle\langle s^2\rangle + 2\langle s^2\rangle^3 = \frac{8b^6}{9N^3}\mathrm{Tr}[\mathbf{R}^{-1}]^3 = \frac{8b^6}{9N^3}\sum_{i=1}^{N-1}\frac{1}{\lambda_i^3} \tag{6.70}$$

The above formulae can also give the sixth and fourth moments of the radius of gyration through the Rouse matrix describing chain dynamics.

The significance of the present graph-theoretical approach is that the mathematical method has the potential to provide an algorithmic way to calculate high-order moments of the radius of gyration and the relaxation time for any tree-like chain. It is difficult to calculate these values with the usual statistical methods because of the great difficulty of evaluating the distribution function.

The characteristic polynomial $\phi(z)$ can be rewritten in terms of the eigenvalues of the Rouse matrix as [15]

$$\varphi(z) = \prod_{i=1}^{N-1}\left(1 + \frac{\gamma}{\lambda_i}z\right) \tag{6.71}$$

Making use of the relation

$$\ln\varphi(z) = \sum_{i=1}^{N-1}\ln\left(1+\frac{\gamma}{\lambda_i}z\right) = \sum_{i=1}^{N-1}\left[\frac{\gamma}{\lambda_i}z - \frac{1}{2}\left(\frac{\gamma}{\lambda_i}\right)^2 z^2 + \frac{1}{3}\left(\frac{\gamma}{\lambda_i}\right)^3 z^3 - \cdots\right] \tag{6.72}$$

the sums of reciprocal powers of the eigenvalues can be readily determined from the following equation:

$$(n-1)!\sum_{i=1}^{N-1}\left(\frac{\gamma}{\lambda_i}\right)^n = (-1)^{n-1}\frac{\partial^n}{\partial z^n}\varphi(z)|_{z\to 0} \tag{6.73}$$

Using (6.31), we have

$$\sum_{i=1}^{N-1} \tau_i^n = \frac{(-1)^{n-1}}{(n-1)!} \left( \frac{N\varsigma_0}{4k_B T} \right)^n \frac{\partial^n}{\partial z^n} \varphi(z)|_{z \to 0} \tag{6.74}$$

The coefficient of $z$ in the characteristic polynomial $\varphi(z)$ can be related to the Wiener index of the molecular graph $G$, that is, the total sum of the elements of its distance matrix. Consequently, any high-order coefficient in $\varphi(z)$ for a tree-like graph, or any coefficient of the characteristic polynomial for its line graph, has the potential to be a new topological index. The high-order Wiener indices can be defined from the coefficients as [66]

$$W_k \equiv (k-1)! \left( \frac{2}{3} \right)^{k-1} N^k \sum_{i=1}^{N-1} \left( \frac{1}{\lambda_i} \right)^k \tag{6.75}$$

It is apparent that the first index $W_1$ becomes the original Wiener index $W$. The high-order Wiener indices can be used for calculating the higher moments of the radius of gyration and mechanical relaxation spectrum of branched chains.

The $n$ th moments of the total sum of the relaxation times can be expressed using the $n$th order Wiener index:

$$\sum_{i=1}^{N-1} \tau_i^n = \frac{2}{3(n-1)!} \left( \frac{\varsigma_0 b^2}{4Nk_B T} \right)^n W_n \tag{6.76}$$

By making use of (6.69) and (6.70), we can express the fourth and sixth moments of the radius of gyration in terms of the second and third order Wiener indices as follows:

$$\langle s^4 \rangle_0 = \left( \frac{b}{N} \right)^4 (W_2 + W_1^2) \tag{6.77}$$

$$\langle s^6 \rangle_0 = \left( \frac{b}{N} \right)^6 (W_3 + 3W_2 W_1 + W_1^3) \tag{6.78}$$

## 6.5 Applications

This section deals with the relationship between the typical rheological functions and the graph-theoretical descriptors. The zero-shear rate viscosity $\eta_0$ and the steady-state compliance $J_e^0$ [63] are

$$\eta_0 = \frac{ck_B T}{N} \sum_{i=1}^{N-1} \tau_i \tag{6.79}$$

$$J_e^0 = \frac{N}{ck_\mathrm{B}T}\sum_{i=1}^{N-1}\tau_i^2 \Big/ \left(\sum_{i=1}^{N-1}\tau_i\right)^2 \tag{6.80}$$

where $J_\mathrm{e}^0$ is a measure of the elastic energy stored under steady flow. These basic rheological functions can be rewritten using the high-ordered Wiener indices as

$$\eta_0 = \frac{c\zeta_0}{6}\left(\frac{b}{N}\right)^2 W_1 \tag{6.81}$$

$$J_e^0 = \frac{3N}{2ck_\mathrm{B}T}\left(\frac{W_2}{W_1}\right) \tag{6.82}$$

Thus, these rheological functions can be obtained from the topological nature of the corresponding tree-graph. This is a very important result because we can realize the effects of the branching feature such as the branch length, branch number, and branch position on these rheological behaviors.

In addition, the intrinsic dynamic storage $[G']$ and loss $[G'']$ moduli can be obtained using the following relations:

$$[G'] = \sum_{i=1}^{N-1}\omega^2\tau_i^2/(1+\omega^2\tau_i^2), \quad [G''] = \sum_{i=1}^{N-1}\omega\tau_i/(1+\omega^2\tau_i^2) \tag{6.83}$$

where $\omega$ is the frequency of strain oscillation. These dynamic rheological functions can be determined from the relaxation spectra.

As described before, it is possible to obtain the relaxation spectra through (6.38) from the eigenvalues of the adjacency matrix of its line graph. The reduced intrinsic dynamic moduli $[G']$ and $[G'']$ calculated from the relaxation spectra are shown in Figure 6.4 for linear, H-shaped, Y-shaped, and dendrimer chains with a fixed vertex number $N = 94$. In the flow region, $\omega \to 0$, $[G']$ and $[G'']$ tend to $\omega^2\tau_\mathrm{R}$ and $\omega\tau_\mathrm{R}$, respectively, as is obvious from (6.83) where $\tau_\mathrm{R}$ is the highest relaxation time or Rouse time. The Rouse time $\tau_\mathrm{R}$ can be estimated from the smallest eigenvalue of **K** or **R**, or the second smallest eigenvalue of $\mathbf{L}^+$ or **Z**. The order of $\tau_\mathrm{R}$ was in linear > H-comb > Y-star > dendrimer. The higher frequency slope of $[G]$ for the dendrimer was found to be larger than half that for the linear chain. It is evident that the Y-star is almost intermediate between the H-comb and the dendrimer.

## 6.6
## Final Remarks

Throughout this chapter, we have demonstrated that the graph-theoretical approach provides a topological insight into branching in polymer architecture. The problems of the dynamics and statistics of various branched polymers without rings and loops were found to be completely reduced to the problem of the eigenpolynomial of the corresponding tree graphs. This suggests that various ideas and concepts obtained from graph theory can be applied directly to the topological analysis of the architecture of branched polymers. For example, advantages in the formulation due to the topological invariants are that the conformational statistics and dynamics involving LS and intrinsic viscosity of any branched polymer can not only be simply and homologously evaluated, but can also be expressed as a function of topological parameters such as the position of branches and the length of main or side chains. Thus, the significance of the present graph-theoretical approach is its provision of useful descriptors expressing the nature and degree of branching. In particular, it is noteworthy that the present mathematical method has the potential to provide an

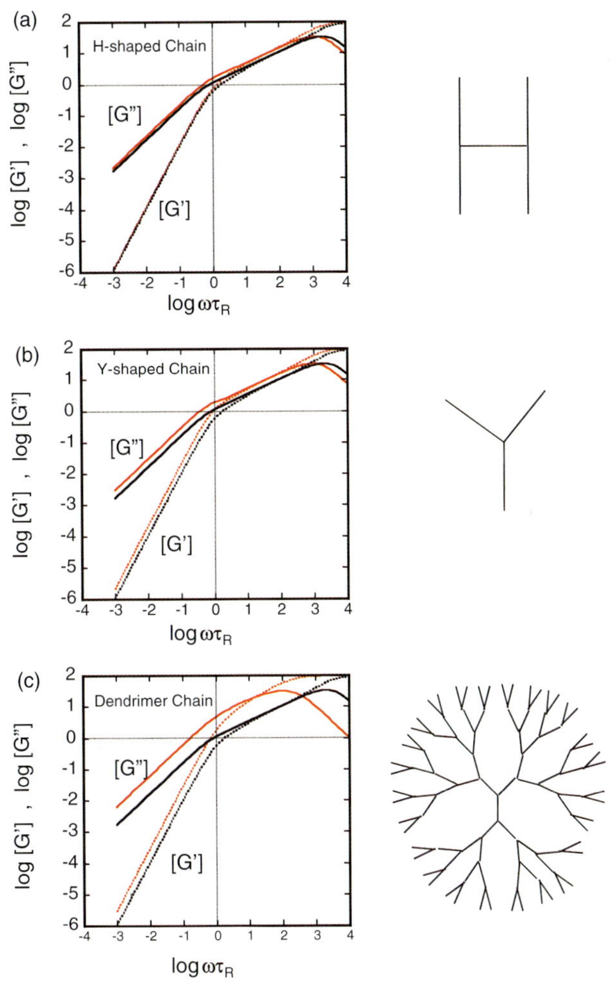


**Figure 6.4** Logarithmic plot of reduced intrinsic moduli $[G']$ and loss $[G'']$ against $\omega\tau_R$ for linear: (a) H-shaped; (b) Y-shaped; and (c) dendrimer chains with $N = 94$, where $\tau_R$ is the maximum relaxation time in Rouse dynamics. The red lines denote the branched chains and the black lines denote the linear chain.

algorithmic way to calculate high-order moments of the radius of gyration and the relaxation time for any tree-like chain.

Additional interest is in the existence of isospectral tree graphs [72]. The isospectral graphs [73–81] are topologically nonidentical (or nonisomorphic) graphs whose nonidentical adjacency matrices give an identical polynomial and an identical set of eigenvalues. The graph-theoretical statement that some molecules have isospectral mates presents a difficulty issue in discriminating their topological structure. We found that the Zimm matrix representing the dynamics of flexible chains is identical with the Laplacian matrix in graph theory. McKay [82] found a mathematical technique for constructing a large number of pairs of Laplacian cospectral tree graphs or cospectral trees with cospectral line graphs. This means that there are many tree-like Gaussian chains with different branched structures showing the same relaxation time spectrum [83].

Currently, only topological descriptor identified in polymer chemistry is the g-factor [7] reflecting the smaller spatial expansion of branched polymers but it is an index with low discriminating power. As demonstrated in this chapter, the graph-theoretical formulation of chain dynamics will make it possible to find more powerful descriptors for leading to the quantitative structure–activity relationships of branched polymers. The topological sense resulting from discrete mathematics such as graph theory may play a central role in molecular design of polymeric materials in future.

## References


1 Burchard, W. (1999) Solution properties of branched macromolecules. *Advances in Polymer Science*, **143**, 113.

2 Ebrahimi, K.G., Takahashi, M., Arai, O., and Masuda, T. (1995) Effects of molecular weight distribution on dynamic viscoelasticity and biaxial extensional flow behavior of polystyrene melts. *Journal of Rheology*, **39**, 1385.

3 Fujimoto, T., Narukawa, H., and Nagasawa, M. (1970) Viscoelastic properties of comb-shaped polystyrenes. *Macromolecules*, **3**, 57.

4 Graessley, W.W., Masuda, T., Roovers, J.E.L., and Hadjichristidis, N. (1976) Rheological properties of linear and branched polyisoprene. *Macromolecules*, **9**, 127.

5 Masuda, T., Ohta, Y., and Onogi, S. (1981) Rheological properties of anionic polystyrenes III. Characterization and rheological properties of four-branch polystyrenes. *Macromolecules*, **4**, 763.

6 McGraph, J.E. (1981) *Anionic Polymerization: Kinetics, Mechanism and Synthesis*, ACS, Washington.

7 Zimm, B.H. and Stockmayer, W.H. (1949) The dimensions and chain molecules containing branches and rings. *Journal of Chemical Physics*, **17**, 1301.

8 Casassa, E.F. and Tagami, Y. (1969) An equilibrium theory for exclusion chromatography of branched and linear polymer chains. *Macromolecules*, **2**, 14.

9 Dawkins, J.V., Maddock, J.W., and Coupe, D. (1970) Gel-permeation chromatography: Examination of universal calibration procedures for polydimethylsiloxane in a poor solvent. *Journal of Polymer Science A-2*, **8**, 1803.

10 Grubisic, Z., Rempp, P., and Benoit, H. (1967) A universal calibration for gel permeation chromatography. *Journal of Polymer Science B*, **5**, 753.

11 Sylvester, J.J. (1878) Chemistry and algebra. *Nature*, **17**, 284.

**12** Cayley, A. (1874) On the mathematical theory of isomers. *Philosophical Magazine*, **67**, 444.

**13** Sylvester, J.J. (1878) On an application of the new atomic theory to graphical representation of the invariants and covariants of binary quantics. *American Journal of Mathematics*, **1**, 64.

**14** Kramers, H.A. (1946) The behavior of macromolecules in inhomogeneous flow. *Journal of Chemical Physics*, **14**, 415.

**15** Rouse, P.E. (1953) A theory of the linear viscoelastic properties of dilute solutions of coiling polymers. *Journal of Chemical Physics*, **17**, 1272.

**16** Bueche, F. (1954) The viscoelastic properties of plastics. *Journal of Chemical Physics*, **22**, 603.

**17** Zimm, B.H. (1956) Dynamics of polymer molecules in dilute solution: Viscoelasticity, flow birefringence and dielectric loss. *Journal of Chemical Physics*, **24**, 269.

**18** Ham, J.D. (1957) Viscosity theory of branched and cross-linked polymers. *Journal of Chemical Physics*, **26**, 625.

**19** Zimm, B.H. and Kilb, R.W. (1959) Dynamics of branched polymer molecules in dilute solution. *Journal of Polymer Science B*, **37**, 19.

**20** Brazhnik, O.D. and Freed, K.F. (1996) Application of graph theory to the statistical thermodynamics of lattice polymers. I. Elements of theory and test for dimers. *Journal of Chemical Physics*, **105**, 837.

**21** Eichinger, B.E. (1980) Configuration statistics of Gaussian molecules. *Macromolecules*, **13**, 1.

**22** Gordon, M. (1979) From Riemanns metric to the graph metric, or applying Occams razor to entanglements. *Polymer*, **20**, 1349.

**23** Kuchanov, S.I., Krolev, S.V., and Payukov, S.V. (1988) Graphs in chemical physics of polymers. *Advances in Chemical Physics*, **72**, 115.

**24** Nitta, K. (1994) A topological approach to statistics and dynamics of chain molecules. *Journal of Chemical Physics*, **101**, 4222.

**25** Harary, F. (1962) The determinant of the adjacency matrix of a graph. *SIAM Review*, **4**, 202.

**26** Harary, F. (1969) *Graph Theory*, Addison-Wesley, MA.

**27** Kirchhoff, G. (1847) Über die Auflösung der Gleichungen, auf welche man bei der Untershchung der linearen Verteilung galvanischer Ströme geführt wird. *Annual Review of Physical Chemistry*, **72**, 497.

**28** Spialter, L. (1963) The atom connectivity matrix and its characteristic polynomial: A new computer-oriented chemical nomenclature. *Journal of the American Chemical Society*, **85**, 2012.

**29** Van Dam, E.R. and Haemers, W.H. (2003) Which graphs are determined by their spectrum? *Linear Algebra and its Applications*, **373**, 241.

**30** Cvetković, D. (2005) Signless Laplacian and line graph. *Bull Cl Sci Math Natur Sci Math*, **131**, 85.

**31** Ugi, T., Marquarding, D., Klusacek, H., Gokel, G., and Gillespie, P. (1970) Chemie und logische Strukturen. *Angewandte Chemie*, **82**, 741.

**32** Wiener, H. (1947) Correlation of heats of isomerization and differences in heats of vaporization of isomers among the paraffin hydrocarbons. *Journal of the American Chemical Society*, **69**, 2636.

**33** Wiener, H. (1947) Structural determination of parafin boiling points. *Journal of the American Chemical Society*, **69**, 17.

**34** Platt, J.R. (1952) Prediction of isomeric differences in paraffin properties. *The Journal of Physical Chemistry*, **56**, 328.

**35** Bonchev, D. and Tranajstić, N. (1977) Information theory, distance matrix, and molecular branching. *Journal of Chemical Physics*, **67**, 4517.

**36** Bonchev, D. and Mekenyan, O. (1980) Topological approach to the calculation of the-electron energy and energy gap of infinite conjugated polymers. *Zeitschrift für Naturforschung*, **35a**, 739.

**37** Bonchev, D., Mekenyan, O., and Polansky, O.E. (1981) Topological approach to the predicting of the electron energy characteristics of conjugated infinite polymers. II. PPP-calculations. *Zeitschrift für Naturforschung*, **36a**, 643.

**38** Bonchev, D., Mekenyan, O., and Polansky, O.E. (1981) Topological approach to the predicting of the electron energy characteristics of conjugated infinite polymers. III. The influence of some structural modifications of polymers. *Zeitschrift für Naturforschung*, **36a**, 647.

39 Bonchev, D., Mekenyan, O., Proti, G., and Tranajstić, N. (1979) Application of topological indices to gas chromatographic data: Calculation of the retention indices of isomeric alkylbenzenes. *Journal of Chromatography*, **176**, 149.

40 Mekenyan, O., Bopnchev, D., and Trinajistić, N. (1980) Chemical graph theory: Modeling the thermodynamic properties of molecules. *International Journal of Quantum Chemistry*, **18**, 369.

41 Mekenyan, O., Dimitrov, S., and Bopnchev, D. (1963) Graph-theoretical approach to the calculation of physicohemical properties of polymers. *European Polymer Journal*, **19**, 1185.

42 Hosoya, H. (1971) Topological index. A newly proposed quantity characterizing the topological nature of structural isomers of saturated hydrocarbons. *Bulletin of the Chemical Society of Japan*, **44**, 2332.

43 Mohar, B. (1993) A novel definition of the Wiener index for trees. *Journal of Chemical Information and Computer Sciences*, **33**, 153.

44 Hosoya, H., Kawasaki, K., and Mizutani, K. (1972) Topological index and thermodynamic properties K. Empirical rules on the boiling point of saturated hydrocarbons. *Bulletin of the Chemical Society of Japan*, **45**, 3415.

45 Lovász, L. and Pelikán, J. (1973) On the eigenvalues of trees. *Period Math Hung*, **3**, 175.

46 Flory, P.J. (1953) *Principles of Polymer Chemistry*, Cornell University, New York.

47 Fischer, E.W., Hahn, K., Kugler, J., Struth, U., and Born, R. (1984) An estimation of the number of tie molecules in semicrystalline polymers by means of neutron scattering. *Journal of Polymer Science*, **22**, 1419.

48 Guttman, C.M., Hoffman, J.D., and DiMarzio, E.A. (1979) Monte Carlo calculation of SANS for various models of semicrystalline polyethylene. *Faraday Discussions of the Chemical Society*, **68**, 297.

49 de Gennes, P.G. (1979) *Scaling Concepts in Polymer Physics*, Cornell University, New York.

50 Debye, P. (1946) The intrinsic viscosity of polymer solutions. *Journal of Chemical Physics*, **14**, 636.

51 Flory, P.J. (1969) *Statistical Mechanics of Chain Molecules*, Hanser, New York.

52 Grosberg, A.Y. and Khokhlov, A.R. (1994) *Statistical Physics of Macromolecules*, AIP, New York.

53 Gottlieb, M. and Bird, B.R. (1976) A molecular dynamics calculation to confirm the incorrectness of the random-walk distribution for describing the Kramers freely jointed beadrod chain. *Journal of Chemical Physics*, **65**, 2467.

54 Widmann, A.H. and Davies, G.R. (1998) Simulation of the intrinsic visocisty of hydrobranched polymers with varing topology 1. Dendric polymers built by sequential addition. *Computational and Theoretical Polymer Science*, **8**, 191.

55 Bonchev, D., Markel, E., and Dekmezian, A.H. (2001) Topological analysis of long-chain branching patterns in polyolefins reciprocal distance matrix. *Journal of Chemical Information and Computer Sciences*, **41**, 1274.

56 Bonchev, D., Markel, E., and Dekmezian, A.H. (2002) Long chain branch polymer chain dimensions: Application of topology to the Zimm–Stockmayer model. *Polymer*, **43**, 203.

57 Sack, R.A. (1953) Mean square radius of randomly coiled molecular chain. *Nature*, **171**, 310.

58 Dobson, G.R. and Gordon, M. (1964) Configurational statistics of highly branched polymer systems. *Journal of Chemical Physics*, **41**, 2389.

59 Kurata, M. and Fukatsu, M. (1964) Unperturbed dimension and translational friction constant of branched polymers. *Journal of Chemical Physics*, **41**, 2934.

60 Peticolas, W.L. (1963) Introduction to the molecular viscoelastic theory of polymers and its applications. *Rubber Chemistry and Technology*, **36**, 1422.

61 Forsman, W.C. (1976) Graph theory and the statistics and dynamics of polymer chains. *Journal of Chemical Physics*, **65**, 4111.

62 Mohar, B. (1991) Eigenvalues, diameter, and mean distance in graphs. *Graphs and Combinatorics*, 7, 53.

63 Ferry, J.D. (1980) *Viscoelastic Properties of Polymers*, 3rd edn, John Wiley & Sons, Inc, New York.

64 Sheridan, P.F., Adolf, D.B., Lyulin, A.V., Neelov, I., and Davies, G.R. (2002) Computer simulations of hyperbranched polymers: The influence of the Wiener index on the intrinsic viscosity and radius of gyration. *Journal of Chemical Physics*, **117**, 7802.

65 Stockmayer, W.H. and Fixman, M. (1953) Dilute solutions of branched polymers. *Annals of the New York Academy of Sciences*, **57**, 334.

66 Nitta, K. (2009) Polymer, nonlinearity, in *Encyclopedia of Complexity and Systems Science* (ed R.A. Meyers), Springer, Berlin, pp. 6833–6855.

67 Flory, P.J. and Fox, T.G. (1954) Treatment of intrinsic viscosities. *Journal of the American Chemical Society*, **73**, 1904.

68 Brinkman, H.C. (1947) A calculation of the viscosity and the sedimentation constant for solutions of large chain molecules taking into account the hampered flow of the solvent through these molecules. *Physica*, **13**, 447.

69 Debye, P. and Bueche, A.M. (1948) Intrinsic viscosity, diffusion, and sedimentation rate of polymers in solution. *Journal of Chemical Physics*, **16**, 573.

70 Fixman, M. (1962) Radius of gyration of polymer chains. *Journal of Chemical Physics*, **6**, 306.

71 Nitta, K. (1999) A graph-theoretical approach to statistics and dynamics of tree-like molecules. *Journal of Mathematical Chemistry*, **25**, 133.

72 Collatz, L. and Sinogowitz, U. (1957) Spektren endlicher grafen - Wilhelm Blaschke zum 70. Geburtstag gewidmet. *Abhandlungen aus dem Mathematischen Seminar der Universität Hamburg*, **21**, 63.

73 Balaban, A.T. and Harary, F. (1971) The characterization polyomial does not uniquely determine the topology of a molecule. *Journal of Chemical Documentation*, **11**, 258.

74 Balasubramanian, K. (1982) Spectra of chemical trees. *International Journal of Quantum Chemistry*, **21**, 581.

75 Harary, F., King, C., Mowshowitz, A., and Read, R.C. (1971) Cospectral graphs and digraphs. *Bulletin of the London Mathematical Society*, **3**, 321.

76 Heilbronner, E. and Jones, T.B. (1976) Spectral differences between isospectral molecules. *Journal of the American Chemical Society*, **100**, 6506.

77 Herndon, W.C. and Ellzey, M.L. (1975) Isospectral graphs and molecules. *Tetrahedron*, **31**, 99.

78 Mowshowitz, A. (1972) The characteristic polynomial of a graph. *Journal C of Combinatorial Theory, Series B*, **12**, 177.

79 Randić, M. (1982) On evaluation of the characteristic polynomial for large molecules. *Journal of Computational Chemistry*, **3**, 421.

80 Randić, M. (1985) On the characteristic equations of the characteristic polynomial. *SIAM Journal on Algebraic and Discrete Methods*, **6**, 145.

81 Schwenk., A.J. (1973) Almost all trees are cospectral, in *New Directions in the Theory of Graphs* (ed F. Harary), Academic Press, New York, pp. 155–163.

82 McKay, B. (1977) On the spectral characteristics of trees. *Ars Combinatoria*, **3**, 219.

83 Nitta, K. (2009) On the structural non-identifiability of flexible branched polymers. *Entropy*, **11**, 907.

# 7
# Structural-Similarity-Based Approaches for the Development of Clustering and QSPR/QSAR Models in Chemical Databases

*Irene Luque Ruiz, Gonzalo Cerruela García, and Miguel Ángel Gómez-Nieto*

## 7.1
## Chemical Structural Similarity

Similarity (or similitude from the old French) is a concept hard to define. Many definitions have been proposed, each one oriented to different science, psychology, humanities, and so on fields. However, the general definition states that similarity is a measure of the resemblance between two objects, considering that an object can be anything, being real or not. Therefore, in this context and for any environment or problem, similarity is a "measure" based on a metric capable of generating reproducible values, for which a common representation of the compared objects [1–4] is needed.

The similarity theory defines "similarity" as an empirical method of finding universal relationships between variables that are made dimensionless using appropriate scaling factors. The application of the similarity theory for obtaining a similarity value between two objects requires the representation of the objects through a set of variables "or properties," and those variables, for each object, must be measures in same domain. As many relationship functions can be used, many set of properties can be considered for the representation of the objects, and many empirical methods can be applied, almost infinite definitions of the similarity can be proposed depending on the pursued target. Hence, for the two same objects we can speak about "shaping similarity," "geometrical similarity," "mathematical similarity," "psychological similarity," and so on, regarding the area and goal in where the similarity concept would be applied [5].

In chemistry, similarity is an important concept and it is also difficult to propose a general definition because there are many aspects to be considered, for example, structural aspects, molecular properties, biological activities, reactivity characteristics, and so on. Different approaches to define similarity are based on: (i) the parameterization of the chemical structure by means of various specific property values and with the assistance of mathematical analysis (statistical, geometrical, etc.) or artificial intelligence methods (pattern recognition, neural network, etc.) to establish a numerical model, and (ii) the examination of structural commonalities or differences directly from the structural information of the chemical compounds, usually based on the topological representation of the structure [2, 6–9].

*Statistical Modelling of Molecular Descriptors in QSAR/QSPR*. First Edition.
Edited by M. Dehmer, K. Varmuza, and D. Bonchev

The most generalized definition, probably thanks to its simple computational implementation, is based on the representation of a molecule as a set of number variables or "descriptors," so that a molecule can be considered a point in a multidimensional "descriptor space." These descriptors can be extracted from the whole molecule (e.g., molecular weight, log $D$, number of hydrogen-bond donors, Kier index, dipole moment, BCUT parameters, etc.) derived from the topology or the 3D structure of the molecule. So, similarity between two molecules is some inverse function of the distance between them in the descriptor space, calculated from the frequency of a set of user-defined 2D or 3D substructures or using the "fingerprint" representation of the molecules where descriptors represent only the presence or absence of a substructure [8, 10, 11].

During the last 50 years, the structural information of chemical compounds has been obtained using their graph representation (molecular graph) of their structure, and later applying some mathematical algorithms (that is the procedure for obtaining molecular descriptors). A molecular graph is a nondirected graph $G=(N, E, F)$, where $N$ is the set of nodes representing the atoms, $E$ is the set of edges representing the relationships among atoms, and $F$ is the function that represents the connection type (edge in the graph) among the atoms of the molecular structure [12, 13].

Hereby, a measurement of the chemical similarity can be obtained by comparing the resemblance between the molecular graph representations of the compounds, and therefore, as a measurement of the graph's isomorphism. Two graphs $G_A$ and $G_B$ are isomorphic if there is a bijection $f$ between the set of nodes of both graphs; that is, if we can obtain the adjacency matrix of $G_B$ through a simple permutation of the adjacency matrix of $G_A$, $G_A$ and $G_B$ are not isomorphic and it is necessary to apply a metric for obtaining a similarity measurement [14–16].

Thus, 2D structural similarity considers

- 2D representations of molecules, using a common representation of the molecules through molecular graphs;
- subgraph isomorphism algorithms, used for matching two molecular graphs and extract all the common subgraphs;
- a metric, called similarity index, created for obtaining a similarity value representing the resemblance between the compared molecules.

In a general formulation, similarity measurements can be expressed as Equation 7.1 [17]. The application of this expression involves: (i) a function g(. . .) in charge of transforming the molecular structures $M$ into an amenable molecular representation able to be handled by computers with the aim of matching each pair of molecules of a dataset (molecular graph, molecular descriptor, fingerprint, etc.), and (ii) the proposal of a function $f$(. . .) for deriving similarity measurements used later for chemical database clustering, screening, correlating with experimental activity, and so on:

$$S_{A,B}=f[g(M_A),g(M_B)] \tag{7.1}$$

The properties of function $f$ have been widely studied, and hundreds of metrics can be applied in order to obtain the similarity value [18, 19]. Usually, the similarity value is ranged in the interval [0, 1] (or [−1, 1]), providing a fuzzy measurement of the descriptor spaces or graphs resemblance or a possibility measure for a probabilistic interpretation of the similarity.

### 7.1.1
### Molecular Graph and Structural Similarity

Measurements of chemical structural similarity are commonly related to graph similarity. The similarity value is a measure of the graphs isomorphism when the chemical compounds are represented trough molecular graphs.

Checking whether two graphs $G_A$ and $G_B$ are isomorphic is a difficult computational problem. Different efficient algorithms have been proposed, as the one by Ullman [14], highly efficient when the graphs are very different, but which efficiency diminishes as the graph similarity increases. The problem becomes more complicated when the purpose to find which subgraph of the $G_A$ graph exists in the $G_B$ graph, known as subgraph isomorphism. The detection of commonly induced subgraph isomorphism and especially the detection of the maximum common substructures (MCSs), have been applied to a great variety of fields [15, 20–22].

A graph $G$ is called the MCS of two graphs $G_A$ and $G_B$, if there is no other common subgraph of $G_A$ and $G_B$ that has more nodes in common than $G$ (node-induced subgraph). This contrasts with other definitions where MCS of two given graphs is defined as the common subgraph that contains the maximum number of edges (edge-induced subgraph). Notice that according to this definition, the MCS is not necessarily unique for two given graphs.

The main problem in the detection of the subgraph isomorphism is the fact that it requires a high computational cost [23]. The reduction of its computational cost is still a challenge for researchers for two main reasons: (i) the applicability of these algorithms to any research area, and (ii) the huge increase of information required for any problem (for example, the size of the databases of chemical compounds). Different techniques have been proposed to reduce the computational cost of the subgraph isomorphism: decomposition of the graph into trees and use of back-tracking, forward-checking, looking-ahead algorithms, and so on, optimization techniques involving the use of neural networks, genetic algorithms, simulated annealing, and so on [1, 2, 15, 16, 24]. Most of the proposals are initially based on the building of homomorphic graph representations, generating reduced graphs. Reduced graphs have the inconvenience of lacking structural information although these representations allow the generation of similarity measurements appropriate for a first step in the screening process of large chemical databases [25–31].

### 7.1.2
### Descriptor-Based Structural Similarity

As aforementioned, molecules can be represented as an array of variables or descriptors representing some structural property of the chemical compound in order to extract a structural similarity measurement. Thousands of molecular descriptors have been proposed in the literature, most of them related to some structural or topological property corresponding to the molecular graph that represents the molecule [6, 8, 10–12].

For the calculation of descriptor-based similarity, molecules are represented as Equation 7.2 and the two arrays are compared using mathematical or statistical procedures. However different hints must be considered:

$$\begin{aligned} M_A &= \left(d_1^A, d_2^A, d_3^A, \ldots, d_n^A\right) \\ M_B &= \left(d_1^B, d_2^B, d_3^B, \ldots, d_n^B\right) \end{aligned} \qquad (7.2)$$

- The number and order of descriptors has to be the same in arrays $M_A$ and $M_B$.
- The magnitude of the descriptors must be normalized.
- It is mandatory to eliminate relationships between descriptors in order to diminish the number of variables, and a previous statistical analysis must be performed.

Fingerprints are a special case of descriptor-based similarity, representing the molecules as arrays of 1s and 0s showing the presence or absence in the molecule of some predefined substructure or path [32, 33]. Structural similarity based on fingerprints has shown excellent results for many QSPR/QSAR studies. However, although similarity values can be obtained with high performance, this technique lacks to solve serious problems such as: (i) incomplete representation of the molecular structure (no all paths and substructures are represented in the binary array), (ii) overlapping representation of the substructures (different substructures can be represented by the same variable), and (iii) the absence of the representation of substructure relationships.

### 7.1.3
### Combining Structural Similarity Approaches

Graph-based similarity measurements generate fine values considering all structural elements (atoms and bonds) and relationships in the molecules; however, this approach requires a high computational cost many times more expensive for the analysis of large chemical databases than for smaller ones. Moreover, graph-based similarity generates inconsistent results when comparing datasets with similar molecules. For instance, 1,2-clorobencene, 1,3-clorobencene, and 1,4-clorobencene are molecules with the same number and type of atoms and edges but with their substituent disposed at different positions. Any matching between the three molecules results in the same isomorphism, 1-clorobencene, and therefore all similarity values between any pair of molecules generate the same value. However, these molecules are different; they have different structure, different chemical and physical properties as well as different biological activities.

On other side, descriptor-based similarity measurements are highly efficiently obtained but they use an incomplete representation of the molecules, and the similarity values also frequently produce large degeneracy in later studies (screening, clustering, and QSPR/QSAR) [34, 35].

Recently, investigations have oriented their studies to combine different similarity approaches in order to obtain finer similarity values with an acceptable computational cost [36–40].

These approaches are oriented to: (i) the fusion of different similarity metrics, (ii) the use of different representations of the molecules and the fusion of the similarity measurements obtained through the application of different or not similarity metrics.

The former approach does not solve the degeneracy problems described above, though it has showed excellent results for screening and clustering applications in large chemical databases. The second option deals with solving the degeneracy problems by using different structural representations of molecules and, therefore, by retaining more information of the molecular structure.

### 7.1.4
### Approximate Structural Similarity

Approximate similarity (AS) combines graph-based and descriptor-based representations of the molecules in order to obtain a more accurate similarity measure [17, 36, 37, 41]. The AS concept is based in considering the structural similarity as an incomplete measure of the resemblance between two molecules, so similarity only has into account common aspects of the molecular structures and the remaining structural fragments are not considered. Figure 7.1 shows a clear example, for three molecules with the same number of atoms and bonds. For any pair of these molecules, the isomorphism is the same (structure D) and therefore they have equal similarity values.

However, properties and activities of these molecules are different and when we wish to use similarity values in order to build a predictive model, some degeneracy is obtained. The reason is clear; the classical similarity approach does not consider those fragments that are not included in the isomorphic structure that is different for each molecule.

The AS approach has into account these nonisomorphic fragments (NIFs) [36, 37] in order to build a new similarity metric combining graph-based and descriptor-based similarity approaches with the aim of obtaining more realistic similarity values:

$$AS_{A,B} = f(S_{A,B}, \Gamma_{A,B}, w_{\Gamma}) \tag{7.3}$$

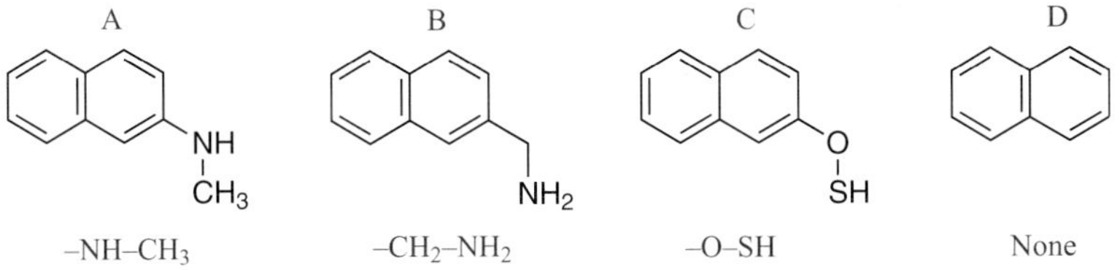


**Figure 7.1** Inconsistent results in graph-based structural similarity.

Expression (7.3) defines mathematically the AS concept. $S_{A,B}$ is any similarity measure (constitutional, fingerprint based, invariant based) obtained by considering isomorphism ($I_{A,B}$) detection and different similarity metrics; $\Gamma_{A,B}$ is the dissimilarity between the substructures that do not form $I_{A,B}$ and is computed by using topological invariants and any distance metric that measures the differences between the descriptor values for the nonisomorphic subgraphs of $G_A$ and $G_B$; $w_\Gamma$ is a weighting factor that adjusts the distance contribution in AS calculation; and finally, *f*() is a mathematical function created for combining $S_{A,B}$, $\Gamma_{A,B}$, and $w_\Gamma$.

Logically, $S_{A,B}$ and $\Gamma_{A,B}$ show opposite meanings: the higher value $S_{A,B}$ shows the greater the similarity between the graphs *A* and *B* is, whereas, the higher value $\Gamma_{A,B}$ shows, the lower the similarity is. In addition, normalization methods are often employed in order to equal similarity and dissimilarity scales and to have the AS distributed within the [0, 1] interval.

Figure 7.2 shows an example of AS for a set of five molecules. In this example, classical similarity is calculated using graph-based similarity considering the MCS as structural information for the isomorphism between molecular graphs. Hence, in the matching between any pairs of molecules an isomorphic fragment corresponding with the MCS is obtained and one or several NIFs are also obtained, which can be observed in Figure 7.2.

Classical similarity (*S*) has been calculated using the Tanimoto index, whereas AS has been obtained with Equation 7.4:

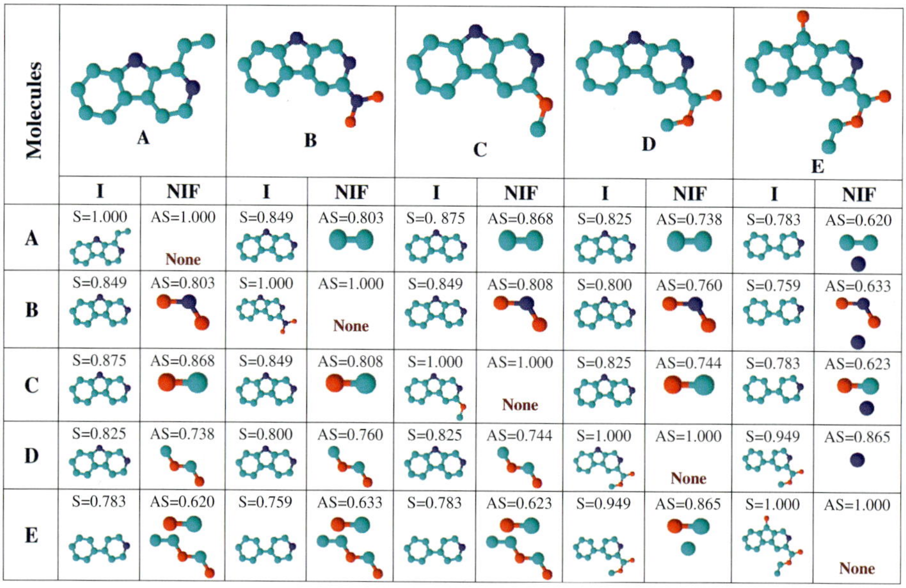


**Figure 7.2** Refinement of approximate similarity against classical similarity.

$$\begin{aligned} AS_{A,B} &= S_{A,B} \times (1 - \Gamma_{A,B} \times w_{A,B}) \\ \Gamma_{A,B} &= |W(\mathrm{NIF}_A) - W(\mathrm{NIF}_B)| \\ w_{A,B} &= \frac{1}{W(A) + W(B)} \end{aligned} \tag{7.4}$$

where $W(\mathrm{NIF}_A)$ and $W(\mathrm{NIF}_B)$ are the values of the Wiener index corresponding with the NIFs of molecules *A* and *B* in the matching, and $W(A)$ and $W(B)$ are the values of the Wiener index of molecules *A* and *B*, respectively.

Approximate similarity models allow the use of diferent metrics and models for the calculation of the classical similarity term (descriptors or graph-based methods, Tanimoto index, Cosine index, etc.). AS also allows the proposal of any model for the calculation of the $\Gamma$ and $w$ terms in the equation. So, the refinement of similarity values obtained for each specific dataset according with its characteristics and the property under study is possible. The main inconvenience of the AS approach is that these terms must be obtained and adjusted experimentally, so a previous analysis of the dataset behavior regarding different descriptors and metrics must be done.

## 7.2
## Clustering Models Based on Structural Similarity

The drugs design is a complex process where scientists discover the relationships between the structure and activity (SAR) of the compounds using data mining methods on molecular fragments. Many machine-learning techniques are applied to prepare the dataset and learn how to classify molecules, for example, variance analysis, linear regression, discriminate analysis, canonical correlation analysis, multidimensional scaling, factor analysis, principal component analysis (PCA), and clustering analysis [42].

The principal objective of clustering is finding the natural groups in a set of objects or patterns [43]. Cluster analysis is defined as [44] "*a statistical classification technique for discovering whether the individuals of a population fall into different groups by making quantitative comparisons of multiple characteristics.*"

From a general point of view, the cluster process includes the following defined steps: (i) pattern representation, (ii) definition of the proximity measure, (iii) decision of the cluster method, and (iv) cluster validation.

The pattern representation (optionally including the extraction and/or selection of characteristics) allows the definition of the number of classes (the number of available patterns) for identifying the subset of the original and more effective characteristics for the clustering process. A good representation of patterns usually improves the results of the clustering process; a bad representation leads to complex clusters whose structure is impossible to understand.

Patterns can be measured as physical objects, for example with a molecule structural characteristic, a molecular descriptor, or molecular similarity values. It is commonly accepted to represent the patterns as multidimensional vectors where each dimension is a single characteristic.

Special care should be taken when a proximity measure is selected for two patterns. The most used metric for continuous characteristics is the Euclidian distance. The

Euclidean distance is commonly used to evaluate the proximity of the objects in a two- or three-dimensional space. It is recommended to be used when the dataset presents compact or isolated clusters. When a lineal correlation exists among the characteristics that represent a pattern, it is more advisable to use other distance metrics as the Mahalanobis distance [45].

The clustering methods can be classified as partitional or hierarchical techniques. Partitional clustering algorithms find all the clusters simultaneously as a partition of data. Hierarchical clustering find clusters recursively in order to create a hierarchical representation, named dendrogram, representing the nested grouping of objects and similarity levels at which groupings change.

One of the firsts and most intuitive cluster method is based on the sphere exclusion theory [46]. The sphere exclusion procedure selects one item from the dataset and excludes from further considerations all items that are within a given similarity/distance threshold to the selected item. This process is repeated until all items have been selected or excluded.

The most popular partitioning algorithm is the *K*-means that has been used in different applications fields for the last 50 years [47]. If we consider a dataset ofobjects $X = \{x_i\}, \quad i = 1, \ldots, n$ represented by *n* d-dimensional points to be classified in *k* clusters $C = \{c_i\}, \quad i = 1, \ldots, k$, the *K*-means algorithm finds such a partition that the squared error between the empirical mean of a cluster and the points in the cluster are minimized. The goal in the *K*-means method is to minimize the sum of squared error over all *k* clusters defined as follows:

$$E = \sum_{k=1}^{K} \sum_{x_i \varepsilon C_k} ||x_i - \mu_k||^2 \tag{7.5}$$

where $\mu_k$ is the mean of de cluster $C_k$.

The procedure steps of *K*-means algorithm are [48]: (1) selection of an initial number of *k* clusters and repetition of steps 2 and 3 until the cluster members are stabilized; (2) creation of a new partition by assigning each pattern/object to its closest cluster center, and (3) computation of the new cluster centers.

The *K*-means algorithm is impacted by the noise. In order to reduce this effect in the *K*-medoid method [43], the clusters are created using the median of the data instead of the mean. Moreover, in the past few years new proposals have appeared related to the *K*-means method. Some of these extensions deal with additional heuristics using a minimum cluster size and merging and splitting clusters [49, 50]. Fuzzy C-means [51, 52] is an extension of *K*-means where each data point can be a member of multiple clusters with a different membership value.

Jarvis–Patrick [53] is other very used partitional cluster method that uses a table of *L* nearest neighbors for each object to be clustered, and then sequentially merges clusters that have at least $L_k$ nearest neighbors in common.

Apart from that, hierarchical clustering algorithms recursively find nested clusters in two ways: (i) considering each data point as a cluster and merging the most similar pair of clusters successively in order to create a cluster hierarchy (bottom-up), and (ii) taking into account all data points as one single cluster and recursively dividing each cluster into smaller ones (top-down). Generally the hierarchical clustering algorithms are variants of the single-link, complete-link, and minimum-variance proposals [54].

Agglomerative techniques are commonly used and three different variants can be chosen depending on how the distance/similarity is defined: single, complete, and average linkage. The single linkage clustering is the simplest hierarchical method. In this case the distance between groups is defined as the distance between the closest pair of objects, and only pairs including one object from each group are considered. In complete linkage clustering, the distance between groups is defined as the distance between the most distant pair of objects, one from each group. And, in average linkage clustering the distance is defined as the average of distances between all pairs of objects, where each pair is made up of one object from each group.

Probably, one of the most popular hierarchical methods is the one proposed by Ward [54]. This method involves an agglomerative clustering algorithm performing a cluster analysis as a variance problem analysis instead of using distance metrics or association measures.

Finally, cluster evaluation refers the need of evaluating the results of the cluster analysis in a quantitative and objective way. Thus, different metrics can be used to evaluate the results of the clustering process [43].

Dunn index [55] is a validation index widely used for clustering algorithms. The Dunn index is a measure that maximizes intercluster distances while minimizing intracluster distances; its larger values indicate the presence of compact and well-separated clusters, so the number of clusters that maximizes the index is taken as the optimal number of clusters.

This validation indices for clustering are expressed as follows (Table 7.1 shows the meaning of the acronym used):

**Table 7.1** Acronyms used in the validation index equations.

| Acronyms | Meaning |
|---|---|
| $n_{\mathrm{CL}}$ | Total number of clusters |
| $d(x,y)$ | Distance between two data elements (objects) |
| $\mathrm{C_i}$ | Reference the $i$th cluster |
| $A$ | Number of pairs with the same class $K_i$ and assigned to the same cluster $C_i$ |
| $B$ | Number of pairs with the same class in $K_i$, but in different clusters $C_i$ |
| $C$ | Number of pairs with different class in $K_i$, but in the same clusters $C_i$ |
| $D$ | Number of pairs with different class $K_i$, that were assigned to a different cluster $C_i$ |
| $\mathrm{Av}_i$ | Average distance of all objects in cluster $c_i$. |
| $\mathrm{An}_i$ | Average distance between and all data items in the closest other cluster (which is defined as the one yielding the minimal) |
| $n_{\mathrm{d}}$ | Number of dimensions |
| $n_{\mathrm{j}}$ | Number of element of $j$th dimension in the whole dataset |
| $\bar{\mathbf{x}}_j$ | The mean of data values of $j$th dimension |
| $x_k$ | Value of the $k$th element in the $i$th cluster for the $j$th dimension |
| $n_{ij}$ | Number of element in the $i$th cluster $j$th dimension |

$$D = \min_{i=1\ldots n_{CL}}\left\{\min_{j=i+1\ldots n_{CL}}\left(\frac{d(c_i, c_j)}{\max_{k=1\ldots n_{CL}}(\mathrm{diam}(c_k))}\right)\right\} \tag{7.6}$$

where: $d(c_i, c_j) = \min_{x\varepsilon c_i, y\varepsilon c_j}\{d(x, y)\}$ and $\mathrm{diam}(c_i) = \max_{x,y\varepsilon c_i}\{d(x, y)\}$

The main disadvantages of the Dunn index are the time it consumes for the index calculation and its sensibility to noise (the maximum cluster diameter can be quite large in a noisy environment). Therefore, several Dunn-like indices have been proposed in order to use different definitions for the cluster distance and diameter [56], as the Davies–Bouldin index [57] that measures the average similarity between each cluster and the most similar to it; small values correspond to compact clusters whose centers are far away from each other; therefore, its minimum value determines the optimal number of clusters [58].

Other validation indices among the most used are Rand [59] and Jaccard. The Rand index is a measurement of the number of pairwise agreements between a clustering $C_i$ and a set of defined classes $K_i$, normalized so that the value ranges between 0 and 1, as follows:

$$R = \frac{a+d}{a+b+c+d} \tag{7.7}$$

The Jaccard index [60] has been commonly applied to assess the similarity between different partitions of the same dataset. The level of agreement between a set of classes $K_i$ and a clustering result is determined by the number of pairs of points assigned to the same cluster in both partitions, and can be obtained as follows:

$$J = \frac{a}{a+b+c} \tag{7.8}$$

The Rand and Jaccard indices produce a result in the range [0, 1], where a value of 1.0 indicates that $C$ and $K$ are identical. A high value for this measure generally indicates a high level of agreement between a clustering and the annotated natural classes [61]. Similar to these indices is the Folks and Mallows index [62].

One of the most used indices for partitional clustering is the Silhouette Width [63] (also named Calinski–Harabasz (CH) index [64]) that is computed as the average Silhouette value over all the data items of a given partition. The Silhouette value for an individual data item, which reflects the confidence in this particular cluster assignment, is computed as follows:

$$S(i) = \frac{An_i + Av_i}{\max(An_i, Av_i)} \tag{7.9}$$

This index takes values in the interval [−1, 1] reflecting the compactness and separation of clusters. A large average Silhouette index denotes a good overall quality of the clustering result, so the optimal number of clusters is the one that gives the largest average Silhouette value [65, 66].

The RMSSTD (root-mean-square standard deviation) index is widely used in hierarchical clustering algorithms, although this index can also be used for evaluating the results of any clustering algorithm. The RMSSTD index [67] is the variance of the clusters, formally defined as follows:

$$\mathrm{RMSSTD} = \sqrt{\frac{\sum_{i=1\ldots n_{CL},j=1\ldots n_d} \sum_{k=1}^{n_{ij}} \left(x_k - \bar{x}_j\right)}{\sum_{i=1\ldots n_{CL},j=1\ldots n_d} \left(n_{ij} - 1\right)}} \tag{7.10}$$

The RMSSTD index measures the homogeneity of the clusters (or the variance of the clusters) at each step of the hierarchical clustering algorithm. A lower RMSSTD value means better clustering, so its minimum determines the optimal number of clusters [68].

On the opposite side, the purpose of the RS index [69] is to measure the dissimilarity of clusters. More formally stated, it measures the degree of homogeneity between groups. The values of RS range, as RMSSTD, from 0 to 1, where 0 means that there is no difference among the clusters and 1 denotes a significant difference. The RS index is calculated as follows:

$$\mathrm{RS} = \frac{\mathrm{SS_t} - \mathrm{SS_w}}{\mathrm{SS_t}} \tag{7.11}$$

where $\mathrm{SS_t} = \sum_{j=1}^{d_i} \sum_{k=1}^{n_j} \left(x_k - \bar{x}_j\right)^2$ and $\mathrm{SS_w} = \sum_{i=1,\ldots,nc,j=1,\ldots,d_i} \sum_{k=1}^{n_j} \left(x_k - \bar{x}_j\right)^2$

### 7.2.1 Clustering of Chemical Databases

Molecular representation, clustering, and classification of chemical databases have great relevance in drug design, QSPR/QSAR, and many chemical application areas. Thus, many researching groups have developed different clustering proposals for each of the clustering stages in order to find most appropriate solutions for medium and large chemical databases.

#### 7.2.1.1 Pattern Representation of Chemicals Structures

One of the most used computational representation of molecules are molecular descriptors: a simple numerical, normally continuous, representation of the structural characteristics of a molecule, including topological, graphical, steric, electrical, and many other molecular and physicochemical properties of a chemical substance.

In this way, molecular databases can be represented as an $n \times d$ matrix, where $n$ represents the number of compounds and $d$ the number of descriptors obtained for each molecule. Using this matrix classical clustering algorithms can be applied [70–72], or also, previously reducing this representational space by applying a PCA over the $n$-dimensional descriptors representation [73].

The main problem to use molecular descriptors in the cluster analysis is that they not necessarily offer a differentiation measure among the molecules, and therefore it is frequent that no similar molecules are assigned to the same cluster.

Fingerprints are another widely used way to represent chemical compounds; these being an array of bits where the presence or absence of structural patterns is represented [33]. There are several types of fingerprints where most have been proposed depending on whether the analysis considers or not predefined patterns, the array size and other structural information, as well as user-defined additional structural keys. Fingerprints generation algorithms take as their source structural information from the molecules (molecular graphs), generating a unique and reproducible array based in the pattern extraction from the molecular graph (from each atom, fragments, paths, and so on), codifying these patterns, and allocating the key pattern generated in the fingerprint array using a hash algorithm.

In cluster analysis, fingerprints can be used in two ways: (i) directly representing each pattern (molecule) as a bit string and applying a selected cluster algorithm over it or (ii) calculating in a preprocessing stage the similarity matrix ($S$) between all the datasets represented through their fingerprints. Thus, the $S(i, j)$ similarity matrix can be used as the proximity metric between patterns in the clustering process. In practice, the clustering based on similarity matrix is the most used [74–76].

Otherwise, the chemical structure can be represented using the graph theory, in this case the molecule is modeled as an undirected graph where the graph vertex represents the atoms, and the graph edges denote the bonds connecting each pair of covalently bonded atoms [77]. Using this representation the proximity metric (for clustering) can be based on the isomorphism concept. Thus, knowing the MCS for two given molecular graphs, we can establish a similarity metric between both graphs and use this distance value in the clustering algorithm.

In general, fingerprints and graph-based representations are more adequate to represent the structural similarities in the clustering process than representational spaces based on molecular descriptors. In [78] both fingerprints and graphs-based molecular representation were compared using several clustering algorithms; this approach suggests that both similarity measures can be effectively used for chemical clustering.

#### 7.2.1.2 Clustering of Chemical Databases

In computational chemistry, clustering methods are broadly applied to arrange similar molecules into the same cluster and dissimilar molecules into different clusters. The result of this classification process is useful when providing a condensed representation of the original dataset and can be employed, for example, in the structure–activity relationships analysis, in the analysis of biological screening data, in molecular docking, and so on [79].

In general, both types of clustering methods (partitional or hierarchical) are applied in the classification of chemical databases.

**Classical Partitional Clustering Applications** Partitional methods are generally much less demanding of computational resources than the hierarchical methods, since only a single partition of the dataset has to be analyzed. Jarvis–Patrick's method has been broadly used in computational chemistry [80]; however, this method has serious limitations when applied to large datasets. The characteristics of the generated clusters are severely associated with the parameters needed for the algorithm execution (the size of the neighbor list and the number of common neighbors needed to form a cluster); according to the selected values, this method tends, in some cases, to create very large but heterogeneous clusters or homogeneous ones but too small.

These problems have been useful for the appearance of new proposals, for example, a clustering method based on exclusion spheres has been proposed [80]. The algorithm considers the following steps: (i) for each element in the dataset the fingerprint is calculated using the daylight application [33]; (ii) the number of neighbors for each molecule in the database is calculated, using the Tanimoto similarity threshold defined as a clustering parameter; (iii) the dataset is sorted in the descending order, so that the compounds with the largest number of neighbors (potential cluster centroids) are placed at the top of the list; (iv) the clustering process is carried out starting with the first compound in the sorted list and its pairwise Tanimoto similarity index to all other compounds is computed. All those molecules with a Tanimoto index above or equal to the value used for the clustering become members of that cluster; each molecule classified in one cluster is flagged and removed from any further comparisons. The process is repeated for all nonflagged molecules down the list.

For validating the proposal, the experiments were carried out with a database containing 5424 molecules, and using a Tanimoto level of 0.8. The method was compared to the Jarvis–Patrick technique considering the same database and creating a very similar number of clusters and singletons. The similarity values for the molecules assigned to a same cluster for Jarvis–Patrick have great variations (from 0.3 to 0.7) depending of the initial clustering parameters. However, for the exclusions sphere method the similarity within the cluster is up to 0.8.

Another alternative used for chemical database clustering is based on the *K*-means method, for example, in [81] a modified version of this algorithm was proposed, this approach also includes a hierarchical representation based on the recursive application of the basic algorithm. The clustering process is carried out as follows: according to the *K*-means two clusters are created, for each cluster a recursive process is carried out where if the maximum distance of a data point to a created virtual mean exceeds the Euclidean distance threshold (clustering parameter), the current cluster is divided again in $k$ new clusters, otherwise the process finishes. The method was tested using four datasets: the first was an artificial dataset, the Fisher's Iris dataset [82], it consists of 150 random samples of flowers from the Iris species setosa, Versicolor, and Virginica. For each species, there are 50 observations for sepal length, sepal width, petal length, and petal width in centimeters. Afterward, two datasets containing small organic molecules (COBRA and MDDR) were used, and finally a SPECS catalog containing 5375 pharmacologically active molecules.

The first validation was applied over the Fisher's Iris dataset where three classes are known, and a PCA analysis for the same dataset was used to contrast the results. It is clear that the tree representations using the *K*-means algorithm are in agreement with the PC projection. The second evaluation shows that for other datasets examples, suitable clusters were created according to the ICE inhibitors in COBRA and glucocorticoid receptor ligands in MDDR.

In chemical applications, it is important that the molecules are assigned to a same cluster, through the application of a certain clustering method, when they have very similar values of chemical, biological, and physical properties. In [73], a fuzzy *K*-means method was proposed, and it was compared to other clustering methods (*K*-means and Ward's) for the prediction of certain molecular properties. Initially, the molecules were represented using the first 45 principal components; these accounting for the 77.6% of the variance in the wide set of molecular descriptors calculated with the Molconn-Z package [83]. In the experiment, two datasets were used. The first consisted in a randomly selected set of 1763 molecules with logP values available. The second was composed by eight groups of molecules with the following composition: 84 Cynotropanes, 311 Diphenylethers, 162 Heterocyclic fluorovinyl nematicides, 272 N-aryl lactam herbicides, 272 imidazolines and oxazolidinediones, 226 strobilurin fungicides, 303 heteroaryl carbinol herbicides, and 710 standard pesticides.

The comparison of the three clustering methods involved simulated property prediction, in which the groupings resulting from a cluster analysis were used to predict the properties of the compounds within each cluster. The predicted property value for a molecule is designed using the mean of the property value for all the other molecules in the same cluster. The set of predicted values is compared with the observed values by calculating the correlation coefficient, a good correlation, and thus an effective clustering, has been denoted by a coefficient value close to 1. As a result, authors described that fuzzy *K*-means showed better correlation values than *K*-means and Ward's methods. Another important conclusion was obtained studying the relationship between the singletons generated by each clustering method, regarding the values considered as outliers. A singleton in fixed clustering is simply a cluster that contains just a single molecule, but in fuzzy clustering a singleton can be a cluster with only one significant membership or a molecule with a membership value close to $1/k$ for all $k$ clusters. It is an indicator that this molecule has not got a relationship to any other cluster. However, in fixed clustering, molecules are grouped together only because they cannot be easily grouped anywhere else. Then, singletons in fuzzed *k*-means provide a simple way of detecting outliers in a dataset.

**Classical Hierarchical Clustering Applications** Hierarchical clustering algorithms are more used for the clustering of chemical databases than partitional ones, especially, in QSAR applications were the molecular dataset has been sorted and classified. An application of Ward's hierarchical clustering method used as a part of the qualitative QSPR/QSAR analysis can be found in [84]. In this work, the hierarchical cluster analysis was performed on the standardized descriptors for a set of 35 compounds, using the Chebychev algorithm to obtain a distance matrix. The results showed a good classification, sorting all the molecules on the basis of their geometrical and electronic similarities/differences into clusters at different hierarchical levels, where compounds belonging to the same group of distinctive activity patterns share chemical structure characteristics. Authors recommended the usage of this approach as an adequate tool to retrieve the meaning from a subset of physicochemical descriptors and for sorting the species in clusters related to particular anticancer profiles.

The visualization of the result of the clustering process is especially important to recognize the trends of some physical properties with respect to certain chemical classes; for this purpose, the conventional hierarchical clustering representation (dendrograms) are commonly used. In [85], a new hierarchical representation method is proposed based on the HeiankyoView representation [86, 87], where the hierarchical data leaf nodes are mapping as colored square icons and nonleaf nodes as rectangular borders. In this approach, Ward's method was applied using the normalized Euclidean distance, each molecule was represented in a space formed by four molecular descriptors (log *S*, Clog *P*, molecular weight, and counts of hydrogen bonding donors). At the end of the clustering process, the molecules can be visualized as equi-shaped icons without overlapping one another in the two-dimensional display space. For validating the proposal, a dataset of 908 compounds was chosen, that ranged from − 10.41 to + 2.06 in terms of the common logarithm (log *S*). HeiankyoView representation clearly indicates that structurally similar compounds tend to have a similar aqueous solubility; this method is appropriate to intuitively understand structure–activity relationships in a large-scale dataset.

In hierarchical clustering, analysis is also quite important to obtain a metric that allows us to evaluate different clustering techniques considering the differences among the generated hierarchical representations (dendrograms). This metric has been useful in chemical clustering applications for comparing results when different groups of descriptors or different similarity coefficients are used. In [88], three different indices for calculating dissimilarities and for analyzing different structural aspects between hierarchical clustering trees are described.

The first index (triples index) looks for all the possible sets of three members (triples) contained in the dataset and contrasts the structural connections of them (maximal 3-subtrees) in the hierarchical tree. The second index, the partition index, builds all the possible nonredundant binary partitions of a dendrogram by deleting its internal links. The third index (cluster index) is based on the contrast of the nontrivial clusters (singletons are not considered) present in the two dendrograms; when this cluster index reaches the value of 0 it means that the contrasted dendrograms have all their clusters in common. For the validation, a set of 1000 molecules was selected from the National Cancer Institute (NCI) database, molecules were represented by 1024-bit fingerprints, and their similarities were calculated using the Tanimoto coefficient. In the analysis, five different grouping methods were applied to the Tanimoto similarity matrix to create five dendrograms. The grouping methods were the following: single, complete, centroid, unweighted average linkages, and Ward's methods.

**Nonclassical and Mixed Clustering Methods** The need to create efficient molecular representations with a low computational cost, and to obtain a high relationship value among the results of the clustering process (regarding to the structural characteristics, the behavior of some biological activity or physic-chemical properties) is a fact that accelerates the creation of new proposals guided to introduce new models for the molecular representation, new clustering algorithms based or mixing some classic methods, or proposing specific methods for chemical applications. For example, in [30] a clustering algorithm for grouping molecules according to their chemotypes is proposed. The algorithm is implemented through the following steps: (i) Pre-clusters are created using a version of the *K*-means algorithm and considering the Tanimoto similarity matrix for the molecular fingerprint representation, (ii) For each cluster the chemotype (sharing a predefined number of rings and atoms) is extracted using a MCS algorithm, (iii) In an iterative procedure, similar chemotypes and singletons are fused to larger chemotypes; singletons that cannot be assigned to any chemotype are grouped using the proportion of overlapping between the molecules, (iv) Representatives from each chemotype and the singletons are used in a second round of the hierarchical *K*-means algorithm to provide a final hierarchical grouping.

This MCS clustering approach provides a superior clustering with reduced heterogeneity in individual clusters in the overall clustering with respect to the classical *K*-means fingerprint version.

Other clustering approaches use the MCS detection over a reduced representation of molecules, for example, the method proposed in [30] where an initial preclustering using the exclusion sphere algorithm on the fingerprint database representation is carried out. Molecules in a cluster are represented as reduced graphs, and are applied as the maximum common edge substructure (MCES) algorithm, where one or more reduced graph cluster representatives are selected. The reduced graph cluster representatives are interpretable in terms of the functional activity and can be mapped directly back to the molecules giving information about the potential activity contained in the cluster.

The MCS computation has also been used at the same time with hierarchical clustering. In [87] a new hierarchical agglomerative clustering algorithm has been proposed; it uses the MCS as the similarity metric, and using an approximate approach at first, it reduces the number of operations to calculate the MCS among all the elements in the dataset. In the first step, the algorithm considers all the molecules in the dataset as singletons, and in the following steps a merged process is carried out where all the elements with the same MCS are assigned to a new cluster in the next level of the hierarchy. At the end of the process, a tree (MCS tree) is created where the root node contain one (or several) MCS common to the whole dataset. This method creates disjointed clusters where a molecule only belongs to a single cluster.

The implementation of this algorithm can be found in [89]; it uses some heuristics to reduce the computational cost, for example, a fingerprint similarity threshold and a minimum number of atoms were established as a mandatory condition for calculating the MCS, as well as the use of two clustering ways: one approximate and other exact.

The clustering method proposed in [90] is based on the representation of each of the molecules in the databases and *n*-dimensional array representing the similarity with the rest of the molecules in the database. The similarity calculation can be obtained using the MCS algorithm proposed in [16]. The clustering algorithm considers two phases: (i) the preprocessing stage where pairwise similarities for the whole database set is calculated, and (ii) the clustering process where the projection of similarities values onto a defined projection space is carried out and clusters are generated. This approach permits the selection of the projection space and the possibility of modifying the number of clusters and therefore the adjustment of the population of the clusters according to the needs of the problem. The results have been compared with the *K*-means method using the same number of clusters in both methods; the proposed methods generate more compact and isolated clusters than the *K*-means method.

A new clustering method based on the representation of the molecular structure using graphs of cycles and chains was proposed in [26]. The algorithm was developed in a two-stage process: (i) Preprocessing: in this phase the representation based on the cyclicity of the molecular structure of the database elements is generated. Three representations were evaluated considering or not the cycles and the chains, and considering or not the size. For these models the corresponding representations based on the equivalence classes were also taken into account, (ii) Processing: once the corresponding representations are built, the generated graphs are compared in order to find the isomorphic graph; in this process the Ullman algorithm is used, the graphs that are equal are assigned to one class or cluster. Two graphs are equal if they have the same number of nodes, of the same type (cycles and chains), with the same label (size, for those representations), and maintain the same set of relationships among them.

As a result of this clustering process, an index is built with the following information: molecule ID, number of nodes in the respective graph representation with information of the nodes corresponding to cycles, nodes corresponding to chains, and the class to which the molecule is assigned. This index allows us to carry out the clustering process having into account the molecule representation based on the reduced graphs generated at different abstraction levels. As several reduced graphs are generated for each molecule considering different molecular information (cycles, chains, or both) several kinds of partitional classifications are built, which can be modeled like hierarchical clustering by the different levels of abstraction considered. Authors refine each level of the hierarchical clustering considering the equivalent classes assignment for each of the types of the reduced graphs.

For the validation, a public domain database of 20677 chemical structures was used, where all the molecules were properly classified in each model for the hierarchical representation, and structural similarities values were calculated for each cluster. The higher discrimination model that generates clusters with low populations has the highest similarity values among the assigned molecules.

Recently, [91] proposes a visual representation of molecular structural hierarchies alongside with activity information. This approach creates a sequential fragmentation for all the molecules in the dataset using the decomposition rules proposed in [92]. The method generates a 2D layout where common fragments are represented with a common depiction motif and operates by searching for a global optimum aesthetic ideal. By traversing the hierarchy from the root descending to the bottom, it is possible to reveal groups of structurally related molecules. These hierarchical representations facilitate the evaluation of the structures, which are represented by the fragment nodes, and a significant amount of structure–activity information can be inferred from the layout.

## 7.3
## QSPR/QSAR Models Based on Structural Similarity

Quantitative structure–activity relationships (QSAR) represent an *in silico* methodology that establishes mathematical functions relating descriptions of biological receptor ligands with their pharmacological behavior. In this way, estimation of ligand activities and parameters, which are only available after synthesizing drugs, can be carried out from structural descriptors obtained before the drug development. So, synthesis processes are optimized with regard to environmental and economical factors.

The attempts to relate the molecular structure to compounds properties and biological activities have their origin in the XX century, were Hansch and Fujita in 1964 [93] established the principles and the foundations of QSAR investigations. These authors proposed that the biological activity of chemicals compounds is the result of the contribution of many factors behaving independently. From the last 50 years, investigations have been focused on finding those parameters related to chemical activity. Thus, thousands of molecular descriptors having into account factors such as molecular and topology features have been proposed, which allow us to build a parametric representation of chemical compounds for its later use in the application of statistical models in the development of 2D-QSAR models. Afterwards, at the end of the XX century new trends in QSAR models proposed the use of three-dimensional parameters (3D-QSAR) based on the extensive calculation of the molecular space and the alignment of the dataset molecules according to a common pharmacophore (COMFA, COMSIA, COMMA). Among the 3D-QSAR methods more recently proposed, molecular quantum similarity measures (MQSM) show an efficient approximation for the prediction of pharmacological and toxicological activities. The MQSM is based on the Maggiora similarity principle [3], but considering similarity measures based on the molecular electronic distribution (i.e., density function and kinetic energy) instead of the topological similarity used by 2D-QSAR methods.

Despite the accurate structure–activity correlations obtained with CoMFA, some shortcomings are often involved, namely requirement of information about the receptor drug interaction in order to align the dataset structures, impossibility of achieving a single alignment when very flexible molecules are modeled, complexity of the optimization of 3D conformation methods based on molecular mechanics, molecular dynamics, and so on [94–97].

Therefore, if we are looking for fast and cheap QSAR methods, some of the above-described drawbacks can be overcome by using 2D methods. Despite losing certain spatial, conformational, and mechanical aspects in the modeling ability, new QSAR methods based on fast and simple tools should be attempted with the aim of enhancing the predictive ability of 2D-QSAR tools. Thus, QSAR models will be characterized by both speed and accuracy.

Independently of the dimensionality of QSAR approaches, the developments of QSAR models imply the proposal of different schemes for: data representation, data refinement, data modeling, and data prediction. The main differences between all QSAR methods is due to the paradigm used for the spatial representation of the dataset and the statistical model used for building the prediction model. Most of the modeling techniques assume a linear relationship between molecular descriptors and the target property, which might be an adequate methodology for many datasets, using traditional statistical methods such as multivariate linear regression, principal component regression, partial least-squares regression (PLSR) with leave-one-out (LOO) validation processes [98].

However, the explosive growth of the amount of structural and biological data and the necessity of developing robust and reversible QSAR models for large and complex datasets has delivered the proposals of nonlinear models based on $k$-nearest neighbors ($k$NN), simulated annealing-partial least squares, support vector machines (SVMs), neural networks, and recently on automated lazy learning QSAR (ALL-QSAR) [99].

The advantages of QSPR/QSAR models are those related to the optimization of chemical processes regarding to residues generation, time, resources involved, and so on. QSAR models are often evaluated by means of analytical parameters such as accuracy, robustness, reversibility, costs (resources involved and time), and so on. Thus, for a chemical point of view, the quality of the model is estimated by how well it performs the mapping between the descriptors and the activity in the training set. This mapping is expressed by the correlation coefficient $R^2$ or the root-mean-square error (RMSE) between the experimental activity and the predicted one, as follows:

$$R^2 = \frac{\sum(\hat{y}_i - \bar{y})^2}{\sum(y_i - \bar{y})^2} \quad \text{RMSE} \sqrt{\frac{\sum(\hat{y}_i - \bar{y})^2}{n}} \tag{7.12}$$

where $y_i$ and $\hat{y}_i$ are the observed and fitting activities for the sample $i$, and $\bar{y}$ is the mean activity over the $n$ samples in the training set.

The validation of the model involves internal and/or external validation. For internal validation, usually cross-validation is used, consisting in removing one sample (leave-one-out) or groups (leave-group-out) of samples in a systematic or random way, for generating a statistical model for the remaining compounds, as well as for using this generated model for the prediction of those previously removed samples. The quality of the model is expressed by the slope, bias, determination coefficient $Q^2$, and standard error in prediction (SEP) calculated as Equation 7.12 where $\hat{y}_i$ are the predicted values in the external validation process.

### 7.3.1 Dataset Selection

Choosing the dataset may be the most important step when building QSAR models. This step consists in selecting the most appropriated, more distinct, and wider number of chemical compounds with a known property or activity that will be used as the training set for building a prediction model later used to predict that property or activity, unknown for another set of molecules and whose objective is to predict their values.

In this process, the training set selection is performed from chemicals databases retrieving the appropriated molecules on a basis of a specific objective criterion. Usually, the structural similarity basic objective is used in the dataset selection. Although similarity measurements are useful for extracting molecules that are structurally close to a given one, it happens many times that similar molecules highly differ in the property/activity values, thus, generating outliers in the validation model. Other approaches based on the use of molecular descriptors fall into the same problems, so it reveals, *a priori*, the existence of a correlation between the property/activity and descriptors values.

The use of high throughput screening (HTS) has widely extended for the identification of objective compounds [100]. HTS usually generates large sets classifying compounds as active or inactive regarding the target property/activity; however, compounds can share few structural characteristics and, therefore, have low structural similarity, complicating the use of the retrieved set for the development of QSAR models.

In order to avoid this inconvenience, some new methodologies have been proposed combining target functions (structural, similarity, and activity) for dataset selection. An example is the proposals based on indirect similarity [101], or binary QSAR technique [102] that correlates compound structures, using molecular descriptors, with a binary expression of the activity (1s and 0s for active and inactive compounds, respectively) and calculates a probability distribution for active and inactive compounds in a training set.

### 7.3.2
### Dataset Representation

Once selected, the compounds for the building of the QSAR model, the next step is to propose a space of representation of these molecules capable of being processed computationally. The most usual representation of the dataset is through a 2D-matrix, with a number of rows equal to the cardinality of the dataset ($M$) and a number of columns equals to the number of variables ($V$) representing those characteristics of the dataset that "we presume" are related to the property/activity to be predicted by the model.

Selection of the variables ($V$) determines the type of the QSAR approach taken. Variables are related to experimentally or theoretically calculated information about the molecules concerning with biological properties (activity, toxicity or bioavailibility, and metabolism), or physicochemical calculated descriptors including parameters accounting for hydrophobicity, topology, electronic, and sterical effects that have been determined empirically or, more recently, by computational methods. So, differences between 2D and 3D-QSAR models deal with the kind of selected descriptors, descriptors derived from molecular structure and topology (2D) and descriptors derived from structural information obtained after the alignment of the molecules of the dataset such as shape, steric, electrostatic, and so on 3D descriptors [103–105].

The high number of experimental and theoretical calculated descriptors that can be obtained from molecules datasets is related to the erroneous idea of generating large representation spaces. Although many descriptors can generate fitting models, recent studies [106] have shown that it conducts to a higher computational effort; the model has a complex mathematical interpretation that can easily be misunderstood and has less applicability.

Other approximations (2D and 3D) are based on the representation of the dataset through a similarity matrix. Here, the number of variables (*V*) is equal to the number of samples (*M*) and each variable represents the similarity between two dataset elements. Different approaches have been used in order to generate similarity matrixes. Fingerprints and graph-based similarity have shown its utility in many studies, building similarity matrixes using one similarity index or combining some of them.

Recent solutions are based on the use of the AS concept in order to build similarity matrixes mixing graphs, fingerprints and descriptor-based similarity values with dissimilarity measurements based on molecular descriptors calculated for NIFs extracted in the matching process between the dataset pairwises. These proposals have the advantages of a low computational cost and a reduced representational space allowing the consideration of any type of structural and molecular information in a unique similarity value between any pair of dataset elements.

### 7.3.3
### Fitting of the Dataset Representation

Before building the model the preparation and fitting of the representational space is necessary. Fitting the dataset representation involves, mainly two steps: (i) data standardization and (ii) data reduction.

Variables (*V*) corresponding to dataset elements representing different molecular properties are usually measured in very different scales. Standardization tends to generate, through the mathematical transformation of the representation space, an autoscalling of the variables with the aim of transforming them in order to perform a comparable contribution. Some, although related, standardization methods are usually used having into account statistical methods in the autoscalling function such as standard deviation, maximum, minimum, mean, and so on.

On the other side, large representational spaces based on the consideration of many variables conduct to computational expensive and complex models, with correlated and few significant variables. Hence, different proposals have been made to cope with fitting the representational space, such as the use of PCA, nonlinear mapping, partial least squares (PLS), neural networks, genetic algorithms, hybrid methods, machine learning, and random elimination descriptors usually [107].

### 7.3.4
### Building and Validation of the QSAR Model

Different statistical techniques can be used for building the QSAR models. Partial least-squares regression (PLSR) also allows us to reduce the original data space, being easier to visualize both trends and influences of the original variables on properties than when the original space is considered. Thus, the study of the number, type, and characteristics of the PLS factors provides scientists with structured information of their multivariate systems. In addition, PLSR considers the variance of both predictors and properties in the building of the reduced space. Thus, this construction leads to better correlations between data and properties. Other techniques based on the reduction of the variables only take into account the predictor variance. For example, principal components regression retains the relevant factors that explain the predictor set. Moreover, PLSR is useful when similarity matrices are used as representational space, so PLSR allows the use of symmetric matrices while others regression techniques, for example, multiple linear regression (MLR), require systems with more objects than predictors.

In this process the dataset is split in two subsets, training and validation, in a recommended proportional relationship of 85/15%. Usually, the validation dataset is arbitrarily selected from the entire dataset (for instance using analysis cluster, similarity/dissimilarity, activity threshold, etc.) [108], but in these strategies information about both training and validation sets is considered. The aim is to consider a fully independent validation set in order to test the predictive capacity of the model. Thus, when the dataset cardinality is enough, high splitting of the validation set in some subsets is proposed and they are used as repetitive external validation sets [103]. Apart from this, Hawkins *et al.* [105] show the problems of the dataset fragmentation for medium and small datasets, proposing the use of all the data in the calibration step and the check of the fitting by cross-validation.

In the internal validation step, cross-validation techniques, usually leave-one-out (LOO), group-leave-out (GLO), and recently Monte Carlo-CV [103, 109] are used in order to consider all the elements of the dataset, evaluating as a result, the determination coefficient ($Q^2$) and the standard error (SECV). Values of $Q^2$ greater than 0.50 and values of SECV lower than the standard deviation of the dataset predicted property are the accepted threshold for proposing a model with a statistical meaning. Shenk and Westerhaus [110] stated that $Q^2 > 0.90$ indicates an excellent precision. If $Q^2$ values are between 0.7 and 0.90, that would mean that the model has a good precision. On the other hand, $Q^2 < 0.70$ indicates that the equation can only be used for screening purposes, which enables distinction between low, medium, and high values for the measured parameter. Finally, if $Q^2 < 0.50$, then the equation only discriminates between high and low property values.

In this step, outliers are also detected. Outliers are dataset elements showing an anomalous behavior not fitting the model. Detection of outliers is checked through the values of the Student statistical variable, calculated as follows:

$$T_i = \frac{y_i - \hat{y}_i}{\text{SECV}} \tag{7.13}$$

where $y_i$ represents the value determined by the PLSR equation in the cross-validation and $\hat{y}_i$ represents the reference value. For each sample, values of $T_i \geq 2.5$ are considered as outliers.

Outliers are classified on statistical and activity parameters, also known as activity cliffs. Maggiora [34] studies the importance of splitting in the QSAR models both type of outliers, and the influence of the space representation in their apparition and detection. While statistical outliers widely depend of the dataset characteristics and the invariance of the similar neighbor's samples, activity cliffs fall into the deviation due to small fragments generating high deviation in the activity or property to be modeled.

The model built in cross-validation is used for the prediction step. In this step, the validation dataset is used, those elements not considered for the building of the model, and the statistical parameters – $r^2$ and standard error in prediction (SEP) – are obtained. Taking into account the SEP value, it is also accepted by the scientific community that the limit for considering equations as robust tools is $1.5 \times$ SECV. Here, values of SEP lower than SECV could reveal overfitting problems. In this case, the analysis of the number of PLS factors is needed and the quality of the model and the data representation could be reconsidered.

The final and, maybe, more important step is the analysis of the predictive capacity of the model. Here, the analysis of $Q^2$ and SECV in validation and $r^2$ and SEP during the validation step is not only important, but also the interpretation of outliers and the scope of the generated model depending on the dataset size and the cardinality and method used in the external validation. Recently, these parameters have been modified [104, 108, 111–113] in order to generate more robust and reproducible QSAR models.

## References


1 Willet, P., Barnard, J.M., and Downs, G. (1998) Chemical similarity searching. *Journal of Chemical Information and Computer Sciences*, **38** (6), 983–996.

2 Willett, P. (1987) *Similarity and Clustering in Chemical information Systems*, Research Studies Press, Letchworth, UK.

3 Johnson, M.A. and Maggiora, G.M. (1990) *Concepts and Application of Molecular Similarity*, Wiley-Interscience, New York.

4 Popelier, P.L.A. (1995) *Molecular Similarity and Complimentarity based on the Theory of Atoms in Molecules* (ed. P.M. Dean), Blackie Academic & Professional, London.

5 Rouvray, D.H. (1975) Similarity in chemistry: Past, present and future. *Topics in Current Chemistry*, **1**, 173.

6 Dehmer, M., Varmuza, K., Borgert, S., and Emmert-Streib, F. (2009) On entropy-based molecular descriptors: Statistical analysis of real and synthetic chemical structures. *Journal of Chemical Information and Modeling*, **49**, 1655–1663.

7 Nikolova, N. and Jaworska, J. (2003) Approaches to measure chemical similarities. A review. *The QSAR and Combinatorial Science*, **22**, 1006–1026.

8 Randic, M. (1998) Topological indices, in *Encyclopedia of Computational Chemistry* (eds P. Schleyer, N.L. Allinger, T. Clark, J. Gasteiger, P.A. Kollman, H.F. Schaefer, and P.R. Schreiner), John Wiley & Sons Ltd, Chichester, UK.

9 Holliday, J.D., Salim, N., Whittle, M., and Willett, P. (2003) Analysis and display of the size dependence of chemical similarity coefficients. *Journal of Chemical Information and Computer Sciences*, **43**, 819–828.

10 Balaban, A.T. (1999) Historical developments of topological indices, in *Topological Indices and Related Descriptors in QSAR and QSPR* (eds J. Devillers, and A.T. Balaban), Gordon and Breach, Amsterdam, The Netherlands.

11 Todeschini, R. and Consonni, V. (2000) *The Handbook of Molecular Descriptors Methods and Principles in Medicinal Chemistry Series* (eds R. Mannhold, H. Kubinyi, and H. Timmerman), Wiley-VCH, New York.

12 Rouvray, D.H. and Balaban, A.T. (1979) Chemical applications of graph theory, in *Applications of Graph Theory* (eds R.J. Wilson and L.W. Beineke), Academic Press, New York.

13 Gross, J. and Yellen, J. (1999) *Graph Theory and Its Applications*, CRC Press, Boca Raton, FL.

14 Ullmann, J.R. (1976) An algorithm for subgraph isomorphism. *Journal of the ACM*, **23** (1), 31–42.

15 Raymond, J.W., Gardiner, E.J., and Willet, P. (2002) Heuristic for similarity searching of chemical graphs using a maximum common edge subgraph algorithm. *Journal of Chemical Information and Computer Sciences*, **42** (2), 305–316.

16 Cerruela García, G., Luque Ruiz, I., and Gómez-Nieto, M.A. (2004) Step-by-step calculation of all maximum common substructures through a constraint satisfaction based algorithm. *Journal of Chemical Information and Computer Sciences*, **44** (1), 30–41.

17 Urbano Cuadrado, M., Luque Ruiz, I., and Gómez-Nieto, M.A. (2006) A steroids QSAR approach based on approximate similarity measurements. *Journal of Chemical Information and Modeling*, **46**, 1678–1686.

18 Carbo-Dorca, R. and Mezey, P.G. (1996) *Advances in Molecular Similarity*, vol. 1, JAI Press, London.

19 Leach, A.R. and Gillet, V.J. (2003) *An Introduction to Chemoinformatics*, Springer, Berlin.

20 Bunke, H., Foggia, P., Guidobaldi, C., Sansone, C., and Vento, M. (2002) A comparison of algorithms for maximum common subgraph on randomly connected graphs. *Lecture Notes in Computer Science*, **2396**, 123–131.

21 Cerruela García, G., Luque Ruiz, I., and Gómez-Nieto, M.A. (2003) A New Algorithm to Obtain All Maximum Common Subgraphs in Molecular Graphs Using Binary Arithmetic and Constraints Satisfaction Model. Proceedings of International Conference of Computational Methods in Sciences and Engineering, Kastoria, Greece, September, pp. 156–159.

22 Cerruela García, G., Luque Ruiz, I., and Gómez-Nieto, M.A. (2005) A constraint satisfaction based algorithm for the calculation of structural similarity through maximum overlapping common substructures. *MATCH – Communications in Mathematical and in Computer Chemistry*, **53**, 237–251.

23 Garey, M.R. and Johnson, D.S. (1979) *Computers and Intractability: A Guide to the Theory of NP-Completeness*, W. H. Freeman, San Francisco.

24 Chen, L. and Robien, W. (1992) MCSS: A new algorithm for perception of maximal common substructures and its applications to NMR spectral studies: 1. The Algorithm. *Journal of Chemical Information and Computer Sciences*, **32**, 501–506.

25 Cerruela García, G., Luque Ruiz, I., and Gómez-Nieto, M.A. (2004) Representation of the molecular topology of cyclical structures by means of cycle graphs: 1. Extraction of topological properties. *Journal of Chemical Information and Computer Sciences*, **44**, 447–461.

26 Luque Ruiz, I., Cerruela García, G., and Gómez-Nieto, M.A. (2004) Representation of the molecular topology of cyclical structures by means of cycle graphs: 2. Application to clustering of chemical databases. *Journal of Chemical Information and Computer Sciences*, **44**, 1383–1393.

27 Luque Ruiz, I., Cerruela García, G., and Gómez-Nieto, M.A. (2004) Representation of the molecular topology of cyclical structures by means of cycle graphs: 3. Hierarchical model of screening of chemical databases. *Journal of Chemical Information and Computer Sciences*, **44**, 1903–1911.

28 Gillet, V.J., Willett, P., and Bradshaw, J. (2003) Similarity searching using reduced graphs. *Journal of Chemical Information and Computer Sciences*, **43**, 338–345.

29 Takahashi, Y., Sukekawa, M., and Sasaki, S. (1992) Automatic identification of molecular similarity using reduced-graph representation of chemical structure. *Journal of Chemical Information and Computer Sciences*, **32**, 639–643.

30 Gardiner, E.J., Gillet, V.J., Willett, P., and Cosgrove, D.A. (2007) Representing clusters using a maximum common edge substructure algorithm applied to reduced graphs and molecular graphs. *Journal of Chemical Information and Modeling*, **47**, 354–366.

31 Barker, E.J. (2004) Chemical Similarity Searching Using Reduced Graphs. Ph.D. Thesis, University of Sheffield, Sheffield, United Kingdom.

32 Liu, R. and Zhou, D. (2008) Using molecular fingerprint as descriptors in the QSPR study of lipophilicity. *Journal of Chemical Information and Modeling*, **48**, 542–549.

33 Daylight Chemical Information Systems, Inc. http://www.daylight.com (accessed November 10, 2010).

34 Maggiora, G.F. (2006) On outliers and activity cliffss – why QSAR often disappoints. *Journal of Chemical Information and Modeling*, **46**, 1535–1535.

35 Diudea, M.V. (2001) *QSPR/QSAR Studies by Molecular Descriptors*, Nova Science Publishers, Huntington, NY.

36 Urbano Cuadrado, M., Luque Ruiz, I., and Gómez-Nieto, M.A. (2007) QSAR models based on isomorphic and nonisomorphic data fusion for predicting the blood brain barrier permeability. *Journal of Computational Chemistry*, **28**, 1252–1260.

37 Luque Ruiz, I., Urbano Cuadrado, M., and Gómez-Nieto, M.A. (2007) Data fusion of similarity and dissimilarity measurements using wiener-based indices for the prediction of the NPY Y5 receptor antagonist capacity of benzoxazinones. *Journal of Chemical Information and Modeling*, **47**, 2235–2241.

38 Ginn, C.M.R., Willett, P., and Bradshaw, J. (2000) Combination of molecular similarity measures using data fusion. *Perspectives in Drug Discovery and Design*, **20**, 1–16.

39 Salim, N., Holliday, J., and Willett, P. (2003) Combination of fingerprint-based similarity coefficients using data fusion. *Journal of Chemical Information and Computer Sciences*, **43**, 435–442.

40 Chen, J., Holliday, J., and Bradshaw, J. (2009) A machine learning approach to weighting schemes in the data fusion of similarity coefficients. *Journal of Chemical Information and Computer Sciences*, **49** (2), 185–194.

41 Urbano Cuadrado, M., Luque Ruiz, I., and Gómez-Nieto, M.A. (2006) Refinement and use of the approximate similarity in QSAR models for benzodiazepine receptor ligands. *Journal of Chemical Information and Modeling*, **46**, 2022–2029.

42 An, A. and Wang, Y. (2001) Comparisons of classification methods for screening potential compounds. Proceedings of the IEEE International Conference on Data Mining (ICDM '01) (eds N. Cercone, T.Y. Lin, and X. Wu), pp. 11–18.

43 Jain, A.K. (2010) Data clustering: 50 years beyond *K*-means. *Pattern Recognition Letters*, **31** (8), 651–666.

44 Merriam-Webster Online Dictionary. Cluster analysis. http://www.merriam-webster-online.com (accessed October 10, 2010).

45 De Maesschalck, R., Jouan-Rimbaud, D., and Massart, D.L. (2000) The mahalanobis distance. *Chemometrics and Intelligent Laboratory Systems*, **50**, 1–18.

46 Wooton, R., Cranfield, R., Sheppy, G.C., and Goodford, P.J. (1975) Physicochemical activity relationships in practice. 2. Rational selection of benzenoid substituents. *Journal of Medicinal Chemistry*, **18**, 607–613.

47 MacQueen, J. (1967) Some methods for classification and analysis of multivariate observations, in *Fifth Berkeley Symposium on Mathematics. Statistics and Probability*, University of California Press, California, pp. 281–297.

48 Jain, A.K. and Dubes, R.C. (1988) *Algorithms for Clustering Data*, Prentice-Hall, Englewood Cliffs, NJ

49 Ball, G. and Hall, D. (1965) *ISODATA, a Novel Method of Data Analysis and Pattern Classification Technical Report NTIS AD 699616*, Stanford Research Institute, Stanford, CA, pp. 129–137.

50 Forgy, E.W. (1965) Cluster analysis of multivariate data: Efficiency vs. Interpretability of classifications. *Biometrics*, **21**, 768–769.

51 Dunn, J.C. (1973) A fuzzy relative of the ISODATA process and its use in detecting compact well-separated clusters. *Journal of Cybernetics*, **3**, 32–57.

52 Backer, E. (1978) *Cluster Analysis by Optimal Decomposition of Induced Fuzzy Sets*, Delft University Press, The Netherlands.

53 Jarvis, R.A. and Patrick, E.A. (1973) Clustering using a similarity measure based on shared near neighbors. *IEEE Transactions on Computers*, **22** 1025–1034.

54 Ward, J.H. (1963) Hierarchical grouping to optimize an objective function. *Journal of the American Statistical Association*, **58**, 236–244.

55 Jain, A.K., Murty, M.N., and Flynn, P.J. (1999) Data clustering: A review. *ACM Computing Surveys*, **31** (3), 264–323.

56 Theodoridis, S. and Koutroubas, K. (1999) *Pattern Recognition*, Academic Press, New York.

57 Davies, D.L. and Bouldin, D.W. (1979) Cluster separation measure. *IEEE Transactions on Pattern Analysis and Machine Intelligence*, **1** (2), 95–104.

58 Bolshakova, N. and Azuaje, F. (2003) Cluster validation techniques for genome expression data. *Signal Processing*, **83** (4), 825–833.

59 Rand, W.M. (1971) Objective criteria for the evaluation of clustering methods. *Journal of the American Statistical Association*, **66**, 846–850.

60 Jaccard, P. (1912) The distribution of flora in the alpine zone. *New Phytologist*, **11**, 37–50.

61 Yeung, K.Y. and Ruzzo, W.L. (2001) Principal component analysis for clustering gene expression data. *Bioinformatics (Oxford, England)*, **17** (9), 763–774.

62 Fowlkes, E. and Mallows, C. (1983) A method for comparing two hierarchical clustering. *Journal of the American Statistical Asociation*, **78**, 553–569.

63 Rousseeuw, P.J. (1987) Silhouettes. A graphical aid to the interpretation and validation of cluster analysis. *Journal of Computational and Applied Mathematics*, **20**, 53–65.

64 Calinski, R.B. and Harabasz, J. (1974) A dendrite method for cluster analysis. *Communications in Statistics*, **3**, 1–27.

65 Kaufman, L. and Rousseeuw, P.J. (1990) *Finding Groups in Data: An Introduction to Cluster Analysis*, John Wiley & Sons, Inc, New York, pp. 53–65.

66 Chen, G., Jaradat, S.A., Banerjee, N., Tanaka, T.S., Ko, M.S.H., and Zhang, M.Q. (2002) Evaluation and comparison of clustering algorithms in analyzing ES cell gene expression data. *Statistica Sinica*, **12**, 241–262.

67 Sharma, Subhash (1996) *Applied Multivariate Techniques*, John Wiley & Sons, Inc, New York

68 Halkidi, M., Batistakis, Y., and Vazirgiannis, M. (2001) On clustering validation techniques. *Intelligent Information Systems Journal*, **17** (2–3), 107–145.

69 Legány, C., Juhász, S., and Babos, A. (2006) En Proceedings of the 5th WSEAS International Conference on Artificial Intelligence, Knowledge Engineering and Data Bases, World Scientific and Engineering Academy and Society (WSEAS), Madrid, Spain, pp. 388–393.

70 Scsibrany, H., Karlovits, M., Demuth, W., Müller, F., and Varmuza, K. (2003) Clustering and similarity of chemical structures represented by binary substructure descriptors. *Chemometrics and Intelligent Laboratory Systems*, **67** (2), 95–108.

71 Liu, W. and Johnson, D.E. (2009) Clustering and its application in multi-target prediction. *Current Opinion in Drug Discovery and Development*, **12** (1), 98–107.

72 Rnnar, S. and Andersson, P.L. (2010) A novel approach using hierarchical clustering to select industrial chemicals for environmental impact assessment. *Journal of Chemical Information and Modeling*, **50** (1), 30–36.

73 Holliday, J.D., Rodgers, S.L., Willett, P., Chen, M.Y., and Mahfouf, M. (2004) Clustering files of chemical structures using the fuzzy *K*-means clustering method. *Journal of Chemical Information and Computer Sciences*, **44** (3), 894–902.

74 Cross, S., Baroni, M., Carosati, E., Benedetti, P., and Clementi, S. (2010) FLAP: GRID molecular interaction fields in virtual screening. Validation using the DUD data set. *Journal of Chemical Information and Modeling*, **50** (8), 1442–1450.

75 Culp, M., Johnson, K., and Michailidis, G. (2010) The ensemble bridge algorithm: A new modeling tool for drug discovery problems. *Journal of Chemical Information and Modeling*, **50** (2), 309–316.

76 Wild, D.J. and Blankley, C.J. (2000) Comparison of 2D fingerprint types and

hierarchy level selection methods for structural grouping using ward's clustering. *Journal of Chemical Information and Computer Sciences*, **40** (1), 155–162.
77 Raymond, J.W. and Willett, P. (2002) Maximum common subgraph isomorphism algorithms for the matching of chemical structures. *Journal of Computer-Aided Molecular Design*, **16**, 521–533.
78 Raymond, J.W., Blankley, C.J., and Willett, P. (2003) Comparison of chemical clustering methods using graph and fingerprint-based similarity measures. *Journal of Molecular Graphics and Modelling*, **21** (5), 421–433.
79 Downs, G.M. and Barnard, J.M. (2002) Clustering methods and their uses in computational chemistry. *Reviews in Computational Chemistry*, **18**, 1–40.
80 Butina, D. (1999) Unsupervised data base clustering based on daylight's fingerprint and Tanimoto similarity: A fast and automated way to cluster small and large data sets. *Journal of Chemical Information and Computer Sciences*, **39** (4), 747–750.
81 Böcker, A., Derksen, S., Schmidt, E., Teckentrup, A., and Schneider, G. (2005) A hierarchical clustering approach for large compound libraries. *Journal of Chemical Information and Modeling*, **45** (4), 807–815.
82 Fisher, R.A. (1936) The use of multiple measurements in taxonomic problems. *Annals of Eugenics*, **7**, 179–188.
83 Molconn-Z is available from eduSoft at http://www.eslc.vabiotech.com/.
84 Pablo, D., Dans, E., and Coitio, E. (2009) Density functional theory characterization and descriptive analysis of cisplatin and related compounds. *Journal of Chemical Information and Modeling*, **49** (6), 1407–1419.
85 Yamashita, F., Itoh, T., Hara, H., and Hashida, M. (2006) Visualization of large-scale aqueous solubility data using a novel hierarchical data visualization technique. *Journal of Chemical Information and Modeling*, **46** (3), 1054–1059.
86 Itoh, T., Takakura, H., Sawada, A., and Koyamada, K. (2006) Hierarhical visualization of network intrusion detection data. *IEEE Computer Graphics and Applications*, **26** (2), 40–47.
87 Vargyas, M., Papp, J., Csizmadia, F., Csepregi, S., Papp, A., and Vadasz, P. (2008) Maximum Common Substructure Based Hierarchical Clustering. First presented at the 8th International Conference on Chemical Structures, 1–5 June.
88 Restrepo, G., Mesa, H., and Llanos, E.J. (2007) Three dissimilarity measures to contrast dendrograms. *Journal of Chemical Information and Modeling*, **47** (3), 761–770.
89 Chemaxon http://www.chemaxon.com. (last accessed, Nov 2010).
90 Luque Ruiz, I., Cerruela Garcia, G., and Gomez-Nieto, M.A. (2005) Clustering chemical databases using adaptable projection cells and MCS similarity values. *Journal of Chemical Information and Modeling*, **45** (5), 1178–1194.
91 Clark, A.M. (2010) 2D depiction of fragment hierarchies. *Journal of Chemical Information and Modeling*, **50** (1), 37–46.
92 Schuffenhauer, A., Ertl, P., Roggo, S., Wetzel, S., Koch, M.A., and Waldmann, H. (2007) The scaffold tree–visualization of the scaffold universe by hierarchical scaffold classification. *Journal of Chemical Information and Modeling*, **47** (1), 47–58.
93 Hansch, C. and Fujita, T.F. (1964) $\rho$-$\sigma$-$\pi$ Analysis. A method for the correlation of biological activity and chemical structure. *Journal of the American Chemical Society*, **86**, 148–152.
94 Cramer, R.D. III, Paterson, D.E., and Brunce, J.D. (1998) Comparative molecular field analysis (COMFA). 1 Effect of shape on binding of steroids to carrier proteins. *Journal of the American Chemical Society*, **110**, 5959–5967.
95 Klebe, G., Abraham, U., and Mietzner, T. (1994) Molecular similarity indices in a comparative analysis (CoMSIA) of drug molecules to correlate and predict their biological activity. *Journal of Medicinal Chemistry*, **37**, 4130–4146.
96 Silverman, B.D. and Platt, D.E. (1996) Comparative molecular moment analysis (CoMMA): 3D-QSAR without molecular superposition. *Journal of Medicinal Chemistry*, **39**, 2129–2140.

97 Rum, G. and Herndon, W.C. (1991) Molecular similarity concepts. 5. Analysis of steroid–protein binding constants. *Journal of the American Chemical Society*, **113**, 9055–9060.

98 Esbensen, K.H. (ed.) (2002) *MultiVariate Data Analysiss in Practice*, Camo Process AS, Oslo.

99 Zhang, S., Golbraikh, A., Oloff, S., Kohn, H., and Tropsha, A. (2006) A novel automated lazy learning QSAR (ALL-QSAR) approach: Method development, applications, and virtual screening of chemical databases using validated ALL-QSAR models. *Journal of Chemical Information and Modeling*, **46**, 1984–1995.

100 Gao, H., Williams, C., Labute, P., and Bajorath, J. (1999) Binary quantitative structure-activity relationship (QSAR) analysis of estrogen receptor ligands. *Journal of Chemical Information and Computer Sciences*, **39**, 164–168.

101 Wale, N., Watson, I.A., and Karypis, G. (2008) Indirect similarity based methods for effective scaffold-hopping in chemical compounds. *Journal of Chemical Information and Modeling*, **48**, 730–741.

102 Stanford, School of Medicine. http://helix-web.stanford.edu/psb99/Labute.pdf (last accessed July 2010).

103 Dimitrov, S., Dimitrova, G., Pavlov, T., Dimitrova, N., Patlewicz, G., Niemela, J., and Mekenyan, O. (2005) A stepwise approach for defining the applicability domain of SAR and QSAR models. *Journal of Chemical Information and Modeling*, **45**, 839–849.

104 Shi, L.M., Fang, H., Tomg, W., Wu, J., Perkins, R., Blair, R.M., Branham, W.S., Dial, S.L., Moland, C.L., and Sheenan, D.M. (2001) QSAR models using a large diverse set of estrogens. *Journal of Chemical Information and Computer Sciences*, **41**, 186–195.

105 Hawkins, D.M., Basak, S.C., and Shi, X. (2001) QSAR with few compounds and many features. *Journal of Chemical Information and Computer Sciences*, **41**, 663–670.

106 Nicolotti, O. and Carotti, A. (2006) QSAR and QSPR studies of a highly structured physicochemical domain. *Journal of Chemical Information and Modeling*, **46**, 264–276.

107 Zheng, W. and Tropsha, A. (2000) Novel variable selection quantitative structure–property relationship approach based on the *k*-nearest-neighbor principle. *Journal of Chemical Information and Computer Sciences*, **40**, 185–194.

108 Consonni, V., Ballabio, D., and Todeschini, R. (2009) Comments on the definition of the Q2 parameter for QSAR validation. *Journal of Chemical Information and Modeling*, **49**, 1669–1678.

109 Hawkins, D.M., Basak, S.C., and Mills, D. (2003) Assessing model fit by cross-validation. *Journal of Chemical Information and Computer Sciences*, **43**, 579–586.

110 Shenk, J.S. and Westerhaus, M.O. (1996) *Calibration the ISI Way. In Near Infrared Spectroscopy: The Future WaVes*, NIR Publications, Chischester, UK, pp. 198–202.

111 Schüürmann, G., Ebert, R.U., Chen, J., Wang, B., and Kühne, R. (2008) External validation and prediction employing the predictive squared correlation coefficient test set activity mean vs training set activity mean. *Journal of Chemical Information and Modeling*, **48**, 2140–2145.

112 Golbraikh, A. and Tropsha, A. (2002) Beware of q2!. *Journal of Molecular Graphics & Modelling*, **20**, 269–276.

113 Konovalov, D.A., Llewellyn, L.E., Vander Heyden, Y., and Coomans, D. (2008) Robust cross-validation of linear regression QSAR models. *Journal of Chemical Information and Modeling*, **48**, 2081–2094.

# 8
# Statistical Methods for Predicting Compound Recovery Rates for Ligand-Based Virtual Screening and Assessing the Probability of Activity

*Martin Vogt and Jürgen Bajorath*

## 8.1
## Introduction

Ligand-based virtual screening methods that are based upon similarity measures of molecules are among the classical approaches to compound database mining [1–4]. A very popular approach is to represent molecules as fingerprints and determine molecular similarity by calculating the similarity of the fingerprints in a pairwise manner. Fingerprints are bit string representations where individual bit positions correspond to specific features in a molecule. It should be noted that individual bit-feature correspondence usually does not apply to hashed or folded fingerprint formats. While many popular similarity measures such as the Tanimoto coefficient [1] are mostly applied on binary representations, in general, similarity measures can be defined for arbitrary numerical representations. The Bayesian statistical framework presented in the following will be formulated for this general application, but special consideration will be given to binary fingerprints.

Typically, virtual screening methods are evaluated and benchmarked in a retroactive manner. Regardless of the types of methods that are investigated, this is most often done by considering sets of known active compounds that are divided into a reference or training set, sometimes referred to as "baits," and a test set, usually referred to as potential "hits," that are then added to a background-compound database. The ability of identifying as many of these hits as possible in a virtual screen is regarded as a measure of success of a method in benchmark situations [2]. Conventionally, hit and recovery rates of active compounds are determined as performance measures in retrospective virtual screening trials [2, 5]. Hit rates are calculated by determining the number of correctly identified active molecules in database selection sets of a given size. In biological screening, hit rates refer to the fraction of active molecules within an experimentally tested compound collection. In virtual screening, recovery rates monitor the fraction of potential hits that are retrieved in database selections of increasing size. This so-called cumulative recall of active database compounds is a popular measure of virtual screening performance [5, 6]. In many instances, cumulative recall curves are drawn for up to about

*Statistical Modelling of Molecular Descriptors in QSAR/QSPR.* First Edition.
Edited by M. Dehmer, K. Varmuza, and D. Bonchev

1% of a source database as a graphically intuitive measure of screening performance [6]. For practical virtual screening applications, where database compounds are preselected for experimental evaluation, the calculation of hit rates is straightforward, but recovery rates could only be determined, if all database compounds would be assayed; otherwise, it would of course not be known how many active molecules the screening database might contain, if any. Thus, in contrast to hit rates, recovery rates can rarely be determined to evaluate the performance of "real life" virtual or sequential screening projects [2]. However, in benchmark situations, determining compound recall is essential to evaluate the specificity of virtual screening calculations, that is, the ability to select active compounds and distinguish them from false positives.

Given the importance of compound recall as a criterion for computational screening performance, the question whether compound class-specific recovery rates could also be predicted is of considerable interest. A prediction of the recovery rate would provide a relative measure for the potential success of a virtual screen, yielding an assessment of the fraction of active compounds contained in a database that would be expected to be recovered. Such a prediction, however, does not provide any indication of how many active compounds might be contained in the database. The basic underlying idea is the following: because compound collections are typically projected into (possibly very high dimensional) chemical descriptor spaces for virtual screening applications, systematic differences in the descriptor value distributions of database compounds and active reference compounds can be exploited to assess the potential for success of a virtual screening. This means that by comparing descriptor value distributions of reference and database compounds, one would like to assess what fraction of active compounds could be retrieved from the database in a virtual screen, provided the chemical features of reference and active database compounds are indeed similar (as one should expect if they share the same biological activity).

The following considerations are based on Bayesian principles where the likelihood of activity is estimated based on the value distributions of descriptors [7]. In contrast to probabilities, likelihood estimations do not yield absolute probabilities of activity. Instead, they merely allow a relative ranking of the compounds with respect to their probability of being active. The Kullback–Leibler (KL) divergence [8] can be used to quantify the differences in value distributions of active and inactive compounds and can thus be related to the expected performance of a virtual screen by training a linear model relating KL-divergence to expected recall.

In the final section of this chapter, alternative approaches for predicting activity will be discussed that might also be utilized to predict the virtual screening performance. For example, instead of predicting recovery rates, a conceptually distinct approach aims at quantitatively estimating the probability of activity for individual compounds. Such an approach, of course, would not only provide a ranking of database compounds in the order of decreasing probability of activity, but also a numerical estimate of the number of molecules expected to be active in a given selection set of database compounds.

## 8.2 Theory

First, we introduce a Bayesian approach to virtual screening. This approach can be applied to continuous, discrete, or binary descriptors such as fingerprints, which results in a specific fingerprint-based weighting scheme for virtual screening. It is shown that the KL-divergence from information theory is a suitable measure for quantifying differences between feature value distributions and that it can be used to assess the discriminatory power of features for predicting the activity of molecules. Then a practical calculation protocol is presented to predict the expected performance of virtual screening trials.

### 8.2.1 Bayesian Approach to Virtual Screening

Bayesian modeling and Bayesian naïve classifiers are currently among the most popular methods used for different types of applications in the context of virtual screening [9–13] and compound classification [14–16]. Bayesian modeling produces an estimate of the likelihood that compounds exhibit a desired property such as a target-specific activity. The key aspect of Bayesian modeling is the interpretation of descriptors as random variables that follow different value distributions depending on a certain property or specific activity. Thus, active compounds are expected to show different distributions than inactive ones for descriptors that are relevant for the activity. Typically, molecules are projected into high-dimensional chemical space representation. In order to efficiently analyze such high-dimensional data representations, the assumption is made that the dimensions have conditionally independent distributions, which plays a crucial role in the Bayesian approach. The quality of a Bayesian model will largely depend on the accurate modeling of the descriptor distributions and on the validity of the independence assumption. The independence assumption is also called the *naïvety* assumption, which will hardly ever be fully met. Nevertheless, Bayesian models have proven to be rather successful in a variety of virtual screening and other chemoinformatics practical applications.

Descriptor distributions have to be estimated from training data. In contrast to similarity search methods where single template searches are feasible, estimates of descriptor value distributions cannot be obtained from individual molecules. Discrete descriptors like individual fingerprint bits can be estimated using frequency counting, which is usually combined with some form of Laplacian correction, because the number of training data points tends to be relatively small.

Formally, let **A** describe a property like a specific biological activity and let $A$ be a set of reference compounds having this property. If the compounds $c \in A$ are represented by fingerprints $\mathbf{v}^c$, where $\mathbf{v} = (v_i)_{i=1,\ldots,m}$ is a fingerprint with $m$ positions, then the individual positions $v_i$ can be treated as Bernoulli-distributed random variables and the success probability $\Pr(v_i^c = 1 | c \in \mathbf{A})$ for a bit being set on for a compound with property **A** can be estimated from the reference data $A$:

$$p_i^{\mathbf{A}} = \frac{\#\{c \in A | v_i^c = 1\}}{N_A} \approx \Pr(v_i^c = 1 | c \in \mathbf{A}) \tag{8.1}$$

where $\#\{c \in A | v_i^c = 1\}$ is the number of compounds in $A$ where bit $i$ of the fingerprint $\mathbf{v}$ is set on and $N_A = \#\{c \in A\}$ is the total number of compounds in set $A$. It is usually sensible to apply a form of correction to the estimates $p_i^{\mathbf{A}}$ in order to avoid degenerate estimates if all bits are set on or all bits are set off in the reference set and to avoid overfitting to a usually relatively small reference set. A common approach is to apply an $m$-estimate correction [17], similar to a Laplacian correction, by incorporating the *background* frequency $p_i^{\mathbf{B}}$ of position $i$ being set on into the estimate:

$$\hat{p}_i^{\mathbf{A}} = \frac{N_A p_i^{\mathbf{A}} + m p_i^{\mathbf{B}}}{N_A + m} \tag{8.2}$$

Here, $m$ is a scaling factor determining the weight of the background frequency in the corrected estimate. For practical purposes, setting $m$ to 1 usually produces meaningful results. The background frequency can be estimated from the compound library $B$ used for the virtual screen:

$$p_i^{\mathbf{B}} = \frac{\#\{c \in B | v_i^c = 1\}}{\#\{c \in B\}} \approx \Pr(v_i^c = 1 | c \in \mathbf{B}) \tag{8.3}$$

Note that Equations 8.1–8.3 can also be easily adopted for discrete descriptors with more than two possible values.

For continuous data, it is necessary to make some assumptions about the distributions. In the absence of further knowledge, generally assuming the presence of Gaussian distributions has been shown to yield promising results in many applications [11, 18], although this is clearly only an approximation in many instances. If $\mathbf{x} = (x_i)_{i=1,\ldots,n}$ is a representation of molecules by $n$ continuous-valued descriptors that can be modeled by Gaussian distributions, then the means $\mu_i$ and standard deviations $\sigma_i$ can be estimated from the reference set:

$$\mu_i = \frac{1}{N_A} \sum_{c \in A} x_i^c \tag{8.4}$$

$$\sigma_i^2 = \frac{1}{N_A - 1} \sum_{c \in A} (x_i^c - \mu_i)^2 \tag{8.5}$$

Similar to the estimates for binary descriptors a correction can be applied to the estimates by incorporating the sample means $\nu_i$ and standard deviations $\tau_i$ from a background database:

$$\hat{\mu}_i = \frac{N_A \mu_i + m \nu_i}{N_A + m} \tag{8.6}$$

$$\hat{\sigma}_i^2 = \frac{(N_A - 1)\sigma_i^2 + m\tau_i^2}{N_A - 1 + m} \tag{8.7}$$

An alternative to assuming specific types of distributions is to divide continuously valued descriptors into discrete value ranges using binning schemes [18, 19].

As will be explained below, distance measures in chemical descriptor space can be interpreted in a probabilistic way as likelihoods and yield a theoretical foundation for the appropriateness of metrics like the normalized Euclidean distance. The basic approach is to consider the probability $\Pr(c \in \mathbf{A}|\mathbf{x}^c)$ of a compound $c$ represented by the (multi-dimensional) descriptor $\mathbf{x}^c$ to show a desired property $\mathbf{A}$. This probability cannot be estimated directly. Instead, from a set of training compounds known to possess property $\mathbf{A}$ one can estimate the probabilities $\Pr(\mathbf{x}^c|c \in \mathbf{A})$ (or probability densities $p(\mathbf{x}^c|c \in \mathbf{A})$ in the case of continuous distributions), that is, the probability of a compound $c$ to adopt descriptor value $\mathbf{x}^c$, given that the compound has property $\mathbf{A}$, as shown in Equations 8.1–8.7. Both probabilities are related to each other through Bayes' theorem:

$$\Pr(c \in \mathbf{A}|\mathbf{x}^c) = \frac{\Pr(\mathbf{x}^c|c \in \mathbf{A})\Pr(c \in \mathbf{A})}{\Pr(\mathbf{x}^c)} \tag{8.8}$$

The probability $\Pr(\mathbf{x}^c)$ can be estimated from the training data. However, the probability $\Pr(\mathbf{A})$, that is, the probability that a compound has property $\mathbf{A}$, is generally unknown. When ignoring these terms, one is not able to estimate the probability in a quantitatively meaningful way; rather, one obtains a relative measure termed the likelihood $L(c \in \mathbf{A}|\mathbf{x}^c) \propto \Pr(\mathbf{x}^c|c \in \mathbf{A})$. Consequently the likelihood ratio

$$R(\mathbf{x}^c) = \frac{L(c \in \mathbf{A}|\mathbf{x}^c)}{L(c \in \mathbf{B}|\mathbf{x}^c)} = \frac{\Pr(\mathbf{x}^c|c \in \mathbf{A})}{\Pr(\mathbf{x}^c|c \in \mathbf{B})} \tag{8.9}$$

will give a relative likelihood measure of compound $c$ having the desired property $\mathbf{A}$ when compared to compounds belonging to a class $\mathbf{B}$ not having property $\mathbf{A}$. Figure 8.1 shows Gaussian distributions estimated from samples of a hypothetical descriptor for sets of active and inactive compounds. The height of the curves is dependent on the overall probability for a compound to be active (and is artificially increased for visualization purposes).

If compound $c$ is represented by continuous descriptors $\mathbf{x}^c = (x_i^c)_{i=1\ldots n}$ in an $n$-dimensional chemical space and the assumptions of descriptor independence and Gaussian distributions are made, then from

$$L(c \in \mathbf{A}|\mathbf{x}^c) \propto p(\mathbf{x}^c|c \in \mathbf{A}) = \prod_{i=1}^{n} p(x_i^c|c \in \mathbf{A}) = \prod_{i=1}^{n} \frac{1}{\sqrt{2\pi}\hat{\sigma}_i} \exp\left(-\frac{(x - \hat{\mu}_i)^2}{2\hat{\sigma}_i^2}\right) \tag{8.10}$$

where $\hat{\mu}_i$ and $\hat{\sigma}_i$ are the mean and standard deviation of descriptor $i$, the negative log-likelihood can be derived:

$$-\log L(c \in \mathbf{A}|\mathbf{x}^c) \propto \sum_{i=1}^{n} \frac{(x - \hat{\mu}_i)^2}{2\hat{\sigma}_i^2} + \text{const} \tag{8.11}$$

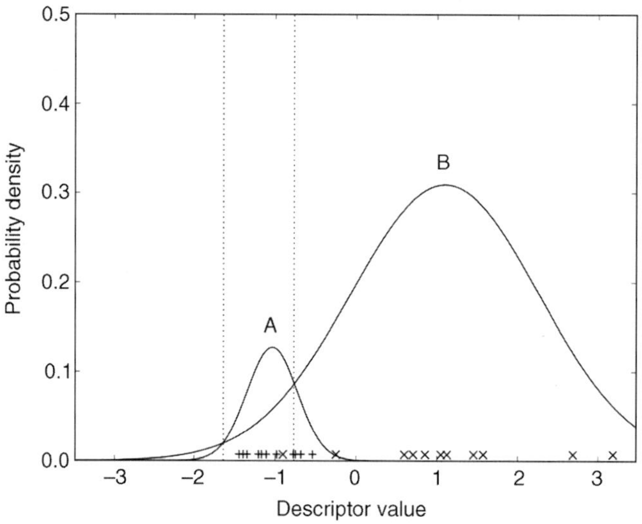


**Figure 8.1** Bayesian screening and estimation of probability densities. The schematic representation shows estimates for Gaussian distributions of a hypothetical descriptor based upon a small number of reference samples of active (curve A) and inactive (curve B) compounds. For illustration purposes, the curves are scaled assuming that 10% of a compound set would be active. The region between the dotted lines indicates the descriptor range for which compounds are more likely to be active than inactive. In practice, the ratio of active to inactive compounds in a database is unknown, but usually very small. Thus, most compounds within the indicated descriptor range will be inactive. However, they are expected to show considerable enrichment in activity compared to a random selection, especially if multiple (uncorrelated) descriptors are combined.

Note that for continuous distributions the conditional probabilities $\Pr(\mathbf{x}^c|c \in \mathbf{A})$ are replaced by the probability density functions $p(\mathbf{x}^c|c \in \mathbf{A})$. Thus, the normalized Euclidean distance [20] in chemical space is related to the assumed Gaussian distributions of the descriptor values. The ability to relate similarity metrics to descriptor value distributions, given the basic assumption of independence, makes it possible to assess the quality of these measures. It should be noted that the likelihood $L(c \in \mathbf{A}|\mathbf{x}^c) \propto \Pr(\mathbf{x}^c|c \in \mathbf{A})$ is only a relative measure of probability. For instance, if $x$ represents a structural feature that is present in 70% of a class of active compounds **A**, it might be an indicator of activity. But if this feature is also present in 90% of the compounds in set **B**, then the probability of activity is about 3.8 times higher for the 10% of the molecules that do not possess this structural feature.

Bayesian classification takes the likelihoods $L(c \in \mathbf{B}|\mathbf{x}^c)$ of compounds not possessing property **A** into account by considering the ratio of these likelihoods:

$$R(c) = \frac{L(c \in \mathbf{A}|\mathbf{x}^c)}{L(c \in \mathbf{B}|\mathbf{x}^c)} = \prod_{i=1}^{n} \frac{\Pr(\mathbf{x}_i^c|c \in \mathbf{A})}{\Pr(\mathbf{x}_i^c|c \in \mathbf{B})} \tag{8.12}$$

Taking the logarithm yields the *log-odds* score for a compound $c$:

$$\log R(c) = \frac{L(c \in \mathbf{A}|\mathbf{x}^c)}{L(c \in \mathbf{B}|\mathbf{x}^c)} = \sum_{i=1}^{n} \left(\log \Pr(\mathbf{x}_i^c|c \in \mathbf{A}) - \log \Pr(\mathbf{x}_i^c|c \in \mathbf{B})\right) \quad (8.13)$$

When one considers the negative of the logarithm, then minimizing the "distance" $-\log R(c)$ is the same as maximizing the odds $R(c)$. Following this approach for (a) an $n$-dimensional continuous descriptor space and (b) an $m$-dimensional binary fingerprint representation yields the following similarity measures:

a)

$$\log R(c) = \frac{1}{2}\sum_{i=1}^{n}\left(\left(\frac{x_i^c - \nu_i}{\tau_i}\right)^2 - \left(\frac{x_i^c - \hat{\mu}_i}{\hat{\sigma}_i}\right)^2\right) + \text{const} \quad (8.14)$$

Here $\hat{\mu}_i$ and $\hat{\sigma}_i$ are the estimated sample mean and standard deviation for descriptor $i$ for a set of training compounds $A$, and $\nu_i$ and $\tau_i$ are, respectively, the sample mean and standard deviation of descriptor $i$ of the background database $B$ not possessing that property:

b)

$$\log R(c) = \sum_{i=1}^{m} v_i \left(\log\frac{\hat{p}_i^{\mathbf{A}}}{p_i^{\mathbf{B}}} - \log\frac{1-\hat{p}_i^{\mathbf{A}}}{1-p_i^{\mathbf{B}}}\right) + \text{const} \quad (8.15)$$

For a fingerprint $\mathbf{v} = (v_i)_{i=1\ldots m}$ the Bayesian approach yields a weighting factor of $\log\frac{\hat{p}_i^{\mathbf{A}}}{p_i^{\mathbf{B}}} - \log\frac{1-\hat{p}_i^{\mathbf{A}}}{1-p_i^{\mathbf{B}}}$ for bit position $i$, where $\hat{p}_i^{\mathbf{A}}$ is the relative frequency of bit $i$ being set on for **A** and $p_i^{\mathbf{B}}$ is the relative frequency of bit $i$ being set on for **B** as estimated from Equations 8.2 and 8.3. Similar weighting schemes for binary fingerprints have been introduced [21, 22] in the context of substructural analysis methods [23].

When searching for active compounds in a large compound database using a relatively small set of active reference structures, the vast majority of database compounds will likely be inactive and only relatively few compounds will also be active. Consequently, the training set for estimating the probability distributions of active compounds consists of the reference structures and for all practical purposes, the distributions of the inactive compounds can be well approximated by considering the total compound database, including potential actives, as they only marginally influence the estimates.

### 8.2.2
### Predicting the Performance of Bayesian Screening

Due to the statistical nature of the approach, the potential success of a virtual screening based on Bayesian modeling relies on the difference in the distribution of descriptors for classes of active and inactive compounds. Simply put, the more the distributions of descriptors differ, the larger the discriminatory power of these descriptors with respect to a given biological activity. A suitable quantitative measure for the difference between distributions is the KL-divergence [8]:

$$D[p(\mathbf{x}|\mathbf{A})||p(\mathbf{x}|\mathbf{B})] = \int p(\mathbf{x}|\mathbf{A}) \log \frac{p(\mathbf{x}|\mathbf{A})}{p(\mathbf{x}|\mathbf{B})} d\mathbf{x} \tag{8.16}$$

The KL-divergence corresponds to the expected score of the log-likelihood-ratio $\log R(c)$ for compound class **A** as can be seen from Equations 8.12 and 8.13. Given estimates for the conditional distributions $p(\mathbf{x}|\mathbf{A})$ and $p(\mathbf{x}|\mathbf{B})$, the KL-divergence can be calculated analytically for normally distributed descriptors

$$D[p(\mathbf{x}|\mathbf{A})||p(\mathbf{x}|\mathbf{B})] = \sum_{i=1}^{n} \left( \log \frac{\sigma_i}{\tau_i} + \frac{\tau_i^2 - \sigma_i^2 + (\nu_i - \mu_i)^2}{2\sigma_i^2} \right) \tag{8.17}$$

and for fingerprints

$$D[\Pr(\mathbf{x}|\mathbf{A})||\Pr(\mathbf{x}|\mathbf{B})] = \sum_{i=1}^{n} \left( p_i^{\mathrm{A}} \log \frac{p_i^{\mathrm{A}}}{p_i^{\mathrm{B}}} + (1-p_i^{\mathrm{A}}) \log \frac{1-p_i^{\mathrm{A}}}{1-p_i^{\mathrm{B}}} \right) \tag{8.18}$$

In practice, Equations 8.16–8.18 are used to analyze the fitness of chemical descriptor spaces for virtual screening. Thus, the KL-divergence can be used to assess the importance of individual descriptors for the detection of activity for specific biological targets. It is therefore possible, by considering only the most discriminating descriptors, to select low-dimensional descriptor representations of molecules individually for virtual screening on specific compound activtiy classes [18].

Importantly, the relation of the KL-divergence to the log-likelihood ratio can be exploited to establish a quantitative relationship between the KL-divergence and the expected performance of virtual screening calculations [24, 25]. This provides the basis for predicting the performance of virtual screening trials. The performance of a virtual screening trial can be measured as the ratio of the number of active compounds retrieved in the selected set to the total number of actives in the compound database (i.e., the recall rate). In a first step, virtual screening benchmark trials are performed using a number of different activity classes from a database such as the MDDR [26] or ChEMBL [27]. For these classes, the KL-divergence is calculated from training sets and the recall rates of active compounds from a database are determined using the Bayesian models derived from the training sets. In a second step, a linear regression model relating the logarithm of the KL-divergence to the recall rate is derived to predict recall rates. The following section describes in detail an exemplary protocol that can be followed in order to establish such a quantitative relationship.

### 8.2.3
### Practical Prediction of Compound Recall

In the following, we present a detailed protocol of how to apply the methodology discussed above for practical virtual screening purposes using fingerprints. Calculations for continuous descriptors follow the same route.

For a screening database, a fingerprint, and a set of compound activity classes, the following steps are required in order to generate a regression curve relating the KL-divergence to recall.

1) Fingerprint calculation and bit frequency determination:
   Initially, the fingerprints for all compounds in the database and the activity classes are calculated. Then the probabilities $p_i^{\mathrm{B}}$ for each bit position $i$ are estimated from the relative frequencies according to Equation 8.3.
2) Determination of the data points for the regression curve:
   For each activity class, a number of randomized virtual screening trials are performed and the ratio of active compounds retrieved among a fraction of top-ranked database is determined. For practical applications, one should consider on the order of 100 trials per activity class where randomized subsets of about 10 compounds are used as reference molecules and the remaining compounds are added to the database as potential hits.
   - For each trial, one estimates the probabilities $p_i^{\mathrm{A}}$ for each bit position from the 10 reference compounds using the relative frequencies and corrects these values using the $m$-estimate according to Equation 8.2. Here, a value of 1 for $m$ yields meaningful results in many practical applications.
   - One then determines the KL-divergence according to Equation 8.18.
   - A virtual screen is performed using the 10 reference compounds and the recall of the active compounds is determined in, for example, the top 1% of the database (i.e., one determines the ratio of hits within the top-ranked 1% of the database).
3) The linear regression curve is derived:
   Step 2 yields about 100 data points consisting of a KL-divergence and a recall rate per activity class. For the pooled data from all activity classes, one determines a linear regression model by relating the logarithm of the KL-divergence to the recall rate.
4) Prediction of compound recall for an activity class that was not included in the regression analysis:
   For a set of active compounds, one determines the relative frequency distributions of each feature and applies the $m$-estimate correction according to Equation 8.2. Then the KL-divergence according to Equation 8.18 is determined. Finally, the regression curve derived in step 3 can be utilized to predict the recall performance for a given fingerprint.

It should be noted that expected recall rates are predicted under the assumption that active compounds are contained in the database that are similar to reference compounds in the utilized feature spaces. Furthermore, no information is available if the database indeed contains compounds having the desired activity. The estimated recall rate only predicts the percentage of active compounds that might be expected to be retrieved for the specific activity class and fingerprint, if such compound would be available in the database. However, if a fingerprint yields low predicted recall, it would not be a suitable search tool to retrieve such molecules. Rather, one would choose a fingerprint that produces high-predicted recall. If active compounds are contained in the database, this fingerprint would have a high probability of retrieving them. The practical relevance of this approach is that we can estimate for any given set of available active compounds, a fingerprint, and a screening database, how successful a virtual screen in principle might be and thus select those search tools that are most promising for a given virtual screening task.

### 8.2.4
### Exemplary Results

The protocol described above was applied to predict recall rates for Bayesian screening campaigns using a set of 142 continuous numerical descriptors implemented in the Molecular Operating Environment (MOE) [28] and, alternatively, MACCS keys, a fingerprint consisting of 166 substructural features [29]. Figure 8.2 shows the two linear regression models for the different descriptor sets built on the basis of 40 compound activity classes taken from the MDDR database. These models were then used to predict the recovery rates for seven other activity classes (represented by colored symbols in Figure 8.2) that were not included in model derivation. Recovery rates were predicted for selecting the top 1000 compounds from a screening database containing 5.8 million molecules [30].

Figure 8.2 shows that for both numerical and fingerprint descriptors, recovery rates observed in test calculations were overall well predicted for different activity classes by the regression models. It should also be noted that the choice of different descriptors often leads to significant changes in the virtual screening performance on a given activity class, which is also evident in Figure 8.2. This, again, illustrates the practical relevance of the recall prediction approach. For example, for the "blue square" activity class in Figure 8.2, the MACCS fingerprint is a much better choice than numerical descriptors, leading to much higher observed and predicted recall performance. Similar preferences for one or the other descriptor are also observed for other activity classes. In a practical prospective virtual screening application, the observed recall rate would of course be unknown, but for the "blue square" class, the predictions would strongly suggest using the fingerprint as a descriptor for a prospective application. These comparative tests can be carried out for a set of available active reference compounds for different methods and molecular representations in order to determine the most promising combination on the basis of predicted compound recall. If the screening database changes, these test calculations should be repeated.

## 8.3
## Alternative Approaches to the Prediction of Compound Recall

The KL-divergence approach discussed above is based on quantifying the differences in values distributions of sets of active and inactive compounds. Ideally, it requires a representative set of active reference compounds under the assumption that potentially active compounds in a database have descriptor values consistent with the

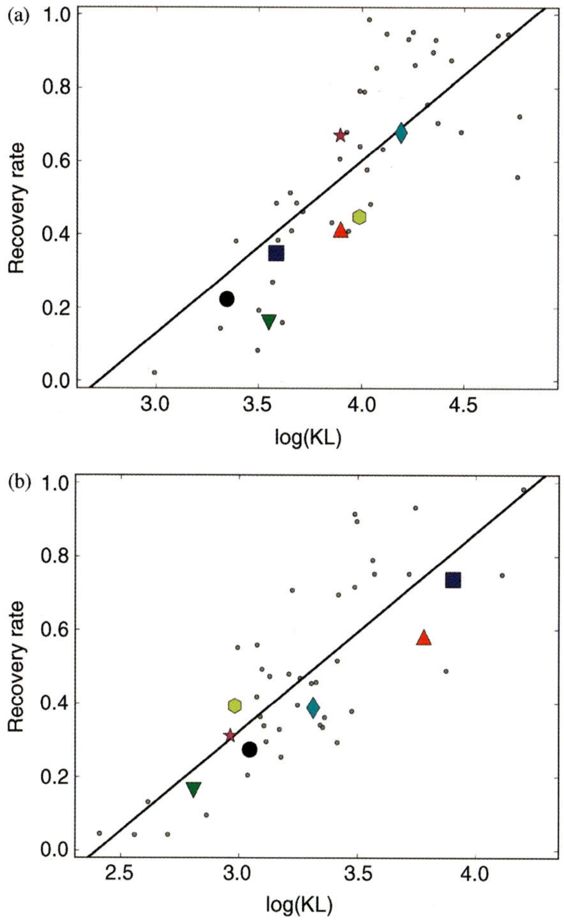


**Figure 8.2** Estimation of compound recall rates. The graphs show the relation between Kullback–Leibler divergence (KL-divergence) and the recall rate of active compounds from a screening database. A total of 40 activity classes (small gray dots) were utilized to establish a linear relationship between the logarithm of KL-divergences and observed recall rates for the top 1000 compounds of a database containing ~5.8 million molecules. A linear regression model was derived and seven other activity classes (colored symbols) were used to predict recovery rates on the basis of calculated KL-divergences using the linear model and evaluate the predictions. The measured recall rates for these seven activity classes are represented using different symbols. (**a**) Shows the regression models and Bayesian screening results for 142 continuous-valued descriptors and (**b**) the corresponding results for the MACCS fingerprint. The data have been taken from [30].

estimated value distributions. Because the approach is based on principles of Bayesian modeling, it conceptually differs from classical similarity searching, at the heart of which is the definition of a similarity metric that quantifies the similarity of a pair of molecules, given a chosen molecular representation. In similarity searching, pairwise comparisons are carried out between one or more available reference molecule and all database compounds yielding database rankings in the order of decreasing similarity to reference molecule(s). Similarity metrics such as the popular Tanimoto coefficient are frequently based upon fingerprint representations of binary features. One fundamental question in this context is to what extent activity can be inferred from the similarity value. In other words, in order to successfully search for active rather than only chemically similar compounds, one must establish a relation between fingerprint similarity and activity similarity. A popular rule of thumb has been adopted from a publication by Patterson *et al.* [31] who reported the analysis of "neighborhood behavior," which showed that 85% of compounds having a Tanimoto coefficient value of at least 0.85 when compared to know active compounds using their so-called unity fingerprints were also active. However, it should be noted that Tanimoto coefficient values are greatly dependent on the fingerprints that are used [32] and not transferable from one to another descriptor. Thus, such rules have limited generalization potential. In a recent effort, Muchmore *et al.* [33] developed a framework for quantitatively relating Tanimoto similarity scores of different fingerprints to activity probabilities. The probabilities were calculated based upon statistical analysis of pairwise comparisons of molecules from a database of 66 000 compounds with varying activity annotations for 23 different targets combined with sampling of random pairs from another in-house database. The probabilities of activity were determined from the similarity scores by fitting a sigmoidal curve to data points relating similarity values to the percentage of active compounds. This led to a quantitative assessment of the confidence that a compound was active. Confidence estimates from different fingerprints were combined using principles from belief theory [34, 35] as a framework for data fusion. Properly calibrated curves for these probabilities could also be used in a prospective manner to assess the expected success of a virtual screen, that is, by determining how many active compounds would be expected to be found in a selection set of a given size. Crucial in this context is the correct calibration from training data relating similarity scores to the probability of activity. However, establishing structure and activity similarity relationships on the basis of belief theory represents another attractive concept for the prediction of virtual screening performance.

## 8.4 Conclusions

Predicting biological activity from chemical structure is a central theme in chemoinformatics and computational medicinal chemistry. Virtual screening is a widely used approach for activity prediction that attempts to identify novel active compounds in large databases. A variety of in part rather different methodologies have been developed or adapted for ligand- and also target structure-based computational database screening. Ligand-based virtual screening generally extrapolates from known active compounds to identify novel hits. One of the premier ligand-based virtual screening techniques is similarity searching, for which many different molecular descriptors and search methods are available. The performance of virtual screening methods is typically evaluated in retrospective benchmark calculations, which often overestimate virtual screening hit and recall rates and usually do not scale with results observed in practical applications. Hence, the estimation of practical virtual screening performance has been a longstanding goal that has, however, been difficult to achieve from a methodological point of view. Recently, first attempts have been made to develop statistical methods capable of predicting virtual screening performance in a prospective manner that are based on principles of probabilistic modeling and information theory. Our discussion has focused on approaches for activity and compound recall prediction that utilize Bayesian principles and the KL-divergence method. These approaches essentially attempt to derive probabilities of specific biological compound activities based on the systematic comparison of descriptor value distributions in known active reference and screening database compounds. Despite their inherent statistical approximations, promising results have been obtained in these activity predictions. We have shown that linear models can be calibrated that quantitatively relate calculated KL-divergence to observed virtual screening recall performance for many different activity classes. Applying these models, it has then been possible to predict the recall performance for other sets of known active reference compounds, given search tools, and screening databases. A basic assumption underlying this approach is that chemical features of known reference and unknown active database compounds are similar. These test calculations also make it possible to select the best-performing molecular representations and search methods for practical virtual screening applications. It is important to note that the recall prediction models we have discussed estimate the fraction of active compounds in a database that is expected to be recovered in a virtual screening campaign using a specific search tool, provided the database does contain compounds having the desired activity. Of course, if it does not, no active compounds can be identified. Given the promising result of recall predictions obtained thus far, it is anticipated that other concepts will also be developed or adapted to bridge between molecular and activity similarity and predict the ability of different molecular representations and computational screening tools to identify new active compounds in large databases.

## References


1 Willett, P., Barnard, J.M., and Downs, G.M. (1998) Chemical similarity searching. *Journal of Chemical Information and Computer Sciences*, **38**, 983–996.

2 Bajorath, J. (2002) Integration of virtual and high-throughput screening. *Nature Reviews. Drug Discovery*, **1**, 882–894.

3 Willett, P. (2005) Searching techniques for databases of two- and three-dimensional chemical structures. *Journal of Medicinal Chemistry*, **48**, 4183–4199.

4 Willett, P. (2006) Similarity-based virtual screening using 2D fingerprints. *Drug Discovery Today*, **11**, 1046–1053.

5 Edgar, S.J., Holliday, J.D., and Willett, P. (2000) Effectiveness of retrieval in similarity searches of chemical databases: A review of performance measures. *Journal of Molecular Graphics & Modelling*, **18**, 343–357.

6 Sheridan, R.P. and Kearsley, S.K. (2002) Why do we need so many chemical

similarity search methods? *Drug Discovery Today*, **7**, 903–911.

7 Duda, R.O., Hart, P.E., and Stork, D.G. (2000) *Pattern Classification*, 2nd edn., Wiley-Interscience, New York, NY, USA, pp. 20–83.

8 Kullback, S. (1997) *Information Theory and Statistics*, Dover Publications, Mineola, NY, USA, pp. 1–11.

9 Xia, X., Maliski, E., Gallant, P., and Rogers, D. (2004) Classification of kinase inhibitors using a Bayesian model. *Journal of Medicinal Chemistry*, **47**, 4463–4470.

10 Bender, A., Mussa, H.Y., Glen, R.C., and Reiling, S. (2004) Molecular similarity searching using atom environments, information-based feature selection, and a naïve Bayesian classifier. *Journal of Chemical Information and Computer Sciences*, **44**, 170–178.

11 Vogt, M., Godden, J., and Bajorath, J. (2007) Bayesian interpretation of a distance function for navigating high-dimensional descriptor spaces. *Journal of Chemical Information and Modeling*, **47**, 39–46.

12 Vogt, M. and Bajorath, J. (2008) Bayesian screening for active compounds in high-dimensional chemical spaces combining property descriptors and molecular fingerprints. *Chemical Biology & Drug Design*, **71**, 8–14.

13 Watson, P. (2008) Naïve Bayes classification using 2D pharmacophore feature triplet vectors. *Journal of Chemical Information and Modeling*, **48**, 166–178.

14 Sun, H. (2005) A naive Bayes classifier for prediction of multidrug resistance reversal activity on the basis of atom typing. *Journal of Medicinal Chemistry*, **48**, 4031–4039.

15 Nidhi, Glick, M., Davies, J.W., and Jenkins, J.L. (2006) Prediction of biological targets for compounds using multiple-category Bayesian models trained on chemogenomics databases. *Journal of Chemical Information and Modeling*, **46**, 1124–1133.

16 Klon, A.E., Lowrie, J.F., and Diller, D.J. (2006) Improved naïve Bayesian modeling of numerical data for absorption, distribution, metabolism and excretion (ADME) property prediction. *Journal of Chemical Information and Modeling*, **46**, 1945–1956.

17 Berthold, M. and Hand, D.J. (2007) *Intelligent Data Analysis: An Introduction;*, Springer, Berlin, Heidelberg, Germany, pp. 245–246.

18 Vogt, M. and Bajorath, J. (2008) Bayesian similarity searching in high-dimensional descriptor spaces combined with Kullback–Leibler descriptor divergence analysis. *Journal of Chemical Information and Modeling*, **48**, 247–255.

19 Labute, P. (1999) Binary QSAR: a new method for the determination of quantitative structure–activity relationships. *Pacific Symposium on Biocomputing*, **99**, 444–455.

20 Godden, J. and Bajorath, J. (2006) A distance function for retrieval of active molecules from complex chemical space representations. *Journal of Chemical Information and Modeling*, **46**, 1094–1097.

21 Ormerod, A., Willett, P., and Bawden, D. (1989) Comparison of fragment weighting schemes for substructural analysis. *Quantitative Structure-Activity Relationships*, **8**, 115–129.

22 Wilton, D.J., Harrison, R.F., Willett, P., Delaney, J., Lawson, K., and Mullier, G. (2006) Virtual screening using binary kernel discrimination: Analysis of pesticide data. *Journal of Chemical Information and Modeling*, **46**, 471–477.

23 Cramer, R., Redl, G., and Berkoff, C. (1974) Substructural analysis. A novel approach to the problem of drug design. *Journal of Medicinal Chemistry*, **17**, 533–535.

24 Vogt, M. and Bajorath, J. (2007) Introduction of an information-theoretic method to predict recovery rates of active compounds for Bayesian *in silico* screening: Theory and screening trials. *Journal of Chemical Information and Modeling*, **47**, 337–341.

25 Vogt, M. and Bajorath, J. (2007) Introduction of a generally applicable method to estimate retrieval of active molecules for similarity searching using fingerprints. *ChemMedChem*, **2**, 1311–1320.

**26** Molecular Drug Data Report (2005) MDL Information Systems Inc., 14600 Catalina Street, San Leandro, CA, USA.

**27** ChEMBL http://www.ebi.ac.uk/chembldb/index.php.

**28** Molecular Operating Environment (MOE) (2007) Vers. 2007.09 Chemical Computing Group Inc., 1255 University Street, Montreal, Quebec, Canada, H3B 3×3, http://www.chemcomp.com.

**29** MACCS Structural Keys, Symyx Technologies Inc, Sunnyvale, CA, USA, http://www.symyx.com.

**30** Vogt, M., Nisius, B., and Bajorath, J. (2009) Predicting the similarity search performance of fingerprints and their combination with molecular property descriptors using probabilistic and information-theoretic modeling. *Statistical Analysis and Data Mining.*, **2**, 123–134.

**31** Patterson, D.E., Cramer, R.D., Ferguson, A.M., Clark, R.D., and Weinberger, L.E. (1996) Neighborhood behavior – a useful concept for validation of molecular diversity descriptors. *Journal of Medicinal Chemistry*, **39**, 3049–3059.

**32** Vogt, M., Stumpfe, D., Geppert, H., and Bajorath, J. (2010) Scaffold hopping using two-dimensional fingerprints: True potential, black magic, or a hopeless endeavor? Guidelines for virtual screening. *Journal of Medicinal Chemistry*, **53**, 5707–5715.

**33** Muchmore, M., Debe, A.D., Metz, J.T., Brown, S.P., Martin, Y.C., and Hajduk, P.J. (2008) Application of belief theory to similarity data fusion for use in analog searching and lead hopping. *Journal of Chemical Information and Modeling*, **48**, 941–948.

**34** Dempster, A.P. (1968) A generalization of Bayesian inference. *Journal of the Royal Statistical Society Series B-Statistical Methodology*, **30**, 205–247.

**35** Shafer, G. (1976) *A Mathematical Theory of Evidence*, Princeton University Press, Princeton, NJ, USA.

# 9
# Molecular Descriptors and the Electronic Structure

*Bögel Horst*

## 9.1
## Introduction

The predictions of chemical properties of molecules on the basis of structural features is one of most challenging tasks, especially the investigation of those molecules not yet known but to guess that they may have interesting features of potentially drugs or other helpful properties (rational drug design). Such information may guide future experimental investigations to create molecules which combine a bunch of helpful properties. The investigation of structure–property relations hires a second benefit, to create a more systematic picture of the molecular world. Until now we know more than 50 million chemical compounds over 61 million biosequences [1], for most of these compounds we know only a few data, many compounds are mentioned in the literature only once. But, in principle, each of these molecules could be characterized by more than 1000 different properties. These data could be arranged as a huge so-called molecular matrix (50 000 000 $\times$ 1000 entries) [2], which is until now only spare populated. Experimentalists are not necessarily interested to fill in the huge blank areas in that matrix; chemical research is driven by more interesting motivations, but computational informatics may therefore be convenient.

The structure of molecules is an important piece of information; structural thinking is the most important and successful concept to understand properties and behavior of compounds and to predict those for unknown substances before they are investigated experimentally.

This concept is not only very useful in chemistry and physics, but also in biology, pharmaceutics, medicine, and material sciences if the interpretation is based on structural information about molecules, solids, surfaces, and in complex biological systems.

In philosophical sense, structure and dynamics of a system cannot be separated, but from cognitive reasons one restricts often on structural aspects independent of dynamical aspects.

*Statistical Modelling of Molecular Descriptors in QSAR/QSPR.* First Edition.
Edited by M. Dehmer, K. Varmuza, and D. Bonchev

One of the most important questions related to structure–property relationships is: Which part of the structural data is in direct relation to the observed properties? As stated by Hamming in 1962: "*The purpose of computing is insight, not numbers*" [3].

## 9.2 The Structure of Molecules

### 9.2.1 General Remarks

Usually if we talk about molecules, we may have in mind a three-dimensional arrangement of the constituent atoms or sometimes as simplified picture without memorizing all the structural details; often, we remember structures as two-dimensional drawings as seen on textbooks, which reflect only the type of atoms and the type of bonds between them.

General structural information about molecules can be represented in very different ways depending on the purpose of using in chemoinformatics [4]. The simplest description arranges the connected atoms as a string (linear notation), among those the Wiswesser Line Notation, also referred to as WLN, invented by Wiswesser [5] already in 1949 and the ROSDAL string (Beilstein) were used to describe the structure of organic molecules in the early times of computer applications. SMILES (simplified molecular input line entry specification) [6], InChI (IUPAC International Chemical Identifier) [7] and InChIKey, sometimes referred to as a hashed InChI, is a fixed length (25 character) condensed digital representation of the InChI, are designed to provide a standard and human-readable way to encode molecular information and to facilitate the fast search for such molecular information in documents, databases, and on the web.

Somewhat more detailed are data based on the topology of the structures; they contain data about the type of atoms and the type of bonds they are connected with. In the Connection Table (CT) for each atom (element symbol) and its neighbors and types of the bond they are connected with are listed; the type of bonds is usually restricted to single, double, triple, or aromatic bonds (indicated by the characters: −, =, #).

Topology of molecular structures can also be represented by graph theory in the form of matrixes or tables; for reasons of simplicity, the hydrogen atoms are often not considered explicitly (hydrogen-depleted graph). A graph consists of a given number of vertices (atoms) and a number of edges (bonds) between them. There are many applications based on graphs; in our context, they could be used for generating topological indices as so-called 2D-descriptors, which are very popular in structure–property relationships and in QSAR.

The experimental 3D structure of molecules is derived from X-ray diffraction and collected in CSD [8] and PDB [9] databases or by spectroscopic investigations; these experimental data cover an immense funds of structural information. Generalized information extracted from those databases had been used to adjust empirical force fields for fast generation of reliable 3D structures. These structural models can be refined by simultaneous optimization of all structural degrees of freedom normally based on the calculated gradient vectors (forces acting on the constituent atoms and guiding to the equilibrium structure). Useful procedures of molecular mechanics are incorporated in many software packages for doing molecular modeling, QSAR, and molecular design.

The quality of the computed equilibrium structures depends on the used force field parameter set and of course on the chemical compound itself. If there are normal bonding patterns in the molecule, the quality is often more reliable than those from much more expensive quantum mechanical computations. More advanced molecular modeling software enables us also to use more expensive quantum chemical methods such as semiempirical SCF-MO (e.g., MOPAC) or *ab initio* calculations (GAUSSIAN), including the popular DFT (density functional theory) procedures or have interfaces for offline computations.

## 9.2.2
## Structure Coding

Structural information can comprise data of different levels of representation:

- 1D – linear notation: for example, brutto formula ($C_2H_5OH$ for ethanol), WLN, or SMILES
- 2D – topology (adjacency matrix, bond electron matrix, CT, etc.)
- 3D – Cartesian or internal coordinates (bond length, valence angles, and dihedral angles)
- 4D – structure and dynamics (e.g., trajectory data of a molecular dynamics simulation)

SMILES strings use a simple and comprehensive chemical language in which molecules and reactions can be specified using ASCII characters for coding atoms and bonds. SMILES are very compact and should be used in canonicalized form (canonical labeling of a graph), which is independent of the numbering of atoms taken by the user.

Matrices for 2D representation are convenient to calculate path lengths and *trough bond* distances and to derive topological indices.

The 3D representation of molecules is the basis for molecular graphics, to inspect the shape and distances between groups and fragments and to carry out computations. Most of the molecular modeling software can handle different formats for storing 3D structural data (*.pdb, *.mol, *.sdf, etc.) and convert to the desired representation. For molecules of pharmaceutical interest, one has to check for possible isomeric forms (esp. tautomers), conformations, and protonation states (typically at pH = 7.4, which represents biological media).

### 9.2.3 Structural Features

One of the most general classifications of chemical compounds is that it may belong to a homologs series (e.g., alkanes, alcohols, amino acids, etc.). All members of such a series have some features in common and show a given tendency in their properties. Linear free energy relations (LFERs) [10] can be used for property predictions.

Similarity is an important feature for comparing molecules, but there is no unique schemata for calculating the similarity. In general, similar compounds should have similar properties; this idea is one criterion for searching of potentially drugs.

If we want to characterize molecular structures, we can choose from the special structure, the electronic structure, and the energetic point of view (Table 9.1).

Molecular similarity – as a paradigm – contains many implicit and explicit assumptions, in particular, with respect to the type of chemical structures and their properties. Molecular similarity has a context that both defines and limits its use. Many rational drug design efforts are based on this principle, which states that structurally similar compounds are more likely to exhibit similar properties. Similarity methods are employed in a wide range of applications successfully, for example, virtual screening, estimation of absorption (A), distribution (D), metabolism (M), excretion (E), and toxicity (Tox), which is summarized to ADME/Tox and prediction of physicochemical properties (biological activity, solubility, partitioning, etc.). The software PASS (prediction of biological activity spectra for substances) [11] uses accumulated information about drugs to predict possible drug activities.

In general, the molecular similarity concept is most often applied when mechanistic knowledge of the system is sparse. It means one of the advantages of using

**Table 9.1** Characteristics of the 3D molecular structure.

| **Geometric** position of the nuclei | **Electronic** "position" of the electrons | **Energy** levels of discrete energy |
|---|---|---|
| Cartesian coordinates | Electron densities | Energy of molecular orbitals |
| Internal coordinates | Atomic charges | Frontier orbitals: HOMO and LUMO |
| Shape of the molecule | Polarity | Atomization energy |
| Volume of the molecule | Dipole moment (vector) | Heat of formation |
| Surface of the molecule | Polarizability (tensor) | Total energy |
| Iso-surfaces (Connolly) | Electronegativity | Binding energy |
| Surface areas for given values | Hard and soft (HSAB) | Dissociation energy |
| | Molecular electrostatic potential | Ionization energy |
| | Molecular lipophilic potential | Electron affinity |
| | Donor/acceptor numbers | Local feature values (local IP) |
| | Atom populations (e.g., Mulliken) | |
| | Bond populations | |

logp

| ID | FORMULA | WEIGHT | SMILES | Cluster 1 ID |
|---|---|---|---|---|
| 117 | C10H8 | 128,17 | C1=CC=C2C=CC=C2C=C1 | 1 |
| 189 | C13H11NO3 | 229,24 | NC1=CC=C(C(O)=C1)C(=O)OC2=CC=C... | 1 |
| 123 | C12H8S | 184,26 | S1C2=CC=CC=C2C3=C1C=CC=C3 | 1 |
| 122 | C12H8N2 | 180,21 | C1=CC=C2N=C3C=CC=CC3=NC2=C1 | 1 |
| 10 | C8H6S | 134,2 | S1C=CC2=C1C=CC=C2 | 1 |
| 67 | C9H7N | 129,16 | C1=CC=C2N=CC=CC2=C1 | 1 |
| 68 | C10H8 | 128,17 | C1=CC=C2C=CC=CC2=C1 | 1 |
| 64 | C9H7NO | 145,16 | OC1=C2N=CC=CC2=CC=C1 | 1 |
| 66 | C9H7N | 129,16 | C1=CC=C2C=NC=CC2=C1 | 1 |
| 130 | C12H10S2 | 218,34 | S(SC1=CC=CC=C1)C2=CC=CC=C2 | 1 |
| 128 | C10H8O3S | 208,24 | O[S++]([O-])([O-])C1=CC=C2C=CC=CC2... | 1 |
| 63 | C8H6N2 | 130,15 | C1=CC=C2N=NC=CC2=C1 | 1 |
| 131 | C12H10 | 154,21 | C1=CC=C(C=C1)C2=CC=CC=C2 | 1 |
| 177 | C13H10N2 | 194,24 | N1C2=C(C=CC=C2)N=C1C3=CC=CC=... | 1 |
| 124 | C13H9N | 179,22 | C1=CC=C2N=C3C=CC=CC3=CC2=C1 | 1 |
| 192 | C15H11ClN2O | 270,72 | ClC1=CC=C2NC(=O)CN=C(C3=CC=CC... | 1 |
| 125 | C14H10 | 178,23 | C1=CC2=CC=C3C=CC=CC3=C2C=C1 | 1 |
| 8 | C8H6O | 118,14 | O1C=CC2=C1C=CC=C2 | 1 |
| 126 | C14H10 | 178,23 | C1=CC=C2C=C3C=CC=CC3=CC2=C1 | 1 |
| 71 | C7H8N4O3 | 196,17 | CN1C(=O)NC(=O)C2=C1NC(=O)N2C | 2 |
| 72 | C6H6N4O3 | 182,14 | CN1C(=O)NC2=C(NC(=O)N2)C1=O | 2 |
| 69 | C7H8N4O3 | 196,17 | CN1C(=O)NC2=C(N(C)C(=O)N2)C1=O | 2 |
| 70 | C7H8N4O3 | 196,17 | CN1C(=O)N(C)C2=C(NC(=O)N2)C1=O | 2 |
| 79 | C7H8N4O2 | 180,17 | CN1C=NC2=C1C(=O)N(C)C(=O)N2 | 2 |
| 78 | C7H8N4O2 | 180,17 | CCN1C(=O)NC(=O)C2=C1N=CN2 | 2 |
| 81 | C5H5N5O | 151,13 | NC1=NC2=C(N=CN2)C(=O)N1 | 2 |
| 80 | C7H8N4O2 | 180,17 | CN1C=NC2=C1C(=O)NC(=O)N2C | 2 |
| 74 | C6H6N4O3 | 182,14 | CN1C(=O)NC2=C1C(=O)NC(=O)N2 | 2 |
| 73 | C6H6N4O3 | 182,14 | CN1C(=O)NC(=O)C2=C1NC(=O)N2 | 2 |
| 76 | C6H6N4O2 | 166,14 | CN1C(=O)NC(=O)C2=C1N=CN2 | 2 |
| 75 | C6H6N4O2 | 166,14 | CN1C(=O)NC2=C(NC=N2)C1=O | 2 |
| 135 | C8H10N4O2 | 194,19 | CCCN1C(=O)NC(=O)C2=C1N=CN2 | 2 |
| 2 | C5H4N4O2 | 152,11 | O=C1NC(=O)C2=C(N1)N=CN2 | 2 |
| 141 | C8H8Cl2IO3PS | 413 | CO[P+]([S-])(OC)OC1=CC(Cl)=C(I)C=C1Cl | 3 |
| 142 | C8H8BrCl2O3PS | 366 | CO[P+]([S-])(OC)OC1=CC(Cl)=C(Br)C=C... | 3 |
| 138 | C8H9ClNO5PS | 297,66 | CO[P+]([S-])(OC)OC1=CC=C(C=C1Cl)[N... | 3 |

**Figure 9.1** Result from a MCS clustering of compounds of a log *P* database with 200 compounds.

molecular similarity methods is that no external knowledge is necessary. The calculated molecular similarity is extensively used in applications such as virtual screening, property prediction (QSAR), synthesis design, and chemical database clustering (Figure 9.1).

Besides similarity, the commonality is helpful to classify chemical compounds only on the basis of their structures. A series of compounds can be clustered according to their most common substructures (MCSs), which can be shown as a tree using Marvin [12] (Figure 9.2).

As a main result from this clustering, one recognizes that different substructures may have different properties; here in Figure 9.3, the first cluster with 20 compounds (bicyclic heteroaromatic compounds) has log *P* values in the range 1.33–0.33, while Cluster No. 2 with nine compounds (with phenyl-$SO_2$ substructure) has values from −0.09 to 2.74 as you can see from the distributions (histograms on the right

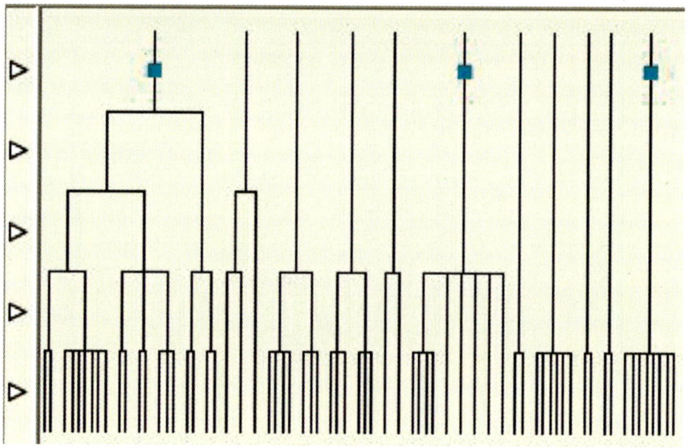

Figure 9.2 Tree of that MCS clustering.

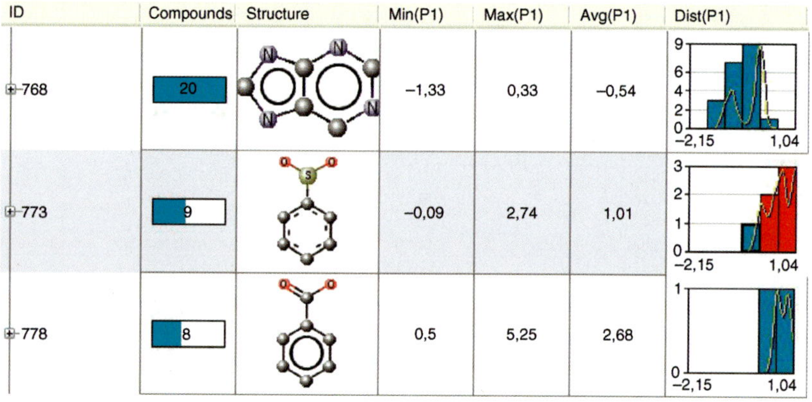


Figure 9.3 The three "marked" clusters of figure 9.2.

side). But similar substructures of Cluster No. 2 and Cluster No. 3 (with phenyl-$CO_2$ substructure) have similar substructures and quite similar log *P* values.

### 9.2.4 Structure and Energy

The energy is the most important quantity to characterize a molecule to show its stability in comparison to isomeric structures. Figure 9.4 shows the dependence of energy (*E*) on a bond distance. In order to compute those quantities, one applies the Born–Oppenheimer approximation, which is the separation of nuclear and electron motion for solving the Schrödinger equation.

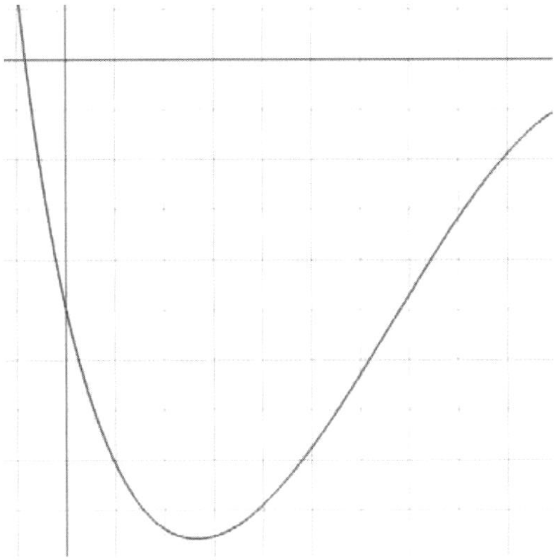

**Figure 9.4** Typical potential energy curve.

The gradient ($g_i$) can be calculated in order to optimize structures; the gradient component corresponds to the negative of the force ($F$). In the minimum of this curve, the gradient vanishes; this characterizes the equilibrium structure:

$$g_i = \frac{\mathrm{d}E}{\mathrm{d}Q_i} = -\mathrm{F} \tag{9.1}$$

Optimization of the structure of a molecule is done by simultaneous minimization of all gradient components, this means minimization of the total energy of the molecule with respect to all atomic positions. Such an optimization leads to a so-called local minimum on the energy hypersurface depending on the chosen start structure. A systematic or stochastic search for other energy minima and the global energy minimum is not trivial and becomes more difficult with increasing number of atoms and the degree of freedoms.

## 9.3 The Electronic Structure

The electronic structure is a very complex quantity; it represents the electron distribution within the molecule and determines the shape and the surface of the molecule and all its chemical behavior and the interaction with other species or the environment (solvent). In all experiment with molecules, we have to deal with the electronic structure (see Table 9.2) especially if one has in mind chemical reactivity and reaction mechanism.

**Table 9.2** Some observations and features of the electronic structure.

| Observation | Feature of the electronic structure |
|---|---|
| X-ray diffraction pattern | Electron density |
| IR vibrational spectra | Force constant |
| NMR-spectra: chemical shift and spin–spin coupling | Electron density and overlap density |
| UV spectra: band position and oscillator strength | Energy of states and transition moment |

From X-ray or electron diffraction experiments, one can derive electron density contour lines, but these diffraction data originate mainly from positions of higher electron density; these are the so-called inner-electrons. The chemical bonding takes place not from the inner-electrons but from the valence electrons, which can arrange between nuclei to form chemical bonds.

The direct way to get complete information about the electronic structure is quantum mechanical calculations based on the Schrödinger equation. The quality of those computations depends on the approximations of the Hamiltonian and the ansatz for the wavefunction (LCAO-MO and basis set in Hartree–Fock and DFT calculations).

The eigenvalues and eigenfunctions (wavefunctions) can be obtained for a given 3D structure of the molecule by solving the Schrödinger equation with some approximations incorporated in the Hamiltonian (in general, one uses an one-electron Hamilton operator and a self-consistent field (SCF) procedure to take into account the influence of other electrons. From these "best" one-electron wavefunctions ($\psi$-orbitals), one can construct an antisymmetric product wavefunction ($\Psi$) of the electronic state of the molecule. The wavefunction contains all necessary information about the electronic structure of the molecule. Any expectation value ($Q$) results from applications of its corresponding operator ($\mathcal{Q}$) and the wavefunction:

$$Q = \int \Psi \mathcal{Q} \Psi \mathrm{d}\tau \tag{9.2}$$

Some properties of molecules can be directly calculated as observables ($Q$), such as the dipole moment by using the corresponding operator ($e^* R$), the kinetic or potential energy by using the parts of the Hamiltonian, but we do not go into the fundamentals of computational chemistry [13].

The quality of the expectation value depends only on the approximation of the wavefunction, but there is no general trend that more expensive computations lead to better results.

Other possibilities of quantities from quantum chemical calculations are based on derivatives of the total energy; some examples are listed in Table 9.3 (here, we have restricted to the first (1) and second (2) derivatives with respect to some external "perturbation."

**Table 9.3** Properties calculated as derivatives of the quantum mechanical energy[a].

| *F* | *B* | *I* | *Q* | *N* | Property |
|---|---|---|---|---|---|
| 1 | | | | | Electric dipole moment ($\mu$) (vector) |
| 2 | | | | | Electric polarizability ($\alpha$) (tensor) |
| | 1 | | | | Magnetic dipole moment |
| | | 1 | | | Hyperfine coupling constant |
| | | | 1 | | Force (gradient of the energy) (vector) |
| | | | 2 | | Force constants (Hessian matrix) |
| | | | | 1 | Electronegativity |
| | | | | 2 | Hardness |

a) *F*, external electric field; *B*, external magnetic field; *I*, internal magnetic field of the spin; *Q*, structural degree of freedom; *N*, number of electrons

It should be mentioned that there are even higher derivatives and of course mixed derivatives, for example, the infrared intensity is given by Equation 9.3:

$$I \propto \left(\frac{\partial \mu}{\partial Q}\right)^2 \propto \left(\frac{\partial^2 E}{\partial Q \partial F}\right)^2 \tag{9.3}$$

The quality of the predicted properties depends only on the accuracy of the calculated wavefunction.

Some remarks about electronegativity ($\chi$), the concept was introduced pure empirically by Pauling and later by Mulliken [14] introduced the orbital electronegativity as the arithmetic means of ionization energy IE) and electron affinity (EA)

$$\chi = 1/2(\mathrm{IP} + \mathrm{EA}) \tag{9.4}$$

The electronegativity of orbitals, atoms, fragments, and molecules is very convenient schemata to guess the change in the electron densities during formation of bonds. Atoms with higher electronegativity tend to attract electron density from their neighbors, but electronegativity of an atom is not a fixed quantity as shown in textbooks. Electronegativity changes with the electron density. In Figure 9.5, the energy of an atom with respect to the number of electrons (*N*) is shown. The gradient $dE/dN$ is the slop of this curve, which is the electronegativity (see Table 9.3), and corresponds to the chemical potential. This dependence is widely used to calculate atomic charges in molecules by the partial equalization of orbital energy (PEOE) [15] algorithm.

As shown above, the electronegativity is the relation between the energy and the atomic charge (see Figure 9.5). Electronegativity is often used as an atomic feature, but fragments and molecules can be treated in the same way. The concept of electronegativity is derived from the energy of the system, which is the most decisive quantity for the interpretation and prediction of chemical changes. The second

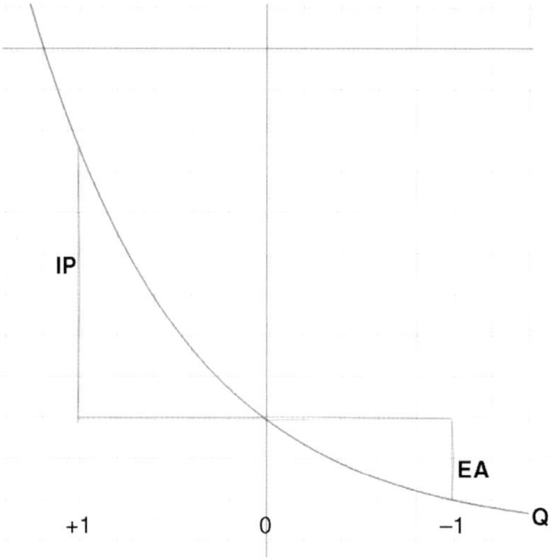


**Figure 9.5** The energy of an atom depending on its charge.

derivative of the energy with respect to the number of electrons is the hardness, which was introduced in chemistry in former times as an empirical concept.

The electronic structure is the most important quantity of the molecular structure. In any experimental investigation, we have to deal with the electronic structure. The geometrical structure results from the electronic structure. This can be seen from, for example, UV-spectra; in ground and excited states, we have different energies and different electronic structure, which results in different geometrical structures.

## 9.4 Dividing Molecules in Atoms and Bonds

### 9.4.1 Bonding in Molecules

There is a question: Is a molecule just the sum of the atoms and the bonds between them? From theoretical point of view, we have the atoms and their electrons can make attractive and repulsive interactions.

Very simply spoken, the reason that molecules may be more stable than the separated atoms is that the electrons in a molecule are attracted not only by their own nuclei but also by some other nuclei. This can be achieved if the electrons prefer position between the atoms (bonding region). In contrast to bonding orbitals, there are even antibonding orbitals, and if those will be occupied with electrons, these will weaken the bond. This leads to the conclusion that there exists an optimal value of electron density between atoms depending on the nature of the atoms and we have to take into account that electrons show general repulsion to other electrons because of their common negative charge.

### 9.4.2 Energy Partitioning

Really the quantum mechanically calculated energy of a molecule can be divided into one- and two-center energies, which summarize to the total energy. This finding is good understanding of traditional chemistry that similar structures (fragments) have similar energies. We will show that the electronic structure is a representative of the energy. Such a relationship is important because the energy of a system is the fundamental quantity if one compares two structures. The relationship between the electronic structures as a representative of the energy of a molecule can be demonstrated by using the well-known Hückel–MO [16] method, which is one of the simplest molecular orbital methods only based on topology. This covers some similarities to the graph theory and topological descriptors [17]. Therefore, the HMO method is only based on molecular skeleton, there are no through space interactions between nonbonded atoms. Only a few simplifying approximations are introduced by Hückel to solve the Schrödinger equation (matrix eigenvalue problem).

The total energy of a molecule can be divided into contributions of atoms and bonds without additional assumptions.

### 9.4.3 Energy and the Hückel Approach

Some of the essential approximations of the HMO method (Equations 9.5–9.9) are as follows:

- The Hamiltonian is simply a sum of one-electron Hamiltonians of electron *i*.
- The wavefunction (MO) is a linear combination of atomic orbitals (LCAOs).
- Atomic one-center integrals depend only on the type of atom, Coulomb integral.
- Bonds are characterized by resonance integrals.
- Overlap is neglected in the secular equations.

$$\mathcal{H} = \sum\nolimits_{i=1}^{n} \mathcal{H}_i \tag{9.5}$$

$$\psi_i = \sum\nolimits_r c_{ir}\varphi_r \tag{9.6}$$

$$\alpha_r = \int \varphi_r \mathcal{H} \varphi_r \mathrm{d}\tau \tag{9.7}$$

$$\beta_{rs} = \int \varphi_r \mathcal{H} \varphi_s \mathrm{d}\tau \tag{9.8}$$

$$S_{rs} = \int \varphi_r \varphi_s \mathrm{d}\tau = \delta_{rs} = 0(\text{for } r \neq s) \tag{9.9}$$

The orbital energy $E_i$ can be evaluated from the corresponding wavefunction $\psi_i$:

$$E_i = \int \psi_i \mathcal{H} \psi_i \mathrm{d}\tau \tag{9.10}$$

Introducing the LCAO representation for the MOs gives

$$E_i = \sum_r c_{ir}^2 \int \varphi_r \mathcal{H} \varphi_r \mathrm{d}\tau + 2\sum_r \sum_{<s} c_{ir} c_{is} \int \varphi_r \mathcal{H} \varphi_s \mathrm{d}\tau \tag{9.11}$$

The total energy $E$ of all the $n$ electrons in the molecule is just the summation over all orbitals with their occupation ($\varrho_i = 2$, 1, or 0 electrons):

$$E = \sum_{i=1}^{n} \varrho_i E_i = \sum_i \varrho_i \sum_r c_{ir}^2 \alpha_r + 2\sum_i \varrho_i \sum_r \sum_{<s} c_{ir} c_{is} \beta_{rs} \tag{9.12}$$

If we introduce the definitions of electron density ($q_r$) and bond order ($p_{rs}$) as elements of the $P$-matrix, we get Equation 9.15:

$$q_r = \sum_i \varrho_i c_{ir}^2 \quad \text{–electron density of atom } r \tag{9.13}$$

$$p_{rs} = \sum_i \varrho_i c_{ir} c_{is} \quad \text{–bond order for bond between atoms } r \text{ and } s \tag{9.14}$$

### 9.4.4
### Energy Components of Atoms and Bonds

So we have split the total energy ($E$) into a summation only over atomic and bond contributions; these contributions are proportional to the electron density and bond order, respectively. The total energy or, in other words, the stability of a molecule is a representative of the distribution of electrons (electronic structure). The electron density of an atom (respectively bond) represents its energetic contribution to the molecule's energy:

$$E = \sum_r q_r \alpha_r + 2\sum \sum_{r<s} p_{rs} \beta_{rs} \tag{9.15}$$

$$E = \sum_r E_r + 2\sum \sum_{r<s} E_{rs} \tag{9.16}$$

$$E = \sum_r \mathrm{E}(\text{Atom}, r) + 2\sum_{r<s} \mathrm{E}(\text{Bond}, r\text{–}s) \tag{9.17}$$

In this context, it should be mentioned that interpretations of reaction mechanisms and the speed of chemical reactions are often explained with orbital electron densities or atomic charges. The electron density of each atom in a molecule represents the energy level of that atom and similarly the bond order indicates energy level of its bond. This could be useful to support empirical incremental schemata in chemistry.

In more advanced MO methods, the total energy can even be expressed in terms of one- and two-center terms (energy partitioning [18]), but not as simple as it was shown for the Hückel–MO method, but there are more terms depending on the more complex Hamiltonian.

This type of energy partitioning corresponds to the conventional chemical intuition that the molecule comprises atoms and bonds and these components have a special electronic structure, which is directly related to energy components which summarize to the total energy of that molecule.

### 9.4.5
### Perturbation Treatment of the Electronic Structure

We can understand the interaction or a chemical reaction between molecules causes a given "perturbation" of the atoms at the reaction center (e.g., change of the hybridization). In the above equations causes such an atomic change some deviation of the Coulomb and Resonance Integrals from the unperturbed value.

In order to predict the local sensitivity of a molecule, the first-order perturbation treatment is very helpful. If any change at the reaction center $\Delta\alpha_r$ (in the electronic state of an atom $r$) within the molecule happens, so the energy change $\Delta E$ of the molecule is proportional to the electron density $q_r$ of that atom. If some change in the bonding occurs, the bond order $p_{rs}$ is a direct measure for the energy changes to be observed:

$$\frac{\partial E}{\partial \alpha_r} = q_r \tag{9.18}$$

$$\frac{\partial E}{\partial \beta_{rs}} = 2p_{rs} \tag{9.19}$$

In the same way, the second-order perturbation (second derivative of the energy with respect to changes) can be used to derive valuable quantities, the so-called self-polarizabilities [19].

If we take into account a certain "perturbation" of an atom in a molecule, the energetic effect is greater for centers of high electron density.

The atomic charge is a very convenient descriptor in QSAR: the reason is that not only that such atomic charge interacts with some atomic centers of an approaching molecule in the sense of the coulomb interaction, but also the electron density of an atom characterizes the atom energetically.

### 9.4.6
### Thermodynamic Equilibrium

Here, we have discussed the stability on the basis of energy; this is very convenient, but in the sense of thermodynamics, the existence of two systems in equilibrium state, given by the Gibbs free energy ($\Delta G$) as in Equation 9.21 gives the decision of which one is more preferred (equilibrium constant $K$).

$$\Delta G = \Delta H - T\Delta S \tag{9.20}$$

$$\Delta G = -RT\ln(K) \tag{9.21}$$

Often one considers only the energy, this implies that we do not take care of the temperature ($T$) of the system and of change in entropy ($S$); $R$ is a gas constant.

### 9.4.7
### Model of "Atom in Molecules"

Since the charge density describes the distribution of electrons in real space, it is a physically measurable quantity. Consequently, when used as a basis for the discussion of chemistry, the charge density allows for a direct physical picture and interpretation. For more expensive investigation of the bonding situation of the calculated electron distribution in a molecule Baders atoms in molecules (AIM) [20] procedure, the so-called topological analysis of the electron density is recommended. It characterizes regions of atoms and bond critical points as characteristic points of the spatial electron density distribution in the 3D molecular structure. An example of using AIM is shown for a Li–allyl complex [21] structure in Figure 9.6. The electronic structure of the complex was calculated using DFT with a moderate basis set. Figure 9.6 shows the nonclassical bonds between Li-ion and the three carbon atoms which form a

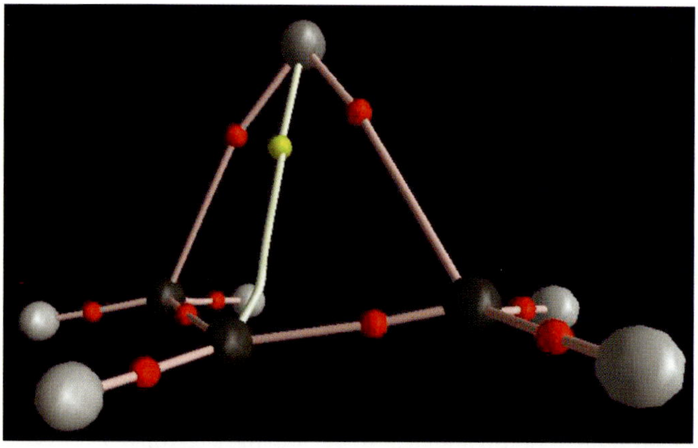

**Figure 9.6** Li–allyl complex with *bond critical points* between the atoms (in red and yellow).

$\pi$-complex structure. The AIM procedure provides information about the extension of atomic regions and would be useful to investigate similarity between molecules.

## 9.5
## Structure and Dynamics

### 9.5.1
### Molecular Flexibility

Even at the temperature of 0 K, the atoms in molecules move around their equilibrium positions, the movement increases with rising temperature. The molecules show certain flexibility and may exist in different conformations and isomers (e.g., tautomers). An ensemble of molecules shows a broad range of kinetic energies described by the statistics of a Maxwell–Boltzmann distribution depending on the temperature. With increasing temperature, we can observe more intense internal rotations and other kinetic processes like chemical reactions.

Molecular dynamics (MD) addresses the problem of solving the classical equations of motion for a system of $N$ atoms. With computational methods, we can simulate single molecules or an ensemble of particles. There are some strategies of running MD simulations, including those, where the temperature can be changed (heating or cooling) for certain periods of time (e.g., simulated annealing).

### 9.5.2
### Molecular Dynamics Simulation

As an example of a MD simulation on a small molecule, the results are shown for ethanol at a constant temperature of 290 K with time step of $t = 0.002$ ps and duration of 100 time steps. Additional conditions are given in Figure 9.7; an interface of the MOE [22] modeling software.

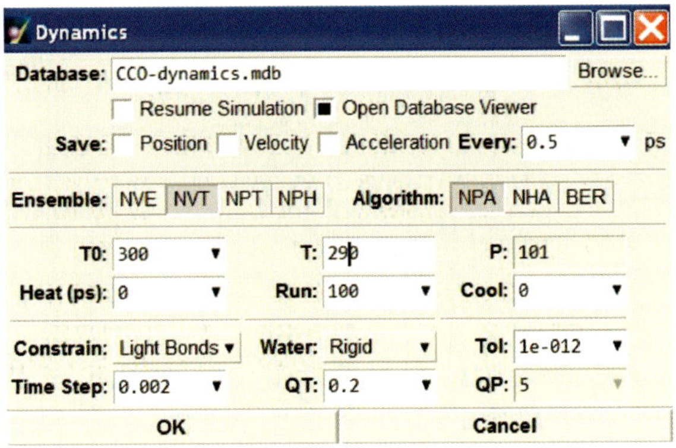


**Figure 9.7** Parameters of the MD simulation.

### 9.5.3
### Conformational Space

To get an impression of the results from such a MD simulation, we have depicted three geometrical parameters from the results of the trajectory database (Figure 9.8). The –C–C–O valence angle (in color: blue) shows only small fluctuations about the equilibrium value of about 109°, the H–C–C–O torsional angle (red) as well has only small fluctuations about 60° and the black curve shows some conformational changes of the C–C–O–H angle (Figure 9.9) starting at about 60°.

This dynamical behavior follows from the profiles of the potential energy hypersurface. In Figure 9.10, the potential energy curve for dihedral angle C–C–O–H (the internal rotation about the C–O bond) shows three minima, the deepest energy at 180° (marked with red line), and two equivalent conformations at 60° and –60°. In Figure 9.11, the internal rotation about the C–C bond is shown; here the barriers are about 3.5 kcal/mol, whereas the barrier about the C–O bond is only 1.5 kcal/mol. So we have observed at low temperatures ($T = 290$ K) only rotations about the C–O bond.

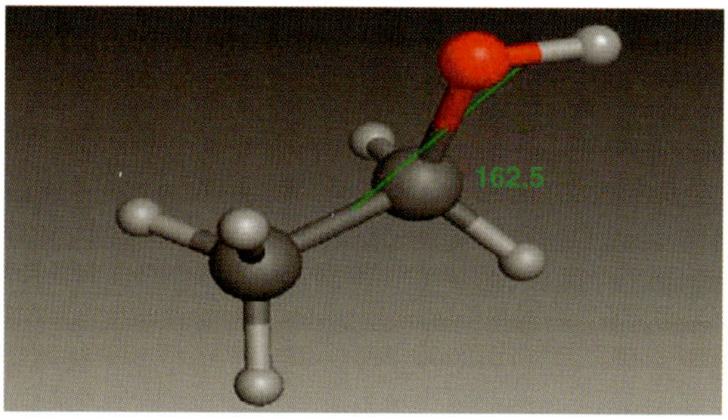


**Figure 9.8** Ethanol molecule and C–C–O–H torsional angle.

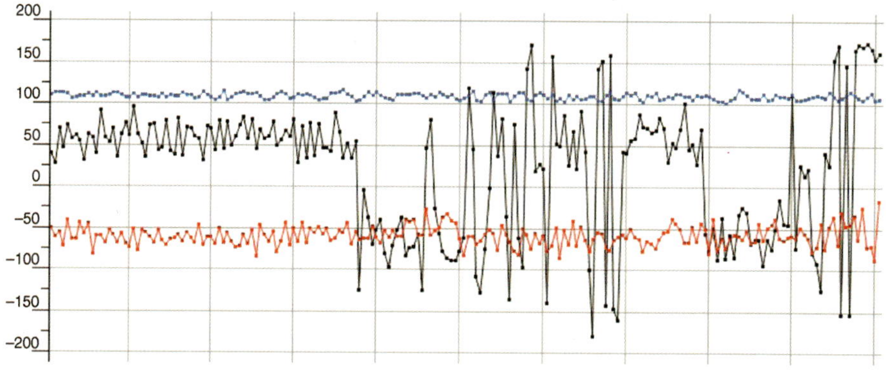


**Figure 9.9** Some results from the MD simulation of ethanol molecule ($x$-axis is time).

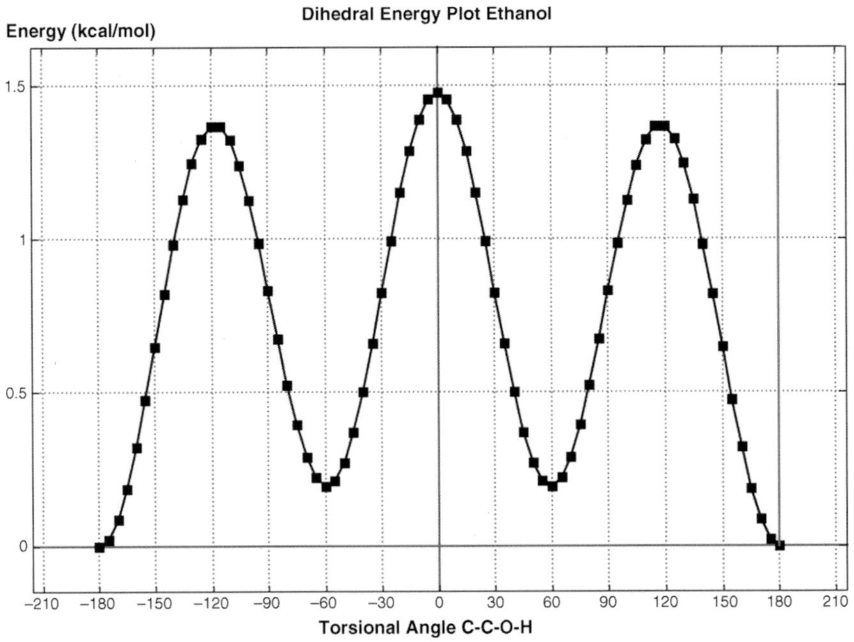


**Figure 9.10** Calculated torsion potential C—C—O—H.

If we take these two internal rotations together, we get the potential energy surface shown in Figure 9.12 as contour plot (energy minima are shown in pink color). In the middle of Figure 9.12 is the global minimum (both angles are 180°) surrounded by eight local minima of higher energies. Here, we have chosen the ethanol as a small molecule, which has only these two torsional degrees of freedom; in applications of drug design, we have to deal with much larger molecules with high-dimensional hypersurface.

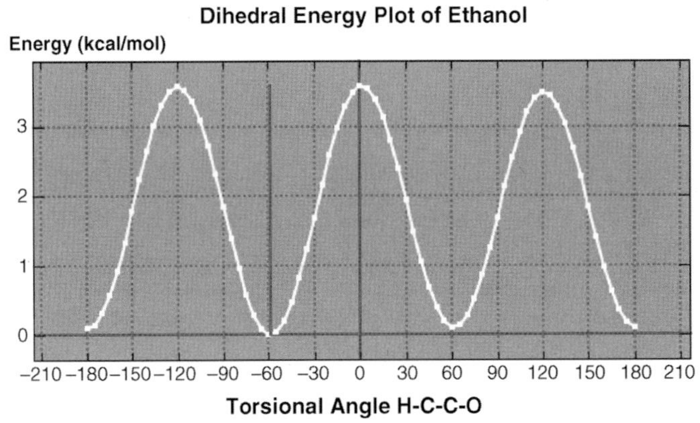


**Figure 9.11** Calculated torsion potential H—C—C—O.

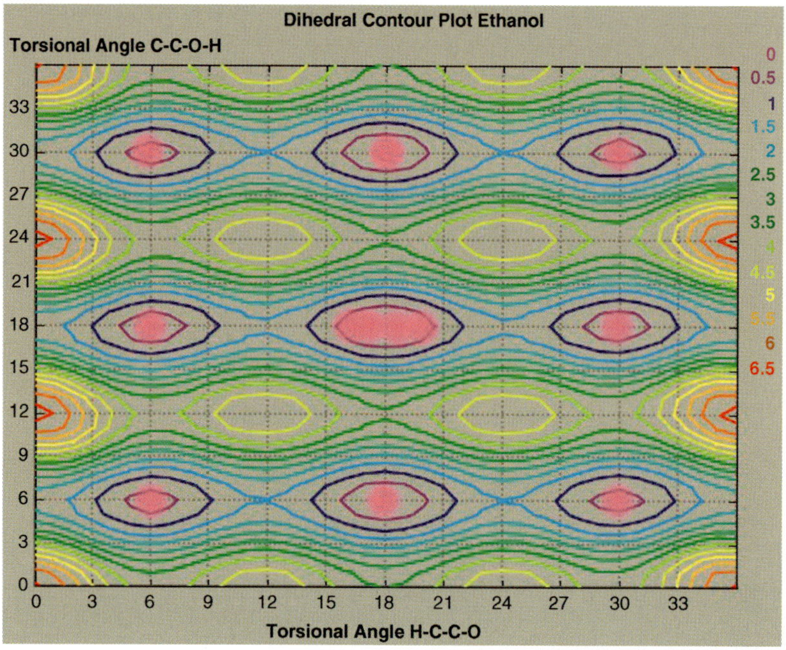


**Figure 9.12** Contour plot of ethanol molecule (energies in kcal/mol with step = 0.5).

MD simulations are very useful to study the flexibility of molecules and search for preferred conformations in molecules with many degrees of freedom and to study the immovability of drug molecules in receptor pockets or in ion channels.

## 9.6
## Structure and Properties

### 9.6.1
### Structure Property Relationships

Mathematical approaches of linking chemical structure and activity/property of chemical compounds in a quantitative manner started already 40 years ago by Hansch [23] and have been developed to powerful tools in drug discovery [24].

If we want to investigate structure–property relationships, we should have a closer look to what type of property we are interested in.

### 9.6.2
### Type of Molecular Properties

The microscopic properties should have a stronger relationship to the structure of a separated single molecule; some of the properties can be directly calculated quantum mechanically from wavefunction as an expectation value or as derivatives of the energy. Microscopic properties are independent of the temperature.

In the case of macroscopic properties, we have an ensemble of identical molecules and we have to take into account statistics (partition functions) handled in statistical thermodynamics, where one can go from microscopic to macroscopic properties. In some of these properties, we have strong intermolecular interactions (e.g., fusion point or boiling point). By the way, these interactions originate of course from the electronic structure of a single molecule.

Drug-like properties and properties for biological activity are quite more complex than the above-mentioned properties and their measurements are often not exact reproducible. The counterparts (organism, living cell, etc.) where the action of the molecule is observed are not identical and we have matrix, solvent, or environmental influences and modification (e.g., hydrolysis) and metabolism.

Two parallel approaches for QSAR are predominant in the literature: one guided by mechanistic methods (including read-across) and another by the use of statistical methods. To bridge the gap between these two approaches and to show some differences is the aim of this paper. The basic assumption for all molecules based on a very general hypothesis is that similar molecules have similar activities. This principle is also called structure–activity relationship (SAR).

Applications in chemo- and bioinformatics utilize many ideas, concepts, and methods to find relationships between molecules. Of great interest to the pharmaceutical industry is the elucidation of a set of chemical/physical descriptors modulating the relationship between the molecular structure and pharmacological activity that could be used to predict activity based solely on the chemical structure.

### 9.6.3
### Molecular Commonality and Similarity

Thereby the concept of "similarity," where molecules may be grouped according to their biological effects or physicochemical properties, has found extensive use in drug discovery. In principle, there are two different approaches for investigating structure–property relationships. The first starts with a broad collection of descriptors ($D_i$) and uses statistical methods to find out which should be considered and optimizes the equation for describing the property ($P$), where $P$ and $D_i$ are vectors of the size of the molecule dataset. The second approach is based on an intuitive selection of descriptors.

### 9.6.4
### Multilinear Regression

The number of the selected descriptors is the dimension of the space to be investigated. The application of the multiple linear regression method calculates mathematical models (QSAR) by performing standard multivariable regressions

**Table 9.4** Types of molecular properties and computational methods.

| **Molecular properties** | | |
|---|---|---|
| Microscopic | Macroscopic | Biological activity |
| **Structure** | | |
| Single molecule | Ensemble of molecules | Drug, organism, cell, and receptor |
| 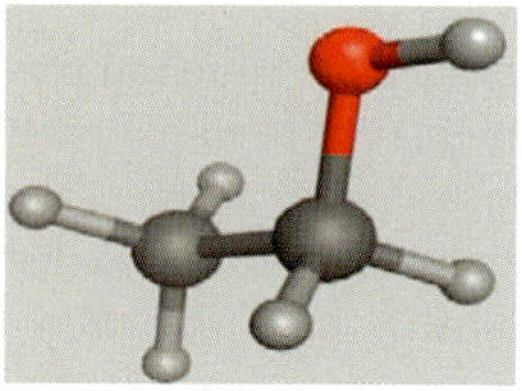 | 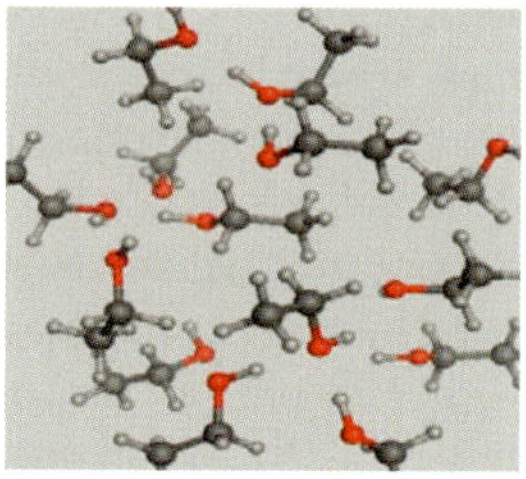 | 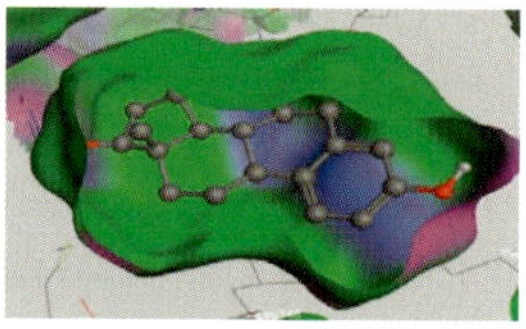 |
| **Methods** | | |
| Quantum mechanics SCF-MO, *ab initio*, DFT | Quantum mechanics FF-molecular mechanics | Molecular modeling QSAR |
| Interpretation of the wavefunction | Molecular dynamics statistical thermodynamics | Similarity, commonality statistical models |
| **Properties** | | |
| Dipole moment molecular spectra | Boiling point enthalpy, entropy | IC50, LD50 $pK_i$ |

using multiple variables in a single equation. When one uses multiple linear regression, one assumes that the variables are independent (not correlated). Also, the number of descriptors (independent variables) initially considered should not be more than one-fifth the number of compounds in the training sets, to exclude the possibility of chance correlations.

Therefore, one calculates a correlation matrix between all descriptors and the property (activity) field in order to select those descriptors, which show good correlation to the property but do not have remarkable intercorrelation within the descriptor set:

$$\mathcal{P} = a0 + a1D1 + a2D2 + a3D3 + \cdots \tag{9.21}$$

There are some helpful criteria for dividing the experimental dataset into a training and a test set. The validation of the derived mathematical model (the parameters $a_i$ in Equation 9.21) is an important step in QSAR and provides an indication of the reliability of the model, which can be used to predict activities for substances not contained in training or test set. One of the simplest validation is the cross-validation (X-validation) with leave-one-out (LOO), which is convenient for small datasets. The cross-validation process repeats the regression many times on subsets of the data.

Usually each molecule is left out in turn, and the $R^2$ is computed using the predicted values of the missing molecules (the cross-validated $R^2$).

Structure–property relations are fundamentally based on the electronic structure; this is also the important reason for bonding forces and the arrangement of atoms, their dynamics, and chemical behavior (including intermolecular interactions and chemical transformations).

A descriptor reduces some of the structural information into a single number. Some of them represent the whole molecule and others represent only individual atoms or fragments. Sometimes it is not trivial to elucidate which part of the information is really covered by the descriptor and how useful it is to describe an interesting property. Those computationally generated descriptors could be augmented by experimental data, but this is often not feasible due to their unavailability if experimentally unknown molecules are in the investigated dataset (Table 9.4). Here, the physicochemical properties and structures are expressed by numbers; one can derive a mathematical relationship (model) or quantitative structure–activity relationship, between the two. The mathematical model can then be used to predict the biological activities of other chemical structures. In order to improve structure–property relationships, new descriptors are being invented from time to time.

### 9.6.5
### Selection of Molecular Descriptors

There are several types of descriptors: constitutional, topological, geometrical, quantum chemical, functional groups, chemical, experimental quantities and so on. In Table 9.5, we have summarized some descriptors, which can be chosen in the MOE molecular modeling package [22]. Until now there is an enormous number of definitions of QSAR descriptors, a good survey over more than 3000 can be found in the two volume book of Todeschini [25], and the Dragon software [26, 27] is able to calculate 4885 descriptors of different classes. A review about quantum chemical descriptors [19] explains some of the theoretically based descriptors.

## 9.7
## Modeling of Physicochemical Properties of the Isomers of Hexane

As an example for macroscopic properties, we chose the isomers of hexane, which have the same brutto formula ($C_6H_{14}$) and exist in a series of five skeleton isomers (Table 9.6). All of these isomers have different chemical properties. Here, we want to demonstrate that simple descriptors can be used to describe the trend of the experimental properties. First, we use the Wiener index, which was first used as topological index derived from a distance matrix [28] to describe the paraffin boiling points. It is called the branching index (it is an integer number), which is calculated from summation of the elements of the distance matrix. We use multiple linear

**Table 9.5** Some of the descriptors which could be automatically calculated (out of 333 in the MOE package).

| CODE | CLASS | DESCRIPTION |
|---|---|---|
| AM1_dipole | i3D | Dipole moment |
| AM1_E | i3D | Total energy (kcal/mol) |
| AM1_Eele | i3D | Electronic energy (kcal/mol) |
| AM1_HF | i3D | Heat of formation (kcal) |
| AM1_HOMO | i3D | HOMO energy (eV) |
| AM1_IP | i3D | Ionization potential (kcal/mol) |
| AM1_LUMO | i3D | LUMO energy (eV) |
| ASA | i3D | Water-accessible surface area |
| ASA+ | i3D | Positive-accessible surface area |
| ASA− | i3D | Negative-accessible surface area |
| ASA_H | i3D | Total hydrophobic surface area |
| ASA_P | i3D | Total polar surface area |
| CASA+ | i3D | Charge-weighted positive surface area |
| CASA− | i3D | Charge-weighted negative surface area |
| DASA | i3D | Absolute difference in surface area |
| DCASA | i3D | Absolute difference in charge-weighted areas |
| Density | i3D | Mass density (AMU/$\mathring{A}^3$) |
| Dipole | i3D | Dipole moment |
| E | i3D | Potential energy |
| E_ang | i3D | Angle bend energy |
| E_ele | i3D | Electrostatic energy |
| E_nb | i3D | Nonbonded energy |
| E_oop | i3D | Out-of-plane energy |
| E_so1 | i3D | Solvation energy |
| E_stb | i3D | Stretch-bend energy |
| E_str | i3D | Bond stretch energy |
| E_strain | i3D | E minus energy of local minimum |
| E_tor | i3D | Torsion energy |
| E_vdw | l3D | van der Waals energy |
| FASA+ | i3D | Fractional positive surface |
| FASA− | i3D | Fractional negative accessible surface area |
| FASA_H | i3D | Fractional hydrophobic surface area |
| FASA_P | i3D | Fractional polar surface area |
| FCASA+ | i3D | Fractional charge-weighted positive surface area |
| FCASA− | i3D | Fractional charge-weighted negative surface area |
| Glob | i3D | Molecular globularity |
| MNDO_dipole | i3D | Dipole moment |
| MNDO_E | i3D | Total energy (kcal/mol) |
| MNDO_Eele | i3D | Electronic energy (kcal/mol) |

regression (MLR) as model and start from a limited amount of molecular descriptors, which were widely proven to have mechanistic interpretation. The selected experimental properties are the heat of formation (HoF) (H-exp_kcal), the heat of vaporization (H-vap_kcal), and the boiling point (BP_K) [29].

**Table 9.6** Database of the hexane isomers with SMILES notation and experimental data.

| | Mol | SMILES | Name | H-exp_kcal | H-vap_kcal | BP_K |
|---|---|---|---|---|---|---|
| 1 | 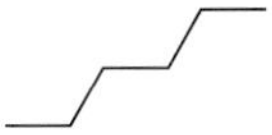 | cccccc | Hexane | −39.9400 | 7.5840 | 341.9000 |
| 2 | 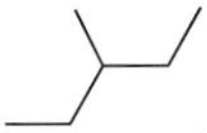 | CCC(C)CC | 3-Methylpentane | −41.0200 | 7.2830 | 336.4000 |
| 3 | 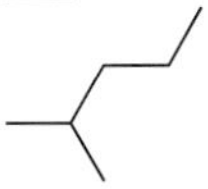 | CCCC(C)C | 2-Methylpentane | −41.6600 | 7.1900 | 334.0000 |
| 4 | 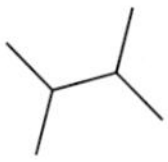 | CC(C)C(C)C | 2,3-Dimethylbut | −42.4900 | 7.0100 | 331.2000 |
| 5 | 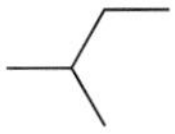 | CCC(C)(C)C | 2,2-Dimethylbut | −44.3500 | 6.6750 | 322.9000 |

As the first property we chose the stability of the isomers (property = HoF) and got with the "Wiener index" (Weiner path), a good correlation (Figure 9.13) with $R^2 = 0.834$ and with a similar descriptor "chi1," the molecular connectivity index (so-called Randic connectivity index), we got $R^2 = 0.959$ (Figure 9.14).

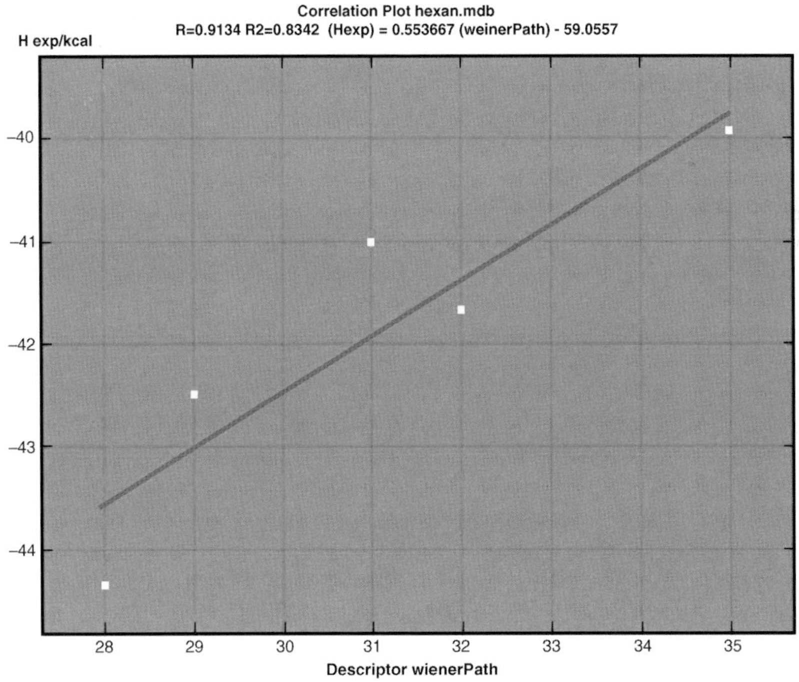


**Figure 9.13** Hexane heat of formation and Wiener index.

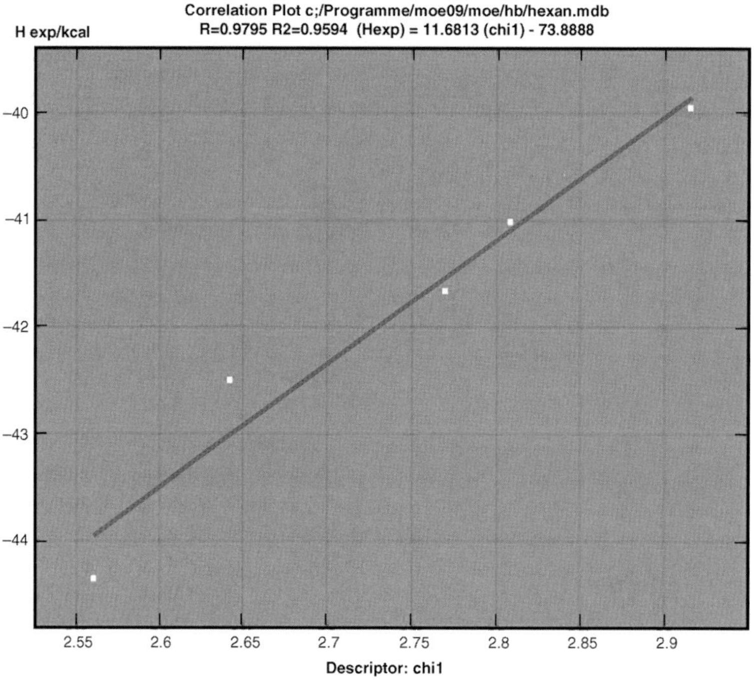


**Figure 9.14** Heat of formation and chi1.

The branching of the isomers should be counted by the number of rotatable C—C bonds. Two indices, the number of rotatable bond (b_1rotN), and the Zagreb (Zagreb) index provided the best correlations (Figures 9.15 and 9.16).

The reason for this dependence is the general observation that spherical structures, which have, for example, higher branching or smaller number of rotatable bonds have the smallest surface in comparison to other isomers (see Figures 9.20 and 9.21) and they are more stable in comparison to linear structures.

The calculated correlation matrix (Table 9.7) shows the relationship between properties and descriptors. We see high correlation between these three experimental values and some molecular descriptors as well. The experimental properties are fully correlated and have high correlation to nearly all the selected descriptors and even the descriptors are strong correlated among each other. From that reason, a combination from those descriptors is not useful for QSAR.

The experimental HoF can be compared with the calculated energies. For the isomers, we have the optimized 3D structures calculated with an empirical force field (Merck molecular force field, MMFF94) and got the steric energy (descriptor no. 4: FF E kcal) $R^2 = 0.916$ (Table 9.7) and the semiempirical SCF-MO calculated HoF (descriptor no.11: AM1_HF) shows $R^2 = 0.856$. The correlation plots are shown in Figures 9.17 and 9.18.

Both calculated energies have good correlation to the experimental HoF, but the calculated energies show both the wrong trend of stability. Theoretically the *n*-hexane

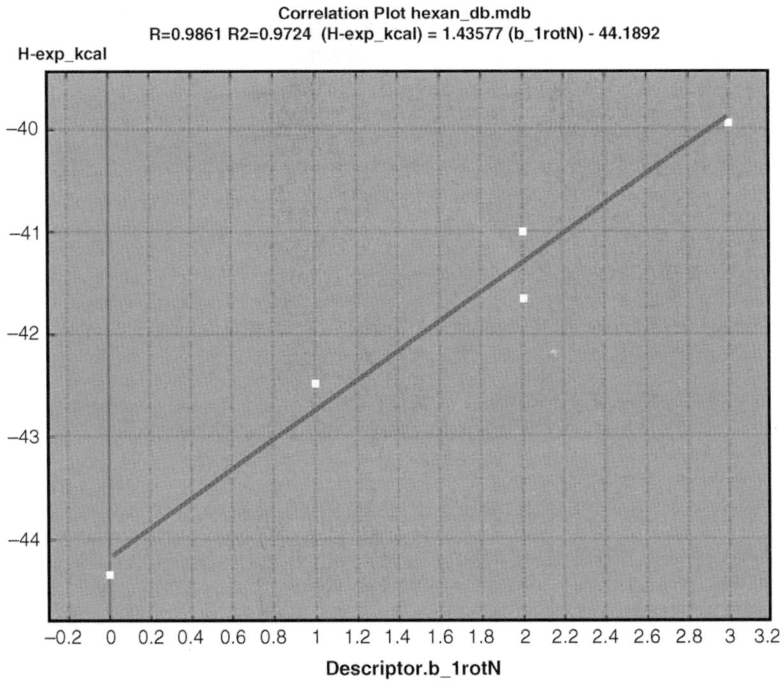


**Figure 9.15** HoF versus number of rotatable bonds.

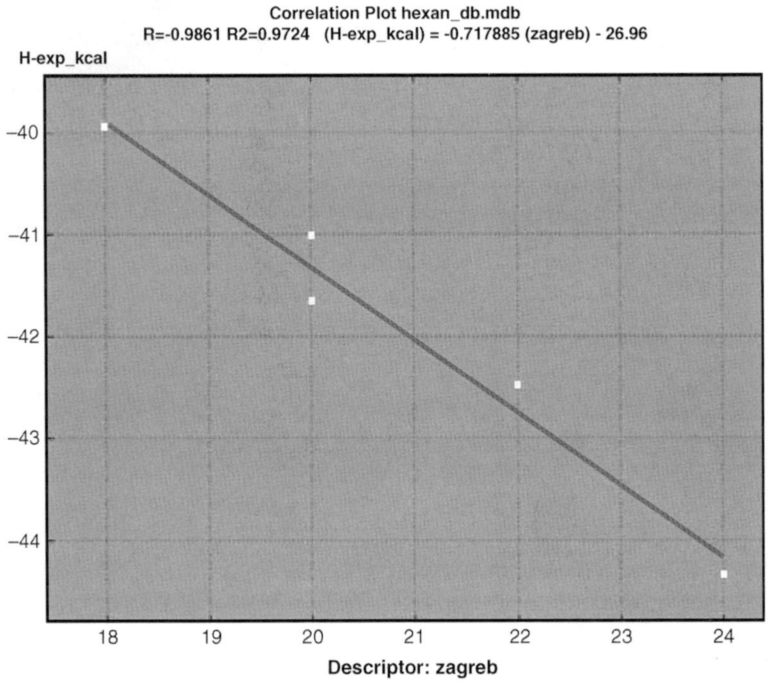


**Figure 9.16** HoF with the Zagreb index.

**Table 9.7** Correlation matrix between properties and descriptors of the hexane database.

| | 1 | 2 | 3 | 4 | 5 | 6 | 7 | 8 |
|---|---|---|---|---|---|---|---|---|
| 1. Hexp | 100 | 100 | 100 | 82 | 99 | −95 | 98 | 95 |
| 2. Hvap | 100 | 100 | 100 | 87 | 99 | −97 | 99 | 97 |
| 3. BP | 100 | 100 | 100 | 83 | 98 | −94 | 97 | 95 |
| 4. ASA | 82 | 87 | 83 | 100 | 89 | −94 | 88 | 94 |
| 5. b_1 rotN | 99 | 99 | 98 | 89 | 100 | −98 | 99 | 99 |
| 6. chi0 | −95 | −97 | −94 | −94 | −98 | 100 | −99 | −99 |
| 7. chi1 | 98 | 99 | 97 | 88 | 99 | −99 | 100 | 98 |
| 8. E_sol | 95 | 97 | 95 | 94 | 99 | −99 | 98 | 100 |
| 9. Kier2 | 94 | 97 | 94 | 96 | 96 | −99 | 97 | 98 |
| 10. log *P* (o/w) | 77 | 82 | 77 | 93 | 83 | −93 | 88 | 88 |
| 11.PEOE_VSA_HYD | −92 | −89 | −92 | −64 | −90 | 79 | −84 | −84 |
| 12. pmi | 87 | 91 | 88 | 99 | 92 | −97 | 93 | 96 |
| 13. vdw_area | −92 | −89 | −92 | −64 | −90 | 79 | −84 | −84 |
| 14. vsurf_G | 89 | 92 | 89 | 99 | 94 | −98 | 94 | 97 |
| 15. ASA_H | 84 | 88 | 86 | 97 | 89 | −90 | 86 | 92 |
| 16. rgyr | 89 | 92 | 89 | 99 | 94 | −98 | 94 | 97 |
| 17. vsurf_S | 87 | 91 | 87 | 99 | 93 | −98 | 94 | 97 |

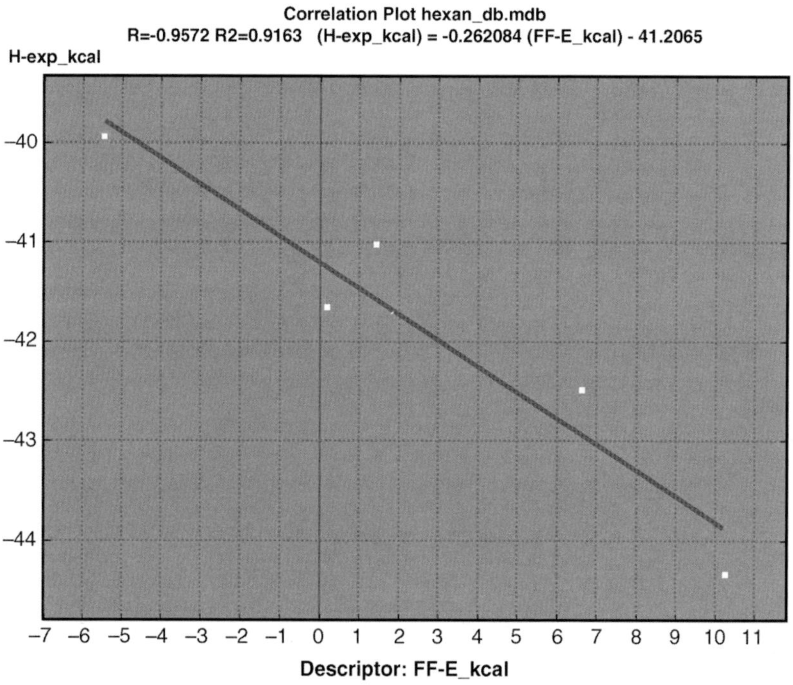


**Figure 9.17** Experimental heat of formation and calculated FF energy.

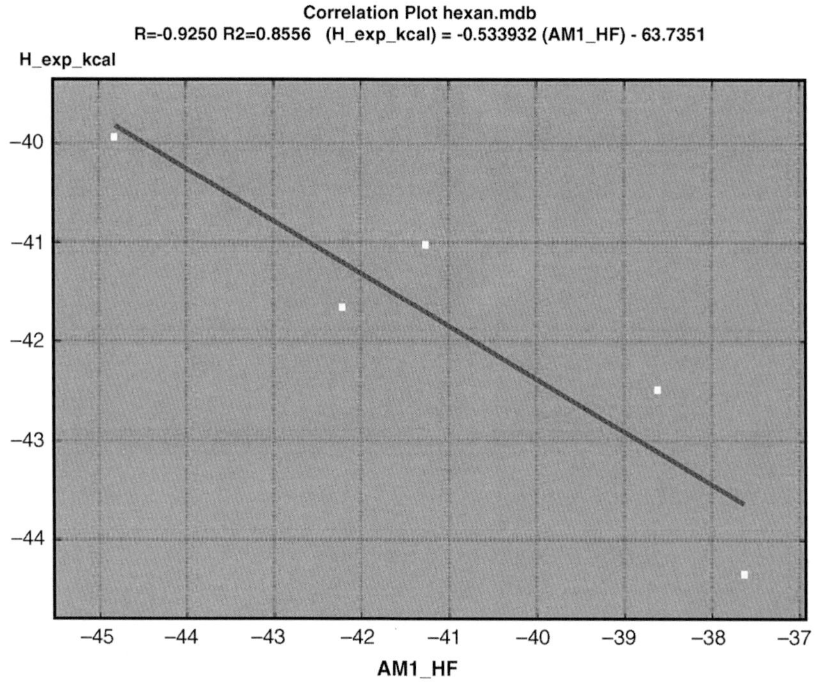


**Figure 9.18** Experimental heat of formation and quantum chemical calc. HoF (AM1 Hamiltonian).

should be the most stable isomer having the lowest energy; this error for disfavoring branched isomers can even be observed with elaborated DFT calculations [30]. This mistake opens our attention not only to trust a good correlation values, but also to analyze in terms of quality and also for mechanistic interpretability of modeling descriptors.

The property boiling point (BP_K) can be described with the same descriptors of Table 9.8 with quite similar accuracy. The correlation to the number of rotatable bond (b_1rotN) is shown in Figure 9.18 with $R^2 = 0.968$.

The reason for this decreasing boiling point with lower numbers of rotatable bonds resp. increasing branching is caused by the interaction potential between molecules (Figure 9.19). Spherical structures have a smaller contact surface. The CPK models of the two isomers of *n*-hexane and 2,2-dimethyl-butane are shown as dimer arrangements in Figures 9.20 and 9.21. The isomers should be in general more stable and should have lower boiling points the more spherical their structure is. At this point, we have neglected the dynamical behavior that chain-molecules like *n*-hexane show at room temperature conformations of crowded structures. From this explanation, we observe that the boiling point varies in the same way as the heat of vaporization (Figure 9.22).

Since the point of contact for all drugs lies inevitably with the molecular surface, a descriptive model of the molecular surface would be useful, which we will describe later.

**Table 9.8** Correlation of heat of formation (Hexp) of the hexane isomers by calculated descriptors.

| No. | Descriptor | $R^2$ | Class | Meaning of the descriptor |
|---|---|---|---|---|
| 1 | b_1rotN | 0.972 | 2D | Number of rotatable single bonds |
| 2 | zagreb | 0.972 | 2D | Zagreb index |
| 3 | chi1 | 0.959 | 2D | Atomic connectivity index (order 1) |
| 4 | FF E kcal | 0.916 | 3D | Force field energy of the opt. structure |
| 5 | E_sol | 0.911 | i3D | Solvation energy |
| 6 | chi0 | 0.898 | 2D | Atomic connectivity index (order 0) |
| 6 | Kierflex | 0.877 | 2D | Molecular flexibility |
| 8 | balabanJ | 0.876 | 2D | Balaban averaged distance sum connectivity |
| 9 | Kier2 | 0.876 | 2D | Second kappa shape index |
| 10 | VSA | 0.860 | i3D | Van der Waals surface area |
| 11 | AM1_HF | 0.856 | i3D | Heat of formation (kcal/mol) AM1-Hamiltonian |
| 12 | vdw_area | 0.850 | 2D | Van der Waals surface area ($A^2$) |
| 13 | Weiner path | 0.834 | 2D | Weiner path number |
| | Log *P* (o/w) | | 2D | Log of the octanol/water partition coefficient |
| | PEOE_VSA_HYD | | 2D | VDW surface areas of hydrophobic ($\|q_i\|<0.2$)atoms |
| | Pmi | | i3D | Principal moment of inertia |
| | vsurf_G | | i3D | Surface globularity |
| | ASA_H | | i3D | Total hydrophobic surface area |
| | Rgyr | | i3D | Radius of gyration |
| | vsurf_S | | i3D | Interaction field surface area |

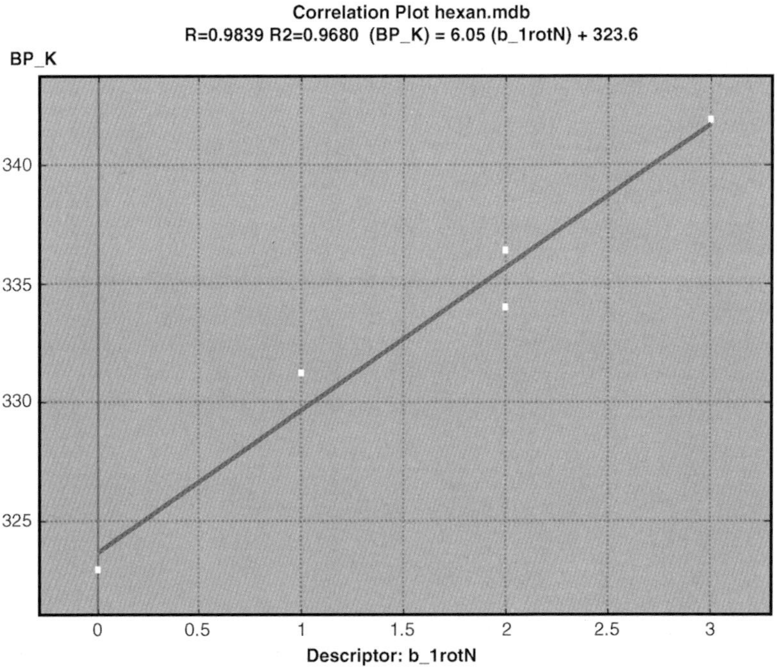


**Figure 9.19** Correlation of the boiling point with the number of rotatable bonds.

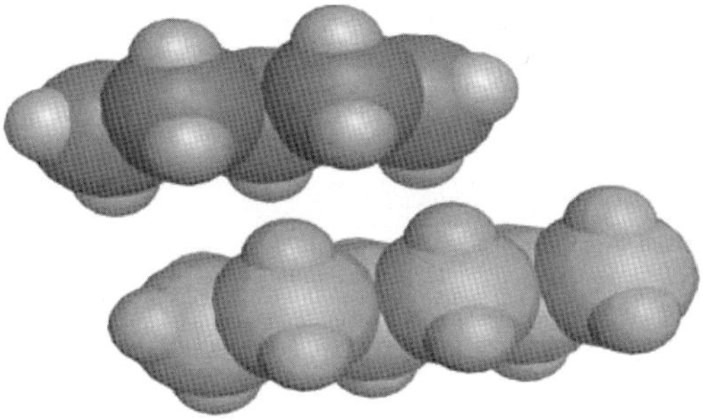

**Figure 9.20** Two molecules of *n*-hexane (No. 1).

The nature of this surface is electronic and quantum mechanical methods are those which describe the electronic structure of the molecule. The branching of the isomers is counted by the number of rotatable C–C bonds or by the branching index, both descriptors show good correlation to all the three experimental properties of the hexanes. Furthermore, we should expect some interrelations between these three properties.

From inspection of the correlation matrix, one realizes that between the three experimental properties a strong relationship exists (first three lines in Table 9.7). The correlation between these three experimental values and some molecular descriptors is very high (more than 80%). The experimental properties are fully correlated and have high correlation to nearly all the selected descriptors but even the descriptors are strongly correlated among each other (Figure 9.23). From that

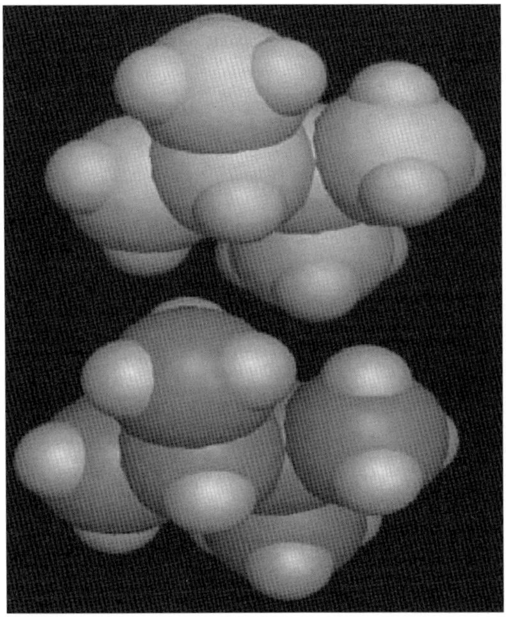

**Figure 9.21** Dimer of 2,2-dimethylbutane (No. 5).

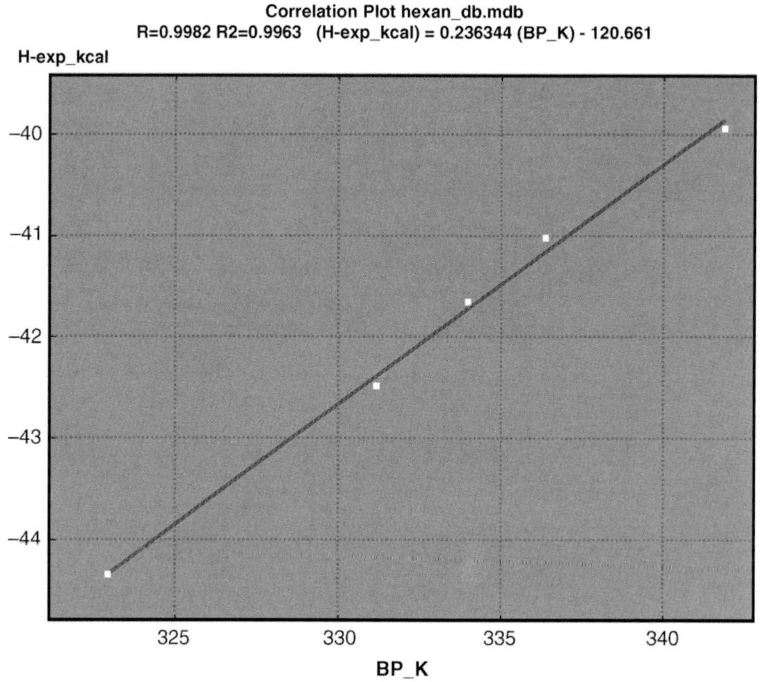


**Figure 9.22** Correlation between the boiling point and the heat of formation.

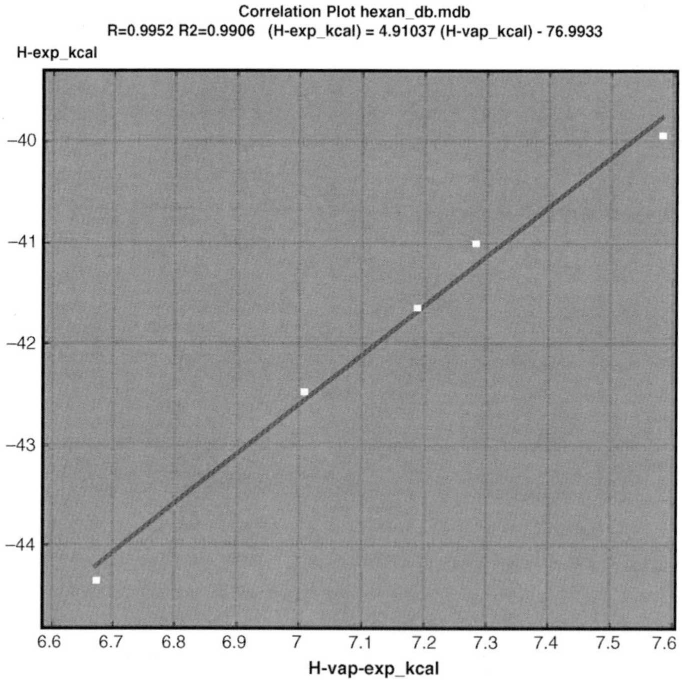


**Figure 9.23** Correlation of heat of vaporization and heat of formation.

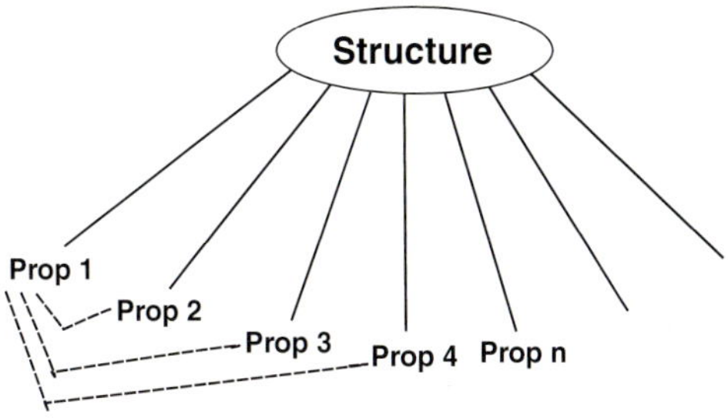


**Figure 9.24** Structure of a molecule and its properties.

reason, a combination from those correlated descriptors is not useful in MLR investigations for QSAR.

The good relationship between the three experimental properties is a good opportunity to remember that all properties of a molecule depend on the same electronic structure; this implies that some properties are interrelated, which is shown schematically in Figure 9.24. This intercorrelation between different properties of the same compound opens new general aspects in searching for QSPR. Once again "structure" means the electronic structure, which is essential for the observed properties.

## 9.8 Modeling of the Proton Affinity

### 9.8.1 Proton Affinity of Pyridines

#### 9.8.1.1 Data and Mechanism

Proton affinity and bases and acids properties in general are important quantities of molecules having biological activity, especially nitrogen containing basic centers; these can be exist in their neutral or protonated form in dependence of the pH value of the surrounding medium; for example, pyridine (Figure 9.25):

$$H_3O^+ + \text{pyridine} \rightarrow [\text{H}-\text{pyridine}]^+ + H_2O \tag{9.22}$$

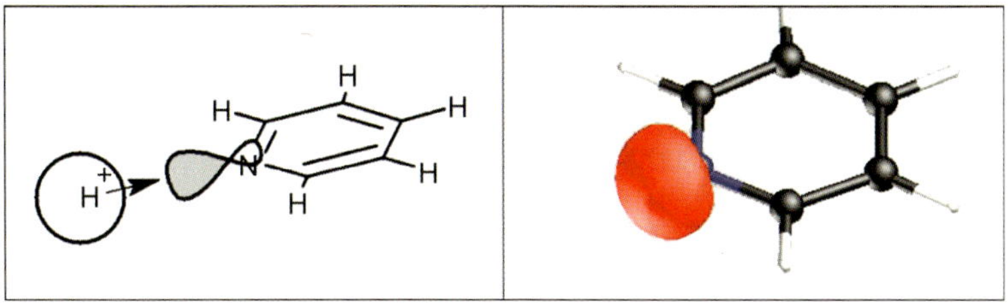


**Figure 9.25** The protonation of the pyridine molecule (a) and the shape of the lone-pair orbital (LMO) at nitrogen (b).

If this process takes place in the gas phase, we do not have any influence of the solvatation. The experimental quantity is the proton affinity (PA); the data taken from [31]. The more negative this value is, the more preferred is the protonated form (right of the equation). In order to predict the PA of a series of 13 mono- and disubstituted pyridine compounds (Figure 9.26), we developed some structural

| | mol | PA | name |
|---|---|---|---|
| 1 |  | 222.3000 | pyridine |
| 2 |  | 225.5000 | 3-me-py |
| 3 |  | 226.4000 | 4-me-py |
| 4 | 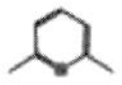 | 230.2000 | 2,6-dime-py |
| 5 | 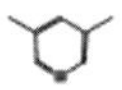 | 228.3000 | 3,5-dime-py |
| 6 | 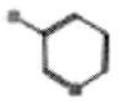 | 215.6000 | 3-F-py |
| 7 | 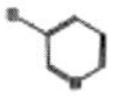 | 215.9000 | 3-Cl-py |
| 8 | 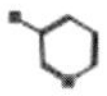 | 228.1000 | 3-amino-py |
| 9 | 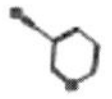 | 209.6000 | 3-cyano-py |
| 10 | 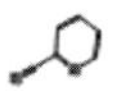 | 208.6000 | 2-cyano-py |
| 11 |  | 210.5000 | 4-cyano-py |
| 12 | 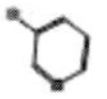 | 222.2000 | 3-hydoxy-py |
| 13 | 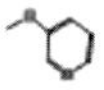 | 225.3000 | 3-meoxy-py |

**Figure 9.26** Database of the pyridine compounds ($n = 13$).

models based on the mechanism of this interaction of a proton with the aromatic heterocycle. In order to derive a QSAR mathematical model, it would be necessary to compute some molecular descriptors to fit to a QSAR model.

In the following, we want to have a closer look for choosing those molecular descriptors, which could be derived from quantum mechanical calculations of the electronic structure. Starting first with a simple approach, we will improve the prediction model step by step. A semiempirical molecular orbital package MOPAC [32] was used to calculate the used descriptors. The aim of this comparative study of mechanistically relevant and statistically relevant QSAR models is demonstrated with the following models I–VI for the proton affinity.

#### 9.8.1.2 **Model I**

We assume that the electrostatic interaction between the proton and the negatively charged nitrogen dominates the interaction energy. In order to derive this descriptor $Q_N$, we have carried out quantum chemical SCF-MO calculations with optimizing the structure with the well-known MOPAC package using the AM1-Hamiltonian. For all compounds, the atomic charge of this $N$ atom is negative, so it would attract the proton. But there is no correlation between the PA and $Q_N$ (Figure 9.27). The reason may result from wrong assumption about the interaction mechanism that the coulomb forces should favor the proton affinity or one has to ask, if the AM1 calculated semiempirical charges do not represent the real charges.

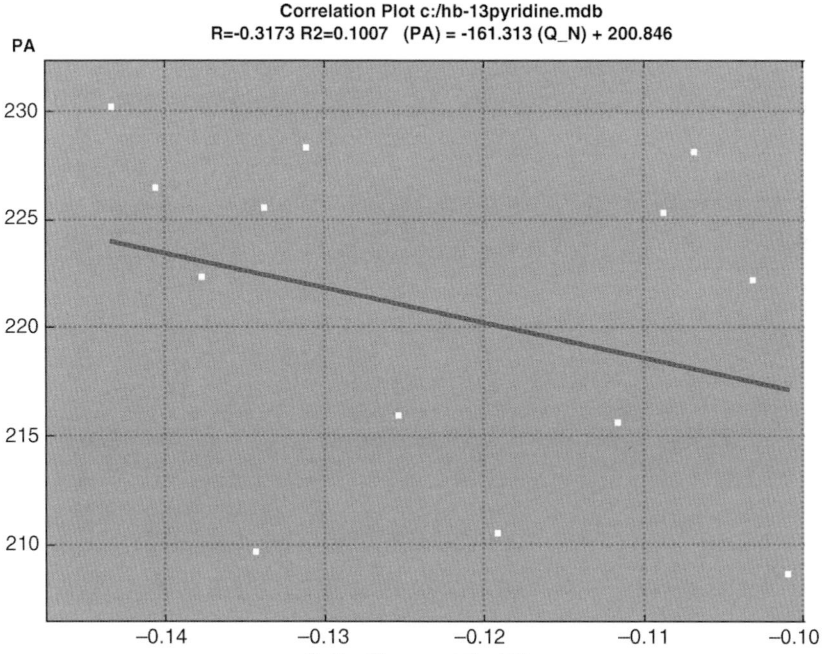


**Figure 9.27** Correlation plot of the PA in dependence of atomic charge at the nitrogen.

We do not want to go into the details of such MO calculations [13] and the inherent quality, but we want to try a next mechanistic model number II.

#### 9.8.1.3 Model II

This model II is based on mechanistic assumption that the proton will interact with its empty 1s orbital. The proton should have a strong attraction to electrons located at the aniline molecule. This corresponds to an electrophilic attack to aniline; therefore, the ionization energy of the molecule (acting as a donor) should be a good measure for the proton affinity. Lower ionization energies should favor the protonation. From the quantum chemical MO calculation, the first ionization potential (IP) is the negative of the energy of the highest occupied molecular orbital (HOMO), which is described by Koopmans' theorem [33].

$$IP = -E_{\text{HOMO}} \tag{9.23}$$

The correlation of PA with the energy of the HOMO (am1_homo) of the aniline compounds in shown in Figure 9.28. It is of remarkable quality ($R^2 = 0.695$), we have chosen the energy of the HOMO because an electrophilic attack of the proton is attracted by electron density especially in molecular orbitals of higher energy.

In order to verify the above assumption, we have to take a look to the overlap between the acceptor orbital (proton 1s orbital) and the donor-MO of aniline, which is

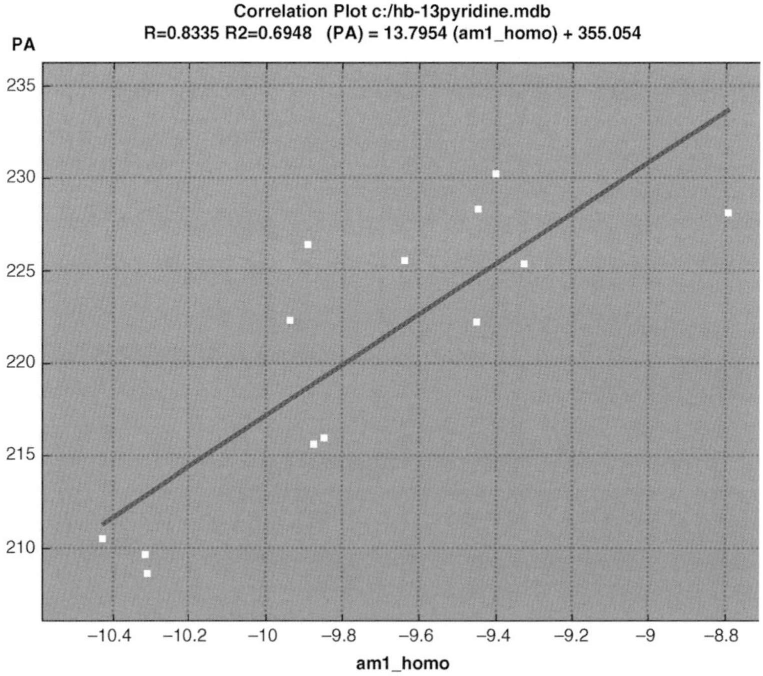


**Figure 9.28** Correlation of the PA with $E$(HOMO).

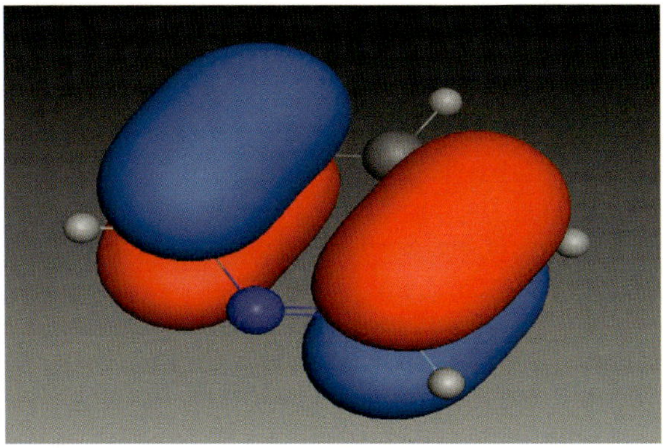

**Figure 9.29** The HOMO of aniline.

necessary for forming a new bond. The symmetry of this HOMO of the aniline molecule is shown in Figure 9.29.

The aniline molecule and even the protonated species belongs to the symmetry point group $C_{2v}$; the HOMO is of irreducible representation $a_2$ and has no overlap to the approaching proton 1s orbital, which belongs to $a_1$, such a process will be symmetry-forbidden. Both orbital do not have the same symmetry for forming a chemical bond because of symmetry restrictions and furthermore this HOMO does not have any probability (local density) at the nitrogen atom. The question arises: Why we have such a good correlation of the HOMO to the PA? The answer may probably the fact that the next lower MO, the HOMO-1 (see Figure 9.30) has the right symmetry ($a_1$) for the orbital interaction to the proton and that these occupied orbitals behalf similar in their energy levels ($E_{\mathrm{HOMO}} = -9.936$ eV resp. $E_{\mathrm{HOMO-1}} = -10.654$ eV) in this series of similar anilines. In these unsaturated aromatic heterocyclic compounds, the HOMO belongs to the $\pi$-system and the electrons in those MOs are not located at the Nitrogen in direction to the approaching proton. So,

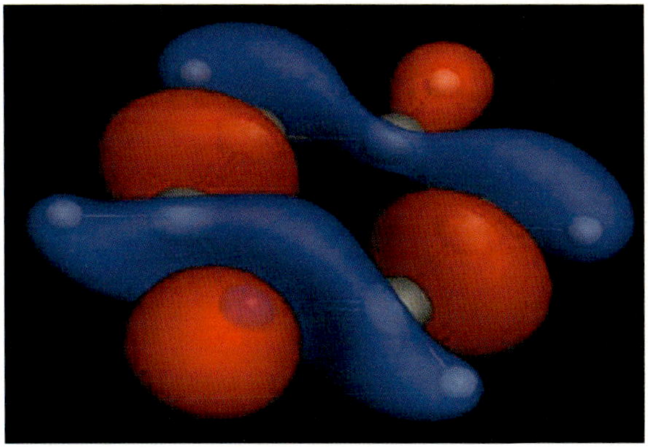

**Figure 9.30** The (HOMO-1) MO of aniline.

we can construct a new mechanistic model III, which locates electrons of the aniline in the desired direction to the proton.

#### 9.8.1.4 Model III

The lone-pair at the nitrogen atom, which belongs to the $\sigma$-electron system is close to the molecular plane of the planar ring. The calculated MOs are so-called canonical orbital, which fulfill the symmetry and are usually distributed over the whole molecule.

These canonical MOs of a molecule can be transformed (unitary transformation) into localized MOs [34], which consist only of two- and one-center orbitals, the so-called localized molecular orbitals (LMOs), which are occupied each with two electrons corresponding to the well-known picture of valence formulas with sigma-bonds and lone pairs. One of the goals of calculating LMOs is to create just those MOs, which are similar and transferable between fragments in different compounds.

The transformation of canonical molecular orbitals to orbital that are equivalent to the chemical picture of localized bonding does not change any expected value. Since all predictable properties depend only on the total electron density, the localization does not change these densities. One usually used criterion for the localization is to maximize the self-repulsion of the two electrons that occupy the LMO, which results in located one- and two-center LMOs (Figure 9.31).

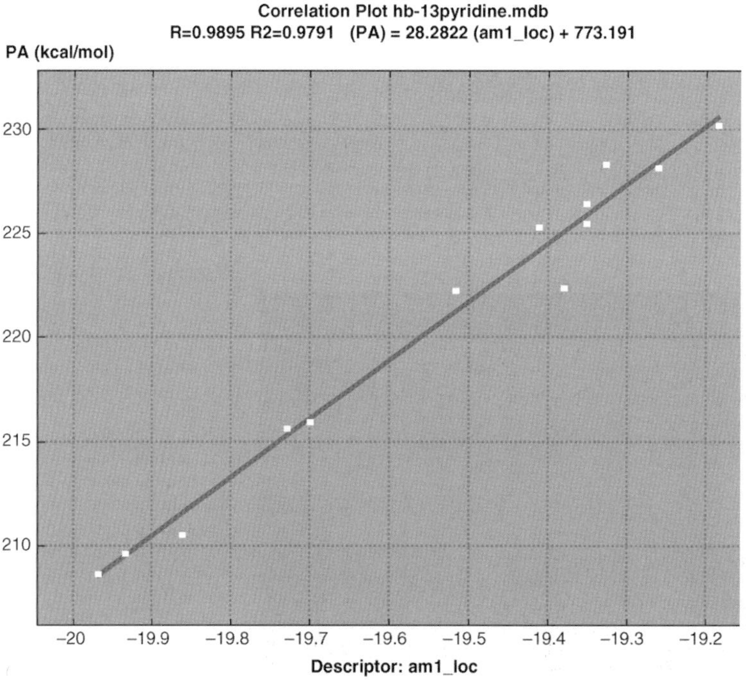


**Figure 9.31** Proton affinity and calculated $E$(LMO) (called am1_loc).

#### 9.8.1.5 Model IV

We want to improve the previous model II. From the point of view of perturbation theory, the interaction of an unoccupied MO (proton) and an occupied MO leads to an expression where the energy difference between interacting orbitals is in the denominator. The energy of the empty 1s orbital of the proton has $E_{1s} = 0.000$, so we tried the reciprocal of $E$(HOMO) value (taken from model II) of the aniline, calculated with MOPAC using the AM1 Hamiltonian. The correlation plot (Figure 9.32) with $R^2$ =0.664 is a little bit lower that means a "better" mechanistic descriptor does not necessarily leads to better correlations; the reason may be the same as discussed in model II, the different symmetry of both interacting orbitals.

#### 9.8.1.6 Model V

Next, we want to introduce a well-known concept in chemical reactivity of hard and soft acids and bases (HSABs) [35]. This empirical concept was later derived by Klopman and Hudson [36, 37] by means of the perturbation treatment between two approaching molecules resulting in a three term equation (Equation 9.24).

Term 2 domination in the case of "charge controlled" interactions (hard – hard combination) and term 3 describes the "orbital controlled" mechanism (soft – soft combination), usually in the last term, the interaction between the frontier orbitals the HOMO of the donor and the LUMO of the acceptor molecule is a good approximate.

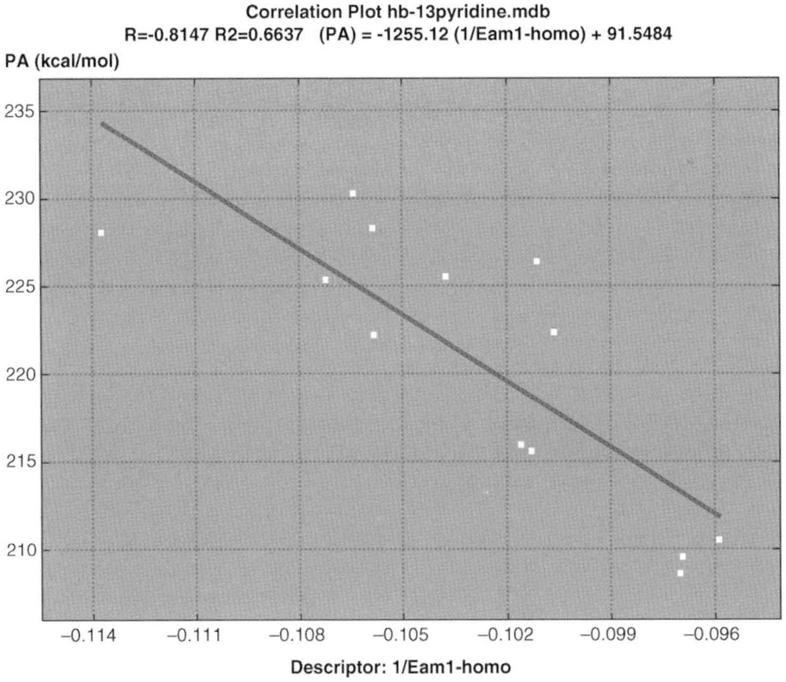


**Figure 9.32** The proton affinity and the descriptor 1/$E$(HOMO).

This is the perturbation of the energy between two interacting molecules A and B and can be evaluated by the well-known Klopman–Hudson equation (Equation 9.24):

$$\Delta E = \underbrace{\sum_{ab} (q_a + q_b)\beta_{ab} S_{ab}}_{\text{Term 1}} + \underbrace{\sum_{A<B} \frac{Q_A Q_B}{\varepsilon R_{AB}}}_{\text{Term 2}} + \underbrace{\sum_{i}^{n}\sum_{j}^{m} \left(\varrho_i - \varrho_j\right) \frac{\left(\sum_r \sum_s c_{ir} c_{js} \beta_{rs}\right)^2}{E_i - E_j}}_{\text{Term 3}} \quad (9.24)$$

Molecule 1 consists of atoms $A$, atomic orbitals $a$, and MOs $i$; molecule 2 consists of atoms $B$, atomic orbitals $b$, and MOs $j$:
where

| | |
|---|---|
| $q$ | electron densities at atomic orbital |
| $\beta$ | resonance integral |
| $S$ | overlap integral |
| $Q$ | atomic charge |
| $\varepsilon$ | dielectric constant |
| $R$ | diatomic distance |
| $\varrho$ | occupation number of MOs (0, 1, or 2 electrons) |
| $c$ | LCAO coefficient (indices: MO AO) |
| $E$ | eigenvalue (MO energy) |

The first term describes the global repulsion between the electron clouds of both molecules: one assumes that this term behaves nearly independent in a series of similar compounds and will be not discussed in detail. The further explanation is based on two other terms. The second term corresponds to the pair-wise coulomb interaction between atoms of both molecules and the third term represents the interaction of occupied MOs of one molecule (acting as donor) and unoccupied MOs of the other molecule (acting as acceptor).

In analogy to model IV, we use here the descriptor $1/E$(LMO) of the lone pair at the nitrogen atom (Figure 9.33) and found $R^2 = 0.979$ much better than model IV with $R^2 = 0.664$. The improved quality results from a better mechanistic descriptor to simulate the chemical process shown in Figure 9.25.

#### 9.8.1.7 **Model VI**

This is the last MLR for the proton affinity of anilines. We want to consider both, the second and the third term of the Klopman–Hudson equation (Equation 9.24) by the descriptors $Q(N)$ (the atomic charge of the nitrogen) and $1/E$(LMO) (LMO energy of the lone-pair electrons at the nitrogen atom), respectively.

The estimated normalized MLR model is

$$\mathrm{PA/SD} = -46.45708 - 1.00902 \times 1/E(\mathrm{LMO})/\mathrm{SD} + 0.05368 \times \mathrm{Q(N)/SD} \quad (9.25)$$

where SD is the standard deviation of the resp. values results in $R^2 = 0.981$ which is of excellent quality. The relative importance of the descriptors is 1.000 for $D1 = 1/E$ (LMO) and 0.053 for $D2 = Q(N)$.

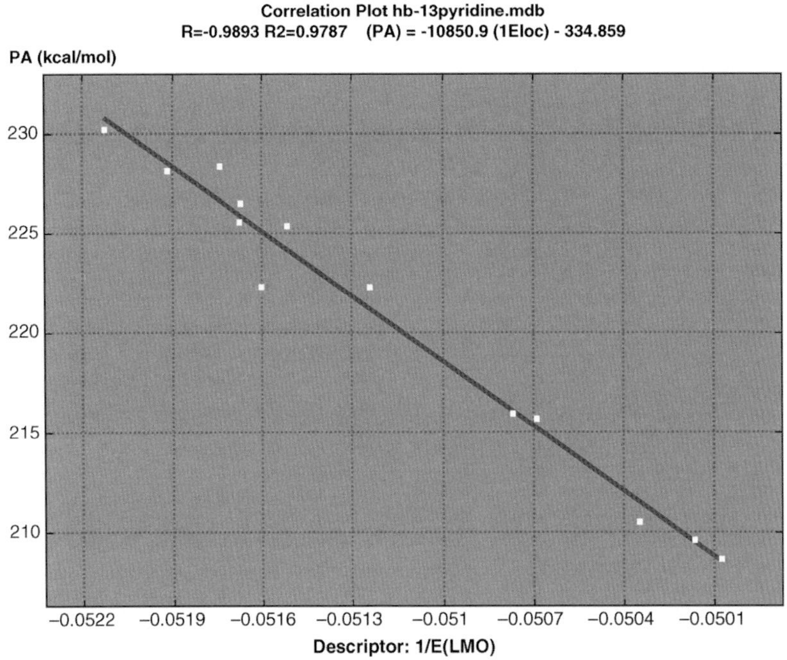


**Figure 9.33** The proton affinity and the descriptor $1/E$(LMO); LMO is the lone pair at nitrogen.

Finally, we carried out the validation of this model by X-validation which results in $X - R^2 = 0.970$, which is not essentially reduced from the $R^2$. The X-validated predicted proton affinities are shown in Figure 9.34.

The use of the LMO energy of the electrons of the lone pair at the nitrogen atom – $E$(LMO) – guaranties for good correlations (Table 9.9). The contribution of the atomic charge of the nitrogen atom can be neglected.

### 9.8.2 Basicity of N-Heterocyclic Aromatics

In order to show the usefulness of this descriptor LMO, we have built a larger database of 54 *N*-containing heterocyclic aromatic compounds. The dataset consists of three different chemical classes: 10 imidazole, 14 pyridine, and 30 pyrazole compounds. Experimental data of PA are taken from [31]. We have the dataset restricted to heterocycles with only one "lone-pair" nitrogen atom, which is part of a conjugated cycle, just to avoid isomeric products. In some cases, we have replaced missing experimental p$K_b$ values by calculated ones using an average of calculations by the ACD/Labs [38] and ChemAxon [12] procedure.

As in the previous model VI, a two-descriptor model was successful to describe the PA. From the correlation matrix (Table 9.10), the $1/E$(LMO) from quantum chemical

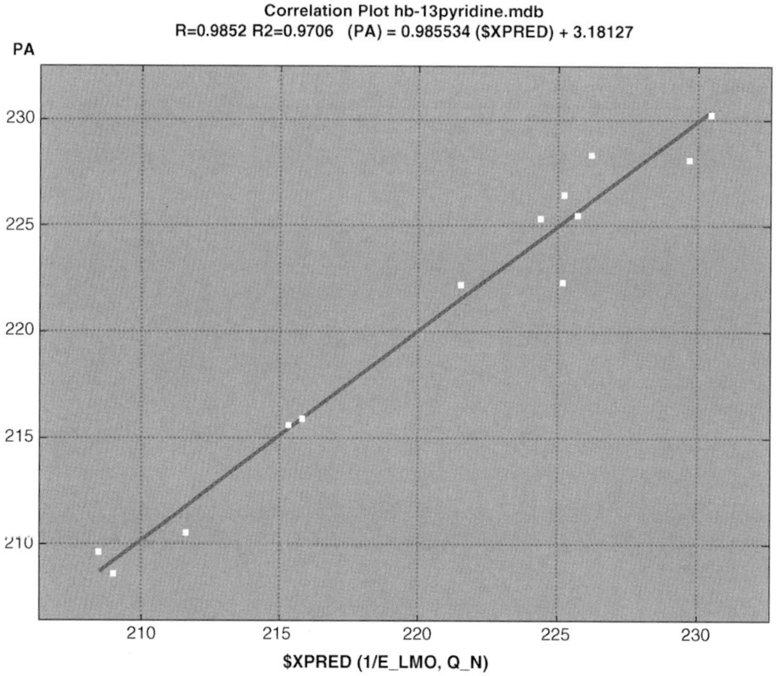


**Figure 9.34** X-validated predictions of PA (2 descriptor model).

**Table 9.9** Comparison of the used models.

| **Model no.** | **I** | **II** | **III** | **IV** | **V** | **VI** |
|---|---|---|---|---|---|---|
| Descriptor | Q(*N*) | *E*(HOMO) | *E*(LMO) | 1/*E*(HOMO) | 1/*E*(LMO) | 1/*E*(LMO),Q(*N*) |
| $R^2$ | 0.101 | 0.695 | 0.979 | 0.664 | 0.979 | 0.981 |

**Table 9.10** Correlation matrix of the proton affinity and useful descriptors [44].

| | **1** | **2** | **3** | **4** | **5** | **6** | **7** |
|---|---|---|---|---|---|---|---|
| 1. PA (kcal/mol) | 100 | −30 | 69 | −67 | −68 | 38 | 36 |
| 2. homo_ohne_PON | −30 | 100 | 28 | −28 | −28 | −38 | −18 |
| 3. Loca-Energie a.u.ohne_PON | 69 | 28 | 100 | −100 | −100 | −13 | −3 |
| 4. Zn1 +_Klopman_ohne_PON_au | −67 | −28 | −100 | 100 | 100 | 14 | 3 |
| 5. 1/loca_msindo | −68 | −28 | −100 | 100 | 100 | 14 | 3 |
| 6. apol | 38 | −38 | −13 | 14 | 14 | 100 | 95 |
| 7. *S* log *P* | 36 | −18 | −3 | 3 | 3 | 95 | 100 |

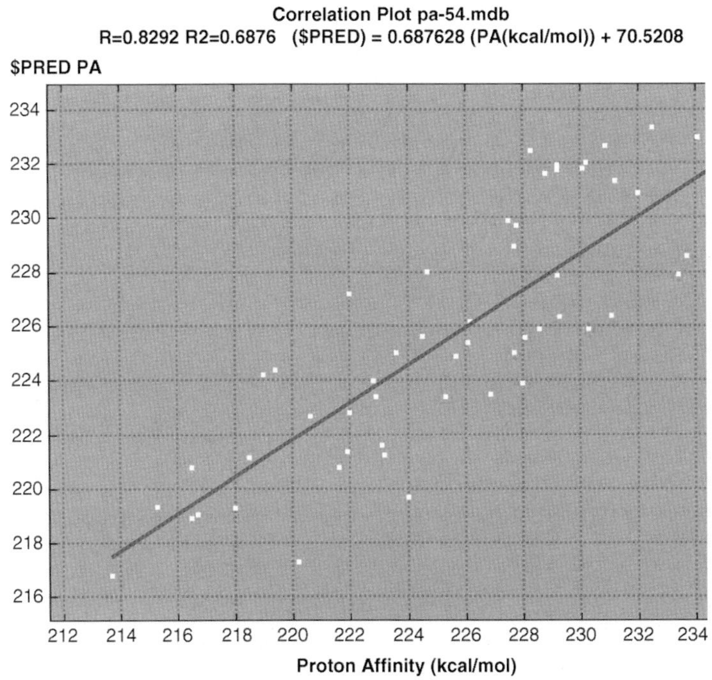


**Figure 9.35** Predicted PA versus exp. PA (imidazoles are marked).

calculations using the MSINDO[39] program and apol, the sum of the atomic polarizabilities calculated with MOE, were selected.

The derived normalized QSAR model ($n = 54$ aromatic *N*-heterocycles, $R^2 = 0.688$) shows that the Klopman front orbital term is very convenient, and the relative importance of both descriptors is 1: 0.645. The predicted PA is shown in Figure 9.35:

$$\text{PA/SD} = 18.03596 - 0.74414{*}1/\text{loca\_msindo/SD} + 0.48023{*}\text{apol/SD} \tag{9.26}$$

The correlation for the subdataset of imidazoles results in a better correlation ($n = 10$, $R^2 = 0.913$) shown in Figure 9.36.

## 9.9 Molecular Surface Properties

Many of the molecular properties especially those macroscopic properties and biological activities depend on the contact of the molecule with its environment. Since the point of contact arrives from the molecular surface, which is the outer contour of the electronic structure, a descriptive model of the molecular surface is

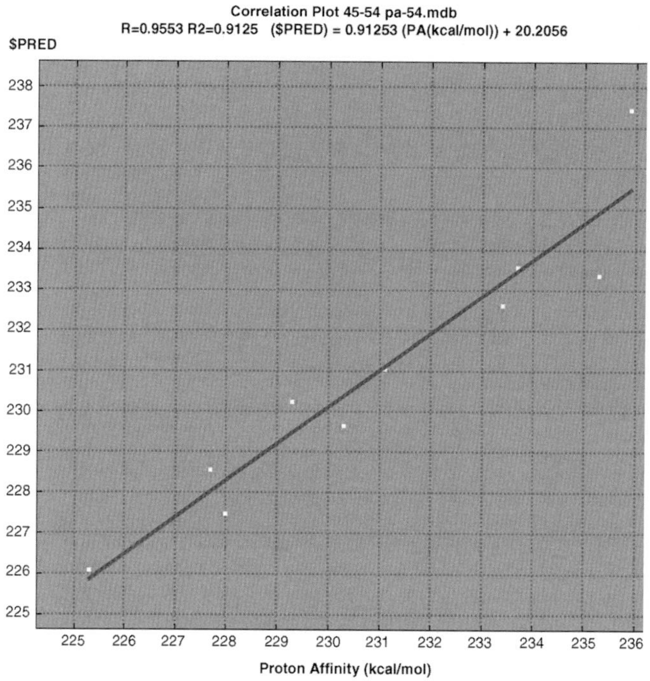


**Figure 9.36** Predicted PA of substituted imidazole molecules.

needed. Molecular orbital (MO) surfaces represent the spatial electronic distribution of a selected orbital in a molecule. According to the Fukui's frontier orbital theory, the highest occupied and the lowest unoccupied molecular orbitals (HOMO and LUMO) are crucial in predicting the local chemical reactivity (electrophilic or nucleophilic mechanism).

In drug design, those 3D models become an interesting possibility to construct those structures, which fits corresponding regions of receptor and drug. The three-dimensional visualization of molecules, calculated iso-surfaces, and color-coded surfaces allows us to inspect certain features of the electronic structure by means of molecular graphics.

Besides descriptor-based models, the 3D visualization of molecules, iso-surfaces, and color-coded surfaces allows us to inspect certain features of the electronic structure. For taking the structure of the receptor site into account, one can construct those drug structures, which fits corresponding regions of receptor, because it combines spherical and electronic features, which fit for optimal orientation and interaction.

With color-coded surfaces showing the surface distribution of a molecular property such as electrostatic or hydrophobic potential, local hardness and softness[40], local ionization potential or electron affinity [41], Fukui function [42] and so on, it also allows one to easily compare different molecules visually. The 3D-QSAR has great predicting ability for new drug candidates, because those descriptors have relevance

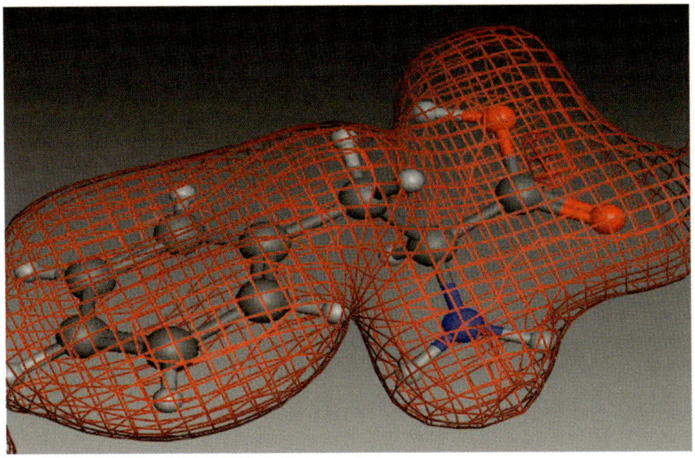

**Figure 9.37** Electron density distribution (iso-surface as mesh) of phenylalanine.

to the calculated physicochemical properties partially based on the electronic structure of the molecules.

In the following, the amino acid phenylalanine is shown in different 3D models. The electron density surface (Figure 9.37) itself represents the shape of the molecule, but it is not convenient to realize special features of surface properties.

More sense is to check Figures 9.38–9.41, which show the amino acid phenylalanine in different volume rendering views.

Very useful is the Connolly surfaces [43], so-called water-accessible surface, which can be colored with calculated quantities. Figure 9.40 shows the molecular electrostatic potential (MEP), and the molecular lipophilic potential (MLP) is shown in Figure 9.41. In order to see the molecular skeleton, the z-clipping mode was applied to cut-off objects in the foreground.

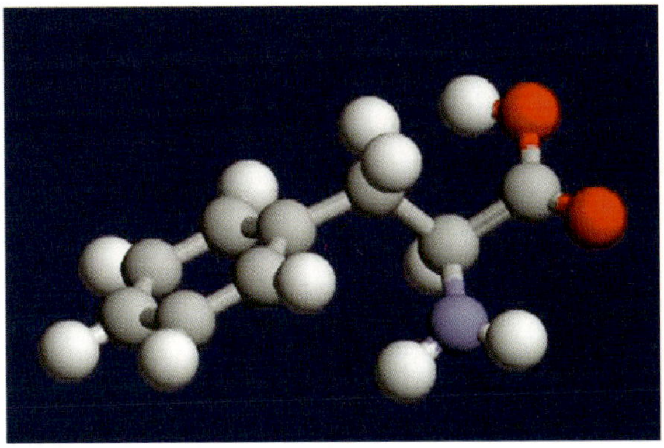

**Figure 9.38** Phenylalanine "ball-stick" model.

**Figure 9.39** Phenylalanine CPK model.

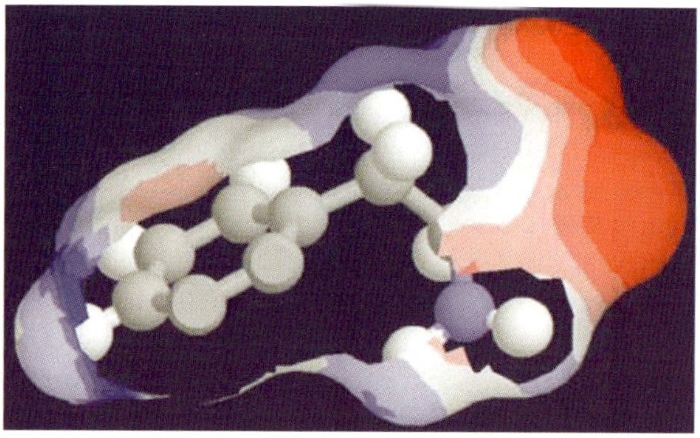

**Figure 9.40** Molecular electrostatic potential (Z-clip).

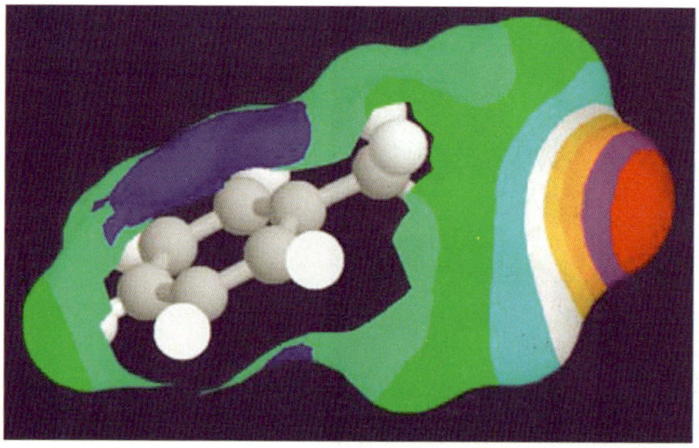

**Figure 9.41** Molecular lipophilic potential.

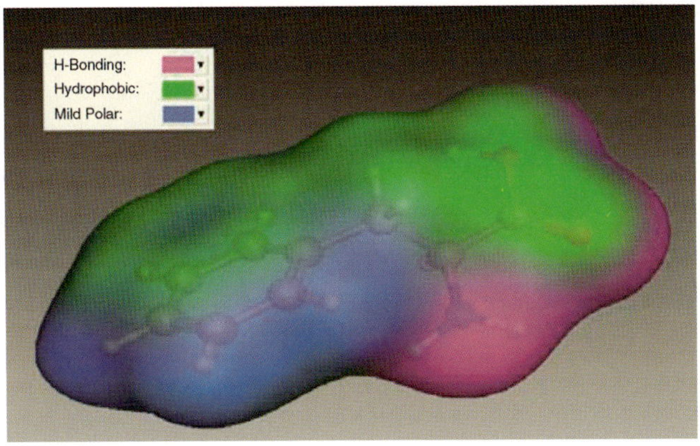


**Figure 9.42** H-bonding and hydrophobic areas.

Besides MEP and MLP, some other local quantities, such as hydrogen bonding and hydrophobic (Figure 9.42) or interaction potentials (Figure 9.43), are helpful to realize possible interactions to surrounding molecules.

Molecular dynamics simulations can be used to detect local interaction centers; here, we see a specific arrangement of the water molecules at the polar amino-acid group, whereas the hydrophobic part of the phenyl group shows no strong interaction to water molecules. Figure 9.44 shows what we could expect from MEP, hydrogen bonding, and interaction potential (Figures 9.40, 9.42, and 9.43).

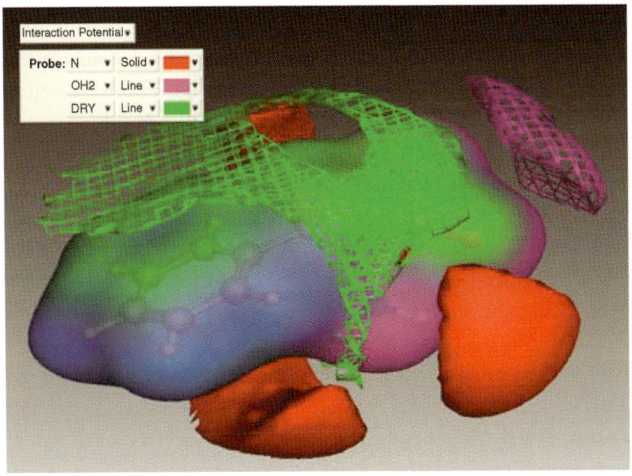


**Figure 9.43** Interaction potential surrounding the structure.

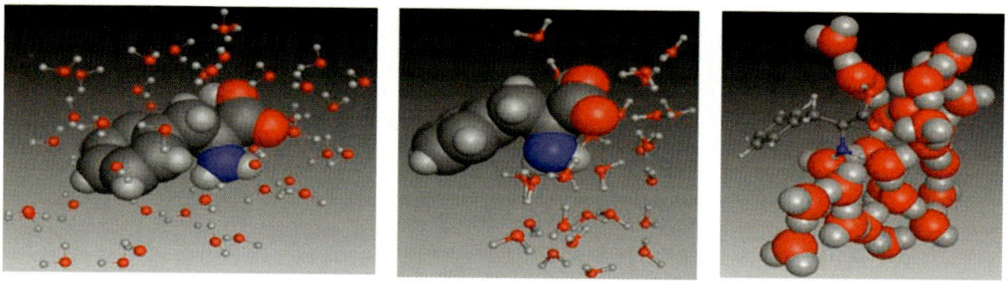

**Figure 9.44** MD simulation of phenylalanine with a water shell (initial distribution, relaxed distribution, and hydrogen bonding waters shown in CPK).

## 9.10 Conclusions

Computational chemistry offers different methods from empirical, semiempirical, and *ab initio* level of approximation. The most recent molecular modeling software contains a user-friendly graphical user interface and convenient visualizations, interfaces for databases, computational methods, and statistical tools. For huge data collections, a batch operating mode and automatic ranking in hit lists enables the user to concentrate on the refinement of structural models and searching for mechanistic interpretations.

Descriptors derived from the calculated electronic structure or from molecular orbitals are often very suitable to model experimental data and for property predictions. They have the possibility for mechanistic interpretation and transferability of models to other classes of compounds. The modeling "experiments" and computer simulations should give some insights into the behavior of molecules and their interactions. These investigations should not only produce new structures and new data but also to improve our knowledge.

The experimental proton affinity of a series of nitrogen heterocyclic compounds was successfully predicted by the calculated LMO of the lone-pair electrons in the sense of Klopmans perturbation treatment of chemical reactivity. These nitrogen-containing compounds are substructures for scaffold-based drug design. On the bases of this investigation of the proton affinity, we want to continue the research to the interaction of nitrogen lone-pair electrons to transitions metals ions in enzymes to act as an inhibitor.

The QSAR studies on the five structural isomers of hexane have revealed that there also exist beneficial relationships between the different properties of the same molecule, which open new chances on the elaboration of general structure–property relationships.

The scientific community in bioinformatics relies on a huge and rapidly growing number of data collections and software which can be freely accessed and modified. Scientists profit from concepts of open data, open access, and public license.

## References


1 CAS – Chemical Abstracts Service, a Division of the American Chemical Society *Columbus, OH* 43202 *U.S.A.* [online] http://www.cas.org/.

2 Murray-Rust, P., Glen, R.C., Rzepa, H.S., and Stewart, J.J.P., (2003) Proc. of UK e-Science All Hands Meeting, pp. 802–809.

3 Hamming, R.W. (1962) *Numerical Methods for Scientists and Engineers*, McGraw-Hill, New York.

4 Gasteiger, J. and Engel, T. (eds) (2003) *Chemoinformatics – A Textbook*, Wiley-VCH Verlag GmbH & Co. KGaA, Weinheim.

5 Wisswesser, W.J. (1982) *Journal of Chemical Information and Computer Sciences*, **22**, 88–93.

6 Weininger, D. (1988) *Journal of Chemical Information and Computer Sciences*, **28**, 31–36.

7 InChI – International Chemical Identifier [online], http://www.inchi.info/.

8 CSD – Cambridge Structural Database (2010) [online], http://www.ccdc.cam.ac.uk/products/csd/.

9 PDB – Protein Data Bank [online], http://www.pdb.org/pdb/home/home.do.

10 Platts, J.A., Butina, D., Abraham, M.H., and Hersey, A. (1999) *Journal of Chemical Information and Computer Sciences*, **39**, 835–845.

11 Poroikov, V.V., Filimonov, D.A., Ihlenfeldt, W.A., Gloriozova, T.A., Lagunin, A.A., Borodina, Y.V., Stepanchikova, A.V., and Nicklaus, M.C. (2003) *Journal of Chemical Information and Computer Sciences*, **43**, 228–236.

12 Marvin V 5.2 (2009) [ online], http://www.chemaxon.com/products/marvin/.

13 Jensen, F. (2007) *Introduction to Computational Chemistry*, John Wiley & Sons Ltd., Chichester, UK

14 Mulliken, R.S. (1934) *Journal of Chemical Physics*, **2**, 782–793.

15 Gasteiger, J. and Marsili, M. (1980) *Tetrahedron*, **36**, 3219–3228.

16 Heilbronner, E. and Bock, H. (1968) *Das HMO-Modell und seine Anwendung*, Bd. 1–3 Verlag Chemie, Weinheim.

17 Bogel, H., Dettmann, J., and Randic, M. (1997) *Croatica Chemica Acta*, **70**, 827–840.

18 Gordon, M.S. and England, W. (1973) *Journal of the American Chemical Society*, **95**, 1753–1760.

19 Karelson, M., Lobanov, V.S., and Katritzky, A.R. (1996) *Chemical Reviews*, **96**, 1027–1044.

20 Bader, R.F.W. (1990) *Atoms in Molecules: A Quantum Theory*, Oxford University Press, Oxford, UK, [ online] http://www.chemistry.mcmaster.ca/aim/.

21 Fedorovsky, M. and Bögel, H., Chemgapedia [online] http://www.chemgapedia.de/.

22 MOE – Molecular Operating Environment (2009) [ online], http://www.chemcomp.com.

23 Hansch, C. (1969) *Accounts of Chemical Research*, **2**, 232–239.

24 Leach, A.R. (2001) *Molecular Modelling – Principles and Applications*, 2nd edn, Pearson Prentice-Hall, New York.

25 Todeschini, R. and Consonni, V. (2009) *Molecular Descriptors for Chemoinformatics*, vol. 2, Wiley-VCH, Weinheim.

26 Dragon 6 [online], http://www.talete.mi.it/products/dragon_description.htm.

27 Mauri, A., Consonni, V., Pavan, M., and Todeschini, R. (2006) *MATCH*, **56**, 237–248.

28 Wiener, H. (1947) *Journal of the American Chemical Society*, **69**, 17–20.

29 Linstrom, P.J. and Mallard, W.G. (2010) [online], http://webbook.nist.gov.

30 Grimme, S. (2006) *Angewandte Chemie*, **118**, 4571–4575.

31 Hunter, E.P.L. and Lias, S.G. (1998) *Journal of Physical and Chemical Reference Data*, **27**, 244.

32 Stewart, J.J.P. (1990) *Journal of Computer-Aided Molecular Design*, **4**, 1–45 [online] http://openmopac.net/.

33 Koopmans, T. (1934) *Physica (Elsevier)*, **1**, 104–113.

34 Edmiston, C. and Ruedenberg, K. (1963) *Reviews of Modern Physics*, **35**, 457–464.

35 Pearson, R.G. (1986) *Proceedings of the National Academy of Sciences of the United States of America*, **83**, 8440–8441.

36 Klopman, G. and Hudson, R.F. (1967) *Theoretica Chimica Acta*, **8**, 165–174.

**37** Hudson, R.F. and Klopman, G. (1967) *Tetrahedron Letters*, 1103–1108.

**38** ACD [online], http://www.acdlabs.com/products/phys_chem_lab/pka/exp_table.html.

**39** Bredow, T., Geudtner, G., and Jug, K. (2001) *Journal of Computational Chemistry*, **22**, 861–887.

**40** Boegel, H., Hansen, S., and Laube, U. (1997) *Journal of Chemical Information and Computer Sciences*, **37**, 316–319.

**41** Clark, T. (2010) *Journal of Molecular Modeling*, **16**, 1231–1238.

**42** Michalak, A., De Proft, F., Geerlings, P., and Nalewajski, R.F. (1999) *Journal of Physical Chemistry A*, **103**, 762–771.

**43** Connolly, M.L. (1983) *Journal of Applied Crystallography*, **16**, 548–558.

**44** Entzian, A., Bögel, H., Buchholz, M., and Heiser, U. (2010) *Journal of Chemical Information*, **2**, P44.

# 10
# New Types of Descriptors and Models in QSAR/QSPR

*Christian Kramer and Timothy Clark*

## 10.1
## Introduction

In this chapter, we will concentrate on novel information-rich descriptors intended not only for quantitative structure–property relationships (QSPRs), but also for quantitative structure–activity relationships (QSARs). We will concentrate on descriptors that are derived from quantum mechanical calculations and that do not rely heavily on the 2D molecular structure. Many such descriptors are derived from local properties calculated at a molecular surface, so that we can consider two main aspects: the local properties themselves and the techniques used to derive descriptors from them. The goal of just about all QSARs and QSPRs is to describe intermolecular interactions. We, therefore, need local properties that describe the individual components of these interactions. These have been reviewed in detail [1], but will be described here once more in order to set the scene for the individual types of descriptors.

Intermolecular interactions are conventionally described in terms of four components [2]:

1) Pauli (electron–electron) repulsion
2) Coulomb (electrostatic) interactions
3) Dispersion (van der Waals) attraction
4) Donor–acceptor (Lewis acid-base) interactions.

In conventional classical treatments of molecules (e.g., in force fields and similar treatments for docking, scoring, QSAR, etc.), the Pauli repulsion and the dispersion interaction are combined into a single Lennard–Jones [3] or Buckingham [4] potential, although soft-core variations (i.e., ones with a less steep rise in the Pauli repulsion term) are now becoming popular for free-energy perturbation calculations [5]. Coulomb interactions are generally considered to play a dominant role in determining intermolecular interactions because of their strength in the gas phase and their long range. Their importance decreases in polar solutions, but QSAR and QSPR have

*Statistical Modelling of Molecular Descriptors in QSAR/QSPR*. First Edition.
Edited by M. Dehmer, K. Varmuza, and D. Bonchev

traditionally relied heavily on predominantly Coulomb-based models. Unfortunately, the common atom-centered monopole (net atomic charge) model often does a very poor job of reproducing the molecular electrostatic potential (MEP) around the molecule [6], so that one of the major advantages of descriptors based on quantum-mechanical treatments is that they can reproduce the MEP around the molecule more accurately than simple classical models. However, it is quite remarkable that the classical models (and many based on quantum mechanics) completely neglect the Lewis acid–base (electron donor–acceptor) interactions. This neglect is not based on systematic studies that reveal such interactions to be unimportant, but rather on the assumption that classical Lennard–Jones and Coulomb interactions are adequate for describing intermolecular interactions.

## 10.2 Local Properties

### 10.2.1 Molecular Electrostatic Potential

The idea of local properties (i.e., properties with single characteristic values at any given point around the molecule) is best illustrated by the MEP [7]. The MEP is defined as the energy of interaction between the molecules and a unit positive charge assuming that the molecule is not polarizable. In a classical atom-centered multipole model, the MEP is defined as

$$V(\mathbf{r}) = \sum_{i=1}^{natoms} \frac{q_i}{|\mathbf{R}_i - \mathbf{r}|} \tag{10.1}$$

where $V(\mathbf{r})$ is the MEP at position $\mathbf{r}$, $q_i$ is the charge assigned to atom $i$ at position $\mathbf{R}_i$, and *natoms* is the number of atoms in the molecule. Within a quantum-mechanical treatment, this expression becomes

$$\mathrm{V}(\mathbf{r}) = \sum_{i=1}^{natoms} \frac{\mathrm{Z}_i}{|\mathbf{R}_i - \mathbf{r}|} - \int \frac{\varrho(\mathbf{r}')\mathrm{d}\mathbf{r}'}{|\mathbf{r}' - \mathbf{r}|} \tag{10.2}$$

where $Z_i$ is the nuclear charge of atom $i$ and $\varrho(\mathbf{r}')$ is the electron density at position $\mathbf{r}'$. The important relationship between Equations 10.1 and 10.2 is that, if we were able to partition the electron density of the molecule into "atomic electron densities" in a unique way, we could express the MEP as an atom-based-model:

$$V(\mathbf{r}) = \sum_{i=1}^{natoms} \left( \frac{Z_i}{|\mathbf{R}_i - \mathbf{r}|} - \int \frac{\varrho_i(\mathbf{r}')\mathrm{d}\mathbf{r}'}{|\mathbf{r}' - \mathbf{r}|} \right) \tag{10.3}$$

where $\varrho'_i$ is the electron density due to atom $i$ at point $\mathbf{r}'$. Thus, Equation 10.1 subsumes the charge of the nucleus and the integral over the electron density due to the atom into a single point charge at the nucleus, making it equivalent to Equation 10.4:

$$V(\mathbf{r}) = \sum_{i=1}^{natoms} \frac{Z_i - \varrho_i(\mathbf{r}')\mathrm{d}\mathbf{r}'}{|\mathbf{R_i} - \mathbf{r}|} \approx \sum_{i=1}^{natoms} \frac{q_i}{|\mathbf{R_i} - \mathbf{r}|} \tag{10.4}$$

There are two problems with the point-charge approach: the electron density assigned to each atom is reduced to a point charge at the nucleus and the assignment of electron density to individual atoms.

The first problem leads to errors because the electron density assignable to each atom has a finite extent and is not usually isotropic. An atom with a lone pair, for instance, has more electron density in the direction of the lone pair than in any other. However, the electrostatic potential due to a point charge is the same in every direction, so that the anisotropy represented by the lone pair cannot be reproduced by the atom-centered multipole model. This deficiency can be removed by using an electrostatic model in which electrostatic multipoles are used at the positions of the atoms and possibly at other points such as bond centers [8]. Such models are, however, often derived directly from the electron density, although atom-centered multipoles are used in sophisticated force fields [9, 10].

The second problem is that there is no unique way to partition the electron density and assign it to individual atoms. The atoms-in-molecules (AIM) approach provides a rational basis for dividing up the electron density, but only one of many. As a consequence, the net atomic charge also cannot be defined uniquely and is not observable experimentally. An alternative way to define net atomic charges is not to divide up the electron density, but rather to fit the charges so that they can reproduce the MEP at an array of points around the molecule as much as possible. A variety of schemes is available to determine such potential-derived charges [11].

### 10.2.2 Electron Density

The electron density is instrumental in determining the Pauli repulsion between two atoms or molecules. It has therefore been used to define one type of molecular surface, the so-called isodensity surface [12]. This is simply the 3D electrodensity contour at a standard value that approximates accepted van der Waals' radii (usually 0.001 e $\mathrm{Bohr}^{-1}$). Thus, the electron density can be used to define the molecular size and shape.

### 10.2.3 Local Polarizability

The dispersion interaction is related to the polarizability by the London equation [13–16]. A local definition of polarizability is, therefore, useful to help describe van der Waals' attraction between molecules [17]. Generally, it is both difficult and time-consuming to calculate molecular polarizabilities [18], but an economical variational technique for semiempirical molecular-orbital calculations was introduced by Rivail [19–21] and later parameterized to give very accurate molecular electronic polarizabilites [22]. This technique can be partitioned (arbitrarily) to give an additive polarizability model [23], on the basis of which a local polarizability $\alpha(\mathbf{r})$ can be defined [24].

### 10.2.4
### Local Ionization Energy and Local Electron Affinity

Early attempts to define local properties to describe the electron donating and accepting abilities of molecules were based on Fukui's frontier molecular orbital (FMO) approximation [25] and were therefore named Fukui functions [26]. However, the FMO approximation is only appropriate for molecules that have frontier orbitals that are well separated energetically from all others, and so is not appropriate, for instance, extensively conjugated systems. Sjoberg *et al.* [27] therefore introduced the local average ionization energy, $I(\mathbf{r})$, which takes all occupied orbitals into account:

$$I(\mathbf{r}) = \frac{\sum_{i=1}^{nocc} -\varepsilon_i \varrho_i(\mathbf{r})}{\sum_{i=1}^{nocc} \varrho_i(\mathbf{r})} \tag{10.5}$$

where $\varepsilon_i$ is the eigenvalue (energy) of orbital *i*, *nocc* is the number of occupied orbitals, and $\varrho_i(\mathbf{r})$ is the electron density due to orbital *i* at position **r**. Thus, the local average ionization energy is essentially a Koopmans theorem ionization potential weighted by the electron density assignable to the individual occupied orbitals at any point in space. $I(\mathbf{r})$ calculated using density-functional theory (DFT) has proven to be a powerful property for describing a number of phenomena [28–31].

The concept of the local average ionization energy can be extended to that of a local electron affinity $A(\mathbf{r})$ simply by summing over virtual orbitals, rather than occupied ones [24]:

$$A(\mathbf{r}) = \frac{\sum_{i=nocc+1}^{norbs} -\varepsilon_i \varrho_i(\mathbf{r})}{\sum_{i=nocc+1}^{norbs} \varrho_i(\mathbf{r})} \tag{10.6}$$

where *norbs* is the total number of orbitals. Note that Equation 10.6 is only appropriate for minimal basis sets. For methods that include polarization functions or other additional orbitals, an intensity-filtering technique must be used to obtain sensible values for $A(\mathbf{r})$ [32].

$I(\mathbf{r})$ and $A(\mathbf{r})$ can be combined [24] to give the local electronegativity, $\chi(\mathbf{r})$, and the local hardness, $\eta(\mathbf{r})$, using definitions analogous to Mulliken [33, 34] and Pearson [35], respectively.

## 10.3
## Descriptors Derived from Local Properties

### 10.3.1
### PEST Methodology

In 2001, Breneman and coworkers introduced the property-encoded surface translator (PEST) descriptors to represent properties on 3D surfaces in a compact way [36]. Surface autocorrelations have been used before, for example by Gasteiger *et al.* [37], but the PEST implementation of Breneman and coworkers has become more popular and is still being developed further. The basic idea of the PEST methodology is to represent a 3D surface by its autocorrelation graph, that is, measuring the distance between the surface points. Each 3D object has a shape-specific distribution of distances between the surface points with similar surfaces having similar distances. Practically, it is sufficient to measure distances between 100 000 randomly selected surface point pairs. For three prototype surfaces of ball, cube, and open cylinder each with maximum diameter of length 2, the distance distribution (50 000 pairs are sufficient for convergence) histograms are shown in Figure 10.1.

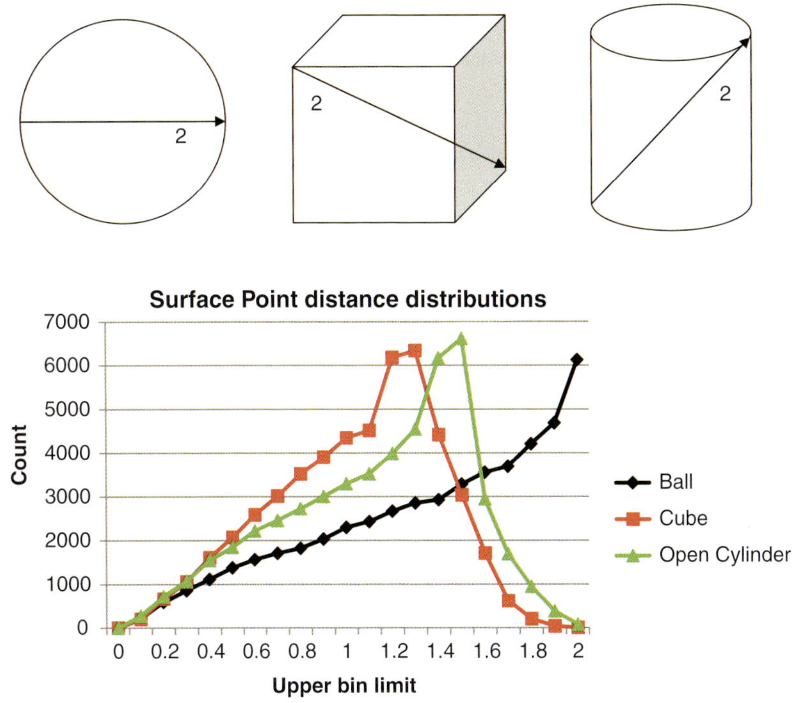


**Figure 10.1** Autocorrelation graphs for prototypical shapes. Calculated for 50 000 random surface point samples, maximum diameter = 2 and bin width = 0.1.

Different properties like the MEP or $I(\mathbf{r})$ can be calculated for each surface point and the distance histograms can be augmented by chemical properties. In the PEST methodology, this is practically implemented by binning the range of distances and the range of local properties. In a recent study, Breneman *et al.* published the PESD (property-encoded shape distribution) descriptors for predicting ligand-binding affinities. These descriptors are based on the shape histograms described above and lipophilicity and the electrostatic potential. They used 24 bins with 1 Ångstrøm bin width for the distances, nine bins for the electrostatic potential and 14 bins for lipophilicity. Since pairs of points are considered, this gives a set of $24 \times 9 \times 9 + 24 \times 14 \times 14 = 6648$ descriptors.

The PEST descriptors have proven very competitive in several models: prediction of protein HPLC retention times, glass transition temperatures of proteins, HIV reverse transcriptase ligand-based QSAR, clustering of protein binding sites and prediction of ligand-binding affinities [38–41].

## 10.4
## MEP as Descriptor for Hydrogen-Bonding Strengths

Kenny has shown that the electrostatic potential 0.55 Å from the hydrogen atom on a vector pointing outward along the hydrogen–heteroatom connection correlates with $R^2 = 0.55$ with experimental binding constants of hydrogen-bonded complexes [42]. This is remarkable in so far as in reality the interaction between two monomers is influenced by many other factors, so we can assume that the correlation of the MEP with the hydrogen-bonding strength is probably even higher. Therefore, the MEP can be used directly as an indicator of hydrogen-bonding strength.

Kenny found that the correlation of the MEP of other points closer to the van der Waals surface is lower. This is interesting because the local properties close to the van-der-Waals surface are usually considered to be important. The van der Waals radius for hydrogen is difficult because of the lack of experimental data, especially for strongly polar hydrogens capable of acting as H-bond donators. In a recent analysis of crystal structure and neutron diffraction data, Taylor found that there are especially short contacts for hydrogen-bond donors, supporting a very small van der Waals radius for polar hydrogens [43].

## 10.5
## ParaSurf (Politzer–Murray) Descriptors

A molecular surface (of any sort) is commonly described as a series of triangles (tesserae) that share edges in order to form the surface. As these triangles are generally of approximately the same size, their corners represent an acceptably uniform distribution of points on the surface. Politzer *et al.* [44, 45] first used the variance of the MEP values calculated at the surface points as a molecular descriptor. A particularly useful extension of this idea is to use the separate variances of all points with positive and negative MEP, so that a so-called balance parameter, $\nu$ can be defined as

$$\nu = \frac{\sigma_+^2 \cdot \sigma_-^2}{\left(\sigma_+^2 + \sigma_-^2\right)^2} \tag{10.7}$$

where $\sigma_+^2$ and $\sigma_-^2$ are the variances in the MEP for all points with positive and negative values, respectively. The balance parameter helps describe the juxtaposition of positive and negative areas on the molecular surface, which can interact with each other in order, for instance to increase the boiling point.

The concept introduced by Politzer–Murray *et al.* [45] is not limited to the MEP as local property and has been extended to derive analogous descriptors for $\alpha(\mathbf{r})$, $I(\mathbf{r})$, $A(\mathbf{r})$, $\chi(\mathbf{r})$, and $\nu(\mathbf{r})$ [46]. These descriptors are available in the program ParaSurf™ [47]. In addition to the variance, the skew and kurtosis of the property distributions have proven to be useful descriptors. It is particularly noteworthy that adding descriptors about the electron donor/acceptor properties (i.e., those based on $I(\mathbf{r})$ and $A(\mathbf{r})$) provides significant extra information that is not encoded by traditional 2D or 3D descriptors or by those based only on the MEP.

## 10.6 4D: Conformational-Ensemble-based Descriptors

3D QSA(P)R descriptors depend on the geometry of the compounds. Since most practically interesting chemicals such as drugs, agrochemicals, and polymers contain flexible bonds, it is necessary to guess which geometry to use for the models. For drugs, this is nearly impossible if the binding pocket is unknown since the conformation with the lowest energy in solution is in most cases not identical with the conformation in the binding pocket. In order to account for flexibility, several 4D QSAR methods have been developed. The fourth dimension consists of the ensemble of structures.

In 1997, Hopfinger and coworkers proposed a scheme for a 4D ligand-based QSAR extension [48] to 3D grid-based methods such as CoMFA [49] (comparative molecular field analysis) or CoMSIA [50] (comparative molecular similarity analysis). In both approaches each ligand is placed in a grid and at each vertex, the specific interaction energy (Lennard–Jones and electrostatic potential for CoMFA and interaction with a hydrophobic, negatively charged, positively charged or aromatic probe for CoMSIA) is calculated. After obtaining the grid for each ligand, the grids are overlaid to give a maximum similarity in each grid point. For each grid point, the interaction energies are used as descriptors in a partial least-squares (PLSs) fitting algorithm.

Hopfinger *et al.* proposed replacing the single conformation-dependent grid with potentials averaged over an ensemble of ligands. They obtained the ensemble from an MD simulation of the docked ligand in the binding pocket. Different conformations for one ligand are overlaid within the binding site and the potentials for different pharmacophore atom types are averaged over all conformations. The 4D QSAR paradigm has proven useful in several examples [51–53].

Several other approaches have been developed to incorporate the fourth dimension of ligand flexibility within a QSAR or pharmacophore model. Martin *et al.* have described a number of receptor-based pharmacophore models for kinases using their autoshim methodology [54]. Several different docking poses are generated for each ligand. The descriptors are calculated according to the fit of the ligand in the pharmacophore points defined by the receptor. Initially, the regression model is trained with a randomly selected pose for each ligand. In the next step, the highest scored pose for each ligand from the first model is selected to train the next model. This is performed iteratively until the model and the training set converge.

We recently published a 4D model for $\log P_{ow}$, the negative logarithm of the distribution coefficient between octanol and water [55]. The public $\log P_{ow}$ dataset collections are probably the largest ones available for QSA(P)R models. In contrast to all other datasets, they contain a large number ($\sim$ 1.500) of sample molecules with no rotatable bond. This allows conformationally unbiased 3D QSPR models to be generated. Using the binned ParaSurf descriptors described above for this dataset, we developed $\log P_{ow}$ models with excellent performance on the test set ($R^2 = 0.92$, $MUE = 0.36$), which is probably as good as the experimental error allows. For an initial test set of three drugs with rotatable bonds, we were able to show that the $\log P_{ow}$ predictions improve for Boltzmann-weighted ensembles compared to the single conformations. We are currently investigating this topic further.

## 10.7
## Proper Validation/Generation of QSA(P)R Models

In real-life applications, QSA(P)R models can be used for two different purposes: The first is to identify factors that influence some target property. For example, a QSPR model for aqueous solubility based on molecular fragments will very likely find that acid-containing fragments increase the solubility. So if the aqueous solubility of some molecules were to be increased, one suggestion would be to add fragments that contain acid groups. The same can be found for H-bond donor capabilities: if one particular H-bond of a molecule was to be made stronger, the molecule should be substituted with fragments that increase the MEP at the donating hydrogen. QSAR models can thus be used to analyze and illuminate physicochemical relationships.

The second purpose of QSA(P)R models is to predict target values, which will then further be used in selection or prioritization of compounds or for hypothesis generation. For example, if a measured biological activity of a pharmaceutically active compound is much lower than expected and the predicted solubility is also very low, the activity might be solubility limited. Further optimization of this compound might, therefore, be performed by increasing solubility, which can also be characterized by a QSPR model.

For both purposes, it is important to estimate the reliability of the model. The quality of the predictions is usually assessed with an independent external test set.

A random subset is completely left out of the modeling procedure (including any descriptor selection) and later used to assess the model. It should be noted that this is usually the only reliable assessment of the quality of a model. Internal quality measurements are often highly inflated due to overtraining, especially when advanced methods such as random forests, support vector machines, or neural networks are used for model building. As large an external test set as possible should be used to measure the quality of the model since the all important quality measures fluctuate strongly for smaller datasets. Therefore, techniques such as bagging and bootstrapping have been developed [56]. In bagging, a random test subset is repeatedly split off the overall dataset, a model is generated based on the remaining training set, and the left-out test set is predicted based on the final submodel. In the end, the final predictions for each compound are obtained by averaging the individual test-set predictions. This approach has the advantage that the entire available dataset can be used for both building the model and assessing its quality in an unbiased manner. The authors of this chapter hope that in the future more and more QSA(P)R models will be built using bagging or a flavor thereof to obtain better estimates of the true quality of the models. At present, QSA(P)R models are very often published for which the performance on the internal dataset is high and that for the randomly selected validation set much lower. This is a clear indicator of overtraining.

For multiple linear regression models, the *F*-value can be used as a measure of quality [57]. The *F*-value can be used to compare the performance of a model to that which would be obtained using random descriptors. Using a significance level of 95%, a model is significant if its $R^2$ is higher than the 95% of those that would be obtained using random descriptors. This is a very useful measure, because it allows a direct assessment of the quality of model without cross-validation. However, the *F*-values must be calculated properly. In 2005, Livingstone and Salt showed that the *F*-values usually used by the QSAR community are incorrect because they do not take the descriptor pool size into account [58]. The problem basically is that with increasing size of the pool of descriptors available for modeling, the probability that one random descriptor out of the pool is significantly correlated with the target value rises. Statistically, it makes a difference whether a descriptor is selected from a pool of 10 or 10 000 descriptors. Standard *F*-values used for significance assessment are only calculated for a given number of samples $n$ and number of descriptors in the model $p$. For QSAR models however, it is also important to take the size $k$ of the pool from which the descriptors were selected into account. We have, therefore, developed a new formula for three different significance levels (0.90, 0.95, and 0.99) to calculate the *F*-value depending on $n$, $p$, and $k$ [59]. These equations approximates the simulated *F*-values very well and can be used directly in any multiple linear regression scheme.

We have implemented a stepwise multiple linear regression algorithm using bagging and the *F*-value adjusted for descriptor-pool size that gives very good MLR models in our hands. The descriptor-pool size-adjusted *F*-value is used to stop descriptor inclusion when the descriptors have no significant correlation with the target value. In cross-validation experiments, the models are usually better than

PLS models because they include fewer descriptors that only represent noise. Further, the difference between $R^2$ and RMSE between the training and test data is usually very low, indicating that the models are not overtrained and that all descriptors make significant contributions to the predictions. The importance of the individual descriptors in these models can be assessed by two different criteria: the coefficients and the number of times a certain descriptor is selected for one of the submodels. The higher the absolute coefficient of the standardized descriptor, the more it contributes to the overall prediction and thus the higher is its importance. Further, the more often one specific descriptor is selected, the higher again is its significance. The *X*% significance criterion cannot exclude all random descriptors—it can only be used to exclude those descriptors that apparently have no significant contribution to the target data. However, if the 95% significance level is used, 5 of 100 random descriptors are still found to be significant. Since we use the bagging strategy and always create at least 50 single models, we can identify and remove many of the remaining random descriptors, because they are only selected in a few of the bagging models. This way we can build very stable and robust QSA(P)R models, which are at the same time clearly interpretable.

## 10.8 Conclusions

We have presented a set of novel descriptors that can be calculated from the evaluation of the quantum mechanical wavefunction for any molecule. The advantage of this set is that it should be physically complete, that is, all relevant types of molecular interactions should be possible to model based on these descriptors. They have a direct physical interpretation and they have successfully been used in QSAR and QSPR studies. We have also presented a statistical workflow for generating multiple linear regression models that gives models where every descriptor used is significantly correlated with the target activity. In our hands, this workflow has always outperformed all other methods for deriving multiple linear regression models with descriptor selection.

We are presently extending the range of QSAR and QSPR models based on the quantum mechanical descriptors and we develop new ways of dealing with the complexity occurring when the descriptor values are calculated for every point on the molecular surface.

## Acknowledgments

We thank many colleagues in Erlangen, Basel, and Portsmouth and the computational chemistry groups at Boehringer-Ingelheim, Biberach, and Novartis, Basel, for many useful discussions. C.K. thanks the Novartis Institutes for BioMedical Research for a Presidential PostDoc Fellowship.

## References


1 Clark, T., Byler, K.G., and de Groot, M.J. (2008) Biological communication via molecular surfaces, in *Molecular Interactions – Bringing Chemistry to Life* (Proceedings of the International Beilstein Workshop, Bozen, Italy, May 15–19, 2006), Logos Verlag, Berlin, pp. 129–146.

2 Stone, A.J. (1997) *The Theory of Intermolecular Interactions*, Oxford University Press, Oxford.

3 Lennard-Jones, J.E. (1924) On the determination of molecular fields. *Proceedings of the Royal Society of London. Series A, Mathematical and Physical Sciences*, **106** (738), 463–477.

4 Buckingham, R.A. (1938) The classical equation of state of Gaseous Helium, Neon and Argon. *Proceedings of the Royal Society of London. Series A, Mathematical and Physical Sciences*, **168**, 264–283.

5 Steinbrecher, T., Mobley, D.L., and Case, D.A. (2007) *Journal of Chemical Physics*, **127**, 214108.

6 Politzer, P., Murray, J.S., and Concha, M.C. (2008) *Journal of Molecular Modeling*, **14**, 659.

7 Politzer, P. and Truhlar, D.G. (eds) (1981) *Chemical Applications of Atomic and Molecular Electrostatic Potentials*, Plenum Press, New York, NY.

8 Stone, A.J. and Alderton, M. (1985) Distributed multipole analysis methods and applications. *Molecular Physics*, **56** (5), 1047–1064.

9 Devereux, M., Plattner, N., and Meuwly, M. (2009) Application of multipolar charge models and molecular dynamics simulations to study stark shifts in inhomogeneous electric fields. *Journal of Physical Chemistry A*, **113**, 13199–13209.

10 Plattner, N., Bandi, T., Doll, J.D., Freeman, D.L., and Meuwly, M. (2008) MD simulations using distributed multipole electrostatics: Structural and spectroscopic properties of CO- and methane-containing clathrates. *Molecular Physics*, **106**, 1675–1684.

11 Merz, K.M. Jr. (1992) Analysis of a large data base of electrostatic potential derived atomic charges. *Journal of Computational Chemistry*, **13**, 749–767.

12 Bader, R.F.W., Carroll, M.T., Cheeseman, J.R., and Chang, C. (1987) *Journal of the American Chemical Society*, **109**, 7968.

13 Eisenschitz, R. and London, F. (1930) *Zeitschrift Fur Physik*, **60**, 491.

14 London, F. (1930) *Zeitschrift Fur Physik*, **63**, 245.

15 London, F. (1931) *Zeitschrift für Physikalische Chemie (B)*, **11**, 222.

16 London, F. (1937) *Transactions of the Faraday Society*, **33**, 8.

17 Martin, B. and Clark, T. (2006) A dispersion treatment for NDDO-based semiempirical MO-techniques. *International Journal of Quantum Chemistry*, **106**, 1208–1216.

18 See, for instance, Davies, D.W. (1988) An *ab initio* configuration interaction for the polarizability of CsF. *Molecular Physics*, **63**, 1071–1076.

19 Rinaldi, D. and Rivail, J.-L. (1974) *Theoretica Chimica Acta*, **32**, 243.

20 Rivail, J.-L. and Rinaldi, D. (1976) Comptes Rendus, Serie B. *Sciences Physiques*, **299**, 283.

21 Rinaldi, D. and Rivail, J.-L. (1973) *Theoretica Chimica Acta*, **32**, 57.

22 Schürer, G., Gedeck, P., Gottschalk, M., and Clark, T. (1999) Accurate parametrized variational calculations of the molecular electronic polarizability by NDDO-based methods. *International Journal of Quantum Chemistry*, **75**, 17.

23 Martin, B., Gedeck, P., and Clark, T. (2000) An additive NDDO-based atomic polarizability model. *International Journal of Quantum Chemistry*, **77**, 473–497.

24 Ehresmann, B., Martin, B., Horn, A.H.C., and Clark, T. (2003) Local molecular properties and their use in predicting reactivity. *Journal of Molecular Modeling*, **9**, 342–347.

25 Fukui, K., Yonezawa, T., and Nagata, C. (1957) *Journal of Chemical Physics*, **26**, 831.

26 Parr, R.G. and Yang, W. (1989) *Density Functional Theory in Atoms and Molecules*, Oxford University Press, New York.

27 Sjoberg, P., Murray, J.S., Brinck, T., and Politzer, P.A. (1990) *Canadian Journal of Chemistry*, **68**, 1440.

28 Politzer, P.A., Murray, J.S., Grice, M.E., Brinck, T., and Ranganathan, S. (1991) *Journal of Chemical Physics*, **95**, 6699.

29 Politzer, P.A., Murray, J.S., and Concha, M.C. (2002) *International Journal of Quantum Chemistry*, **88**, 19, and references therein.

30 Hussein, W., Walker, C.G., Peralta-Inga, Z., and Murray, J.S. (2001) *International Journal of Quantum Chemistry*, **82**, 160.

31 Murray, J.S., Abu-Awwad, F., and Politzer, P.A. (2000) *Journal of Molecular Structure (Theochem)*, **501–502**, 241.

32 Clark, T. (2010) The local electron affinity for non-minimal basis sets. *Journal of Molecular Modeling*, **16**, 1231–1238.

33 Mulliken, R.S. (1934) *Journal of Chemical Physics*, **2**, 782.

34 Mulliken, R.S. (1955) *Journal of Chemical Physics*, **23**, 1833.

35 Pearson, R.G. (1991) *Chemtracts – Inorganic Chemistry*, **3**, 317.

36 Mazza, C.B., Sukumar, N., Breneman, C.M., and Cramer, S. (2001) Prediction of protein retention in ion-exchange systems using molecular descriptors obtained from crystal structure. *Analytical Chemistry*, **73**, 5457–5461.

37 Wagener, M., Sadowski, J., and Gasteiger, J. (1995) Autocorrelation of molecular surface properties for modeling corticosteroid binding globulin and cytosolic Ah receptor activity by neural networks. *Journal of the American Chemical Society*, **117** (29), 7769–7775.

38 Breneman, C.M., Sundling, C.M., Sukumar, N., Shen, L., Katt, W.P., and Embrechts, M.J. (2003) New developments in PEST shape/property hybrid descriptors. *Journal of Computer-Aided Molecular Design*, **17**, 231–240.

39 Ladiwala, A., Rege, K., Breneman, C.M., and Cramer., S.M. (2005) A priori prediction of adsorption isotherm parameters and chromatographic behavior in ion-exchange systems. *Proceedings of the National Academy of Sciences of the United States of America*, **102**, 11710–11715.

40 Das, S., Kokardekar, A., and Breneman., C.M. (2009) Rapid comparison of protein binding site surfaces with property encoded shape distributions. *Journal of Chemical Information and Modeling*, **49**, 2863–2872.

41 Das, S., Krein, M.P., and Breneman., C.M. (2010) Binding affinity prediction with property-encoded shape distribution signatures. *Journal of Chemical Information and Modeling*, **50**, 298–308.

42 Kenny, P.W. (2009) Hydrogen bonding, electrostatic potential, and molecular design. *Journal of Chemical Information and Modeling*, **49**, 1234–1244.

43 Taylor, R. (2011) Short nonbonded contact distances in organic molecules and their use as atom-clash criteria in conformer validation and searching. *Journal of Chemical Information and Modeling*, **51**, 897–908.

44 Politzer, P., Lane, P., Murray, J.S., and Brinck, T. (1992) Investigation of relationships between solute molecule surface electrostatic potentials and solubilities in supercritical fluids. *The Journal of Physical Chemistry*, **96**, 7938–7943.

45 Murray, J.S. and Politzer, P. (1998) Statistical analysis of the molecular surface electrostatic potential: An approach to describing noncovalent interactions in condensed phases. *Journal of Molecular Structure Theochem*, **425**, 107–114.

46 Ehresmann, B., de Groot, M.J., Alex, A., and Clark, T. (2004) New molecular descriptors based on local properties at the molecular surface and a boiling-point model derived from them. *Journal of Chemical Information and Computer Sciences*, **44**, 658–668.

47 ParaSurf'10, Cepos InSilico Ltd, Kempston, UK (2010) (www.ceposinsilico.com/products/parasurf.htm).

48 Hopfinger, A.J., Wang, S., Tokarski, J.S., Jin, B., and Albuquerque, M. (1997) Construction of 3D-QSAR models using the 4D-QSAR analysis formalism. *Journal of the American Chemical Society*, **119**, 10509–10524.

**49** Cramer, R.D. III, Patterson, D.E., and Bunce., J.D. (1988) Comparative molecular field analysis (CoMFA): 1. Effect of shape on binding of steroids to carrier proteins. *Journal of the American Chemical Society*, **110**, 5959–5967.

**50** Klebe, G., Abraham, U., and Mietzner, T. (1994) Molecular similarity indices in a comparative analysis (CoMSIA) of drug molecules to correlate and predict their biological activity. *Journal of Medicinal Chemistry*, **37**, 4130–4146.

**51** Hopfinger, A.J., Wang, S., Tokarski, J.S., Jin, B., Albuquerque, M., Madhav, P.J., and Duraiswami, C. (1997) Construction of 3D-QSAR models using the 4D-QSAR analysis formalism. *Journal of the American Chemical Society*, **119**, 10509–10524.

**52** Pan, D., Liu, J., Senese, C., Hopfinger, A.J., and Tseng, Y. (2004) Characterization of a ligand–receptor binding event using receptor-dependent four-dimensional quantitative structure–activity relationship analysis. *Journal of Medicinal Chemistry*, **47** (12), 3075–3088.

**53** Santos-Filho, O.A. and Hopfinger, A.J. (2006) Structure-based QSAR analysis of a set of 4-hydroxy-5,6-dihydropyrones as inhibitors of HIV-1 protease: An application of the receptor-dependent (RD) 4D-QSAR formalism. *Journal of Chemical Information and Modeling*, **46** (1), 345–354.

**54** Martin, E.J. and Sullivan., D.C. (2008) Surrogate AutoShim: Predocking into a universal ensemble kinase receptor for three dimensional activity prediction, very quickly, without a crystal structure. *Journal of Chemical Information and Modeling*, **48**, 873–881.

**55** Kramer, C., Beck, B., and Clark, T. (2010) A surface-integral model for $\log P_{OW}$. *Journal of Chemical Information and Modeling*, **50**, 429–436.

**56** Breiman, L. (1996) Bagging predictors. *Machine Learning*, **24**, 123–140.

**57** Fisher, R.A. (1924) On a distribution yielding the error functions of several well known statistics. *Proceedings of the International Congress of Mathematicians*, **2**, 805–813.

**58** Livingstone, D.J. and Salt, D.W. (2005) Judging the significance of multiple linear regression models. *Journal of Medicinal Chemistry*, **48**, 661–663.

**59** Kramer, C., Tautermann, C.S., Livingstone, D.J., Salt, D.W., Whitley, D.C., Beck, B., and Clark, T. (2009) Sharpening the toolbox of computational chemistry: A new approximation of critical *F*-values for multiple linear regression. *Journal of Chemical Information and Modeling*, **49**, 28–34.

# 11
# Consensus Models of Activity Landscapes

*José L. Medina-Franco, Austin B. Yongye, and Fabian López-Vallejo*

## 11.1
## Introduction

The analysis of structure–activity relationships (SAR) of a set of compounds with measured biological activity is a central topic in medicinal chemistry and drug design [1, 2]. Although SAR studies of small-to-medium size data sets are frequently performed by medicinal chemists on a routine basis with no need for computational approaches, the systematic study of the SAR of large data sets requires the application of automated methods. Computational approaches applied to SAR analysis can be divided in two major categories, namely, predictive and descriptive methods [3]. Predictive methods are meant to anticipate the biological activity of molecules before synthesis and/or biological evaluation in order to focus the experimental efforts – along with the associated time and costs – on the most promising compounds. Predictive methods include, but are not limited to, two-dimensional (2D) and three-dimensional (3D) quantitative structure–activity relationships (QSAR), pharmacophore modeling, and machine learning approaches [4–9]. In contrast to predictive methods, descriptive approaches are designed to access, visualize, and to help understand the data obtained from biological screening or chemical optimization [3]. *Activity landscape* methods, as they currently stand, are descriptive approaches that are increasingly being used to systematically characterize the SAR of large data sets [1, 3, 10, 11]. It has been largely recognized that understanding the activity landscape and early detection of *activity cliffs* (chemical compounds with high similar structure but unexpectedly very different biological activity [12]) can be crucial to the success of predictive computational models [13, 14]. Similarly, characterizing the activity landscape will be critical in future ligand-based virtual screening campaigns, for example, for selecting appropriate reference molecules for similarity-based virtual screening [15, 16].

The activity landscape has been defined as "any representation that integrates the analysis of the structural similarity and potency differences between compounds sharing the same biological activity" [3]. The activity landscape has also been conceptualized as the multidimensional space resulting from the addition of

*Statistical Modelling of Molecular Descriptors in QSAR/QSPR.* First Edition.
Edited by M. Dehmer, K. Varmuza, and D. Bonchev

biological activity as another dimension to the chemical space of a compound data set [10]. The activity landscape has been compared to rolling hills or continuous SAR where small changes in the molecular structure are associated with small changes in activity [12]. Such SARs are attractive for molecular-similarity analysis that is based on the similarity-property principle [17, 18]. A discontinuous SAR or rugged activity landscape, however, is populated with molecules with small changes in structure but large changes in activity (i.e., activity cliffs) [12]. Such landscapes are common in lead optimization. It can also happen that structurally diverse compounds have similar activity, which is the basis of scaffold or R-hopping [19–23]. Additionally, active regions with wide variations in chemical structure but small variations in biological activity may suggest different binding modes or sites or may reveal the effect of additional mechanisms such as the interaction with membranes that are not typically considered in several modeling approaches [13]. Understanding activity cliffs in compound data sets is an essential part of the study of SAR. Activity cliffs indicate specific structural features that are crucial for molecular activity. In addition, activity cliffs may appear as apparent statistical outliers in QSAR [12] leading to erroneous conclusions. Three different types of SAR are illustrated in Figure 11.1 that shows the chemical structures and biological activity ($IC_{50}$) of compounds tested against the parasite *Trichomonas vaginalis* [24–26]. The SAR of these compounds has been recently characterized [27]. Figure 11.1a shows an example of a pair of compounds with similar biological activity and very similar chemical structure, for example, continuous SAR. Figure 11.1b shows an example of a pair of molecules with similar activity but different chemical structure, for example, scaffold or R-hopping [23]. Figure 11.1c illustrates an example of an activity cliff: compounds with very different biological activity (more than three orders of magnitude) but very similar chemical structure (the only difference is a $CF_2$ group).

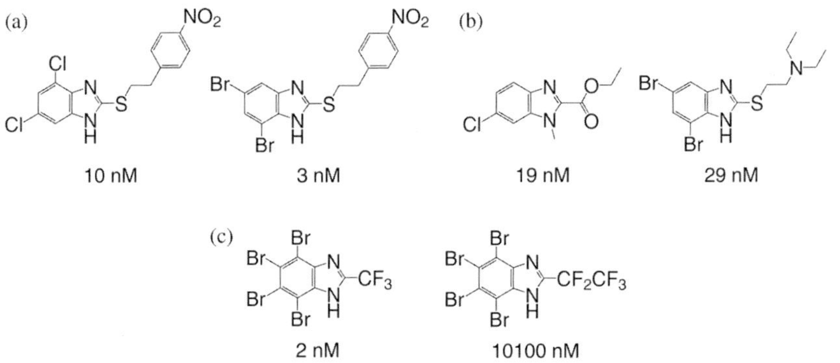


**Figure 11.1** Pair of compounds tested with *T. vaginalis* illustrating different types of SAR: (a) continuous SAR: molecules with similar biological activity and very similar chemical structures; (b) scaffold or R-hopping: molecules with similar activity but different chemical structures; and (c) activity cliff or discontinuous SAR: compounds with very different activity but very similar chemical structures.

The systematic description of the SAR of compound data sets has frequently revealed the complex nature of their underlying SAR; the landscape of several compound classes is often *heterogeneous* with the presence of continuous and discontinuous regions. An example is the data set tested with *T. vaginalis* shown in Figure 11.1 [27].

There is an increasing interest to develop quantitative activity landscape methods [3, 10]. There are two major sources of information to derive such models. One is the biological activity values that are obtained from the experiment, with the inherent experimental errors. The second source of information is the compound similarity relationships that define the chemical space [3]. The compound similarity relationships largely depend on the molecular representations that are chosen, and this topic is an area of intense research in our and other research groups. This chapter focuses on the advances of the development of activity landscape methods (Section 11.2) with particular emphasis on the current approaches to address the dependence of activity landscape on structural representation (Section 11.3).

## 11.2 Characterization of the Activity Landscape

The goal of activity landscape modeling of molecular data sets is to help rationalize the underlying SAR identifying key compounds and structural features for further exploration. The concept of activity landscape is strongly associated with the basic relationships between molecular structure and biological activity. While predictive SAR methods such as pharmacophore modeling and traditional QSAR/QSPR focus on specific molecular descriptors or arrangements of substructures or functional groups associated with activity, descriptive activity landscape models rely on the "similarity property principle," that is, similar structures should have similar biological properties [17, 18] and employ whole-molecular similarity measures [3].

Although the concept of activity landscape is simple in nature, the systematic characterization of activity landscape modeling is a challenging task and it is a research field under intense investigation. Two major difficult tasks in landscape modeling are the strong influence of molecular representation to characterize molecular similarity, and the complex nature of several landscapes involving a combination of smooth and rugged regions [28]. Quantitative efforts to characterize the activity landscapes have been developed until recently. Current graphical and numerical methods of activity landscape modeling are reviewed in several excellent publications [3, 10, 11] and are summarized in Table 11.1. These methods include the structure–activity relationship index (SARI) [29, 30], structure–activity landscape index (SALI) and SALI graphs [13, 14], structure–activity similarity (SAS) maps [23, 31, 32], netowork-like similarity graphs (NSG) [30, 33], and combinatorial analog graphs (CAG) [34]. SARI and NSG have been used to detect molecules with small changes in structure but large changes in selectivity (*selectivity cliffs*) [33]. NSG has been used to study multitarget SAR resulting from compounds tested with two or more targets [3]. Such models have enabled the visual representation and analysis

**Table 11.1** Methods to characterize the activity landscape.

| Approach | Basis | Applications | Refs. |
|---|---|---|---|
| Structure–activity relationship index (SARI) | Combine normalized continuous and discontinuous (for compounds with structure similarity and potency difference above a predefined cut-off) scores in a set of compounds or discontinuity of a compound | Global and local SAR characteristics and detect activity cliffs | [29, 30] |
| Structure–activity landscape index (SALI) | Potency-guided connection of nodes (individual compounds) if they form an activity cliff greater than a predefined SALI score | Detect activity cliffs | [13, 14] |
| Structure–activity similarity (SAS) maps | Depict pairwise structure and activity similarities. Activity similarities are represented as normalized potency difference. Data points denote compound pairs | Detect activity cliffs, continuous and discontinuous local SARs. Consensus activity cliffs for multiple structural representations | [23, 27, 31, 32] |
| Network-like similarity graphs (NSG) | Similarity–potency relationships. Connect nodes (individual compounds) if their similarity values are higher than a predefined cut-off. Color gradient reflects potency. Nodes are sized based on contributions of the corresponding compounds to the local SAR discontinuity | Explore global and local SAR patterns. Detect activity and selectivity cliffs. | [30, 33] |
| Multidimensional scaling (MDS) | Project chemical reference space onto a 2D surface. Color gradient reflects potency levels | Detect activity cliffs, continuous and discontinuous local SARs | [37] |
| Dual Activity Difference (DAD) maps | Pairwise comparison of the activity differences of a set of compounds across two targets. Each data point represents a pair of compounds, colored according to defined zones in the plots or structural similarity | Investigate activity, selectivity profile of a set of compounds against two targets, including selectivity cliffs | [27] |
| Combinatorial analog graphs (CAGs) | Utilize local similarity based on R-group substitution patterns to organize analog series in a hierarchical tree. Nodes represent compound subsets with substitutions at specified sites, while edges link subsets with one or more substitutions sites in common. Nodes are also annotated with normalized discontinuity scores | Study local SARs of closely related analogs of a single compound in advanced stages of drug discovery | [34] |
| Similarity-potency trees (SPT) | Hierarchical representation of compounds in a data set with increasing dissimilarity from a single reference compound down to a similarity threshold. Color gradient reflects potency but without node scaling | Describe local SARs | [38, 39] |
| 3D activity landscape | Compound potency included as a third dimension to a 2D projection of chemical space representation | Detect activity cliffs continuous and discontinuous local SARs | [36] |

of *structure–selectivity relationships*. The authors have proposed using multiple structural representations to derive a *consensus model* for activity landscape and identify *consensus activity cliffs* (see the next section) [35]. Recently, 3D representations of the activity landscape were proposed, confirming the existence of consensus activity cliffs and representation-dependent cliffs [36, 37]. Several of these methods are based on individual pairwise comparisons and consider the global network of SAR in the entire data set. In contrast, *compound-centric landscape views* have been proposed defining a single compound as a structural reference. One of these approaches are the similarity-potency trees (SPTs) [38, 39] reviewed recently [3]. A related representation is the *concentric similarity rings* used by the authors to graphically represent local SAR [23].

SAS maps are one of the first graphical methods proposed to explore the activity landscape of a given data set [31, 32]. SAS maps compare systematically structural similarity and activity similarity for all pairs of compounds in a data set. A general form of the SAS map is presented in Figure 11.2. In this map, the activity similarity is represented on the *Y*-axis and molecular similarity is plotted on the *X*-axis. A variant of the SAS maps represents potency difference in the *Y*-axis [35, 36]. For each pair of molecules, activity similarity can be measured as follows:

$$AS_{i,j} = 1 - \frac{|A_i - A_j|}{\max - \min} \tag{11.1}$$

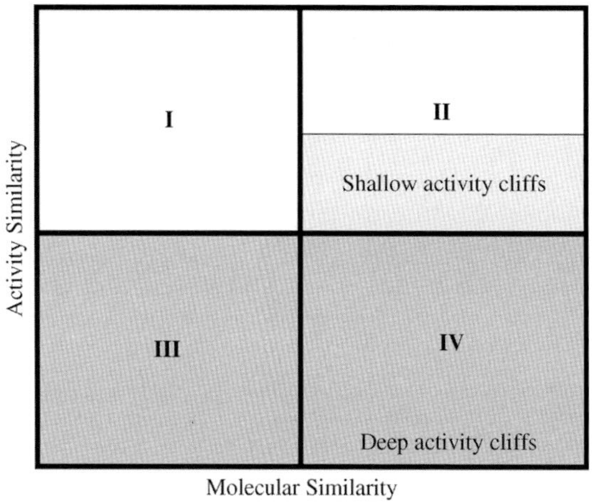


**Figure 11.2** Prototype model of a structure–activity similarity (SAS) map showing four major regions. Region I contains compound pairs that represent scaffold hops, or side-chain hops in the case wherein they differ only via substituents at attachment points. Region II represents compounds with smooth or continuous structure–activity relationships. Region IV contains compounds that represent discontinuities in the structure activity landscape and are termed activity cliffs. Data points in region IV can be distinguished further by small structural changes that result in moderate and dramatic changes for the shallow and deep activity cliffs regions, respectively. Data points in region III represent molecules with both low molecular and activity similarities.

where $A_i$ and $A_j$ are the activities of the $i$th and $j$th molecules, for example, $pIC_{50}$ or $pK_i$ values, and max–min indicate the range of activities in the data set.

SAS maps provide a visual and quantitative characterization of the activity landscape [23, 27, 32]. Four zones can be distinguished in Figure 11.2 labeled as regions I–IV. Data points that fall into region I correspond to pairs of molecules with high activity similarity and low molecular similarity and therefore are associated with regions of scaffold hopping. If the compounds in the data set share the same core scaffold and the differences are only in the attachment points, then region I is associated with side chain or R-hopping [23]. Points plotted in region II denote pairs of molecules with high molecular similarity and high activity similarity. Thus, compounds in this region are in a smooth or continuous SAR landscape. Data points in region III denote pairs of molecules with low molecular similarity and low activity similarity. Region IV identifies pairs of molecules that have high molecular similarity and low activity similarity and therefore corresponds to activity cliffs or discontinuous SAR. Data points that are consistently put in the same region by a number of molecular representations contribute to defining a consensus model of the activity landscape. In order to characterize the SAS maps obtained with different similarity measures, each map can be partitioned by imposing activity and molecular similarity thresholds along the $Y$- and $X$-axis, respectively, and then counting the number of data pairs in the resultant regions I–IV [23, 35]. To define the so-called *deep and shallow activity cliffs*, more than one threshold can be used. This approach has been used to define deep/shallow activity cliffs for a data set of antiparasitic compounds tested against two protozoa [23]. Note that when one or more molecular representations are used for the same set of compounds different ranges of similarity values are usually obtained. In such cases, a normalization criterion has to be employed to compare SAS maps (or another representation of the landscape) across molecular representations [23, 27, 35].

The concept of activity landscape to quantitatively explore SAR can be generalized to other types of landscapes that explore the structure–property relationships (SPR) of compound data sets. One step in this direction is the development of a *flavor landscape* for a comprehensive flavor database [40]. In that work, flavor descriptions of the materials in the database were coded as binary fingerprints and were the basis to compute inter-molecular flavor similarity using the Tanimoto coefficient [41, 42]. Using graphs related to SAS maps, a model of the flavor landscape was generated by means of structure–flavor similarity maps [40].

## 11.3 Consensus Models of Activity Landscape

### 11.3.1 Chemical Space and Molecular Representation

As discussed earlier, activity landscapes can be conceptualized as the multidimensional space resulting from the addition of biological activity as another dimension to chemical space. The chemical space of a compound data set can be viewed "as being analogous to the cosmological universe in its vastness, with chemical compounds populating space instead of stars" [43] or as "the total descriptor space that encompasses all the small carbon-based molecules that could, in principle, be created" [44]. Since the chemical space of a compound data set is strongly influenced by the molecular representation [32, 45, 46], it follows that molecular representation is one of the most critical variables in landscape modeling and a major source of potential inconsistencies in comparing and describing activity landscapes [3]. Figure 11.3 illustrates the drastic effect that different molecular representations can have on the distribution of compounds in chemical space. This figure shows a visual representation of the chemical space of 618 compounds obtained from Torrey Pines (TPIMS) combinatorial libraries (blue) [47–49] retrieved from PubChem [50, 51] and 1731 approved drugs (red) obtained from DrugBank [52] as implemented in the ZINC database [53]. The chemical space was generated using four different fingerprint

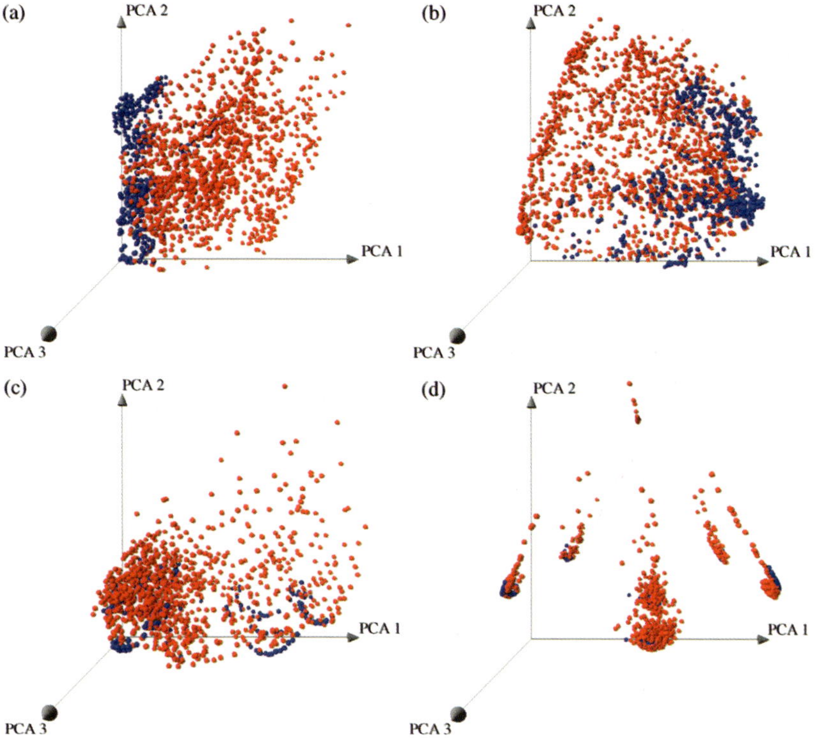


**Figure 11.3** Comparison of the chemical space of 618 TPIMS compounds (blue) and 1731 approved drugs (red). Depiction of a visual representation of the chemical space obtained by PCA of the similarity matrix computed with Tanimoto similarity using (a) MACCS keys, (b) piDAPH3 fingerprints, (c) TGD, and (d) TGT fingerprints. The first three PCs account for 76.1, 62.9, 79.6, and 93.1% of the variance, respectively.

representations, namely: MACCS keys (166 bits), pharmacophore atom triangle (piDAPH3), typed graph distance (TGD), and typed graph triangle (TGT) (see figure legend of Figure 11.3 for details). A recent book chapter by Maggiora and Shanmugasundaram [32] presents an excellent description of how chemical spaces are depicted and how reduced-dimensional chemical spaces are constructed. Although it is possible to conclude that, in general, several TPIMS compounds are structurally different from approved drugs, Figure 11.3 clearly shows that the four chemical space representations are different. Importantly, the neighborhood relationships between a pair of compounds may be different when they are compared in different chemical space representations.

To further illustrate the influence of molecular representation in chemical space, Figure 11.4 shows a visual representation of the chemical space of TPIMS compounds and approved drugs using drug-like properties. These properties are very useful to get information about the *druggability* of compound databases when compared to the property space of drugs [54, 55]. Property space is represented using six scaled molecular properties namely molecular weight (MW), hydrogen bond acceptors (HBA), hydrogen bond donors (HBD), number of rotatable bonds (RB), Log of the octanol/water partition coefficient (*S*log *P*), and topological surface area (TPSA). In contrast to the chemical spaces (Figure 11.3) obtained with molecular fingerprints, Figure 11.4 shows that TPIMS compounds have similar physicochemical properties as approved drugs. Therefore, chemical space and property space give

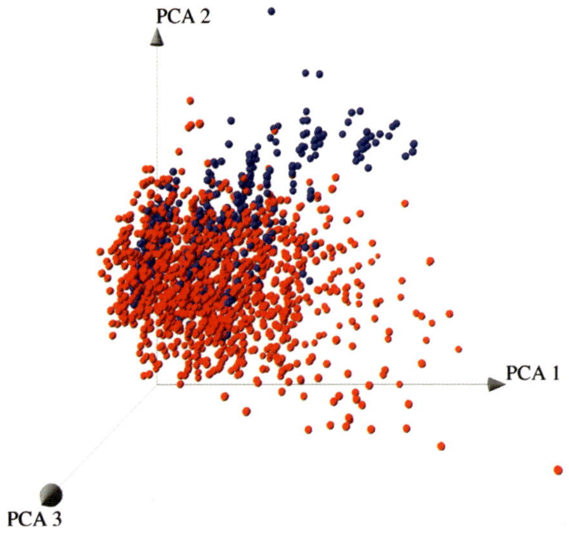


**Figure 11.4** Property space of 618 TPIMS compounds (blue) and 1731 approved drugs (red). The visual representation of the property space was obtained by PCA of six scaled molecular properties (MW, HBA, HBD, RB, SlogP, and TPSA). The first three PCs account for 86.8% of the variance. TPIMS and approved drugs databases were retrieved from PubMed and ZINC database, respectively.

important information about structure and molecular properties; and they are not mutually exclusive.

The dependence of chemical space on molecular representation and hence, the dependence of activity landscape on molecular representation, is rooted in the different nature of the molecular fingerprints, descriptors, and molecular properties used to assess molecular similarity. Different molecular descriptors capture different aspects of molecular structures [56] and there is not always a clear answer what molecular descriptions are *best* to characterize the activity landscape. Recent papers by Bender [56, 57] present a critical analysis of the different nature of molecular descriptors used to assess the similarity between molecules. To illustrate the diverse distribution of pairwise similarity values calculated for the same data set with different molecular representations, Figure 11.5 shows the cumulative distribution function (CDF) of 190 653 pairwise similarities of 618 TPIMS compounds obtained with MACCS keys (166 bits), pharmacophore graph triangle (i.e., graph-based three point pharmacophores) (GpiDAPH3), piDAPH3, pharmacophore atom quadruplet (piDAPH4), TGD, and TGT fingerprints as computed with molecular operating environment (MOE) [58]. In this example, similarity values computed with TGD and TGT fingerprints are shifted toward higher values (median similarities of 0.606 and 0.570, respectively). In contrast, similarities computed with piDAPH4, GpiDAPH3, and piDAPH3 fingerprints are shifted toward lower values (Figure 11.5). It is not surprising that MACCS keys do not have high similarity values since they were developed for speeding up the substructure searching in database and not for similarity analysis. The authors have observed similar different distributions for a

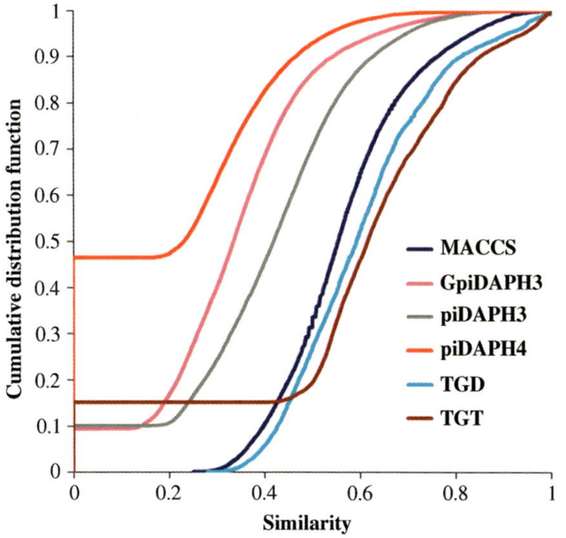


**Figure 11.5** Cumulative distribution functions of 190 653 pairwise structural similarities of 618 TPIMS compounds using six fingerprint representations.

(a)

| | MACCS | GpiDAPH3 | piDAPH3 | piDAPH4 | TGD | TGT |
|---|---|---|---|---|---|---|
| MACCS | 1.00 | | | | | |
| GpiDAPH3 | 0.53 | 1.00 | | | | |
| piDAPH3 | 0.50 | 0.84 | 1.00 | | | |
| piDAPH4 | 0.47 | 0.74 | 0.80 | 1.00 | | |
| TGD | 0.56 | 0.57 | 0.46 | 0.39 | 1.00 | |
| TGT | 0.37 | 0.45 | 0.42 | 0.36 | 0.59 | 1.00 |

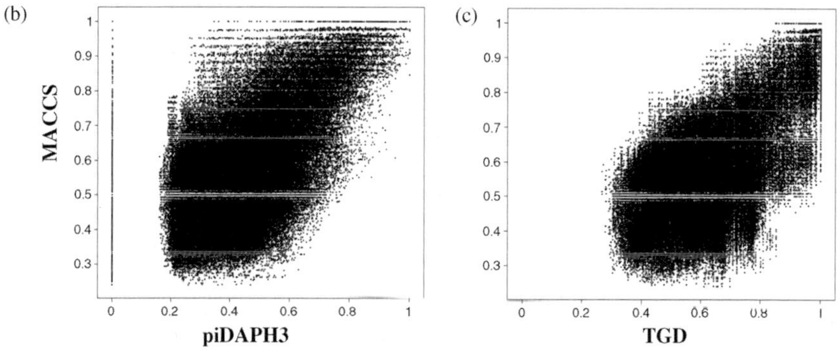


**Figure 11.6** (a) Correlation matrix for the 190 653 pairwise structural similarities of 618 TPIMS compounds for six fingerprint representations; correlation between (b) MACCS keys vs. piDAPH3 and (c) MACCS keys vs. TGD. Each data point indicates a pairwise comparison of 618 TPIMS compounds.

number of 2D and 3D molecular representations obtained for several data sets [23, 27, 35].

The correlation matrix between the six similarity methods for the 190 653 pairwise similarities is shown in Figure 11.6a. The matrix clearly shows the low correlation between several fingerprint representations, for example, between MACCS keys and piDAPH3 (0.5), and MACCS keys and TGD (0.56). Figure 11.6b and 6c further shows the low correlation between MACCS keys/piDAPH3 and MACCS keys/TGD, respectively. The distribution of pairwise similarity values and correlation between different molecular representations can be the basis to identify sets of molecular representations that would lead to similar and redundant landscape models. Similarly, these analyses can be the basis to select a subset of orthogonal, unrelated molecular representations to obtain complementary and consensus models of activity landscapes.

### 11.3.2
### Activity Landscape with Multiple Representations

Common to all activity landscape representations is the depiction of the relationship between molecular similarity and potency of a given data set. In descriptive utilizations of activity landscapes, potency information is derived from experimental measurements, while molecular similarity is determined via computational methods that require numerical descriptions of the compounds [3]. These descriptions are normally achieved employing fingerprint representations. Nonetheless, the authors and other groups have shown that the computed similarity relationships are representation dependent [15, 23, 27, 35, 36]. For instance, compounds that are considered similar using one description might be dissimilar in another representation as illustrated in Figures 11.3 and 11.4. As such, conclusions derived from analyzing activity landscapes will be influenced largely by the choice of the molecular representation that is used to define the chemical space. A consequence of this dependence is the potential presence of *apparent* continuous or discontinuous regions in the activity landscape, for example, *apparent activity cliffs* conceptualized as cliffs identified just for some molecular representations [23, 36]. In order to minimize the effect of the dependence of activity landscape on chemical representation, the authors have proposed the use of multiple molecular representations to model the activity landscape of data sets [23, 27, 35]. The major goal is to identify regions in the landscape that are consistently formed in different reference spaces, that is, to develop *consensus models* of the activity landscape that are consistent with several maximally orthogonal fingerprint representations [23, 27, 35].

Figure 11.7 illustrates the effect that different molecular representations can have on the distribution of compounds in the activity landscape. Figure 11.7 shows the SAS maps of 54 compounds (1431 pairwise relationships) tested against cathepsin B from *Trypanosoma brucei* (*T. brucei*) [59]. The SAS maps were obtained with six 2D and 3D orthogonal fingerprints. The 2D fingerprints were as follows: dendritic, MOLPRINT 2D, MACCS keys encoding 155 patterns from Canvas [60, 61], and TGT from MOE [58]. To evaluate 3D similarities 3-point pharmacophore fingerprints from Canvas (3p) and MOE (piDAPH3) were employed. In addition, different 3D conformations of the compounds in the database were utilized to determine the extent to which input conformation affected activity landscapes. The different 3D input conformations were generated as follows: (1) best whole molecule 3D overlay with a compound that showed best overall activity against three cathepsins (BO), (2) the lowest energy conformer (LE), and (3) input SMILES-generated conformer (IP). Figure 11.7k shows a *consensus SAS map* obtained from combining different molecular representations by means of mean fusion [62, 63]. From the distribution of the data points on the SAS maps, the effect that different molecular representations can have on activity landscapes of the same data set is evident. For the 3D fingerprints (3p and piDAPH3), differences are also observed in the distribution of compound pairs. For the consensus SAS map, the distribution of data points is confined to the middle of the landscape, which captures the low- or high-similarity distributions of the individual representations.

Figure 11.8 shows representative data points in different regions of the landscape obtained from the SAS maps in Figure 11.7, including their potencies, activity, and structure similarities.

The consensus pair, **4** and **54**, representative of data points from region I is shown in Figure 11.8a. The similarity in the potencies of compounds **4** (330 nM) and **54** (220 nM) is reflected in the very high activity similarity score of 0.92, though the

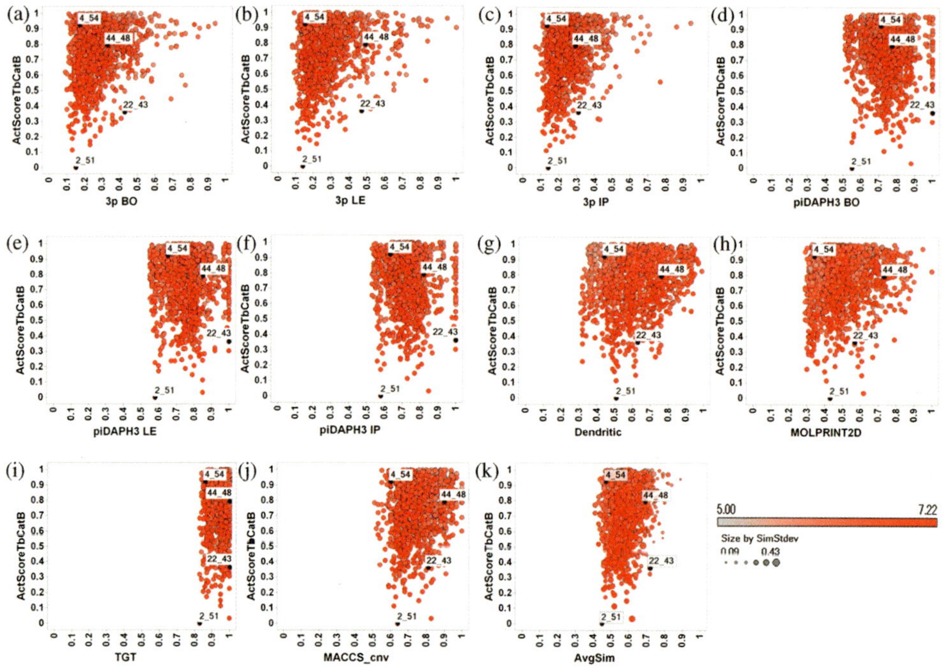


**Figure 11.7** SAS maps of *T. brucei* cathepsin B for several representations of 54 compounds. The panels present the following information: three point pharmacophore implemented in Canvas using (a) best overlay (BO) to a compound with best activity against three cathepsins including *T. brucei* cathepsin B; (b) the lowest energy (LE) conformer of each ligand; and (c) the SMILES-generated input conformer (IP); three point pharmacophore implemented in MOE using (d) best overlay (BO) to a compound with best activity against three cathepsins including *T. brucei* cathepsin B; (e) the lowest energy (LE) conformer of each ligand; and (f) the SMILES-generated input conformer; (g) dendritic; (h) MOLPRINT 2D (i) TGT; (j) MACCS keys implemented in Canvas; (k) Consensus of all the fingerprints, sized by the standard deviation of pairwise similarity. The data points are colored based on the activity of the most active member in a pair. Selected pairs in different regions of the SAS maps are shown. Region I: **4_54**; region II: **44_48**; region III: **2_51;** and region IV: **22_43**.

consensus structure similarity value of 0.47 is low. Compounds **4** and **54** differ at both substituent positions of the purine framework. This pair effectively represents side chain hops because of the shared purine scaffold among all the compounds in this data set. Figure 11.8b portrays a pair of compounds, **44** (540 nM) and **48** (180 nM), in the smooth SAR region of the activity landscape. They display relatively high-consensus structure and activity similarities of 0.69 and 0.79, respectively. The structural difference between **44** and **48** is a *p*-bromobenzyl in **44** as opposed to a 3,4-dichlorobenzyl in **48**. Figure 11.8c illustrates an activity cliff, the pair of compounds, **22** and **43**. The structural difference between this pair is a cyclohexyl and a *m*-trifluorobenzyl group in **22** and **43**, respectively. They share a high structural similarity, yet their potencies are remarkably different: 10 000 nM for compound **22** and 340 nM for compound **43**.

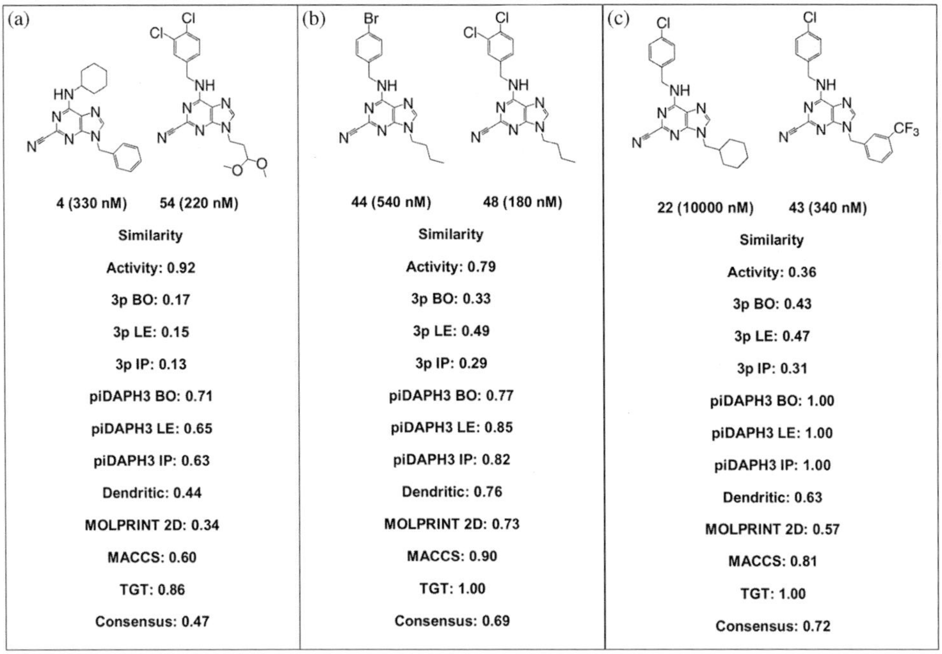


**Figure 11.8** Chemical structures of pairs of compounds from the three informative regions (I, II, and IV) of the SAS maps of *T. brucei*. (a) Data pair involved in side-chain hops I, that is, low structure similarity, but high activity similarity; (b) data pair in the smooth SAR region displaying both high activity and structure similarities; (c) portrays a pair of compounds involved in an activity cliff, that is, high structure similarity but very low activity similarity.

The activity landscape of 48 bicyclic guanidines with reported activity toward the κ-opioid receptor was characterized systematically using 11 2D and 3D structural representations [35]. Despite the inherent conformational issues, the use of 3D structural representations was valuable in that work to distinguish stereoisomers that are crucial in medicinal chemistry, and to characterize the activity landscapes. The activity landscape was explored using graphical representations related to SAS maps. Analyses of the computed similarity values revealed moderate-to-low correlations as well as statistical differences in distributions between the 2D and 3D methods. Nonetheless, quantitative characterizations of the SAS maps identified pairs of compounds that several methods placed in the same regions of the SAS maps, which led to the identification of consensus activity cliffs. Data fusion was also applied for the SALI measure suggesting a *consensus SALI* measure.

More recently, consensus SAS maps have been developed for a data set of 32 benzimidazoles with reported activities against two protozoa: *T. vaginalis* and *G. intestinalis* [23]. In this series of compounds, the consensus models for each target parasite were obtained by combining, using mean fusion [62, 63], the pairwise structure similarities for each compound computed from selected orthogonal fingerprints [23].

One should bear in mind that consensus scoring methods are not meant to "eliminate" data by, for example, disregarding "true" activity cliffs that are not seen by some molecular representations [3]. Consensus models are conceptualized to "prioritize" activity cliffs and other consensus regions in the landscape that are recognized by several molecular representations. There is a high likelihood that consensus regions in the landscape as perceived by computational molecular representations are also recognized by medicinal chemists and other experimental scientists.

In order to measure the number of data points that are consistently put into the same region by different representations, the Degree of Consensus (DoC) has been proposed [35]. The DoC between two methods for each region *R* of a landscape can be computed as follows:

$$DoC^{R}_{m,n} = \frac{Cp_{m,n}}{p_m + p_n - Cp_{m,n}} \tag{11.2}$$

where $Cp_{m,n}$ is the number of *Consensus Pairs* in region *R* (R = I – IV in Figure 11.2 of the SAS map) between methods *m* and *n*; $p_m$ is the number of pairs of molecules assigned by method *m* in region *R*, and $p_n$ is the number of pairs of molecules assigned by method *n* in the same region. Therefore, DoC is a Tanimoto-like expression that indicates the *similarity* that two methods exhibit for assigning molecules to a given region in the activity landscape. To note, DoC depends on the thresholds used to define each region, for example, the thresholds used to differentiate the four quadrants in the SAS maps (Figure 11.2). The DoC measure has been applied to compare the SAS maps of 48 bicyclic guanidines with reported activity toward the κ-opioid receptor [35] and a data set of 32 benzimidazoles with activity against the parasites *T. vaginalis* and *G. intestinalis* [23]. In both case studies, in general, low DoC values were observed in the activity cliffs region of the landscape.

One of the main challenges to quantitatively compare activity landscape models obtained with multiple representations and applying measures such as the DoC is the definition of thresholds that distinguish compounds that are on the borderline between continuous and discontinuous regions of the landscape. One approach that has been proposed is to define the thresholds based on the similarities of the actives compounds in the data set [23, 27, 35]. Another approach that we are currently exploring is the use of statistical measures of the entire data set.

The borderline cases between different regions of the activity landscape give rise to distinguishing deep and shallow activity cliffs [23]. A deep activity cliff has been defined as a pair of similar compounds with a "very large" difference in activity similarity [23]. As such, the deep cliffs are the most dramatic examples of the effect of the small changes of the structure on the activity. In contrast, shallow activity cliffs can be conceptualized as those pairs of compounds with large (but not dramatic) activity difference. In a SAS map, the shallow activity cliffs are located in a twilight zone between regions II and IV in Figure 11.2. This distinction has been helpful to further systematically characterize the activity landscape of the data set of 32 benzimidazoles mentioned above [23]. This concept can be extended to deep/shallow consensus cliffs. Deep/shallow activity cliffs can also be defined in terms not only of the activity similarity or potency difference but also of the structural similarity. Thus, a deep activity cliff can be defined as a pair of compounds with very similar structural similarity and large potency difference. In this case, a shallow activity cliff would be an activity cliff with border line structural similarity.

In the course of characterizing the SAR of compounds screened against *T. vaginalis* and *G. intestinalis*, we introduced the dual activity difference (DAD) map [27]. In this plot, the pairwise potency differences of each target are displayed on the *X*- and *Y*-axes. Structural similarity information can be incorporated by color coding of the data points. DAD maps allow for the simultaneous visualization of structural changes in a series of compounds and associated effects on their potencies against two targets.

Although molecular fingerprints have been frequently used to characterize molecular similarity in landscape modeling, other representations can be used to compute molecular similarity such as properties [3]. We recently introduced property–activity similarity descriptions to analyze the activity landscapes of 32 benzimidazoles against *T. vaginalis* and *G. intestinalis* employing six drug-like properties namely HBA, HBD, MW, SlogP, RB, and TPSA [23]. A relatively high DoC was observed between property and structure similarity. The use of fingerprints and molecular properties can form the basis to construct 3D-SAS maps portraying the relationships structure–property–activity (SPA) similarities. As such, SPA maps provide potential additional benefits to develop consensus models of the activity landscape not only based on structural similarity but also based on property similarity.

In addition to employing molecular physico-chemical properties, we have also investigated activity–docking score similarity using the activities of 54 compounds with available activities against *T. brucei* cathepsin B, human cathepsin H and L (unpublished data). In this study, the activity–property and activity–docking score landscapes had pairs of compounds correctly placed in the respective regions of their SAS maps. Across the three targets, both measures detected active pairs that were categorized by the nonfused consensus model.

As discussed above, 3D molecular representations play a key role in medicinal chemistry. Molecular similarity based on 3D molecular representations has proved to be useful in landscape modeling by distinguishing pairs of compounds that look identical to 2D representations, for example, stereoisomers. However, conformational issues pose the question of what conformer(s) are representative to compute 3D similarity. The conformational landscape of compounds with several RBs is complex and requires the generation of a large number of conformers for exhaustive sampling of conformational space [64]. However, in characterizing activity landscape, a single conformer is often selected for each database compound. As mentioned earlier, it is conceivable that the choice of input conformation could affect the conclusions derived from activity landscapes employing 3D similarity fingerprints. For example, for the 54 compounds tested against cathepsin B from *T. brucei* discussed above (Figures 11.7 and 11.8) [59], to determine whether the different

**Table 11.2** Degree of consensus matrix for the activity cliff region of the SAS map of *T. brucei* cathepsin B (different sets of 3D conformations are employed for the purine analogs).

| | 3p BO[a)] | 3p LE[b)] | 3p IP[c] | piDAPH3 BO | piDAPH3 LE | piDAPH3 |
|---|---|---|---|---|---|---|
| 3p BO | 1.00 | | | | | |
| 3p LE | 0.49 | 1.00 | | | | |
| 3p IP | 0.57 | 0.51 | 1.00 | | | |
| piDAPH3 BO | 0.29 | 0.26 | 0.27 | 1.00 | | |
| piDAPH3 LE | 0.30 | 0.26 | 0.29 | 0.68 | 1.00 | |
| piDAPH3 | 0.26 | 0.23 | 0.26 | 0.61 | 0.71 | 1.00 |

a) BO: Best overlay with a compound that showed best overall activity against all three targets.
b) LE: Lowest energy conformer.
c) IP: Input SMILES-generated conformer.

conformers' representations were complementary, the DoC matrix for the activity cliff region was computed utilizing only data pairs with at least one active compound. Results are summarized in Table 11.2. The off-diagonal elements represent the DoC between two different conformers. The average and median DoC are 0.40 and 0.29, respectively, while values range from 0.23 (between 3p LE and piDAPH3) to 0.71 (between piDAPH3 LE and piDAPH3). The low-to-moderate values indicate that the conformers' representations are identifying activity cliffs that are not present in the landscapes of the other conformers and point to potential benefits of including different 3D conformers in the characterization of activity landscapes.

## 11.4 Conclusions and Future Perspectives

Activity or property landscape modeling is an emerging concept to systematically characterize the SAR/SPR of data sets. Understanding the underlying activity/property landscape of molecular data sets is an essential concept in medicinal chemistry and other disciplines and can be critical to the successful application of predictive models such as QSAR and QSPR. Significant progress has been made to develop models of activity landscapes and a number of graphical and quantitative methods that enables the extraction, analysis, and visualization of the SAR of large data sets have been proposed. A major challenge in activity landscape modeling is the dependence of the SAR with molecular representation. One approach to overcome this dependence is using multiple molecular representations to derive consensus models of the activity landscape, identify consensus activity cliffs, and other consistent regions in the SAR. Consensus representations of the activity landscape have been proposed to obtain robust and comprehensive models of the underlying SAR of compound data sets that can be recognized not only by a statistician and a computational chemist but also by an expert medicinal chemist and/or an experimental scientist. Consensus models of activity landscape have been developed using SAS maps and related graphical representations. However, the basic principle of using multiple representations to generate consensus models can be extended to other methods such as consensus SALI and consensus SARI measures, to name a few.

Currently, activity landscape models are largely used as descriptive methods to visualize and analyze the SAR of screening and optimization data [1]. However, as we proposed earlier, activity landscape models can serve as the basis to develop predictive models [23]. Another challenge that remains to be fully solved is the quantitative comparison of landscapes. An initial step toward this direction is the measure Degree of Consensus. However, it remains to fully explore the quantification considering different thresholds that distinguish borderline regions between continuous and discontinuous regions of the landscape. Similarly, the characterization and quantification of deep and shallow activity cliffs remain to be fully addressed.

## Acknowledgments

The authors are very grateful to Dr. Gerald M. Maggiora (University of Arizona) for motivating discussions and inspiring much of our work on activity landscape modeling. We also acknowledge Dr. Karina Martínez-Mayorga (TPIMS) and M.Sc. Jaime Pérez-Villanueva (UNAM) for helpful insights. We thank Dr. Kendall Byler (TPIMS) for assembling the data set of purine analogs used in this chapter. This work was supported by the State of Florida, Executive Office of the Governor's Office of Tourism, Trade, and Economic Development.

## References

1 Wawer, M., Lounkine, E., Wassermann, A.M., and Bajorath, J. (2010) Data structures and computational tools for the extraction of SAR information from large compound sets. *Drug Discovery Today*, **15**, 630–639.

2 Medina-Franco, J.L., López-Vallejo, F., and Castillo, R. (2006) Diseño de fármacos asistido por computadora. *Educación Química*, **17**, 452–457.

3 Wassermann, A.M., Wawer, M., and Bajorath, J. (2010) Activity landscape representations for structure–activity relationship analysis. *Journal of Medicinal Chemistry*, **53**, 8209–8223.

4 Ooms, F. (2000) Molecular modeling and computer aided drug design. Examples of their applications in medicinal chemistry. *Current Medicinal Chemistry*, **7**, 141–158.

5 Martínez-Mayorga, K. and Medina-Franco, J.L. (2009) Chemoinformatics – applications in food chemistry, in *Advances in Food and Nutrition Research*, vol. 58 (ed. S. Taylor), Academic Press, Burlington, pp. 33–56.

6 Melville, J.L., Burke, E.K., and Hirst, J.D. (2009) Machine learning in virtual screening. *Combinatorial Chemistry and High Throughput Screening*, **12**, 332–343.

7 Ma, X.H., Jia, J., Zhu, F., Xue, Y., Li, Z.R., and Chen, Y.Z. (2009) Comparative analysis of machine learning methods in ligand-based virtual screening of large compound libraries. *Combinatorial Chemistry and High Throughput Screening*, **12**, 344–357.

8 Yang, S.Y. (2010) Pharmacophore modeling and applications in drug discovery: Challenges and recent advances. *Drug Discovery Today*, **15**, 444–450.

9 Tropsha, A. (2010) Best practices for QSAR model development, validation,

and exploitation. *Molecular Informatics*, **29**, 476–488.

**10** Bajorath, J., Peltason, L., Wawer, M., Guha, R., Lajiness, M.S., and Van Drie, J.H. (2009) Navigating structure–activity landscapes. *Drug Discovery Today*, **14**, 698–705.

**11** Peltason, L. and Bajorath, J. (2008) Molecular similarity analysis in virtual screening, in *Chemoinformatics Approaches to Virtual Screening* (eds A. Varnek and A. Tropsha), Royal Society of Chemistry, Cambridge, UK, pp. 120–149.

**12** Maggiora, G.M. (2006) On outliers and activity cliffs–why QSAR often disappoints. *Journal of Chemical Information and Modeling*, **46**, 1535–1535.

**13** Guha, R. and VanDrie, J.H. (2008) Structure–activity landscape index: Identifying and quantifying activity cliffs. *Journal of Chemical Information and Modeling*, **48**, 646–658.

**14** Guha, R. and Van Drie, J.H. (2008) Assessing how well a modeling protocol captures a structure-activity landscape. *Journal of Chemical Information and Modeling*, **48**, 1716–1728.

**15** Eckert, H. and Bajorath, J. (2007) Molecular similarity analysis in virtual screening: Foundations, limitations and novel approaches. *Drug Discovery Today*, **12**, 225–233.

**16** López-Vallejo, F., Caulfield, T., Martínez-Mayorga, K., Giulianotti, M.A., Nefzi, A., Houghten, R.A., and Medina-Franco, J.L. (2011) Integrating virtual screening and combinatorial chemistry for accelerated drug discovery. *Combinatorial Chemistry and High Throughput Screening*, **14**, 475–487.

**17** Johnson, M.A. and Maggiora, G.M. (eds) (1990) *Concepts and Applications of Molecular Similarity*, John Wiley & Sons, Inc, New York.

**18** Bender, A. and Glen, R.C. (2004) Molecular similarity: A key technique in molecular informatics. *Organic and Biomolecular Chemistry*, **2**, 3204–3218.

**19** Schneider, G., Neidhart, W., Giller, T., and Schmid, G. (1999) Scaffold-hopping by topological pharmacophore search: A contribution to virtual screening. *Angewandte Chemie-International Edition*, **38**, 2894–2896.

**20** Zhao, H. (2007) Scaffold selection and scaffold hopping in lead generation: A medicinal chemistry perspective. *Drug Discovery Today*, **12**, 149–155.

**21** Brown, N. and Jacoby, E. (2006) On scaffolds and hopping in medicinal chemistry. *Mini Reviews in Medicinal Chemistry*, **6**, 1217–1229.

**22** Medina-Franco, J.L., Martínez-Mayorga, K., Bender, A., and Scior, T. (2009) Scaffold diversity analysis of compound data sets using an entropy-based measure. *The QSAR and Combinatorial Science*, **28**, 1551–1560.

**23** Pérez-Villanueva, J., Santos, R., Hernández-Campos, A., Giulianotti, M.A., Castillo, R., and Medina-Franco, J.L. (2010) Towards a systematic characterization of the antiprotozoal activity landscape of benzimidazole derivatives. *Bioorganic and Medicinal Chemistry*, **18**, 7380–7391.

**24** Andrzejewska, M., Yépez-Mulia, L., Cedillo-Rivera, R., Tapia, A., Vilpo, L., Vilpo, J., and Kazimierczuk, Z. (2002) Synthesis, antiprotozoal and anticancer activity of substituted 2-trifluoromethyl- and 2-pentafluoroethylbenzimidazoles. *European Journal of Medicinal Chemistry*, **37**, 973–978.

**25** Andrzejewska, M., Yepez-Mulia, L., Tapia, A., Cedillo-Rivera, R., Laudy, A.E., Starosciak, B.J., and Kazimierczuk, Z. (2004) Synthesis, and antiprotozoal and antibacterial activities of *S*-substituted 4,6-dibromo- and 4,6-dichloro-2-mercaptobenzimidazoles. *European Journal of Pharmaceutical Sciences*, **21**, 323–329.

**26** Valdez-Padilla, D., Rodríguez-Morales, S., Hernández-Campos, A., Hernández-Luis, F., Yépez-Mulia, L., Tapia-Contreras, A., and Castillo, R. (2009) Synthesis and antiprotozoal activity of novel 1-methylbenzimidazole derivatives. *Bioorganic and Medicinal Chemistry*, **17**, 1724–1730.

**27** Pérez-Villanueva, J., Santos, R., Hernández-Campos, A., Giulianotti, M.A., Castillo, R., and Medina-Franco, J.L. (2011) Structure-activity relationships of benzimidazole derivatives as antiparasitic agents: Dual activity-difference (DAD) maps. *Medicinal Chemistry Communications*, **2**, 44–49.

28 Peltason, L. and Bajorath, J. (2007) Molecular similarity analysis uncovers heterogeneous structure–activity relationships and variable activity landscapes. *Chemistry and Biology*, **14**, 489–497.

29 Peltason, L. and Bajorath, J. (2007) SAR Index: Quantifying the nature of structure–activity relationships. *Journal of Medicinal Chemistry*, **50**, 5571–5578.

30 Wawer, M., Peltason, L., Weskamp, N., Teckentrup, A., and Bajorath, J. (2008) Structure–activity relationship anatomy by network-like similarity graphs and local structure–activity relationship indices. *Journal of Medicinal Chemistry*, **51**, 6075–6084.

31 Shanmugasundaram, V. and Maggiora, G.M. (2001) Characterizing property and activity landscapes using an information-theoretic approach. 222nd ACS National Meeting, Chicago, IL, United States, CINF-032; American Chemical Society, Washington, DC.

32 Maggiora, G.M. and Shanmugasundaram, V. (2011) Molecular similarity measures, in *Methods in Molecular Biology*, vol. 672 (ed. J. Bajorath), Springer, New York, pp. 39–100.

33 Peltason, L., Hu, Y., and Bajorath, J. (2009) From structure-activity to structure-selectivity relationships: Quantitative assessment, selectivity cliffs, and key compounds. *ChemMedChem*, **4**, 1864–1873.

34 Peltason, L., Weskamp, N., Teckentrup, A., and Bajorath, J. (2009) Exploration of structure–activity relationship determinants in analogue series. *Journal of Medicinal Chemistry*, **52**, 3212–3224.

35 Medina-Franco, J.L., Martínez-Mayorga, K., Bender, A., Marín, R.M., Giulianotti, M.A., Pinilla, C., and Houghten, R.A. (2009) Characterization of activity landscapes using 2D and 3D similarity methods: Consensus activity cliffs. *Journal of Chemical Information and Modeling*, **49**, 477–491.

36 Peltason, L., Iyer, P., and Bajorath, J. (2010) Rationalizing three-dimensional activity landscapes and the influence of molecular representations on landscape topology and the formation of activity cliffs. *Journal of Chemical Information and Modeling*, **50**, 1021–1033.

37 Iyer, P., Wawer, M., and Bajorath, J. (2011) Comparison of two- and three-dimensional activity landscape representations for different compound data sets. *Medicinal Chemistry Communications*, **2**, 113–118.

38 Wawer, M. and Bajorath, J. (2010) Similarity-potency trees: A method to search for SAR information in compound data sets and derive SAR rules. *Journal of Chemical Information and Modeling*, **50**, 1395–1409.

39 Wassermann, A.M., Peltason, L., and Bajorath, J. (2010) Computational analysis of multi-target structure–activity relationships to derive preference orders for chemical modifications toward target selectivity. *ChemMedChem*, **5**, 847–858.

40 Martínez-Mayorga, K., Peppard, T.L. (2011) Yongye, A.B., Santos, R., Giulianotti, M.A., and Medina-Franco, J.L., Characterization of a comprehensive flavor database. *Journal of Chemometrics*, **25**, 550–560.

41 Jaccard, P. (1901) Etude comparative de la distribution florale dans une portion des alpes et des jura. *Bulletin de la Societé Vaudoise des Sciences Naturelles*, **37**, 547–579.

42 Willett, P., Barnard, J.M., and Downs, G.M. (1998) Chemical similarity searching. *Journal of Chemical Information and Computer Sciences*, **38**, 983–996.

43 Lipinski, C. and Hopkins, A. (2004) Navigating chemical space for biology and medicine. *Nature*, **432**, 855–861.

44 Dobson, C.M. (2004) Chemical space and biology. *Nature*, **432**, 824–828.

45 Sheridan, R.P. and Kearsley, S.K. (2002) Why do we need so many chemical similarity search methods? *Drug Discovery Today*, **7**, 903–911.

46 Medina-Franco, J.L., Martínez-Mayorga, K., Giulianotti, M.A., Houghten, R.A., and Pinilla, C. (2008) Visualization of the chemical space in drug discovery. *Current Computer-Aided Drug Design*, **4**, 322–333.

47 Houghten, R.A., Pinilla, C., Appel, J.R., Blondelle, S.E., Dooley, C.T., Eichler, J.,

Nefzi, A., and Ostresh, J.M. (1999) Mixture-based synthetic combinatorial libraries. *Journal of Medicinal Chemistry*, **42**, 3743–3778.

**48** Pinilla, C., Appel, J.R., Borras, E., and Houghten, R.A. (2003) Advances in the use of synthetic combinatorial chemistry: Mixture-based libraries. *Nature Medicine*, **9**, 118–122.

**49** Houghten, R.A., Pinilla, C., Giulianotti, M.A., Appel, J.R., Dooley, C.T., Nefzi, A., Ostresh, J.M., Yu, Y.P., Maggiora, G.M., Medina-Franco, J.L., Brunner, D., and Schneider, J. (2008) Strategies for the use of mixture-based synthetic combinatorial libraries: Scaffold ranking, direct testing, *in vivo*, and enhanced deconvolution by computational methods. *Journal of Combinatorial Chemistry*, **10**, 3–19.

**50** Austin, C.P., Brady, L.S., Insel, T.R., and Collins, F.S. (2004) Molecular biology: NIH molecular libraries initiative. *Science*, **306**, 1138–1139.

**51** This library is freely accessible at http://pubchem.ncbi.nlm.nih.gov (accessed Jan, 2011)

**52** Wishart, D.S., Knox, C., Guo, A.C., Cheng, D., Shrivastava, S., Tzur, D., Gautam, B., and Hassanali, M. (2008) Drugbank: A knowledgebase for drugs, drug actions and drug targets. *Nucleic Acids Research*, **36**, D901–D906.

**53** Irwin, J.J. and Shoichet, B.K. (2005) ZINC – a free database of commercially available compounds for virtual screening. *Journal of Chemical Information and Modeling*, **45**, 177–182.

**54** Vistoli, G., Pedretti, A., and Testa, B. (2008) Assessing drug-likeness – what are we missing? *Drug Discovery Today*, **13**, 285–294.

**55** Singh, N., Guha, R., Giulianotti, M.A., Pinilla, C., Houghten, R.A., and Medina-Franco, J.L. (2009) Chemoinformatic analysis of combinatorial libraries, drugs, natural products, and molecular libraries small molecule repository. *Journal of Chemical Information and Modeling*, **49**, 1010–1024.

**56** Bender, A., Jenkins, J.L., Scheiber, J., Sukuru, S.C.K., Glick, M., and Davies, J.W. (2009) How similar are similarity searching methods? A principal component analysis of molecular descriptor space. *Journal of Chemical Information and Modeling*, **49**, 108–119.

**57** Bender, A. (2010) How similar are those molecules after all? Use two descriptors and you will have three different answers. *Expert Opinion on Drug Discovery*, **5**, 1141–1151.

**58** Molecular Operating Environment (MOE) version (2009) Chemical Computing Group Inc.: Montreal, Quebec, Canada. Available at http://www.chemcomp.com (accessed Jan, 2011)

**59** Mallari, J.P., Shelat, A.A., Obrien, T., Caffrey, C.R., Kosinski, A., Connelly, M., Harbut, M., Greenbaum, D., McKerrow, J.H., and Guy, R.K. (2008) Development of potent purine-derived nitrile inhibitors of the trypanosomal protease TbcatB. *Journal of Medicinal Chemistry*, **51**, 545–552.

**60** Sastry, M., Lowrie, J.F., Dixon, S.L., and Sherman, W. (2010) Large-scale systematic analysis of 2D fingerprint methods and parameters to improve virtual screening enrichments. *Journal of Chemical Information and Modeling*, **50**, 771–784.

**61** Canvas, version 1.3 (2010) *Schrödinger*, LLC, New York, NY.

**62** Willett, P. (2006) Similarity-based virtual screening using 2D fingerprints. *Drug Discovery Today*, **11**, 1046–1053.

**63** Medina-Franco, J.L., Maggiora, G.M., Giulianotti, M.A., Pinilla, C., and Houghten, R.A. (2007) A similarity-based data-fusion approach to the visual characterization and comparison of compound databases. *Chemical Biology and Drug Design*, **70**, 393–412.

**64** Borodina, Y.V., Bolton, E., Fontaine, F., and Bryant, S.H. (2007) Assessment of conformational ensemble sizes necessary for specific resolutions of coverage of conformational space. *Journal of Chemical Information and Modeling*, **47**, 1428–1437.

# 12
# Reverse Engineering Chemical Reaction Networks from Time Series Data

*Dominic P. Searson, Mark J. Willis, and Allen Wright*

## 12.1
## Introduction

Effective tools for the reverse engineering of chemical reaction networks from laboratory-scale time-course concentration data are likely to become of increasing commercial and academic importance. For instance, a model of a reaction network – written as a coupled set of ordinary differential equations (ODEs) describing the dynamic behavior of the system – comprises the central numerical description of the reaction network in modern process simulation and optimization software. Software of this nature is required for numerous reasons, including accurate and economic plant design and process optimization [1] and so methods, tools and procedures for rapidly establishing the reaction pathway from data using as little a priori information as possible are desirable.

In particular, methods that can be applied to data obtained from reaction systems operating far away from chemical or biochemical equilibrium are of interest. This is because batch and semi-batch reactors – rather than continuous stirred tank reactors (CSTRs) – tend to be used in the fine chemical and pharmaceutical industries during the chemical development lifecycle. Furthermore, the increased uptake of high throughput technologies, for example, automated robotic workstations for performing many experiments in parallel, coupled with improved *in situ* sensor technology to provide rich data sets, is likely to provide an increase in the quantity and quality of non-equilibrium experimental reaction data. Currently, this data is generally only used for kinetic fitting – once a kinetic model structure has been postulated – but there exists the potential to extract more useful information, such as reaction network topologies as well as reaction stoichiometries, reaction rates, reaction fluxes, and kinetic models [1–3].

Without suitable prior knowledge, however, the deduction of a fully characterized, interpretable dynamic mathematical description of reaction network from observed time composition data alone is an ill-posed problem. This type of problem – which may be thought of as "reverse engineering" a useful description of a network from data – is usually very difficult to solve uniquely. For instance, when there is little a

*Statistical Modelling of Molecular Descriptors in QSAR/QSPR*. First Edition.
Edited by M. Dehmer, K. Varmuza, and D. Bonchev

priori knowledge of the nature of the entities within the network and when the measured data is corrupted with measurement noise, there are frequently many plausible networks that can explain the data [4, 5].

One of the earliest advances in reaction network inference from non-equilibrium data was made by Bonvin and Rippin [2] who proposed a methodology – called target factor analysis (TFA) – to identify both the number of linearly independent reactions and also to test whether proposed reaction stoichiometries are consistent with the measured time-series concentration data. The principal advantage of TFA is that no knowledge of reaction kinetics or fluxes is required. Once a set of plausible reaction stoichiometries is extracted, the user can then propose suitable ODE kinetic model descriptions and test these against the data. This work has subsequently been developed by a number of workers (e.g., [3, 6–8]).

A contrasting approach is to parameterize a suitable ODE model structure directly from the observed data and to use the resulting model to infer the network properties. One such approach – referred to as the S-system methodology [4, 5, 9–12] – has been investigated extensively in the biosciences for modeling non-equilibrium time-series data (e.g., gene expression microarray data) and forms part of the biochemical systems theoretic (BST) framework for the dynamic modeling and analysis of biological systems. The S-system model structure possesses a number of desirable properties. For instance, experimental time series can be used to estimate the parameters of S-system models that can both approximate the temporal dynamics of a connected network of entities and be used to infer the connective structure of the network. However, an estimated S-system model does not, in itself, provide any direct information about the number or nature of the reaction steps, for example, stoichiometries or reaction orders.

Domain-dependent knowledge can be exploited to narrow the network search space by restricting the form of the ODE model used to explain the dynamic behavior of the chemical reaction network. For instance, elementary chemical and biochemical reactions, occurring in well-mixed, relatively dilute, homogeneous phases – such as may be found in controlled laboratory batch and fed-batch experiments – often obey the law of mass action kinetics. This allows a class of physically interpretable ODE models with pseudo-linear properties to be formulated [13–16]. The pseudo-linear properties of these models allow classical regression techniques to be applied within the ODE model search process.

In this chapter, we describe an approach, closely related to the pseudo-linear approaches described above, that uses a variant of the evolutionary computational method of differential evolution (DE) to search the space of elementary chemical reaction networks. DE [17] is employed to iteratively evolve a population of numerical vectors, each of which encodes the topology of a trial network of elementary reactions. Each trial network has a corresponding stoichiometric matrix, the properties of which may be used to determine whether the network is physically realizable. The objective function for the DE is designed so that the search is biased toward networks that are physically possible. In addition, for each trial network, the rate coefficients – and hence the vector of reaction rates – are identified from the estimated time derivatives of the measured concentration data using multiple linear regression. It is shown, using realistic amounts of noise corrupted time-series data acquired from simulations of several chemical reaction networks in batch experiments, that the approach is effective at recovering the "true" network of elementary reactions and providing good estimates of the associated rate coefficients.

The chapter is structured in the following way. In Section 12.2, mathematical models of elementary reaction networks are discussed in order to provide a definition of the particular network inference problem. In Section 12.3, the numerical representation of possible network topologies is described. In Section 12.4, the objective function used for network search is presented. The procedure used to determine whether trial networks are physically valid is also described for cases where chemical information is either available or not. In Section 12.5, the basic DE method and the self-adaptive variant of DE used to search for reaction networks are described. In Section 12.6, three case studies using simulated chemical reaction systems arepresented. Finally, in Section 12.7 some conclusions and a discussion are presented.

## 12.2
## Problem Definition

A materials balance for $S$ chemical components taking place in $R$ chemical reactions comprising a chemical reaction network can be written for certain classes of chemical reactors – batch, fed-batch (semibatch), and continuous flow stirred tank reactor (CSTRs). These balances are often written to reflect the pragmatic assumptions that the (homogeneous phase) reactor is operating isothermally, is well mixed, and that the overall density of the reaction mixture is not significantly changed by the occurrence of the chemical reactions within the reactor. In this work we will, without loss of generality, consider batch reactions only, that is, there are no materials added or removed during the course of an experiment. In this case, materials balance for each species can be written as

$$\frac{\mathrm{d}[x_i]}{\mathrm{d}t} = f_i \quad i = 1, \ldots, S \tag{12.1}$$

where $[x_i]$ is the molar concentration of species $i$ at time $t$. Equation 12.1 is a set of coupled ODEs that describes the dynamic behavior of the reactive species due to chemical reactions, as represented by the $S$ reaction fluxes $f_i$.

The flux terms $f_i$ are directly linked to the stoichiometries of the $R$ reactions taking place and the kinetic rate terms of these reactions. For instance, consider the $(S \times R)$ stoichiometric matrix $\mathbf{N}$ containing the stoichiometric coefficients for each of the $S$ species in the $R$ reactions:

$$\mathbf{N} = \begin{bmatrix} n_{11} & \cdots & n_{1R} \\ \ldots & \ddots & \ldots \\ n_{S1} & \cdots & n_{SR} \end{bmatrix} \tag{12.2}$$

Here, $n_{ij}$ is the stoichiometric coefficient of the $i$th chemical species in the $j$th reaction. By convention, $n_{ij} < 0$ for a species that undergoes net consumption in a reaction, $n_{ij} > 0$ for a species that undergoes net production and $n_{ij} = 0$ for a species that is either not involved in the $j$th reaction or has no net change in it. The component balances may now be conveniently expressed in terms of **N** and the $R$ individual reaction rates in matrix–vector form:

$$\frac{d\mathbf{x}}{dt} = \mathbf{Nr} \tag{12.3}$$

where **r** is the $(R \times 1)$ dimensional vector of reaction rates and **x** is the $(S \times 1)$ dimensional vector of species concentrations $[x_1], \ldots, [x_S]$ at time $t$. The $R$ reaction rates constituting **r** are, in general, nonlinear functions of the concentrations **x** and linear functions of the rate coefficients. If elementary reactions are assumed, then the form of the $R$ rate terms in **r** is determined uniquely by the reactants in each of the $R$ elementary equations.

The law of mass action kinetics states that the rate of an elementary reaction may be assumed to be directly proportional to the collision frequency of the reactants, and hence the product of the reactant concentrations. Consider a bimolecular elementary chemical reaction with rate coefficient $k$ involving $a$ molecules of species $x_1$ and $b$ molecules of species $x_2$ forming $c$ molecules of species $x_3$ and $d$ molecules of species $x_4$:

$$ax_1 + bx_2 \xrightarrow{k} cx_3 + dx_4 \tag{12.4}$$

The rate $r$ of this reaction at any time, according to the law of mass action, is

$$r = k[x_1]^a[x_2]^b \tag{12.5}$$

For example, consider the hypothetical chemical reaction network comprising four reactive species $x_1, \ldots, x_4$ involved in two elementary reactions with rate coefficients $k_1$ and $k_2$:

$$x_1 + x_2 \xrightarrow{k_1} x_3 + x_4 \quad x_3 \xrightarrow{k_2} x_4 \tag{12.6}$$

The stoichiometric matrix **N** and rate vector **r** for this network are

$$\mathbf{N}^T = \begin{bmatrix} -1 & -1 & 1 & 1 \\ 0 & 0 & -1 & 1 \end{bmatrix} \quad \mathbf{r} = \begin{bmatrix} k_1[x_1][x_2] \\ k_2[x_3] \end{bmatrix} \tag{12.7}$$

Hence, the $S$ ODEs, for known initial conditions of the $[x_i]$, can be used to describe the temporal evolution of the species concentrations:

$$\frac{d\mathbf{x}}{dt} = \mathbf{Nr} = \begin{bmatrix} -k_1[x_1][x_2] \\ -k_1[x_1][x_2] \\ k_1[x_1][x_2] - k_2[x_3] \\ k_1[x_1][x_2] + k_2[x_3] \end{bmatrix} \tag{12.8}$$

Note that any given row of **N** does not uniquely correspond to an elementary reaction, nor does any **N** correspond to a unique set of elementary reactions. For instance, the following reaction network has the same **N** as the network described by Equation 12.6:

$$x_1 + x_2 \xrightarrow{k_1} x_3 + x_4 \quad x_1 + x_3 \xrightarrow{k_2} x_1 + x_4 \tag{12.9}$$

Using the development above, the general problem of inferring a network of elementary reactions from time-series concentration data can be stated as: using only noise-corrupted measurements of **x** sampled from *M* experiments reconstruct one or more plausible elementary reaction networks, that is, the number and nature of the elementary reactions, the corresponding stoichiometric matrix **N**, the rate coefficients, and the reaction rates **r**.

## 12.3
## Reconstruction of Elementary Reaction Networks from Data by Network Search

Srividhya *et al.* [15] described a method to reconstruct (bio)chemical reaction networks from time-series concentration data method by searching the space of networks of elementary unimolecular and bimolecular reactions in order to minimize an objective function that takes into account both the predictive ability and the parsimony of the proposed reaction network. This search is stepwise in nature and is accomplished by adding or removing one reaction at a time from previously high-performing networks of elementary unimolecular and bimolecular reactions. For each network (i.e., set of elementary reactions) evaluated, the corresponding reaction rate terms, multiplied by the appropriate stoichiometric coefficients, are used to set up a constrained regression problem in which the network rate coefficients are computed from the estimated derivatives of the time concentration data using a nonnegative least-squares method. The minimization of an objective function that trades off the mean squared prediction error of the estimated derivatives against the number of reactions in each trial set is used to drive the stepwise search process. The advantages of this approach are that it is automated, nonsubjective, and that it exploits the fact that the rates of change of concentration (the fluxes $f_i$) of the *S* species are linear combinations of the same *R* reaction rate terms. The principal disadvantages of this approach as presented in [15] are that firstly, it uses a stepwise method to search the space of reaction networks. Stepwise methods are essentially local search operators and are likely to be trapped in local minima, especially in noisy, high-dimensional, multimodal search spaces such as that typically encountered in network inference. This appears to be borne out by some of the results presented in [15] in which the stepwise approach fails to recover the correct network for some fairly low-complexity bimolecular reaction networks under noise-free conditions. Secondly, the search method does not penalize physically unrealizable reaction networks. For instance, consider the following trial network:

$$2x_1 \xrightarrow{k_1} x_2 \quad x_1 \xrightarrow{k_2} x_2 \tag{12.10}$$

This is not physically possible, but such a network could nominally satisfy an objective function based on prediction error and network complexity alone. Thirdly, the method makes no provision for the use of a priori information about the reactive species (e.g., molecular weights and chemical composition).

In an attempt to address these issues, a method which takes into account these factors is presented in this work.

### 12.3.1
### Network Search as a Nonlinear Integer Programming Problem

In order to search through the space of reaction networks, it is first necessary to employ a suitable numerical representation of possible network topologies. In this work, the topology of a reaction network containing up to $q_{max}$ reactions is represented by a vector of integers. The search of the network space may then be formulated as the search for the optimal integer vector, that is, an integer programming problem. An optimal solution is that which best satisfies an objective function that balances network complexity and physical validity against how well the network fits the data.

Our integer representation of each of the $q_{\max}$ reactions is simply a description of the identities of the reactants and the products in each reaction. For instance, consider the case where the reactions are restricted to be either unimolecular or bimolecular with a maximum of two reactants and three products.[1)] Each reaction can then be represented by a sequence of five integers. The first two integers represent the identities of the reactants and the remaining three integers represent the identities of the products. A zero is used to "switch off" a reactant or product in a reaction. The purpose of this is to allow the representation of reactions of lower complexity than the specified maximum. Hence, each entry in the integer sequence can take a value in the range 0 to $S$.

The topology of the whole network is represented by concatenating the $q_{\max}$ integer sequences into one vector. This is illustrated in Figure 12.1 where $q_{\max} = 4$ and $S = 3$. The reaction representation rule is also illustrated for the five integers representing reaction 1.

Note that the representation also allows the specification of physically unrealizable reactions, such as reaction 4 in Figure 12.1 which is

$$2x_1 \rightarrow x_1 \tag{12.11}$$

Invalid reactions such as these are interpreted as "no reaction occurring" and are excluded from the trial network. Hence, the actual size $q$ of the network in Figure 12.1 is 3. This is accomplished by examining the columns of the trial network's stoichiometric matrix and removing those that do not contain at least one positive and one negative coefficient. The ability to "switch off" reactions is an important

1) The integer representation used can be easily modified to encompass reaction types of greater or lesser complexity than that illustrated.

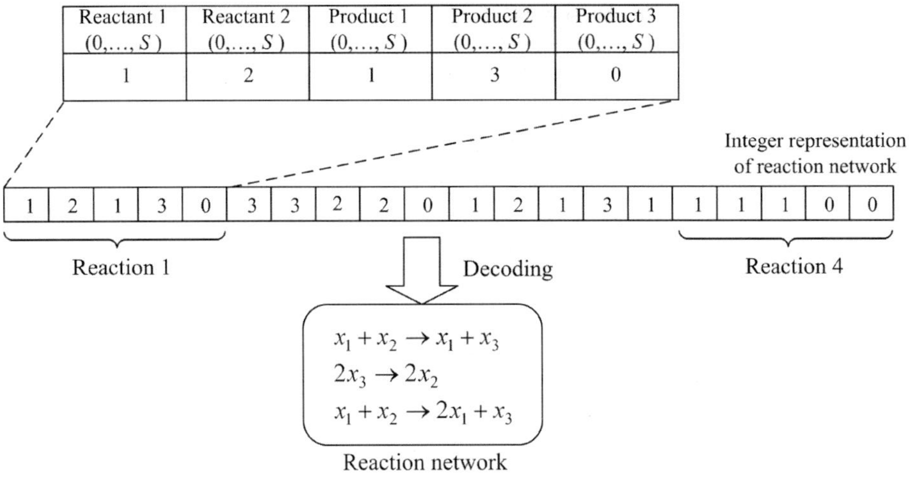


**Figure 12.1** Example of integer vector representation of the topology of a trial reaction network containing $S = 3$ species and a maximum of $q_{max} = 4$ reactions. In this case, reaction 4 is invalid and is excluded from the overall network.

feature because it is not assumed exactly how many reactions $R$ are occurring in the unknown reaction network, but it is assumed that $R < q_{max}$ where $q_{max}$ is set to some "sensible" user defined limit.

While this representation excludes individual reactions that are physically unrealizable, it does not prohibit entire networks with this property. The trial network illustrated in Figure 12.1, for instance, is not physically possible because the third reaction is inconsistent with the first and second reactions. Such networks are not rejected out of hand, however, because any integer programming method employed to search the space of network topologies would require substantial modification in order to search the integer vector space in such a way that only trial solutions corresponding to physically realizable networks are generated. This would then limit the applicability and simplicity of the approach. Instead, we employ a soft constraint to guide the DE-based integer programming method toward physically realizable solutions. This is discussed further in Section 12.4.

### 12.3.2 Estimation of the Rate Coefficients for Trial Reaction Networks

To evaluate how well any trial network topology fits the data, it is necessary to estimate the rate coefficients. However, parameter estimation for each of the corresponding ODE models from time-series concentration data usually requires the use of iterative optimization techniques (e.g., gradient descent methods). This typically involves repeated numerical solution of the ODEs using many trial parameter sets until the simulation closely matches the experimental data (i.e., "kinetic fitting"). A drawback of this approach is that the numerical solution of the ODEs for each trial parameter set for each of the evaluated trial network topologies is – currently – prohibitively computationally expensive.

To avoid the computational expense of numerical integration it is possible – prior to the network search procedure – to estimate the time derivatives for each of the $S$ species from the measured time-series concentration data [18]. This can be achieved by fitting nonlinear "smoothing" functions to the concentration measurements and either numerically or analytically obtaining the derivatives at the required sample times from the fitted function. Examples of this in the literature include the use of forward differencing [15], feedforward artificial neural networks [18], and rational polynomials [16].

Once estimates of the time derivatives are obtained, the rate coefficients for each trial network can be estimated from the data using least-squares methods. This requires the formation of a response vector **y** and a predictor (model design) matrix **X** for each trial network. These can be related to the $(q \times 1)$ vector of the unknown rate coefficients **k** by a linear equation that may be solved using least squares to minimize the vector of errors $\boldsymbol{\varepsilon}$:

$$\mathbf{y} = \mathbf{X}\mathbf{k} + \boldsymbol{\varepsilon} \tag{12.12}$$

Following the development in [15], this is outlined below for a single experiment. Let $[x_i]_n$ represent the sampled concentration of the $i$th species at the $n$th time point during an experiment, where there are $N$ time points, and let $D_{i,n}$ represent the estimated time derivative of the $i$th species concentration at the $n$th time point. An $(N \times q)$ matrix $X_i$ can be formed for each species, the $j$th column of which comprises the unscaled kinetic rate term for the $j$th reaction, evaluated at each time point, multiplied by a coefficient $\nu_{i,j}$. This coefficient is the molecularity of the $i$th species in the $j$th reaction multiplied by $+1$ if the $i$th species is a product in the $j$th reaction, by $-1$ if it is a reactant, and by 0 if it does not take part in the $j$th reaction. For example, in the following trial network, the coefficients are $\nu_{1,1} = -2$; $\nu_{1,2} = 0$; $\nu_{2,1} = 1$; $\nu_{2,2} = -1$; $\nu_{3,1} = 0$; and $\nu_{3,2} = 1$:

$$2x_1 \xrightarrow{k_1} x_2 \quad x_2 \xrightarrow{k_2} x_3 \tag{12.13}$$

The $\mathbf{X}_i$ matrices for this network are shown in Equation 12.14:

$$\mathbf{X}_1 = \begin{bmatrix} -2[x_1]_{n=1}^2 & 0 \\ -2[x_1]_{n=2}^2 & 0 \\ \vdots & \vdots \\ -2[x_1]_{n=N}^2 & 0 \end{bmatrix} \mathbf{X}_2 = \begin{bmatrix} [x_1]_{n=1}^2 & -[x_2]_{n=1} \\ [x_1]_{n=2}^2 & -[x_2]_{n=2} \\ \vdots & \vdots \\ [x_1]_{n=N}^2 & -[x_2]_{n=N} \end{bmatrix} \mathbf{X}_3 = \begin{bmatrix} 0 & [x_2]_{n=1} \\ 0 & [x_2]_{n=2} \\ \vdots & \vdots \\ 0 & [x_2]_{n=N} \end{bmatrix} \tag{12.14}$$

The $\mathbf{X}_i$ are then vertically concatenated to form the $(SN \times q)$ overall predictor matrix **X**. Similarly, the $(SN \times 1)$ response vector **y** is formed from the vertical concatenation of the $S$ vectors $\mathbf{y}_i$ each of which contains the $N$ estimated time derivatives $D_{i,n}$. The extension to $M$ multiple experiments is straightforward and yields an **X** of dimension $(MSN \times q)$ and **y** of dimension $(MSN \times 1)$.

To compute **k**, we use the least-squares normal equation with the Moore–Penrose pseudo-inverse $(\mathbf{X}^T\mathbf{X})^{\#}$ to mitigate problems with co-linearity in **X**:

$$\mathbf{k} = (\mathbf{X}^T\mathbf{X})^{\#}\mathbf{X}^T\mathbf{y} \tag{12.15}$$

In our method, any reactions that yield rate coefficients that are negative or zero are simply deleted from the trial network. The corresponding elements in **k** and columns in **X** are also then deleted. The sum of squares of prediction errors of the estimated derivatives ($\mathrm{SSE}_D$) for the rectified trial network can then be calculated as

$$\mathrm{SSE}_D = (\mathbf{y}-\mathbf{Xk})^T(\mathbf{y}-\mathbf{Xk}) \tag{12.16}$$

In the degenerate case, where there are no reactions remaining in the rectified trial network ($q = 0$), a value of infinity is assigned to the $\mathrm{SSE}_D$.

## 12.4
## Formulation of the Objective Function for Network Search

An objective function based purely on the minimization of prediction errors does not take into account the desired model characteristics. It is preferable to use one of the formal statistical metrics (information criteria; IC) that can be minimized to in order to balance the trade-off between model complexity (in this case, the number of reactions $q$) and how well the model fits the data, for example, Akaike's information criterion (AIC; [19]). However, it is seldom clear which is the "best" criterion to use. A number of ICs were tested and it was found that the Schwarz information criterion (SIC; [20]) appeared to yield the best all round performance. The SIC – sometimes referred to as the Bayesian information criterion – for a trial network can be written as

$$\mathrm{SIC}_D = N_s \ln\left(\frac{\mathrm{SSE}_D}{N_s}\right) + p\ln(N_s) \tag{12.17}$$

where $N_s$ is the sample size and $p = q + 1$ [21].

The SIC measure, however, does not discriminate between physically valid and physically unrealizable reaction networks. To guide the network search process, we propose the use of an additional term in the objective function to penalize trial networks that violate physical constraints. This is achieved my modifying the objective function in Equation 12.17 as follows:

$$\mathrm{SIC}_D = N_s \ln\left(\frac{\mathrm{SSE}_D}{N_s}\right) + (p+\alpha)\ln(N_s) \tag{12.18}$$

where $\alpha = 0$ for physically valid networks and $\alpha = q_{\max}$ for physically invalid networks. The meaning of this is that physically invalid networks are penalized in terms of the SIC measure as if they comprise $q + q_{\max}$ elementary reactions rather than $q$ reactions. Hence, physically unrealizable trial networks are not considered infeasible solutions – but they are strongly penalized. In terms of the evolutionary search process, this allows relatively high-performing partial solutions – although formally invalid – to be explored early on in the expectation that these will be combined with other high-performing partial solutions later in the search process. As the search progresses, the evolutionary process increasingly discriminates in favor of relatively high-performing valid solutions (i.e., those for which $\alpha = 0$). In this case, the modified objective function reverts to the standard SIC.

In order to determine whether a trial network is physically valid, two scenarios are considered. The first scenario is where additional physical information and/or chemical characterization of the reactive species is available. The second scenario is where no additional physical information is available. These will be considered in turn.

### 12.4.1 Physical/Chemical Information Available

Physical and chemical properties of the reaction network that are conserved are referred to as reaction invariant properties. For example, the molecular weight, number of carbon atoms, and number of oxygen atoms of a particular species do not change regardless of the nature and number of the chemical reactions occurring. If $C$ such properties are known for the $S$ species, then a $(C \times S)$ conservation matrix **A** may be formed, where the entry at the $i$th column and $j$th row refers to the value of the $j$th conserved property for the $i$th species. The simplest case of **A** is probably the $(1 \times S)$ vector containing the molecular weights – or the ratios of the molecular weights – of the $S$ species. Due to the conservation of the properties within **A**, the following relationship holds for physically realizable reaction networks [22]:

$$\mathbf{AN} = 0 \tag{12.19}$$

where **N** is the $(S \times q)$ stoichiometric matrix corresponding to the reaction network. Note that this property is independent of the kinetic behavior of the network. If the relationship in Equation 12.19 does not hold for a trial network in the network search process, then $\alpha$ is set to $q_{\max}$ in the search objective function, as indicated in Section 12.3.

### 12.4.2 No physical/Chemical Information Available

If no conserved physical or chemical properties are available for the $S$ species, then a conservation matrix **A** cannot be explicitly formed. However, it is still possible to make use of the relation given in Equation 12.19 even if **A** is unknown. This is because the transpose of any valid conservation matrix $\mathbf{A}^T$ resides in the nullspace of the transpose of the trial network's stoichiometric matrix $\mathbf{N}^T$. This can be seen by rewriting Equation 12.19 as

$$\mathbf{N}^T\mathrm{A}^T = 0 \tag{12.20}$$

Therefore, if $\mathbf{N}^T$ does not have a nullspace, then no conserved physical or chemical properties matrix **A** can exist and thus the network is physically unrealizable. The nullspace of $\mathbf{N}^T$ for any trial network can be calculated by means of the singular value decomposition. In practice, commands for computing nullspaces of matrices are built in to most modern numerical computing software, for example, Mathematica, MATLAB, and so on. Usually the nullspace is computed in the form of a set of orthonormal basis vectors.

There are also trial networks for which the nullspace of $\mathbf{N}^T$ does exist but has rows that contain only zeros. In these cases, a solution of Equation 12.20 is algebraically possible, but the solution does not correspond to a set of physically possible conserved properties. For instance, the following reaction network topology has a nullspace of $[-0.7071\ \ -0.7071\ \ 0]^T$:

$$x_1 \longrightarrow x_2 \quad x_1 \longrightarrow x_2 + x_3 \tag{12.21}$$

The interpretation of this is that **A** exists in the algebraic sense but that it only does so if *all possible* conserved properties for $x_3$ are zero. However, the set of all possible conserved properties includes properties that cannot have zero values, for example, mass. Therefore, such solutions are not physically realizable and so the trial network is penalized in Equation 12.18.

In summary, if the nullspace of $\mathbf{N}^T$ for a trial network either does not exist or contains one or more rows that contain only zeros, then $\alpha$ is set to $q_{max}$ in the network search objective function.

The network topology representation, rate coefficient estimation, and the objective function for the network search problem have been defined in this and previous sections. In the next section, the population-based evolutionary method of differential evolution for searching through the space of networks is discussed.

## 12.5 Differential Evolution for Searching the Space of Reaction Networks

Evolutionary algorithms (EAs) are inherently well suited for many search and optimization problems: they can find near optima in astronomically large, multimodal search spaces and they are robust across many problem classes. Differential evolution (DE) is an EA that was originally conceived as a continuous variable optimization method [17], and although not well very known, studies have shown DE to outperform other EAs for a variety of benchmark problems [23]. Further advantages of DE are that (a) it is compact and very simple to implement, (b) it has relatively few parameters to set, (c) it may easily be modified to operate with discrete, integer, and continuous variables and mixtures thereof, and (d) it can operate both as unconstrained optimizer and as a constrained optimizer with multiple constraints [24].

Before discussing the variant of DE used here, the "basic" DE method for continuous optimization is outlined.[2)]

2) In the optimization literature, this version of DE is referred to as DE/rand/1/bin.

### 12.5.1 Basic DE Optimization Method

Consider a problem involving finding a vector **z** comprising the $n$ continuous parameters $z_1, \ldots, z_n$ with values that minimize an objective function of these parameters $f(\mathbf{z})$. The parameters in **z** generally have upper and lower bounds defined individually by the user for each element of **z**. In order to minimize $f(\mathbf{z})$, DE performs search by simulated evolution on a population of $P$ such trial vectors $\mathbf{z}_{i,g}$ over $G$ generations, where $i = 1, \ldots, P$ and $g = 1, \ldots, G$. The $i$th vector in the gth population contains the $n$ individual trial parameters $z_{i,j,g}$ where $j = 1, \ldots, n$.

The underlying mechanism of DE is that each vector $\mathbf{z}_{i,g}$ in the current population is used to generate a single mutated offspring $\mathbf{m}_{i,g}$. The mutated offspring is then compared directly with its parent on the objective function. If the offspring equals or surpasses its parent, it replaces it in the next generation; otherwise, the parent is retained. Eventually, the evolutionary processes of mutation and selection should drive the population toward the optimum values of the parameter vector **z**. The steps in the DE algorithm are outlined below.

*Step 1. DE initialization*

The algorithm is initialized in the first generation $g = 1$ by randomly generating the $P$ vectors $\mathbf{z}_{i,1}$ within the ranges specified by the upper and lower bounds for each parameter. Each $\mathbf{z}_{i,1}$ is then evaluated against the objective function in order to compute the values $f(\mathbf{z}_{i,1})$.

*Step 2. DE Mutation*

The mutation vector $\mathbf{m}_{i,g}$ is created by replacing a probabilistically determined number of the elements of $\mathbf{z}_{i,g}$ with corresponding elements from an "exploration" or "perturbation" vector $\mathbf{v}_{i,g}$. The perturbation vector itself is created by adding the scaled difference of two randomly selected vectors $\mathbf{z}_{a2,g}$ and $\mathbf{z}_{a3,g}$ from the current population to another randomly selected vector $\mathbf{z}_{a1,g}$ from the current population where $i \neq a1 \neq a2 \neq a3$:

$$\mathbf{v}_{i,g} = \mathbf{z}_{a1,g} + F(\mathbf{z}_{a2,g} - \mathbf{z}_{a3,g}) \tag{12.22}$$

The "scaling factor" $F$ is a user-defined parameter of the DE and is often set in the range $0 < F \leq 2$. The other user-defined parameters are the population size $P$, the number of generations $G$, and the "crossover rate" $CR$. The latter falls in the range $0 \leq CR \leq 1$ and is used to probabilistically determine what proportion of the elements of the current member $\mathbf{z}_{i,g}$ are replaced with elements of the perturbation vector $\mathbf{v}_{i,g}$ when creating $\mathbf{m}_{i,g}$.

The mutated offspring vector $\mathbf{m}_{i,g}$ is formed by means of a binomial crossover mechanism on $\mathbf{z}_{i,g}$ and $\mathbf{v}_{i,g}$ as follows. To determine the $j$th element $m_{i,j,g}$ in $\mathbf{m}_{i,g}$, a random number $a_j$ is generated in the range $0 \leq a_j \leq 1$. If $a_j < CR$, then $m_{i,j,g} = v_{i,j,g}$ otherwise $m_{i,j,g} = z_{i,j,g}$. That is, if the random number is below the crossover rate then a parameter value from the perturbation vector is used and, if not, the parameter value from the current individual is retained. The values in $\mathbf{m}_{i,g}$ are then checked against the upper and lower bound constraints for each parameter. If any element of $\mathbf{m}_{i,g}$ violates the bounds, then it can either be replaced with a random number generated within the bounds or truncated to the exceeded bound value. In this work, the latter option is used.

*Step 3. DE competition and population update*

Each mutated vector is then evaluated against the objective function $f(\mathbf{m}_{i,g})$ and the resulting value compared to that obtained by its parent $f(\mathbf{z}_{i,g})$. If $f(\mathbf{m}_{i,g}) \leq f(\mathbf{z}_{i,g})$, then $\mathbf{m}_{i,g}$ takes the place of $\mathbf{z}_{i,g}$ in the next generation and $\mathbf{z}_{i,g+1} = \mathbf{m}_{i,g}$. Otherwise, the current solution $\mathbf{z}_{i,g}$ is retained and copied to the next generation and $\mathbf{z}_{i,g+1} = \mathbf{z}_i$. The generation counter g is then set to $g + 1$ and the algorithm returns to Step 2.

*Step 4. DE termination*

The above process is repeated until some termination criterion is reached, most commonly after the fixed number of generations $G$ has elapsed. At this point, the vector $\mathbf{z}_{i,G}$ in the final population with the lowest value of the objective function is usually selected as the "best result" of the run. This does not have to be the case, however, as will be seen in Section 12.6.

## 12.5.2
## Self-Adaptive DE with Integer Variables

We use a self-adaptive variant of DE where the evolved parameters vectors are interpreted as integer vectors that encode trial reaction network topologies as described in Section 12.3 using the modified SIC objective function in Equation 12.18. Modifying DE to produce vectors that contain integer elements is simple. The DE algorithm essentially operates as described above on continuous parameter vectors and the only difference is that, prior evaluating a vector $\mathbf{z}_{i,g}$ (or $\mathbf{m}_{i,g}$) against the objective function, the elements of $\mathbf{z}_{i,g}$ (or $\mathbf{m}_{i,g}$) are temporarily rounded to the nearest integer. Hence, by defining appropriate upper and lower bounds for each of the elements in $\mathbf{z}$, the DE can be made to operate as a nonlinear integer search method within specified ranges for each parameter. Note that the underlying vector representation within the DE remains continuous and that the rounded parameter vectors are only used for the purpose of being evaluated by the objective function.

For the integer network representation described in Section 12.3.3, the lower and upper bounds for each continuous parameter $z_j$ in $\mathbf{z}$ are defined such that $(-1/2 + \delta) \leq z_j \leq (S + {}^1\!/_2 - \delta)$ where $\delta$ is a some very small number. This is to ensure that equal lengths of the real number line are rounded correctly to the integers between 0 and $S$.

One of the drawbacks of the version of DE outlined above is that it necessary to determine "good" values of the user-defined control parameters, in particular, the scaling factor $F$ and crossover rate $CR$. The optimal values for these control parameters are, in general, problem dependent and usually a trial and error approach is required to tune them satisfactorily. To counter this, self-adaptive variants of DE exist in which $F$ and $CR$ are themselves evolved along with the parameters required to solve the problem. We use the self-adaptive DE proposed by Brest *et al.* [25] because it has been shown – using a suite of nonlinear benchmark problems – that it equals or outperforms standard DE and other self-adapting DE variants. This self-adaptive modification to the DE method requires only minimal changes and is easy to implement. Each parameter vector in the population $\mathbf{z}_{i,g}$ is linked to corresponding values of $F$ and $CR$, referred to as $F_{i,g}$ and $CR_{i,g}$. In the initial population, the values for $F_{i,g}$ and $CR_{i,g}$ are randomly generated in the ranges $0.1 \leq F_{i,g} \leq 1$ and $0 \leq CR_{i,g} \leq 1$. In Step 2 of the DE algorithm, before creating the mutant offspring $\mathbf{m}_{i,g}$, the values of $F_{i,g}$ and $CR_{i,g}$ are first mutated to new values $\underline{F}_{i,g}$ and $\underline{CR}_{i,g}$. The mutated control parameters are then used instead of fixed values of $F$ and $CR$ to create $\mathbf{m}_{i,g}$. The $\underline{F}_{i,g}$ and $\underline{CR}_{i,g}$ are created using the following probabilistic rules [25].

$$\underline{F}_{i,g} = \begin{cases} F_l + a_1 F_u & \text{if} \quad a_2 < \tau_1 \\ F_{i,g} & \text{otherwise} \end{cases}$$
$$\underline{CR}_{i,g} = \begin{cases} a_3 & \text{if} \quad a_4 < \tau_2 \\ CR_{i,g} & \text{otherwise} \end{cases} \tag{12.23}$$

where the $a_1, \ldots, a_4$ are independent random numbers generated between 0 and 1 each time the computations in Equation 12.23 are carried out and $\tau_1$ and $\tau_2$ are mutation probabilities. The rules have the effect of mutating $F_{i,g}$ to a new value between $F_l$ and $(F_l + F_u)$ with probability $\tau_1$ and mutating $CR_{i,g}$ to a new value between 0 and 1 with probability $\tau_2$. As in [25], we set $\tau_1 = \tau_2 = 0.1$ and $F_l = 0.1$ and $F_u = 0.9$.

If the mutated vector $\mathbf{m}_{i,g}$ outperforms its parent $\mathbf{z}_{i,g}$, then the values of the control parameters in the following generations are updated as $F_{i,g+1} = \underline{F}_{i,g}$ and $CR_{i,g+1} = \underline{CR}_{i,g}$. If not, the original values are kept and $F_{i,g+1} = F_{i,g}$ and $CR_{i,g+1} = CR_{i,g}$.

## 12.6 Network Identification Case Studies

The proposed methods were evaluated using simulations of three hypothetical chemical reaction networks occurring in well-mixed, isothermal batch reactors within an inert solvent. The first reaction network comprises 5 species, the second 6 species, and the third comprises 10 species. Reaction networks 1 and 2 are used to evaluate the methodology when there is no knowledge of any conserved physical properties and reaction network 3 is used to evaluate it when the ratios of the molecular weights of the reacting species are known.

To generate the simulated experimental data, the corresponding ODEs for each network were numerically solved using an accurate adaptive step length solver. Periodically sampled concentration values were then obtained for each species in each experiment. Arbitrary time and concentration units are used.

To simulate the effects of experimental error, the sampled concentration values for each species in each experiment were independently corrupted with additive Gaussian noise. The noise was specified to have a standard deviation equal to 4% of the range of the concentration signal in the experiment. The reaction systems are described below:

*Reaction network 1*

The first hypothetical reaction network comprises five chemical species labeled $x_1, \ldots, x_5$ involved in four reactions with rate coefficients $k_1 = 0.1$, $k_2 = 0.2$,

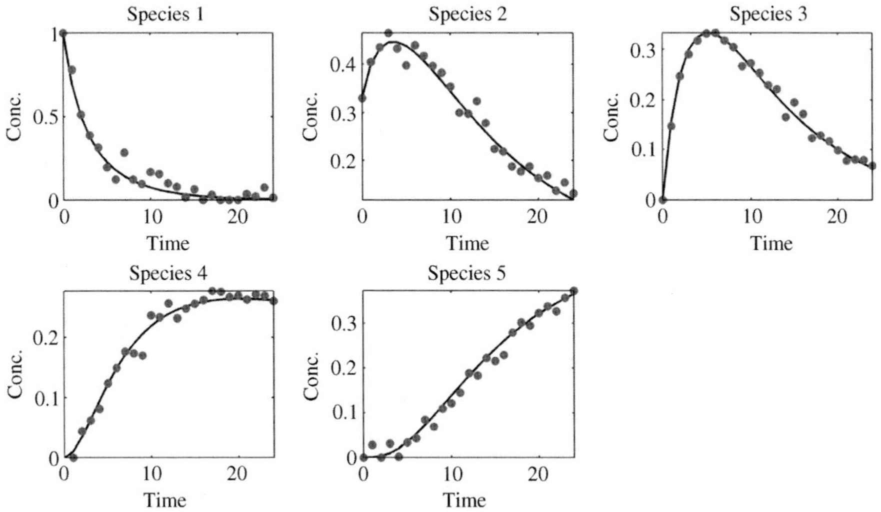


**Figure 12.2** Simulated experimental concentration vs. time data for reaction network 1 (Experiment 2). The noisy sampled values (dots) are plotted against the underlying concentration values (lines).

$k_3 = 0.13$, and $k_4 = 0.3$ as shown in Equation 12.24. The initial reactants are $x_1$ and $x_2$. Four experiments were simulated in which the initial concentrations were as follows: $\{[x_1] = 0.33, [x_2] = 1\}$, $\{[x_1] = 1, [x_2] = 0.33\}$, $\{[x_1] = 1, [x_2] = 1\}$, and $\{[x_1] = 0.75, [x_2] = 1\}$. In each case, the species concentrations were sampled every time unit from 0 to 24 time units:

$$\begin{aligned} &2x_1 \xrightarrow{k_1} x_2 \\ &x_1 \xrightarrow{k_2} x_3 \\ &x_3 \xrightarrow{k_3} x_4 \\ &x_2 + x_4 \xrightarrow{k_4} x_5 \end{aligned} \tag{12.24}$$

Figure 12.2 shows the data acquired from a typical experiment. For each species, the "true" unknown concentration values are plotted against the noisy sampled data.

*Reaction network 2*

The second reaction network comprises six chemical species labeled $x_1, \ldots, x_6$ involved in four chemical reactions with rate coefficients $k_1 = 0.2$, $k_2 = 0.1$, $k_3 = 0.15$, and $k_4 = 0.05$ as shown in Equation 12.25. Note that the fourth reaction is the reverse of the third reaction:

$$\begin{aligned} &x_1 + x_2 \xrightarrow{k_1} x_3 + x_4 \\ &x_2 + x_3 \xrightarrow{k_2} x_5 \\ &x_1 + x_4 \xrightarrow{k_3} x_6 \\ &x_6 \xrightarrow{k_4} x_1 + x_4 \end{aligned} \tag{12.25}$$

The initial reactants are $x_1$ and $x_2$. Again, four experiments were simulated with the following initial concentration values: $\{[x_1]=2.5, [x_2]=2.5\}$, $\{[x_1]=2.5, [x_2]=7.5\}$, $\{[x_1]=7.5, [x_2]=2.5\}$, and $\{[x_1]=10, [x_2]=5\}$. In each case, the species concentrations were sampled every 0.5 time units from 0 to 15 time units.

*Reaction network 3*

The third network comprises 10 chemical species labeled $x_1, \ldots, x_{10}$ involved in six chemical reactions with rate coefficients $k_1=0.35$, $k_2=0.25$, $k_3=0.3$, $k_4=0.4$, $k_5=0.3$, and $k_6=0.1$ as shown in Equation 12.26:

$$\begin{aligned} &x_1 + x_2 \xrightarrow{k_1} x_3 \\ &x_3 \xrightarrow{k_2} x_4 \xrightarrow{k_3} x_5 \xrightarrow{k_4} x_6 \\ &x_4 \xrightarrow{k_5} x_7 + x_8 \\ &x_5 \xrightarrow{k_6} x_9 + x_{10} \end{aligned} \tag{12.26}$$

It is assumed that chemical information, in the form of molecular weights, is available for each of the species in this network and so a $(1 \times 10)$ conservation matrix **A** comprising the ratios of the molecular weights, was formed as follows:

$$\mathbf{A}^{\mathrm{T}} = [1 \quad 3 \quad 4 \quad 4 \quad 4 \quad 4 \quad 1 \quad 3 \quad 2 \quad 2] \tag{12.27}$$

Again, the initial reactants are $x_1$ and $x_2$. For this network, three experiments were simulated with the following initial concentration values: $\{[x_1]=5, [x_2]=5\}$, $\{[x_1]=2, [x_2]=4\}$, and $\{[x_1]=4, [x_2]=2\}$. In each experiment, the species concentrations were sampled every 0.1 time units from 0 to 5 time units.

### 12.6.1
### Estimation of Time Derivatives

Once the simulated experimental time-series concentration data was generated, the time derivative of each species concentration value in each batch was approximated. This was accomplished by fitting a separate second-order rational polynomial of the form in Equation 12.28 to each signal where $t$ is the time value and $[\hat{x}]_t$ is the predicted value of the $i$th species concentration at time $t$:

$$[\hat{x}_i]_t = \frac{p_1 t^2 + p_2 t + p_3}{t^2 + p_4 t + p_5} \tag{12.28}$$

For each signal, the coefficients $p_1, \ldots, p_5$ were estimated from the noisy concentration values $[x_i]_t$ using the Levenberg–Marquardt nonlinear least-squares method. The approximate values of the time derivatives were subsequently obtained at each sampled data point using the analytical derivative of Equation 12.28. For a related S-system identification problem [18], it was observed that, typically, the accuracy of approximated derivative values is poor at the beginning and end of each signal and so, prior to the network inference procedure, we remove the first and last four data points from each signal for each experiment.

### 12.6.2
### DE Settings

Ten runs of the self-adaptive DE with the encoding described in Section 12.3.1 were performed for each reaction network in order to minimize the objective function in Equation 12.18. DE, like other EAs, is a nondeterministic algorithm, and multiple runs are usually required to reliably assess its performance on a particular problem. Each run used a population size $P=400$ and number of generations $G=2000$. Runs were terminated early if the best value of the objective function obtained did not change for 500 generations. The value of $q_{\max}$ (the maximum number of elementary reactions allowed in each trial network) was set to 10. Hence, in each case, the DE searched for an optimal vector containing 50 integers, each of which can take a value between 0 and $S$, for example, for the third reaction system $S=10$ and so the entire search space contains $10^{50}$ reaction networks. All runs were performed using MATLAB v7.1 on a 1.8 GHz Windows PC with 2 Gb of RAM. The time taken for each run averaged at around 10 minutes.

### 12.6.3
### Model Selection Methodology

Instead of choosing the "best" result (i.e., the lowest value of $\mathrm{SIC}_D$) of each DE run as the solution to the network inference problem, we instead performed a further step to choose a solution. This is because the trial network with the lowest $\mathrm{SIC}_D$ is not necessarily the trial network that yields the lowest SIC when its ODEs are numerically integrated and compared directly with the measured concentration data. Errors introduced during the derivative approximation step can result in the DE search finding a trial network with the lowest $\mathrm{SIC}_D$ but not necessarily the lowest SIC when numerically integrated. This is defined as $\mathrm{SIC}_C$:

$$\mathrm{SIC}_C = N_s \ln\left(\frac{\mathrm{SSE}_C}{N_s}\right) + (p+\alpha)\ln(N_s) \tag{12.29}$$

where $\mathrm{SSE}_C$ is the sum of squared errors between the measured concentration data and the numerically integrated prediction of the ODEs corresponding to a trial network, all other terms are identical to those in the DE objective function defined in Equation 12.18. Hence, at the end of each DE, run the ODES corresponding to each trial network in the entire final population were numerically solved and the predicted concentration values evaluated against the experimental data in terms of the $\mathrm{SIC}_C$. The network with the lowest $\mathrm{SIC}_C$ was considered to be the solution of the run, although in practice it would be desirable to consider several of the best performing networks from each run. These could be treated as a working portfolio of plausible reaction networks that could then be examined for chemical plausibility or as the starting point for a set of further experiments to discriminate between network hypotheses.

## 12.6.4
## Results

The results of running the DE 10 times for each of the hypothetical reaction networks are described below:

*Reaction network 1*

Each of the DE runs converged to the same value of $SIC_D = -696.29$ corresponding to the reaction network described by Equation 12.30 with the following estimated rate coefficients $\hat{k}_1 = 0.090$, $\hat{k}_2 = 0.130$, $\hat{k}_3 = 0.130$, $\hat{k}_4 = 0.304$, and $\hat{k}_5 = 0.073$. Note that this network, although close to the true solution, is incorrect. It actually comprises the true network plus an additional fifth reaction (underlined in Equation 12.30) which is stoichiometrically identical to the second reaction but gives rise to different kinetic terms in the corresponding ODEs. This is not a deficiency of the DE search process, however, because the correct network topology actually yields a higher (i.e., worse) value of $SIC_D = -674.650$. It seems likely that errors introduced by the derivative approximation process have led to overfitting whereby a network with an additional incorrect reaction fits the concentration derivative data better than the correct network:

$$\begin{aligned}
&2x_1 \xrightarrow{\hat{k}_1} x_2\\
&x_1 \xrightarrow{\hat{k}_2} x_3\\
&x_3 \xrightarrow{\hat{k}_3} x_4\\
&x_2 + x_4 \xrightarrow{\hat{k}_4} x_5\\
&\underline{x_1 + x_2 \xrightarrow{\hat{k}_5} x_2 + x_3}
\end{aligned} \tag{12.30}$$

However, the application of the model selection methodology outlined in the previous section yielded the correct reaction network topology for each of the DE runs – showing that the correct solution, while not having the lowest $SIC_D$, was in each case present in the final population of each run and has an $SIC_C$ better than that obtained for the incorrect solution above. The estimated rate coefficients for the correct topology are $\hat{k}_1 = 0.090$, $\hat{k}_2 = 0.191$, $\hat{k}_3 = 0.132$, and $\hat{k}_4 = 0.307$ and are close to the true rate coefficients $k_1 = 0.1$, $k_2 = 0.2$, $k_3 = 0.13$, and $k_4 = 0.3$. Hence, it can be concluded that the true reaction network topology has been recovered from the data with satisfactory estimates of the rate coefficients.

*Reaction network 2*

The 10 DE runs converged on solutions with $SIC_D$ values in the range $-768.30$ to $-778.79$. Once again, in each case, the network with the lowest $SIC_D$ of the run did not correspond to the true reaction network. Typically, the solutions found tend to indicate overfitting in that they comprise the correct network with incorrect additional reactions. For example, one such incorrect network is shown in Equation 12.31 with estimated rate coefficients $\hat{k}_1 = 0.2$, $\hat{k}_2 = 0.099$, $\hat{k}_3 = 0.153$, $\hat{k}_4 = 0.053$, and $\hat{k}_5 = 0.009$. Note that the rate coefficients for the correct reactions are close to the true coefficients and that the identified rate coefficient for the incorrect reaction is small in comparison to those for the correct reactions:

$$\begin{aligned}
&x_1 + x_2 \xrightarrow{\hat{k}_1} x_3 + x_4 \\
&x_2 + x_3 \xrightarrow{\hat{k}_2} x_5 \\
&x_1 + x_4 \xrightarrow{\hat{k}_3} x_6 \\
&x_6 \xrightarrow{\hat{k}_4} x_1 + x_4 \\
&\underline{x_3 + x_5 \xrightarrow{\hat{k}_5} 2x_4 + x_5}
\end{aligned} \tag{12.31}$$

Again, the application of the model selection step on the final population of each DE run yielded, in each case, the correct reaction network topology with estimated rate coefficients $\hat{k}_1 = 0.200$, $\hat{k}_2 = 0.099$, $\hat{k}_3 = 0.155$, and $\hat{k}_4 = 0.058$. These are close to the true rate coefficients $k_1 = 0.2$, $k_2 = 0.1$, $k_3 = 0.15$, and $k_4 = 0.05$.

*Reaction network 3*

Unlike the previous hypothetical reaction networks, only 6 of the 10 runs correctly identified the correct network topology after the $SIC_C$ model selection step. This is in contrast with the previous examples in which, in every case, the model selection step yielded the true network topology. This is not too surprising because this network was purposely designed to be a challenging problem in that it contains a "backbone" of unimolecular reactions in series, all with rate coefficients that are similar in magnitude. This means that many of the concentration time-series profiles are highly correlated. Typically, the incorrectly identified networks are very similar to the true network in structure. For example, the incorrect network from run 6 is shown in Equation 12.32 and has estimated rate coefficients $\hat{k}_1 = 0.339$, $\hat{k}_2 = 0.249$, $\hat{k}_3 = 0.267$, $\hat{k}_4 = 0.409$, $\hat{k}_5 = 0.299$, and $\hat{k}_6 = 0.033$. The majority of the true network has been recovered correctly; the only incorrect reaction (underlined) contains the species $x_5$ as the reactant instead of the correct reactant $x_4$. Note that this incorrect reaction network satisfies the mass conservation constraint $\mathbf{AN} = 0$, as did the incorrect networks found in the other DE runs:

$$\begin{aligned}
&x_1 + x_2 \xrightarrow{\hat{k}_1} x_3 \\
&x_3 \xrightarrow{\hat{k}_2} x_4 \xrightarrow{\hat{k}_3} x_5 \xrightarrow{\hat{k}_4} x_6 \\
&\underline{x_5 \xrightarrow{\hat{k}_5} x_7 + x_8} \\
&x_5 \xrightarrow{\hat{k}_6} x_9 + x_{10}
\end{aligned} \tag{12.32}$$

In the 6 of the 10 runs that yielded the correct network topology, the estimated rate coefficients were $\hat{k}_1 = 0.338$, $\hat{k}_2 = 0.249$, $\hat{k}_3 = 0.300$, $\hat{k}_4 = 0.403$, $\hat{k}_5 = 0.299$, and $\hat{k}_6 = 0.100$. Again, these are very close to the true rate coefficients $k_1 = 0.35$, $k_2 = 0.25$, $k_3 = 0.3$, $k_4 = 0.4$, $k_5 = 0.3$, and $k_6 = 0.1$.

A further set of DE runs were performed that did not make use of the molecular weight information in the **A** matrix. In this case, only 1 of the 10 runs yielded the correct network after the model selection step. Of the incorrect networks found in the other nine runs, each was tested and found to violate the $\mathbf{AN} = 0$ conservation constraint. This emphasizes that the network search procedure is enhanced by the incorporation of prior chemical knowledge, especially on "difficult" reaction networks.

## 12.7 Conclusions

Reverse engineering the topology and dynamic behavior of chemical and biological networks from noisy experimental time series data is a hard "inverse" problem. In this chapter, an evolutionary computational method for the identification of chemical reaction networks and the associated rate coefficients has been presented. Three case studies using data from simulated batch reactor experiments were used to demonstrate the technique and it was shown that it is possible to correctly identify the underlying chemical reaction networks and hence their stoichiometries and ODE models.

Advantages of our approach are: (i) it is almost entirely automated in that only a few parameters such as the DE population size $P$, number of generations $G$, and maximum network size $q_{max}$ need to be specified by the user; (ii) it can be applied to non-equilibrium chemical reaction systems, for example, batch and semi-batch reactor experiments; (iii) in contrast to TFA methods, it is not necessary to separately determine either the number of independent reactions or reaction stoichiometries; (iv) the topology of the elementary reactions and the corresponding rate coefficients are estimated simultaneously; (v) a priori chemical information in the form of reaction invariant constraints (e.g., molecular weights) can, when available, be incorporated into the network search procedure in a natural manner; (vi) the use of self-adaptive DE to search the space of elementary reaction networks provides a straightforward, robust optimization method for the estimation of chemical reaction topologies; and (vii) the use of a population-based search makes the subsequent use of a model selection procedure simple to apply.

Currently, there are limitations to the method, principally: (i) it is assumed that concentration measurements are available for all of the species. When this is not true, however, it may be possible to employ existing TFA-related methods [2] for estimating concentration data for unmeasured species; (ii) most chemical and biological networks tend to be sparsely connected and to promote parsimonious solutions, a penalty function (soft constraint) approach was used in this work. It is likely, however, that the use of a multiobjective EA approach (e.g., see [26]) should improve the method by promoting diversity of network complexity in the population; (iii) the current implementation has only been tested on simulated reaction networks comprising elementary reactions and it is as yet unclear how method performs – or could be modified to perform – when presented with observed data from networks containing "complex" reactions, that is, reactions comprising one or more elementary reactions where the intermediates cannot be measured and/or exist only fleetingly as transitional states.

While the advantages of the incorporation of a priori chemical information have been demonstrated, it would be worthwhile considering the incorporation of additional system constraints in order to improve the robustness and reliability of the approach. Energy constraints, as well as those based on mass, could be used to enhance the network search process. For instance, a reaction – at constant temperature and pressure – can only occur if the Gibbs free energy of the reactants exceeds the Gibbs free energy of the products [27].

Future work will attempt to address the points discussed above in addition to testing the method on real experimental data from both "simple" reaction systems and from "complex" oscillating reaction networks such as that described in [28].

Finally, we note that the success of the current approach is dependent on the ability to satisfactorily estimate first derivative values from sufficiently dynamically rich time-series concentration data. The method used here (rational polynomials) is fairly simple but for real data, more robust methods such as that recently described in [29] may be more appropriate.

## Acknowledgment

This work was supported by the UK Engineering and Physical Sciences Research Council (EPSRC).

## References

1 Maria, G. (2004) A review of algorithms and trends in kinetic model identification. *Chemical and Biochemical Engineering Quarterly*, **18** (3), 195–222.

2 Bonvin, D. and Rippin, D.W.T. (1990) Target factor analysis for the identification of stoichiometric models. *Chemical Engineering Science*, **45** (12), 3417–3426.

3 Brendel, M., Bonvin, D., and Marquadt, W. (2006) Incremental identification of kinetic models for homogeneous reaction systems. *Chemical Engineering Science*, **61**, 5404–5420.

4 Kikuchi, S., Tominaga, D., Arita, M., Takahashi, K., and Tomita, M. (2003) Dynamic modelling of genetic networks using genetic algorithm and s-system. *Bioinformatics (Oxford, England)*, **19** (5), 643–650.

5 Noman, N. and Iba, H. (2005) Inference of gene regulatory networks using s-system and differential evolution. GECCO '05: Proceedings of the 2005 Conference on Genetic and Evolutionary Computation. ACM, pp. 439–446.

6 Fotopoulos, J., Georgakis, C., and Stenger, H.G. (1994) Structured target factor analysis for the stoichiometric modeling of batch reactors. Proceedings of the 1994 American Control Conference, 1, pp. 495–499.

7 Harmon, J.L., Duboc, P., and Bonvin, D. (1995) Factor analytical modeling of biochemical data. *Computers and Chemical Engineering*, **19** (12), 1287–1300.

8 Amrhein, M., Srinivasan, B., and Bonvin, D. (1999) Target factor analysis of reaction data: Use of data pre-treatment and reaction–invariant relationships. *Chemical Engineering Science*, **54** (5), 579–591.

9 Savageau, M.A. (1969) Biochemical systems analysis: I. Some mathematical properties of the rate law for the component enzymatic reactions. *Journal of Theoretical Biology*, **25**, 365–369.

**10** Voit, E.O. (2004) The dawn of a new era of metabolic systems analysis. *Biosilico*, **2** (5), 182–189.

**11** Searson, D.P., Willis, M.J., Horn, S.J., and Wright, A.R. (2007) Inference of chemical reaction networks using hybrid s-system models. *Chemical Product and Process Modeling*, **2** (10).

**12** Vilela, M., Chou, I., Vinga, S., Vasconcelos, A.T.R., Voit, E.O., and Almeida, J.S. (2008) Parameter optimization in s-system models. *BMC Systems Biology*, **2**, 35.

**13** Crampin, E.J., Schnell, S., and McSharry, P.E. (2004) Mathematical and computational techniques to deduce complex biochemical reaction mechanisms. *Progress in Biophysics and Molecular Biology*, **86** (1), 77–112.

**14** Searson, D.P., Burnham, S.C., Willis, M.J., and Wright, A.R. (2006) Identification of chemical reaction mechanism from batch process data. Proceedings of the 17th IASTED International Conference on Modelling and Simulation. Montreal, Paper 530-102.

**15** Srividhya, J., Crampin, E.J., McSharry, P.E., and Schnell, S. (2007) Reconstructing biochemical pathways from time course data. *Proteomics*, **7**, 828–838.

**16** Burnham, S.C., Searson, D.P., Willis, M.J., and Wright, A.R. (2008) Inference of chemical reaction networks. *Chemical Engineering Science*, **63** (4), 862–873.

**17** Storn, R. and Price, K. (1997) Differential evolution: a simple and efficient heuristic for global optimization over continuous spaces. *Journal of Global Optimization*, **11** (4), 341–359.

**18** Voit, E.O. and Almeida, J. (2004) Decoupling dynamical systems for pathway identification from metabolic profiles. *Bioinformatics (Oxford, England)*, **20** (11), 1670–1681.

**19** Akaike, H. (1974) A new look at the statistical model identification. *IEEE Transactions on Automatic Control*, **19** (6), 716–723.

**20** Schwartz, G. (1978) Estimating the dimension of a model. *Annals of Statistics*, **6**, 461–464.

**21** Burnham, K.P. and Anderson, D.R. (2002) *Model Selection and Multimodel Inference: A Practical-Theoretic Approach*, 2nd edn, Springer-Verlag.

**22** Aris, R. and Mah, R.H.S. (1963) Independence of chemical reactions. *Industrial and Engineering Chemistry Fundamentals*, **2**, 90–94.

**23** Vesterstroem, J. and Thomsen, R. (2004) A comparative study of differential evolution, particle swarm optimization, and evolutionary algorithms on numerical benchmark problems. Proceedings of the IEEE Congress on Evolutionary Computation, pp. 1980–1987.

**24** Fan, H. and Lampinen, J. (2003) A trigonometric mutation operation to differential evolution. *Journal of Global Optimization*, **27**, 105–129.

**25** Brest, J., Greiner, S., Bošković, B., Mernik, N., and Žumer, V. (2006) Self-adapting control parameters in differential evolution: A comparative study on numerical benchmark problems. *IEEE Transactions on Evolutionary Computation*, **10** (6), 646–657.

**26** Hinchliffe, M. and Willis, M.J. (2003) Dynamic systems modelling using genetic programming. *Computers and Chemical Engineering*, **27** (12), 1841–1854.

**27** Beard, D.A., Liang, S., and Qian, H. (2002) Energy balance for analysis of complex metabolic networks. *Biophysical Journal*, **83** (1), 79–86.

**28** Novakovic, K., Grosjean, C., Scott, S.K., Whiting, A., Willis, M.J., and Wright, A.R. (2007) Achieving pH and Qr oscillations in a phenylacetylene carbonylation reaction using an automated reactor system. *Chemical Physics Letters*, **435** (1–3), 142–147.

**29** Vilela, M., Borges, C.C.H., Vinga, S., Vasconcelos, A.T.R., Santos, H., Voit, E.O., and Almeida, J.S. (2007) Automated smoother for the numerical decoupling of dynamics models. *BMC Bioinformatics*, **8**, 305.

# 13
# Reduction of Dimensionality, Order, and Classification in Spaces of Theoretical Descriptions of Molecules: An Approach Based on Metrics, Pattern Recognition Techniques, and Graph Theoretic Considerations

*George Maroulis*

## 13.1
## Introduction

In recent years, we have witnessed an increasing interest in producing new quantum chemical methods oriented toward a more and more refined view of the molecular world. Coupled to the explosive development of computers and computing technology, this progress has resulted in the production of an amazing wealth of new data. In general, quantum chemical methods are tested over various collections of atomic/molecular properties. The evaluation of their predictive capability is still a major issue in computational quantum chemistry. Ransil's [1] early paper on the subject bears the eloquent title "How good is good agreement with experiment: Evaluating the reliability of quantum-mechanically calculated observables." The possibility of controlling the behavior of methods has been examined early enough [2, 3].

Sometime ago we proposed an information theoretic approach to the classification of quantum chemical methods. The method was initially applied to the evaluation of the quality of the performance of basis sets on an arbitrary collection of molecular properties, with the FH molecule as a test case [4]. It was further tested on OH [5], BH, CH, NH [6], and the triatomic $H_2O$ [7] molecule. It was also used by other researchers on a variety of problems as the performance of various types of basis sets in studies on helium [8, 9] and copper [10]. Later, the method was extended to treat simultaneously a group of molecules and was tested on a collection of properties for the sequence $CH_4$, $NH_3$, $H_2O$, FH, and Ne [11]. A program for the automatic application of this method has been written and made available to the community of computational chemists by Sordo [12].

A most essential element in the evaluation of the performance of quantum chemical methods is the quantification of their relative merit. This quantification always refers to a well-defined collection of atomic/molecular properties. In order to achieve this specific goal, we introduced a new methodology that relies on generalized metrics, graph theory, and pattern recognition techniques [13]. This methodology has been subsequently applied by us [14–16] and other authors [17–19] to various problems.

*Statistical Modelling of Molecular Descriptors in QSAR/QSPR*. First Edition.
Edited by M. Dehmer, K. Varmuza, and D. Bonchev

**Table 13.1** *Ab initio* electric properties for HOOH calculated with basis set [9s6p4d1f/6s4p1d][a)].

| Property | SCF | MP2 | MP3 | DQ-MP4 | SDQ-MP4 | MP4 | CCSD | CCSD(T) |
|---|---|---|---|---|---|---|---|---|
| $\mu_z$ | 0.7279 | 0.6778 | 0.6825 | 0.6818 | 0.6757 | 0.6662 | 0.6763 | 0.6685 |
| $\alpha_{xx}$ | 19.36 | 18.51 | 18.21 | 18.31 | 18.72 | 19.02 | 18.66 | 18.85 |
| $\alpha_{yy}$ | 12.34 | 13.98 | 13.39 | 13.48 | 13.78 | 14.15 | 13.71 | 13.98 |
| $\alpha_{zz}$ | 11.11 | 13.11 | 12.38 | 12.48 | 12.84 | 13.26 | 12.78 | 13.08 |
| $\alpha_{xy}$ | 1.79 | 1.41 | 1.45 | 1.45 | 1.5 | 1.51 | 1.5 | 1.51 |
| $\bar{\alpha}$ | 14.27 | 15.2 | 14.66 | 14.76 | 15.11 | 15.48 | 15.05 | 15.3 |
| $\Delta\alpha$ | 8.31 | 5.58 | 5.95 | 5.96 | 6.06 | 5.97 | 6.06 | 5.97 |
| $\beta_{zzz}$ | −3.7 | −9.1 | −6.2 | −6.6 | −9 | −11 | −8.9 | −10.3 |
| $\beta_{yyz}$ | −6.3 | −4.7 | −5.5 | −5.3 | −5.1 | −5 | −5 | −5 |
| $\beta_{xxz}$ | −2.4 | −4.2 | −3.1 | −3.2 | −4.3 | −5.2 | −4.2 | −4.8 |
| $\beta_{xyz}$ | −0.3 | 3.8 | 1.9 | 2.4 | 3.1 | 3.8 | 2.9 | 3.3 |
| $\beta$ | −7.5 | −10.8 | −8.9 | −9.0 | −11.1 | −12.7 | −10.9 | −12.1 |
| $\gamma_{xxxx}$ | 1327 | 2500 | 1988 | 2132 | 2244 | 2500 | 2143 | 2270 |
| $\gamma_{xxxy}$ | −136 | −124 | −85 | −94 | −126 | −148 | −103 | −111 |
| $\gamma_{xxyy}$ | 449 | 694 | 542 | 577 | 675 | 768 | 642 | 689 |
| $\gamma_{xyyy}$ | 71 | 86 | 59 | 72 | 105 | 124 | 94 | 102 |
| $\gamma_{xyzz}$ | 48 | 75 | 54 | 61 | 83 | 96 | 81 | 91 |
| $\gamma_{xxzz}$ | 569 | 975 | 731 | 788 | 957 | 1097 | 923 | 1000 |
| $\gamma_{yyyy}$ | 877 | 1538 | 1215 | 1275 | 1516 | 1726 | 1514 | 1644 |
| $\gamma_{yyzz}$ | 332 | 596 | 470 | 493 | 587 | 667 | 585 | 634 |
| $\gamma_{zzzz}$ | 1006 | 1970 | 1516 | 1597 | 1941 | 2225 | 1950 | 2143 |
| $\bar{\gamma}$ | 1182 | 2107 | 1641 | 1744 | 2028 | 2303 | 1982 | 2141 |

a) All properties are in atomic units.

**Table 13.2** DFT values for the electric properties of HOOH.

| Property | B3LYP | B3PW91 | mPW1PW91 |
|---|---|---|---|
| $\mu_z$ | 0.6804 | 0.6788 | 0.6824 |
| $\alpha_{xx}$ | 19.22 | 18.88 | 18.84 |
| $\alpha_{yy}$ | 14.07 | 13.87 | 13.73 |
| $\alpha_{zz}$ | 13.12 | 12.92 | 12.75 |
| $\alpha_{xy}$ | 1.55 | 1.53 | 1.54 |
| $\bar{\alpha}$ | 15.47 | 15.22 | 15.11 |
| $\Delta\alpha$ | 6.29 | 6.15 | 6.27 |
| $\beta_{zzz}$ | −10.5 | −10.3 | −9.8 |
| $\beta_{yyz}$ | −5.9 | −5.9 | −5.9 |
| $\beta_{xxz}$ | −6.6 | −6.2 | −5.9 |
| $\beta_{xyz}$ | 2.6 | 2.3 | 2.1 |
| $\bar{\beta}$ | −13.8 | −13.4 | −12.9 |
| $\gamma_{xxxx}$ | 2827 | 2598 | 2484 |
| $\gamma_{xxxy}$ | −132 | −108 | −111 |
| $\gamma_{xxyy}$ | 915 | 847 | 809 |
| $\gamma_{xyyy}$ | 167 | 162 | 152 |
| $\gamma_{xyzz}$ | 150 | 137 | 127 |
| $\gamma_{xxzz}$ | 1350 | 1238 | 1172 |
| $\gamma_{yyyy}$ | 1946 | 1819 | 1740 |
| $\gamma_{yyzz}$ | 769 | 718 | 681 |
| $\gamma_{zzzz}$ | 2547 | 2371 | 2247 |
| $\bar{\gamma}$ | 2678 | 2479 | 2359 |

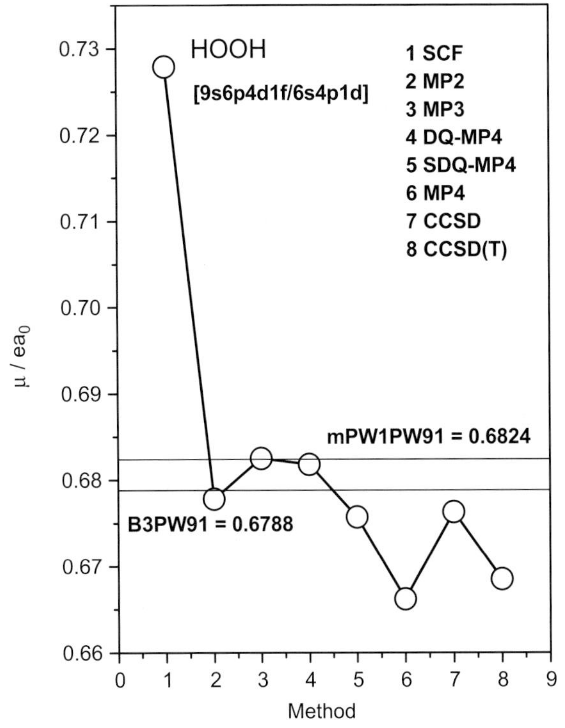


**Figure 13.1** Method dependence of the dipole moment ($\mu_z$) of HOOH.

The aim of this chapter is to show in sufficient detail how this approach is applied to concrete problems in computational quantum chemistry. As a test case, we have chosen the molecule of hydrogen peroxide (HOOH) and its electric properties: dipole moment ($\mu_\alpha$), dipole polarizability ($\alpha_{\alpha\beta}$), and first ($\beta_{\alpha\beta\gamma}$) and second dipole hyperpolarizability ($\gamma_{\alpha\beta\gamma\delta}$). Because of the low symmetry of HOOH, a very large number of independent components are needed to specify the above tensors. We employ conventional *ab initio* methods and widely used density functional theory (DFT) techniques to predict these properties. The relative merit of the quantum chemical methods in the prediction of the electric dipole moment and (hyper)polarizability is analyzed and quantified.

## 13.2 Theory

We present here a few essential points of our approach. Key elements of theory and pertinent technical details about graph theory, pattern recognition, and clustering may be found in the specialized texts of Chartrand and Lesniak [20], Varmuza [21], and Spath [22].

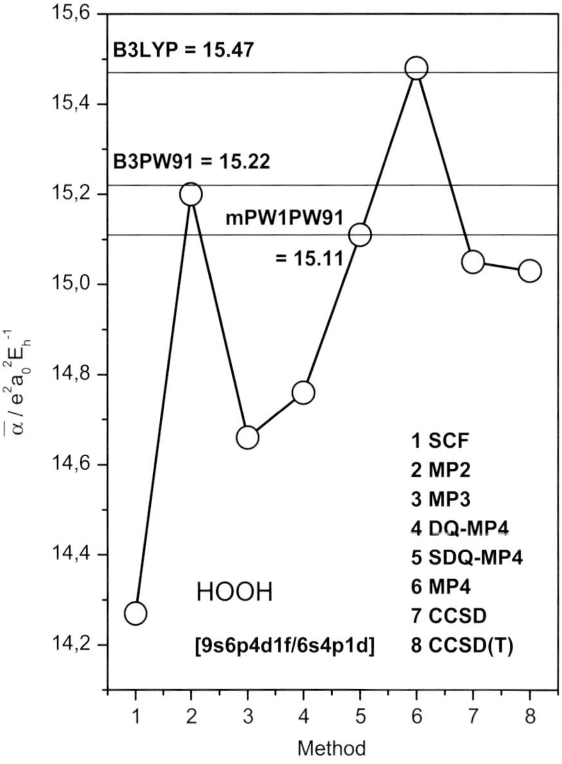


**Figure 13.2** Method dependence of the mean dipole polarizability ($\bar{\alpha}$) of HOOH.

Let us use the symbol $Q_{m\alpha}$ to denote the value of molecular property $\alpha$ calculated with method $m$. $\alpha$ and $m$ take values in the index sets $I_\alpha$ and $I_m$,

$$Q_{m\alpha}, m \in I_m(\text{methods}), \alpha \in I_\alpha(\text{properties}) \tag{13.1}$$

The *theoretical description* $(\text{TD})_i$ is a collection of molecular property values,

$$\text{TD}_i = \{Q_{j\alpha}, \alpha \in I_\alpha\} \tag{13.2}$$

The space of all $\text{TD}_i$ is denoted as TD. We define a distance in the TD space using the generalized Minkowski metric. The *distance* between two $\text{TD}_i$ and $\text{TD}_j$ is defined as

$$D_{ij} = \left( \sum_\alpha \frac{|Q_{i\alpha} - Q_{j\alpha}|^p}{(\max_{ij} |Q_{i\alpha} - Q_{j\alpha}|)^p} \right)^{1/p}, p \geq 1 \tag{13.3}$$

In most applications, we have used the Euclidean metric, that is $p = 2$.

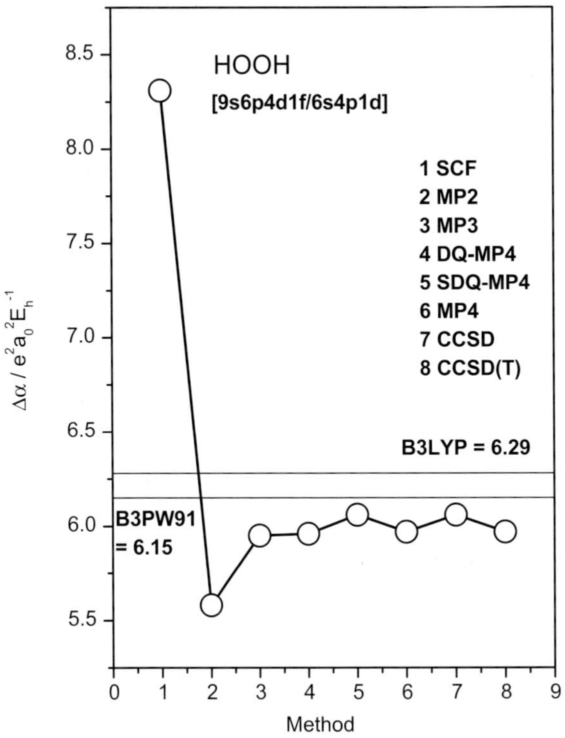


**Figure 13.3** Method dependence of the anisotropy of the dipole polarizability ($\Delta\alpha$) of HOOH.

*Similarity* between two $TD_i$ and $TD_j$ is then defined on the basis of distance/proximity as

$$S_{ij} = 1 - \frac{D_{ij}}{\max\limits_{ij} D_{ij}} \qquad 0 \leq S_{ij} \leq 1 \tag{13.4}$$

Having defined all relevant quantities, we are now in a position to establish a rigorous correspondence between pattern recognition and quantum chemistry. An *object* in pattern recognition corresponds to a *method* in quantum chemistry. Objects are characterized by *features* and methods by well-defined *molecular properties*. A *pattern* is a collection of features, so a *TD* is a collection of *molecular property values*. Last, the *pattern space* is clearly the *space of all TDs*. Schematically:

| Pattern recognition | | Quantum Chemistry |
|---|---|---|
| Object | ↔ | Method |
| Features | ↔ | Molecular property values |
| Pattern | ↔ | TD |
| Pattern space | ↔ | Space of all TDs |

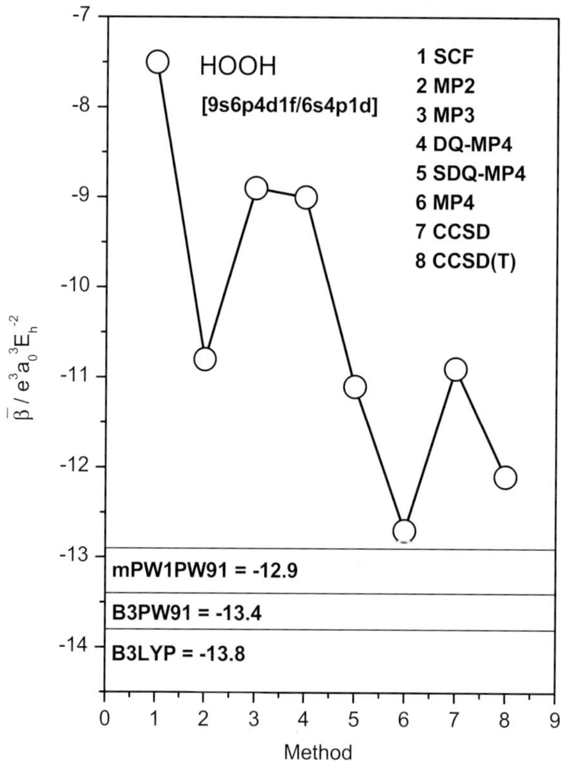


**Figure 13.4** Method dependence of the mean of the first hyperpolarizability ($\bar{\beta}$) of HOOH.

After the definition of distance or proximity ($D_{ij}$) and similarity ($S_{ij}$) in the space of TDs, we advance to the graph theoretic content of our work. We adopt the elements of TD as *vertices* of a *graph* G. We assign to the *edge* defined by $TD_i$ and $TD_j$ a real number, a weight equal to $D_{ij}$. Thus, we are able to construct the *minimum spanning tree* (MST) of the graph G. Subsequently, we obtain various clusterings of TD by removing from the *MST* all edges with weights above a given threshold $D_T$, $D_{ij} \geq D_T$. This method of clustering is called single-linkage cluster analysis.

A distinct advantage of our approach lies in the definition of *MST*, a unique mathematical object characterizing the space of TDs. The order of the problem is determined by the cardinality of $I_m$ and $I_\alpha$ that is the number of methods and molecular properties entering the analysis.

## 13.3
## Methods and Computational Strategy

Detailed presentations of our finite-field approach to the calculation of electric (hyper) polarizabilities may be found in previous work [23]. More specific details are available

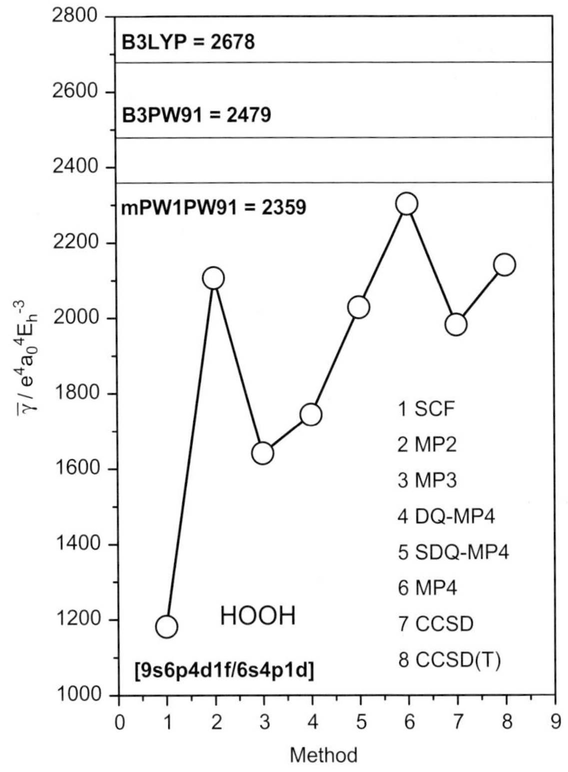


**Figure 13.5** Method dependence of the mean of the second hyperpolarizability ($\bar{\gamma}$) of HOOH.

in our previous work on HOOH [24]. We follow Buckingham's notation and terminology [25] throughout this chapter. The number of independent components needed to specify the electric properties is strictly regulated by symmetry: one for the dipole moment ($\mu_\alpha$), four for the dipole polarizability ($\alpha_{\alpha\beta}$), four for the first dipole hyperpolarizability ($\beta_{\alpha\beta\gamma}$), and nine for the second dipole hyperpolarizability ($\gamma_{\alpha\beta\gamma\delta}$). If we include the energy of the free molecule ($E^0$), we have a 19D problem.

In addition to the Cartesian coordinates, we also calculate the mean ($\bar{\alpha}$) and the anisotropy ($\Delta\alpha$) of the dipole polarizability, the mean first ($\bar{\beta}$), and second ($\bar{\gamma}$) dipole hyperpolarizability.

The *ab initio* methods used in this chapter are

SCF, self-consistent field,
MP2, second-order Møller–Plesset (MP) perturbation theory,
MP3, third-order MP,
DQ-MP4, doubles and quadruples partial fourth-order M,
SDQ-MP4, singles, doubles and quadruples partial fourth-order M,
MP4, complete fourth-order Møller–Plesset perturbation theory,

**Table 13.3** Calculated similarity for all TDs of HOOH calculated with *ab initio* and DFT methods[a].

| | SCF | MP2 | MP3 | DQ-MP4 | SDQ-MP4 | MP4 | CCSD | CCSD(T) | B3LYP | B3PW91 | mPW1PW91 |
|---|---|---|---|---|---|---|---|---|---|---|---|
| SCF | 1 | 0.18463 | 0.38637 | 0.36562 | 0.27770 | 0.11591 | 0.27857 | 0.19247 | 0 | 0.08231 | 0.15365 |
| MP2 | 0.18463 | 1 | 0.61631 | 0.70165 | 0.85642 | 0.74415 | 0.83191 | 0.82573 | 0.48056 | 0.56545 | 0.60803 |
| MP3 | 0.38637 | 0.61631 | 1 | 0.90721 | 0.64019 | 0.42211 | 0.70260 | 0.57978 | 0.25185 | 0.38810 | 0.45338 |
| DQ-MP4 | 0.36562 | 0.70165 | 0.90721 | 1 | 0.72288 | 0.50403 | 0.78314 | 0.65767 | 0.31900 | 0.44968 | 0.51427 |
| SDQ-MP4 | 0.27770 | 0.85642 | 0.64019 | 0.72288 | 1 | 0.77658 | 0.88776 | 0.87555 | 0.54048 | 0.63504 | 0.69150 |
| MP4 | 0.11591 | 0.74415 | 0.42211 | 0.50403 | 0.77658 | 1 | 0.69341 | 0.80215 | 0.65328 | 0.67142 | 0.68895 |
| CCSD | 0.27857 | 0.83191 | 0.70260 | 0.78314 | 0.88776 | 0.69341 | 1 | 0.86643 | 0.48670 | 0.60428 | 0.65982 |
| CCSD(T) | 0.19247 | 0.82573 | 0.57978 | 0.65767 | 0.87555 | 0.80215 | 0.86643 | 1 | 0.58034 | 0.67567 | 0.71380 |
| B3LYP | 0 | 0.48056 | 0.25185 | 0.31900 | 0.54048 | 0.65328 | 0.48670 | 0.58034 | 1 | 0.83086 | 0.77987 |
| B3PW91 | 0.08231 | 0.56545 | 0.38810 | 0.44968 | 0.63504 | 0.67142 | 0.60428 | 0.67567 | 0.83086 | 1 | 0.91946 |
| mPW1PW91 | 0.15365 | 0.60803 | 0.45338 | 0.51427 | 0.69150 | 0.68895 | 0.65982 | 0.71380 | 0.77987 | 0.91946 | 1 |

| Chromatic code | |
|---|---|
| $0.9 \leq S(i,j) \leq 1$ | (red) |
| $0.8 \leq S(i,j) \leq 0.9$ | (blue) |
| $0.7 \leq S(i,j) \leq 0.8$ | (green) |

a) A chromatic code for the most similar performance is also added.

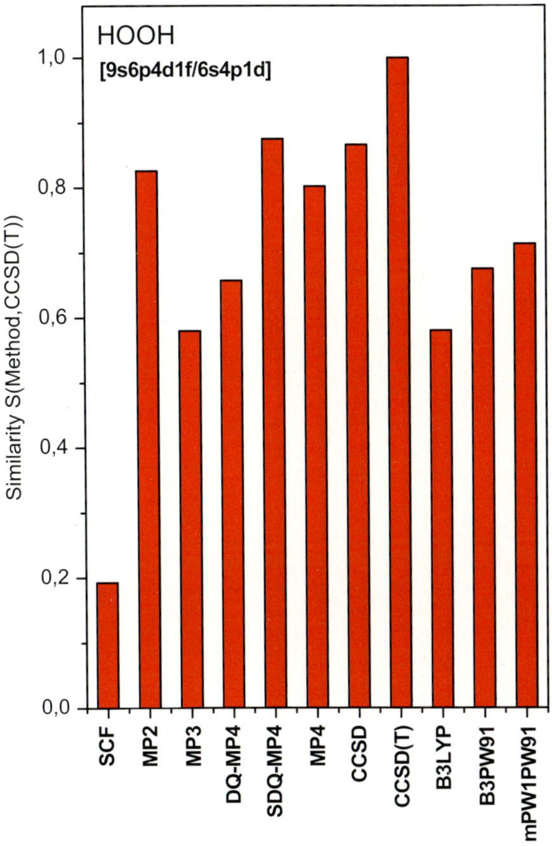


**Figure 13.6** Histogram of the similarity of all methods to the most accurate CCSD(T).

CCSD, singles and doubles coupled cluster, and
CCSD(T), which includes an estimate of connected triples via a perturbational treatment.

The DFT methods are B3LYP, B3PW91, and mPW1PW91. We use the above *ab initio* and DFT methods following closely their implementation in the GAUSSIAN 03 program [26]. Comprehensive presentations of both *ab initio* and DFT methods are available in standard textbooks [27, 28].

All calculations were performed with a large, flexible, purpose-oriented basis set of [9s6p4d1f/6s4p1d]. This basis has been built upon a large [7s4p/4s] substrate of quadruple-zeta quality [29]. The construction of the basis set follows a computational philosophy expanded in sufficient detail in previous work [23].

All calculations have been performed at the theoretical MP2(Full)/cc-pvtz geometry.

Atomic units are used in this chapter. Conversion factors to SI units are as follows: energy, 1 $E_h = 4.3597482 \times 10^{-18}$ J; length, 1 $a_0 = 0.529177249 \times 10^{-10}$ m; $\mu$, 1 $ea_0 = 8.478358 \times 10^{-30}$ cm; $\alpha$, 1 $e^2a_0^2E_h^{-1} = 1.648778 \times 10^{-41}$ $C^2$ $m^2$ $J^{-1}$; $\beta$, 1 $e^3a_0^3E_h^{-2}$

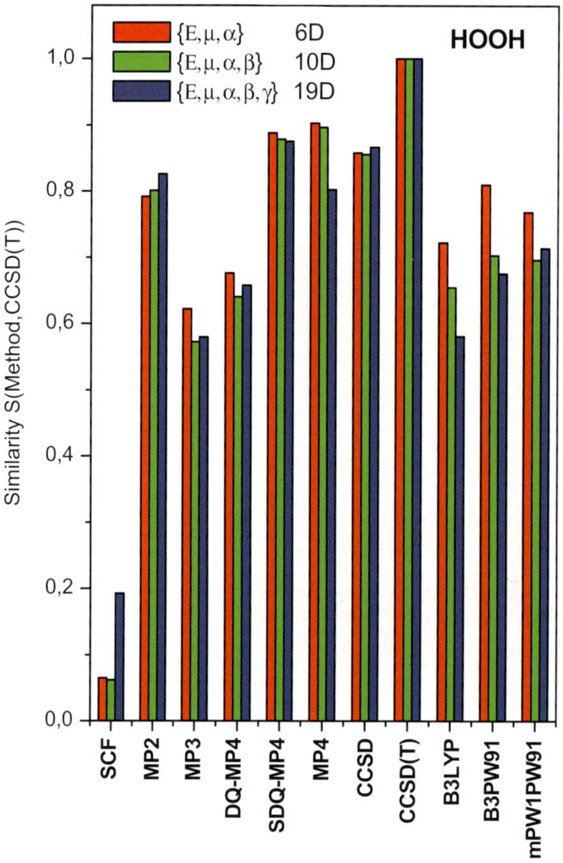


**Figure 13.7** Histogram of the evolution of the similarity *S(Method,CCSD(T))* with the size of the TD (see text).

$= 3.206361 \times 10^{-53}\,\mathrm{C}^3\,\mathrm{m}^3\,\mathrm{J}^{-2}$; and $\gamma$, $1\,e^4 a_0^4 E_\mathrm{h}^{-3} = 6.235378 \times 10^{-65}\,\mathrm{C}^4\,\mathrm{m}^4\,\mathrm{J}^{-3}$. Electric property values are mostly given as pure numbers, that is, $\mu/ea_0$, $\alpha/e^2 a_0^2 E_\mathrm{h}^{-1}$, $\beta/e^3 a_0^3 E_\mathrm{h}^{-2}$, and $\gamma/e^4 a_0^4 E_\mathrm{h}^{-3}$.

## 13.4 Results and Discussion

The calculated values for the electric properties of HOOH are shown in Tables 13.1 and 13.2 for the *ab initio* and DFT methods, respectively. The calculated *ab initio* values are in good agreement with those published earlier and calculated with a less complete basis set. We will not extend the discussion to comparisons with previous results. Interesting recent work on the electric properties of HOOH has been published by Koput [30] and Santiago *et al.* [31]. Our values agree quite well with their findings.

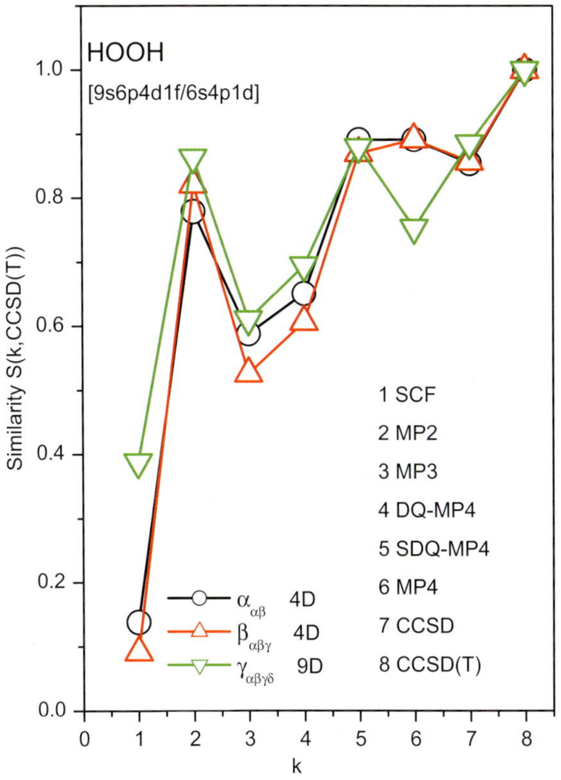


**Figure 13.8** Evolution of the similarity *S(Method,CCSD(T))* for three different property descriptions (see text).

Figures 13.1–13.5 show the method dependence of five fundamental quantities (all of them measurable in principle): dipole moment ($\mu_z$), mean ($\bar{\alpha}$) and anisotropy ($\Delta\alpha$) of the dipole polarizability, and mean first ($\bar{\beta}$) and second ($\bar{\gamma}$) hyperpolarizability. These figures also show the respective calculated DFT values.

For the dipole moment, the agreement between the *ab initio* and DFT values is quite good. The latter are close to the MP2 values but somewhat above the higher level post-Hartree-Fock methods, SDQ-MP4, MP4, CCSD an CCSD(T). For the mean polarizability, both B3PW91 and mPW1PW91 are quite close to the MP2 value. The mPW1PW91 method agrees quite well with the MP4 and CCSD(T) values for this property. As regards the anisotropy of the polarizability, we obtain an interesting picture: the DFT methods are significantly lower than the SCF value and slightly above the post-Hartree-Fock methods.

The predictive capability of the DFT methods in hyperpolarizability calculations is always a highly nontrivial matter. In Figure 13.4, we see a very eloquent picture of the performance of the three selected DFT methods on the calculation of the mean first hyperpolarizability. The DFT values are larger in magnitude than all *ab initio* results. The same is also true for the mean second dipole hyperpolarizability with

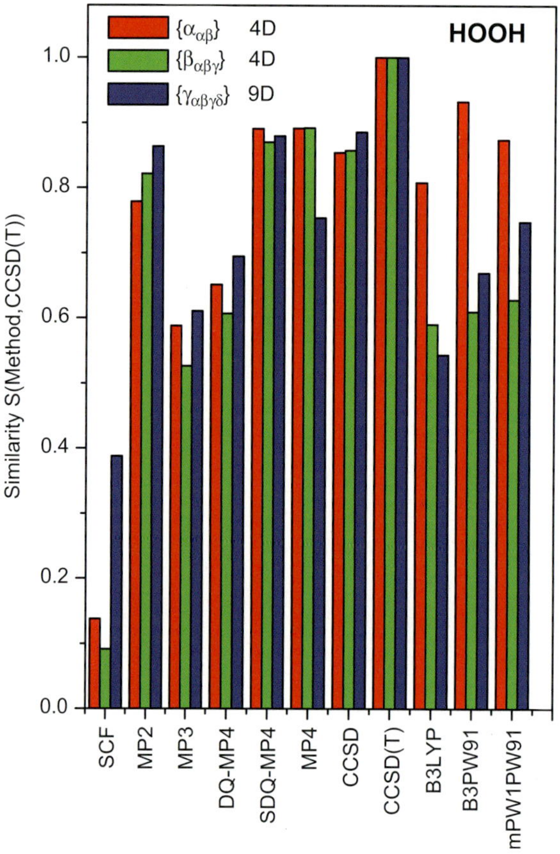


**Figure 13.9** Histogram of the evolution of the similarity *S(Method,CCSD(T))* for three different property descriptions (see text).

the B3LYP method yielding a value significantly above all other *ab initio* and DFT results.

We will now endeavor to propose a quantitative analysis of the above facts.

Table 13.3 presents the calculate similarities for all quantum chemical methods. The similarities refer to 19D TDs that include all properties $\{E^0, \mu_\alpha, \alpha_{\alpha\beta}, \beta_{\alpha\beta\gamma}$, and $\gamma_{\alpha\beta\gamma\delta}\}$. Some fundamental observations first: CCSD(T) is highest level of theory employed in this work so, presumably, it yields the most accurate (or reference) TD. The method closest in performance, as seen from the contents of Table 13.3, is SDQ-MP4. It is clearly the method closest to CCSD(T) with the highest similarity: S(SDQ-MP4, CCSD(T)) = 0.88776. In increasing similarity to CCSD(T), the *ab initio* methods are ordered as:

$$\text{SCF} \rightarrow \text{MP3} \rightarrow \text{DQ-MP4} \rightarrow \text{MP4} \rightarrow \text{CCSD} \rightarrow \text{MP2} \rightarrow \text{SDQ-MP4} \rightarrow \text{CCSD(T)}$$

Of all DFT methods, the most similar to CCSD(T) is mPW1PW91, its similarity to the most accurate one is S(mPW1PW91, CCSD(T)) = 0.71380. Within the DFT methods,

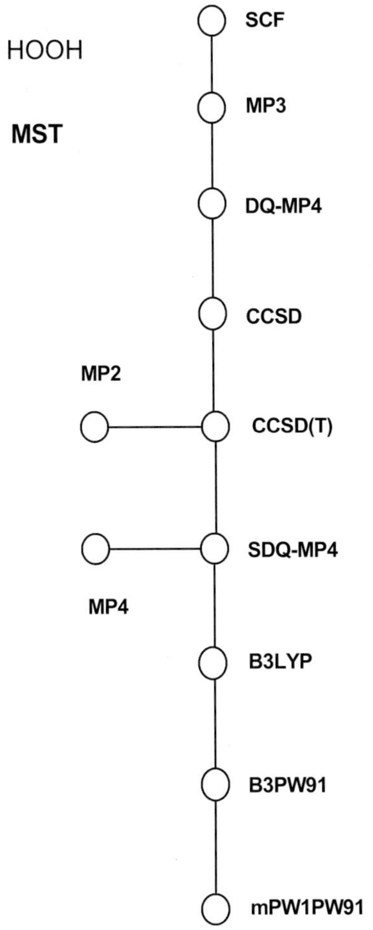


**Figure 13.10** *MST* for the space of TDs.

the mPW1PW91 one is closest to B3PW91 with S(mPW1PW91, B3PW91) = 0.91946. Its similarity to the very popular B3LYP method is significantly lower at S(mPW1PW91, B3LYP) = 0.77987. The similarity of all methods to the most accurate one for the 19D TDs is shown in the histogram of Figure 13.6.

Another question that arises here is the dependence of the calculated similarities on the size of the TD. To examine closer this matter, we have calculated the similarities for reduced size TDs $\{E^0, \mu_\alpha, \alpha_{\alpha\beta}\}$ (6D), $\{E^0, \mu_\alpha, \alpha_{\alpha\beta}, \beta_{\alpha\beta\gamma}\}$ (10D). The respective changes are shown in Figure 13.7 and are most visible for the DFT methods.

Figure 13.8 shows a different analysis, based on reduced TDs referring to distinct properties: dipole polarizability (4D), first (4D), and second (9D) hyperpolarizability. We see that the *ab initio* methods, with the notable exception of the MP4, show reasonable variations from the dipole polarizability to the second hyperpolarizability.

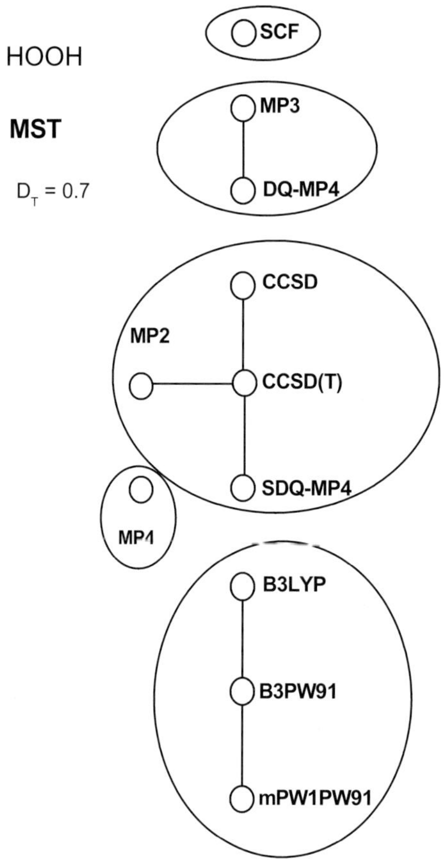


**Figure 13.11** Clustering in the space of TDs.

In Figure 13.9, we have also included the DFT methods. The respective histogram show clearly that the similarities pertaining to the DFT methods decrease significantly passing from $\{\alpha_{\alpha\beta}\}$ to $\{\beta_{\alpha\beta\gamma}\}$ and $\{\gamma_{\alpha\beta\gamma\delta}\}$.

Figure 13.10 shows the MST calculated for 19D TDs. For a threshold of $D_T = 0.7$, we obtain the clustering shown in Figure 13.11. The clustering leads to the following decomposition of the space of TDs

$$TD = C1 \cup C2 \cup C3 \cup C4 \cup C5$$

where

C1 = {SCF}
C2 = {MP3, DQ-MP4}
C3 = {MP4}
C4 = {MP2, SDQ-MP4, CCSD, CCSD(T)}
C5 = {B3LYP, B3PW91, mPW1PW91}.

The most accurate method CCSD(T) is in cluster C4. The DFT methods form a distinct cluster C5.

## 13.5 Conclusions

We have shown that our methodology provides a useful tool for the quantitative analysis of the performance of quantum chemical methods over arbitrary collections of atomic/molecular properties. We have chosen the molecule of HOOH for our analysis. Because of its low symmetry, a large number of independent components are needed to specify the electric polarizability and hyperpolarizability tensors of HOOH. This methodology should be of importance in various lines of research in computational quantum chemistry, with the evaluation of the quality of methods or the design of basis sets for reliable predictions of molecular properties.

## References


1 Ransil, B.J. (1980) *International Journal of Quantum Chemistry*, **17**, 467.
2 Poirier, R.A., Daudel, R., Mezey, P.G., and Csizmadia, I.G. (1980) *International Journal of Quantum Chemistry*, **18**, 715.
3 Poirier, R.A., Daudel, R., and Csizmadia, I.G. (1980) *International Journal of Quantum Chemistry*, **18**, 727.
4 Maroulis, G., Sana, M., and Leroy, G. (1981) *International Journal of Quantum Chemistry*, **19**, 43.
5 Maroulis, G., Sana, M., and Leroy, G. (1984) *Journal of Molecular Structure (THEOCHEM)*, **110**, 182.
6 Maroulis, G., Sana, M., and Leroy, G. (1985) *Journal of Molecular Structure (THEOCHEM)*, **122**, 269.
7 Maroulis, G., Sana, M., and Leroy, G. (1985) *Journal of Molecular Structure (THEOCHEM)*, **121**, 69.
8 Simas, A.M., Thakkar, A.J., and Smith, V.H. Jr. (1983) *International Journal of Quantum Chemistry*, **24**, 527.
9 Robertson, D.D., Smith, V.H. Jr., Simas, A.M., and Thakkar, A.J. (1986) *International Journal of Quantum Chemistry*, **30**, 717.
10 Regier, P.E., Fisher, J., Sharma, B.S., and Thakkar, A.J. (1985) *International Journal of Quantum Chemistry*, **28**, 429.
11 Maroulis, G. (1988) *International Journal of Quantum Chemistry*, **24**, 185.
12 Sordo, J.A. (1998) *Computer Physics Communications*, **113**, 85.
13 Maroulis, G. (1995) *International Journal of Quantum Chemistry*, **55**, 173.
14 Maroulis, G. (1998) *Journal of Chemical Physics*, **108**, 5432.
15 Maroulis, G. (1999) *Journal of Chemical Physics*, **111**, 583.
16 Maroulis, G. (2008) *Journal of Chemical Physics*, **129**, 044314.
17 Xenides, D. (2007) *Journal of Molecular Structure (THEOCHEM)*, **804**, 41.
18 Christodouleas, C., Xenides, D., and Simos, T.E. (2010) *Journal of Computational Chemistry*, **31**, 412–420.
19 Xenides, D., Karamanis, P., and Pouchan, C. (2010) *Chemical Physics Letters*, **498**, 134–139.
20 Chartrand, G. and Lesniak, L. (1986) *Graphs and Digraphs*, Wadsworth, Belmont.
21 Varmuza, K. (1980) *Pattern Recognition in Chemistry*, Springer, Berlin.
22 Spath, H. (1980) *Cluster Analysis Algorithms*, Ellis Horwood, Chichester.
23 Maroulis, G. (2003) *Journal of Chemical Physics*, **118**, 2673. and references therein.
24 Maroulis, G. and Thakkar, A.J. (1992) *Journal of Chemical Physics*, **96**, 6048.

25 Buckingham, A.D. (1967) *Advances in Chemical Physics*, **12**, 107.
26 Frisch, M.J., Trucks, G.W., Schlegel, H.B. *et al.* (2004) *Gaussian 03*, Revision D. 01, Gaussian, Inc., Wallingford CT.
27 Szabo, A. and Ostlund, N.S. (1982) *Modern Quantum Chemistry*, MacMillan, New York.
28 Helgaker, T., Jørgensen, P., and Olsen, J. (2000) *Molecular Electronic-Structure Theory*, Wiley, Chichester, UK.
29 Thakkar, A.J., Koga, T., Saito, M., and Hoffmeyer, R. (1993) *International Journal of Quantum Chemistry*, **S27**, 343.
30 Koput, J. (1996) *Chemical Physics Letters*, **257**, 36.
31 Santiago, E., Castro, M.A., Fonseca, T.L., and Mukherjee, P.K. (2008) *Journal of Chemical Physics*, **128**, 064310.

# 14
# The Analysis of Organic Reaction Pathways by Brownian Processing

*Daniel J. Graham*

## 14.1
## Introduction

Suppose a chemist wishes to convey information to a colleague regarding four molecules. Complete descriptions would comprise the solutions to the Schrödinger equation. If molecule I is constructed of 15 atoms and 42 electrons, the wavefunction $\Psi_I$ depends on $3 \times (15 + 42) = 171$ independent variables. There are three coordinates $x_i$, $y_i$, and $z_i$ for each *i*th electron; there are three coordinates $X_j$, $Y_j$, and $Z_j$ for each *j*th nucleus [1]. The wavefunction is multivariable, to say the least

$$\Psi_I = \Psi_I\left(x_1, y_1, z_1, x_2, y_2, z_2, \ldots, x_{42}, y_{42}, z_{42}, X_1, Y_1, Z_1, X_2, Y_2, Z_2, \ldots, X_{15}, Y_{15}, Z_{15}\right) \tag{14.1}$$

If Molecules II, III, and IV present respective component sets of {14 atoms, 42 electrons}, {15 atoms, 51 electrons}, and {15 atoms, 51 electrons}, it is straightforward to count the variables upon which $\Psi_{II}$, $\Psi_{III}$, and $\Psi_{IV}$ depend. It is not so easy to obtain the wavefunctions.

Clearly, if the chemist elects the most thorough descriptions, there are long stories to tell the colleague. Each wavefunction, while convergent, specifies an infinite number of values. The same statement holds for the probability amplitudes formed by $\Psi_I^2$, $\Psi_{II}^2$, $\Psi_{III}^2$, and $\Psi_{IV}^2$.

Fortunately, there is a short-cut method. If the chemist specifies the formula diagrams, namely

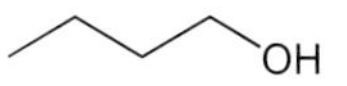


I

II

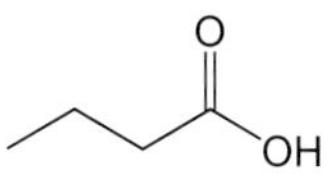


III

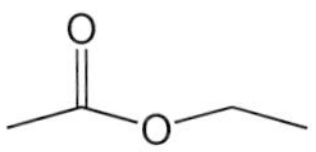


IV

*Statistical Modelling of Molecular Descriptors in QSAR/QSPR*. First Edition.
Edited by M. Dehmer, K. Varmuza, and D. Bonchev

he or she communicates vital information using a handful of digits, letters, and lines. The colleague's response is immediate and far-reaching. He or she sees that the molecules are organic; they host different functional groups but are otherwise closely related. All are four-carbon systems: a primary alcohol, alkyl halide, carboxylic acid, and carboxylic ester. Because of training and experience, the colleague also registers reactions that relate three of the molecules to one another, for example

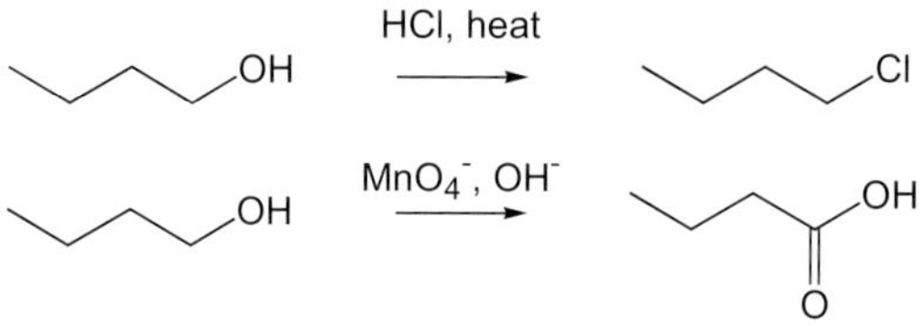


The colleague understands that the carboxylic ester can be derived from the others, but only by a nontrivial reaction series.

How do the most thorough and short-cut methods differ? One answer is that the latter relies on sparse digital code to represent key information. As a consequence, there are no mathematical functions to evaluate, plot, integrate, and so forth. Digital codes are ubiquitous, of course, and long predate the chemical sciences. Yet their significance seems all the more heightened in the age of computers and searchable databases. Chemists will become increasingly adept at portraying molecules via wavefunctions. Communicating with formula diagrams, however, will not become obsolete. They lie indeed at the core of the electronic messages carried by molecules, and their transformations in chemical reactions. As with all physical entities, messages connect with both energy and information [2]. Importantly, the messages carried by molecules can be viewed through a lens furnished by information theory and thermodynamics.

It is the purpose of this chapter to examine archetypal examples of organic molecules and reaction pathways. The principal tool will be Brownian processing; this is a modeling technique which adheres to the shortcut ways for portraying molecules. Brownian processing advances formula diagrams several steps, however, so as to enhance their contact with information theory and thermodynamics. The results include the quantification and correlation of molecular information, and added perspective surrounding transformations, synthetic methods and strategies. The approach benefits from an extensive literature on graph and information theory applied to organic compounds [3–9]. It is indebted as well to the links forged between chemical reactions, algorithms, and computation [10].

This chapter is organized as follows. The next section sets the stage with a few fundamentals regarding information, energy, and computation. The ensuing sections then illustrate the intersection of the fundamentals with select molecules and reactions.

## 14.2
## Electronic Messages, Information, and Energy

Digital systems use finite clusters of discrete signals to convey infinite possible messages. Computer keyboards typically offer $2 \times 47 = 94$ signals; the number is 88 for pianos. Amazingly, nature accommodates digital protocols at the Angstrom scale. The charge interactions of carbon, hydrogen, oxygen, and chlorine atoms are diverse and complex. Yet when assembled in stable compounds, they admit discrete representations via the symbols C–C, C–H, C–O, C=O, C–Cl, O–H, and so forth. There is really no limit to the molecules and, in turn, messages that can be articulated by these symbols. Every assembly presents chemical information of the facts-and-data variety. It is interesting and important to examine the information in quantitative terms. Computers provide helpful guidelines.

The messages of such devices are based on voltages pitched at two levels. A byte is no more than a spatially ordered set of static potentials. For example, (01 001 011) corresponds ideally to (0, 5, 0, 0, 5, 0, 5, 5) V. The word *ideally* acknowledges that matters are not so strict in practice. The low voltages may very well be 0.658 V and the high ones might be 4.82 V. The point is that there are two states that are readily distinguished in the circuitry. Anything close to 0 V is registered as "0"; all potentials in the neighborhood of 5 V are interpreted as "1". Since each constituent has two possible and equally likely states, it contributes one binary digit or bit of information. Information is an additive quantity. Thus, the messages in circuits offer variable amounts, depending on how many constituents are combined: (1001) $\leftrightarrow$ 4 bits, (101 010 010 111) $\leftrightarrow$ 12 bits, (110 111) $\leftrightarrow$ 6 bits, (1 010 101 010 101 010) $\leftrightarrow$ 16 bits, and so on. Messages carry both an information and energy content. If a "1" correspond to $10^{-6}$ amperes of current forced at 5 V for $10^{-9}$ s, then (1001) is paired with energy $(1+0+0+1) \times 5\,\text{V} \times 10^{-6}\,\text{A} \times 10^{-9}\,\text{s} = 10^{-14}\,\text{J}$.

The energies allied with other assemblies are just as easy to compute.

Among the consequences, every message corresponds to a point in a state space definable by energy, information, and other illuminating variables. Computations which transform the message generally relocate the point. A sequence of operations, moreover, marks a pathway that threads the state space. For example, the first two of the above messages can be combined, viz.

1) 1001 $\cup$ 101 010 010 111 $\rightarrow$ 1 001 101 010 010 111

   The two's-complement of the result can then be computed

2) 1 001 101 010 010 111 $\rightarrow$ 0 110 010 101 101 001

   The last message can be added numerically to 110 111:

3) 011 001 010 110 001 + 110 111 $\rightarrow$ 0 110 010 110 100 000

These are arbitrarily selected examples of message transformations. What is important is that they admit plots such as Figure 14.1. The information/energy coordinate for each message has been marked by a filled square. The operations which convert one message into another have been indicated by arrows. Clearly, all transformations are not created equal. The first and third of the above require pairs of messages as input; a single assembly serves as input for the second. In turn, operations 1 and 3 require longer jumps across the state space, plus the coordination of two electronic messages. By comparison, the second operation entails only minor repositioning of the state point.

In the ideal, messages correspond to state points of zero width. In the perfect processor, the transformations are instantaneous and land with 100% accuracy.

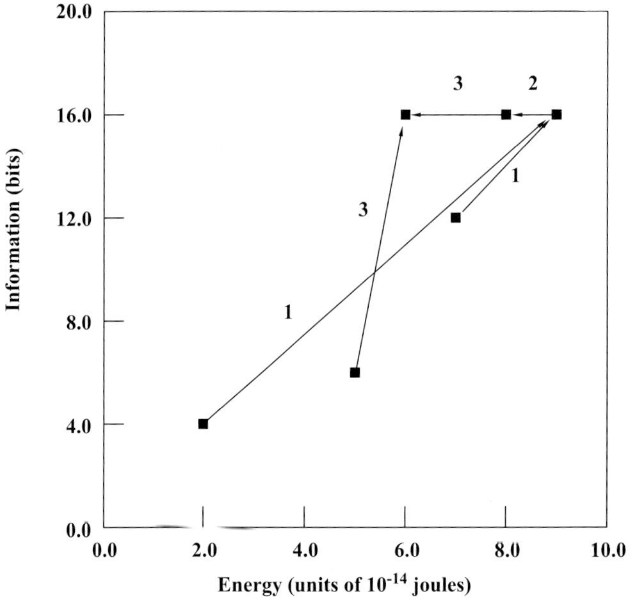


**Figure 14.1** Electronic messages and state space. The information/energy coordinate for each computer message has been marked by a filled square. The transitions which convert messages have been indicated by arrows. The numbers correspond to transitions discussed in text.

Real-life is more complicated, unsurprisingly. Because of circuit noise, flawed components, and the ill-coordination of bytes, state points inescapably express nonzero widths. Further, extraneous states are visited over a series of operations. For example, in a two's-complement transition, if the constituents are not switched in unison, then abundant states are visited to no good purpose in the neighborhoods of the initial and final. Superfluous state sampling is a bane of computation. It increases the likelihood for error, and demands extra time, energy, and programming [11]. By and large, the shortest and surest jumps across state space afford the maximum accuracy and economy. This is one reason that everyday processors operate via fixed-length words (e.g., 32 bit) and low-current, fast-switching components. Such architecture limits the transformations to short, lateral jumps in a coordinate system such as Figure 14.1. State point coordinates are by no means restricted to energy and information. Other variables such as energy density and dispersion, and mutual or correlated information can be used to formulate state spaces. Notably, plots like Figure 14.1 have roots in thermodynamics. Here the state points and pathways are located by temperature, pressure, mass density, heat capacity, and other accessible variables [12].

The messages of computers are readily linked to state points and planes. How do things work for organic molecules?

Let us revisit the compounds of the first section: 1-butanol, 1-chlorobutane, butanoic acid, and ethyl acetate. The structures of these molecules include C–H and C–C constituents. These denote electronic sites which are distinctive by their charge density, bond energy, and location relative to others. If an atom such as helium were to collide with a C—C site of any of the four systems, the same type of electronic message would be registered. Such a message differs from that registered by collision at a C—H site. This is not to imply that the charge distribution and energy are identical for C—C sites, and likewise for C—H. However, the similarities across nature are sufficient to justify the digital labels, C—C and C—H, and to represent the electronic structure in the most elementary terms [13]. Along the same lines, C—C and C—H sites are paired, respectively, with average bond energies of 347 and 414 kJ/mol. One hearkens back to computers, and the shortcut ways for describing molecules. The "1" signals of the former express a spectrum of voltages; the C—C bonds in real molecules demonstrate a range of lengths and energies. However, the natural groupings of potentials in computers, and C—C links in organic compounds, make feasible the digital approaches. In turn, the placement of discrete points in various state planes is as immediate as in Figure 14.1.

Organic molecules pose far greater complexity than computers. One reason is that they typically require more than two labels to differentiate the constituents. 1-Butanol, for instance, requires four labels in a unique assembly: C—H, C—C, C—O, and O—H. The identity and number follow just as readily (i.e., by inspecting the formula diagrams) for butanoic acid, 1-chlorobutane, and ethyl acetate: {C—C, C—H, C=O, C—O, O—H}, {C—C, C—H, C—Cl}, and {C—C, C—H, C=O, C—O}, respectively. Clearly, the uncommon constituents distinguish one molecule from another, for example, 1-butanol and 1-chlorobutane both host C—C and C—H but diverge because of C—O, O—H, and C—Cl. But more than that, the manner in which constituents are linked restricts the electronic sequences that can be communicated and registered in thermal environments. If, say, a helium atom were to collide with the C—O site of 1-butanol, the neighboring O—H site is far more likely to be involved in the subsequent collision, compared with a C—H several bond units away. If the atom were to collide with C=O of ethyl acetate, the ester linkage is significantly more likely to experience the next collision, compared with one of the methyl groups.

Computers and molecules are akin by their digital nature. Their mode and energy for processing follow different tracks, however. Computer messages are logged by voltage readings at strictly controlled time intervals; the energy for doing so originates from direct-current supplies. In contrast, the messages carried by molecules are registered piecewise and erratically by thermal collisions, that is, during and by virtue of Brownian motion. The power is furnished by heat stored in the translational and rotational motion [14].

Formula diagrams offer a century-old model of Angstrom-scale structure. Brownian computation provides a more recent tool by which molecular messages are charted, and their state points located in various coordinate systems [15]. Brownian computation (processing) identifies the most probable sequences of electronic constituents (C—H, C—C, C—O, and so on) registered in a thermal environment such as a gas or liquid. As with formula diagrams, the Brownian algorithm takes the covalent interactions to be overriding. In so doing, the message sequences are those rendered by an unforced, nearest-neighbor random walk over the molecule of interest. Figure 14.2 illustrates a portion of a walk involving 1-butanol. The site or constituent in contact with a hypothetical collider – think helium atom – has been

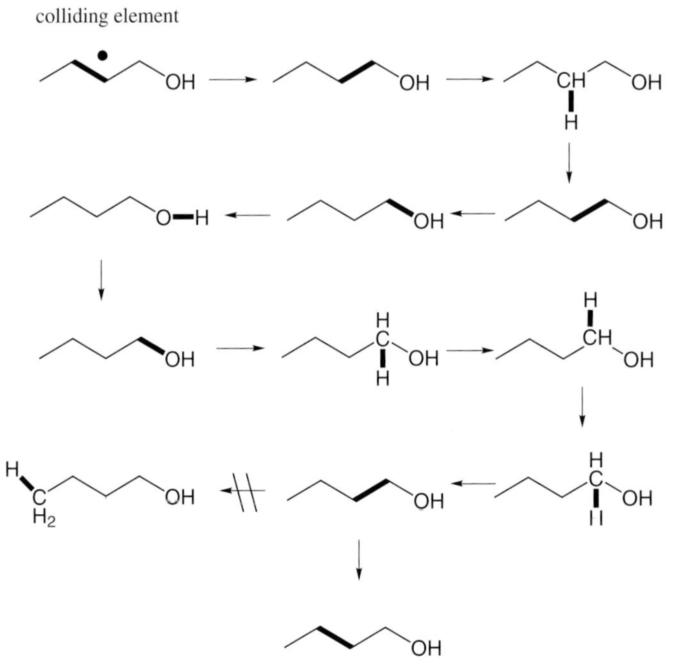


**Figure 14.2** Molecular messages and brownian computation. The sequence of formula diagrams illustrates a portion of a nearest-neighbor random walk involving 1-butanol. The site in contact with a collider such as a helium atom has been indicated using boldface. The walk is recorded using the same labels and bond energies prescribed by the diagram. The walk and digital record tape are unique to the source. The lower, left-most transition is an example of one forbidden by nearest-neighbor selection rules.

indicated using boldface. Notably, the walk is articulated using the same digital labels and covalent bond energies prescribed by the diagram

$$\begin{array}{lllllllll} & (C-C) & (C-C) & (C-H) & (C-C) & (C-O) & (O-H) & \ldots & \\ \ldots & 347 & 347 & 414 & 347 & 351 & 464 & \ldots & \text{kilojoules per mole} \end{array}$$

The sequences have the form of a record tape that appears nonsensical. They are nonetheless unique to the source – the molecule and formula diagram – and shine new light on the chemical function. It is simple to construct these tapes which are typically several million code units in length; one applies computer programs that interface with a formula diagram of interest. In effect, desktop computers draw out the information properties of small-scale Brownian computers. This approach was developed by the author and students and directed to a variety of compounds and chemical reactions [15–19]. It was inspired by Bennett's exposition on the thermodynamics of computation [20]. Bennett's discussion cast a wide net and was not limited to macroscopic systems. Bennett specifically addressed the computational and thermodynamic aspects of DNA transcription. It is universally appreciated that polymerases

**Table 14.1** Examples of first, second, and third order molecular messages with corresponding frequency and energy terms.

| Order | Message | Frequency | Energy (kJ/mole) |
|---|---|---|---|
| 1 | C–H | $f_{\mathrm{C-H}}$ | 436 |
| 1 | C—O | $f_{\mathrm{C-O}}$ | 351 |
| 2 | (C–H)(C–O) | $f_{\mathrm{(C-H)(C-O)}}$ | 436 + 351 |
| 2 | (C—C)(C—C) | $f_{\mathrm{(C-C)(C-C)}}$ | 347 + 347 |
| 3 | (C–H)(C–O)(C–C) | $f_{\mathrm{(C-H)(C-O)(C-C)}}$ | 436 + 351 + 347 |
| 3 | (C—C)(C—O)(O—H) | $f_{\mathrm{(C-C)(C-O)(O-H)}}$ | 347 + 351 + 464 |

register the electronic information of polynucleotide base sites by nearest-neighbor random walks [21]. This view of Angstrom-scale communication can be extended to molecules large and small, on account of their digital nature, and the ubiquity of thermal environments. Such extension is the essence of Brownian processing.

A record tape furnishes a probability function that originates from the formula diagram. In turn, there are several stochastic descriptors which can be established. These include (1) the Shannon information $I_n$ allied with $n$-number of message units, (2) the mutual information $\mathrm{MI}_m$ for an $m$-sized cluster of units, (3) the average energy $E_n$ allied with $n$-number of message units, and (4) the average energy dispersion $\sigma_{En}$ for $n$-size messages. It is straightforward to arrive at these quantities by identifying all the digital sequences logged by the random walk, and by tabulating the occurrence frequencies. Some examples are noted in Table 14.1

The Shannon information at every order follows accordingly

$$I_n = -\sum_i f_i \log_2 f_i \tag{14.2}$$

where the summation is over all $i$ unique messages of order (sequence length) $n$ registered with frequency $f_i$ [15, 22]. The use of base-two logarithms ensures bits to be the appropriate units for $I_n$. The Shannon information quantifies the diversity of electronic messages available from the molecule. Compounds such as methane ($CH_4$) and carbon tetrachloride ($CCl_4$) feature only one type of electronic site, C–H and C–Cl, respectively. These molecules offer zero bits of Shannon information because diversity is absent in their messages. By contrast, common alkaloids such as caffeine and dopamine offer electronic messages rich in diversity, and appreciable Shannon information as a consequence.

The mutual information follows along similar lines. For every $m$-sized cluster, $m \geq 2$, one has:

$$MI_{m=2} = \sum_{ij} f_{ij} \log_2 \left(\frac{f_{ij}}{f_i \cdot f_j}\right) \tag{14.3}$$

$$MI_{m=3} = \sum_{ijk} f_{ijk} \log_2 \left(\frac{f_{ijk}}{f_i \cdot f_j \cdot f_k}\right) \tag{14.4}$$

The expressions for $MI_{m>3}$ follow just as readily. Among other things, the mutual information quantifies the departures from randomness of a molecule's messages; it establishes the degree of pair, triplet, and higher-order correlations [23]. If organic molecules were devoid of internal organization, a frequency such as $f_{ijk}$ would equate trivially with $f_i \cdot f_j \cdot f_k$, and the collision sequences would present zero mutual information. Chemistry would be an inordinately different science if this were the case.

Electronic messages, molecular and otherwise, connect with energy. The average energy at every order obtains via

$$\langle E_n \rangle = \sum_i f_i \, D_i^{(n)} \tag{14.5}$$

The $D_i^{(n)}$ represent the energy affiliated with the $i$th message for a molecule of order $n$. The number of terms for computing $D_i^{(n)}$ equates with $n$, and follows from standard look-up tables [24].

The average dispersion energy $\sigma_{En}$ for messages of length $n$ arrives via

$$\sigma_{En} = \sqrt{\langle E_n^2 \rangle - \langle E_n \rangle^2} \tag{14.6}$$

In modern day computation, the energy and information can be quantified for any size message. Likewise, the themes of Equations 14.2–14.6 can be extended to any order of electronic sequence.

It is apparent that with no upper bound on $n$, there are unlimited Brownian descriptors for a molecule. The more complete the set, the more thorough the characterization of the electronic message space. Fortunately, record tapes admit compression of descriptor sets. The reasons are twofold. One is that $I_n$, $MI_m$, $\langle E_n \rangle$, and $\sigma_{En}$ scale linearly to large degree with order $n$ and cluster size $m$. Secondly, the values of the Brownian descriptors increase with $n$ and $m$. These traits underpin four intensive variables that obtain from a record tape:

$$\xi_I = \left\langle \frac{\Delta I_n}{\Delta n} \right\rangle \quad \xi_{MI} = \left\langle \frac{\Delta MI_m}{\Delta m} \right\rangle \quad \xi_E = \left\langle \frac{\Delta E_n}{\Delta n} \right\rangle \quad \xi_{\sigma E} = \left\langle \frac{\Delta \sigma_{En}}{\Delta n} \right\rangle$$

These are accompanied by uncertainty values based on regression analyses. The results are $\xi_I$, $\xi_{MI}$, $\xi_E$, and $\xi_{\sigma E}$ which, respectively, quantify the average increases in Shannon information, mutual information, energy, and energy dispersion with each electronic site encountered in a thermal collision. These intensive descriptors proceed to locate state points for compounds large and small, natural and synthetic. The statistical uncertainties allied with $\xi_I$, $\xi_{MI}$, $\xi_E$, and $\xi_{\sigma E}$ establish the point widths. In thermodynamics, intensive variables such as temperature and pressure specify locations on phase diagrams. $\xi_I$, $\xi_{MI}$, $\xi_E$, and $\xi_{\sigma E}$ serve an analogous purpose for individual molecules.

Figures 14.3 and 14.4 illustrate data for Molecules I–IV of the introduction. One observes by Figure 14.3 that the average Shannon information per message unit

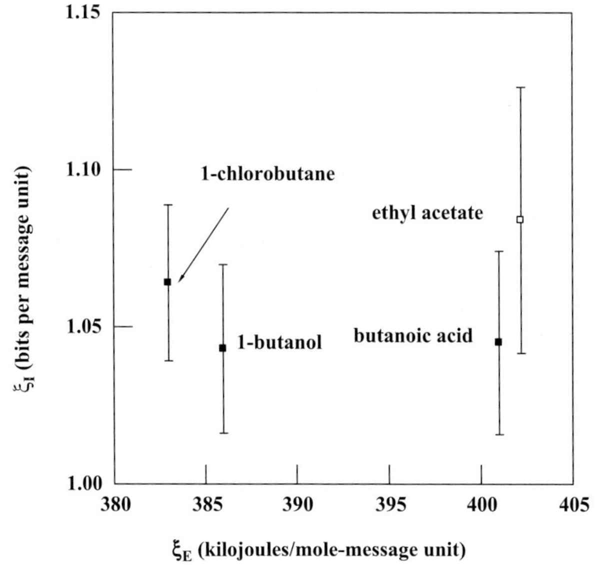


**Figure 14.3** State points and widths. Located are the state points for Molecules I–IV. The coordinates are the intensive descriptors $\xi_I$ and $\xi_E$ calculated via digital record tapes.

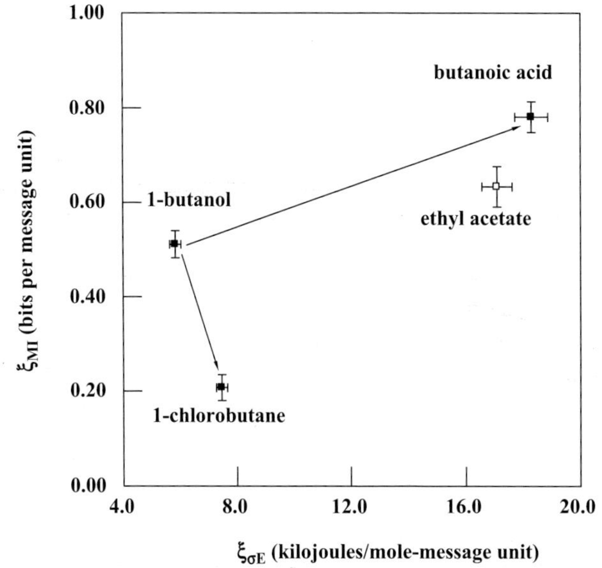


**Figure 14.4** Alternative coordinate systems and state points. Located are the state points for Molecules I–IV in the $\xi_{MI}$ and $\xi_{\sigma E}$ plane. The reactions listed in the introduction connect state points selectively. The point for ethyl acetate (open square) cannot be reached by a single transition originating with any of the companion points.

encountered (measured by $\xi_I$) is low, and indeed falls in the neighborhood of 1 bit. Message-diversity-wise, the contrasts among the four systems are not substantive; the uncertainties surrounding $\xi_I$ largely outweigh the differences. The state space distinctions are more apparent in the $\xi_E$ measures. Here the average energy per message unit registered proves highest for ethyl acetate and lowest for 1-chlorobutane. The $\xi_E$ differences among the four molecules are several percent.

Figure 14.4 illustrates state points in the $\xi_{MI}$ and $\xi_{\sigma E}$ plane. Here the distinctions are acute, based on the electronic message correlations and energy diversity. The state points for the carboxyl systems place in the upper right of the state plane, that is, in the high-correlation-intensity, high-energy-diversity quadrant. Of the four compounds, the alkyl halide rates the lowest in terms of $\xi_{MI}$ and $\xi_{\sigma E}$.

It is then apparent what chemical reactions accomplish. When 1-butanol is converted to a carboxylic acid, both the concentration (intensity) of message correlations and energy diversity are modified in the upward direction. In contrast, substitution of the –OH moiety with a chlorine atom augments the energy diversity only slightly; this is at the expense of the correlation intensity. The transformations of electronic messages are limited. Computer processors are restricted by their hard-wiring. In chemistry, the alcohol, carboxylic acid, and alkyl halide of Figure 14.4 lack the capacity to convert to the ester, among other compounds. In state space terms, the points marked by the filled squares cannot access the open square by a single jump.

## 14.3
## Molecular Messages, Conversions, and State Space Representations

An organic molecule is a source of electronic information and communicates by thermal collisions. A molecule's message repertoire derives from its charge inhomogeneities and their spatial arrangement. Valence-complete systems (e.g., Molecules I–IV) are stable by and large. Thus, the inhomogeneities are frozen under equilibrium conditions, and the available messages are robust. Figure 14.4 shows compounds to be distinguished sharply by the intensity of message correlations and energy diversity. Thus, every library of molecules offers a set of signature points in the $\xi_{MI}$ and $\xi_{\sigma E}$ plane. Further, every reaction offers two or more connected points, while a total synthesis marks a pathway of points over the state space. Obviously, state space plots do not supplant the formula diagrams on which they are based. Rather, they augment the perspective of Angstrom-scale structure, electronic transformations, and synthetic strategies. Molecules are Brownian computers built exceedingly small; state space plots illustrate what the devices really have to offer by way of energy, diversity, and electronic correlations.

The next two figures together furnish a better grasp of the state space formed by $\xi_{MI}$ and $\xi_{\sigma E}$. The points and widths are located in Figure 14.6 for the compounds of Figure 14.5: methane, ethane, cyclohexane, glucose, glycine, aspartic acid, lysine, and ethylene. Note how the point for methane places trivially at the origin. This five-atom assembly lacks charge inhomogeneities and message correlations at the formula

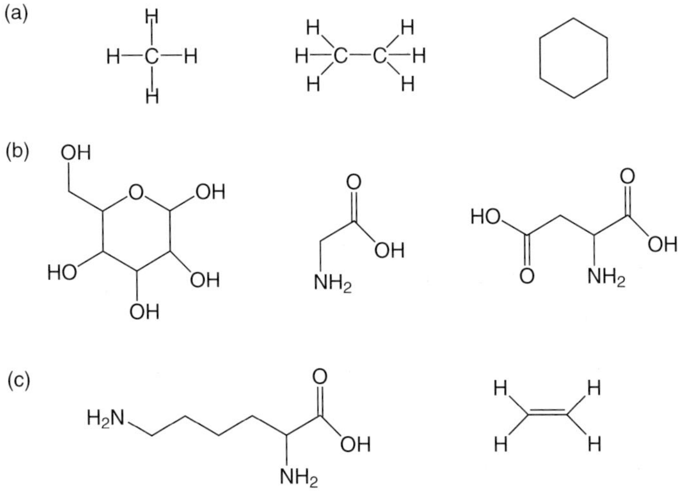


**Figure 14.5** Formula diagrams for diverse molecules. (a) Left-to-right: methane, ethane, and cyclohexane. (b) Left-to-right: glucose, glycine, and aspartic acid. (c) Left-to-right: lysine and ethylene.

diagram level. The inhomogeneities and correlations can be realized only via lower-level descriptions, for instance, appealing to the valence bond orbitals.

Unsurprisingly, the state point for ethane places near that of methane. In the former molecule, the charge inhomogeneities derive from six C—H units linked to a single C—C. The points for the other molecules place much more remotely. The highest "altitude" (i.e., correlated) states are observed for the amino acids: glycine, lysine, and aspartic acid. Cyclohexane presents a low-altitude state at otherwise high dispersion energy. The state point for glucose falls about even with ethane on the $\xi_{\sigma E}$ scale. The carbohydrate is richer, however, in its concentration of electronic correlations. This is the reason the state point lies at modest height in the state space.

The chemical patterns are apparent. The functionalized compounds offer the state points highest in altitude, and furthest from the origin. The saturated hydrocarbons place lowest in Figure 14.6. Cyclohexane is devoid of functional groups. It does, however, offer substantial charge inhomogeneities by the equal mixtures of C—H and C—C.

The inhomogeneities and structure correlations of the Figure 14.6 molecules are stable by nature. At the same time, every compound, methane included, can be chemically activated by heating, contact with a Lewis acid, exposure to ultraviolet light, and so on. Activation renders the charge inhomogeneities fluid. This opens the door for message mutations.

A message mutation derives from the redistribution of charge. It does not transpire randomly, but rather is guided by the natural gradients in a molecule and their correlations. This is important. Inhomogeneities underpin nature's capacity for site-specific activation and mutation. In turn, the correlations restrict the possibilities

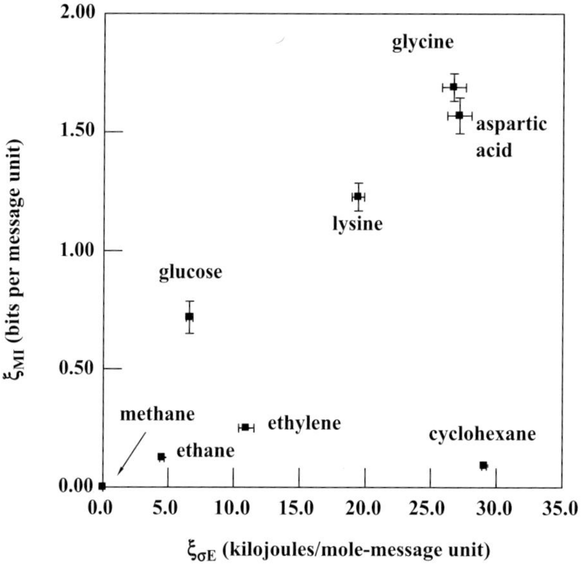


**Figure 14.6** State points and widths for diverse molecules. The state points in the $\xi_{MI}$ and $\xi_{\sigma E}$ plane are marked for the compounds in Figure 14.5.

for the intermediate and terminal products. In effect, charge gradients and correlations provide the fuel and guidance for all activated compounds. By the second law of thermodynamics, for a chemical reaction to transpire, the total entropy must increase [12]. The correlation-intense compounds, such as amino acids and carbohydrates, carry greater wherewithal to direct and restrict the entropy increases, compared with hydrocarbons and other correlation-sparse systems.

In the analysis of reactions by Brownian computation, it is illuminating to use reduced, dimensionless units. The smallest, nonfunctionalized organic molecule with nonzero $\xi_{MI}$ and $\xi_{\sigma E}$ is ethane ($C_2H_6$): $\xi_{MI}$ and $\xi_{\sigma E}$ equal 0.125 bits and 4.53 kJ/mol, respectively (cf. Figure 14.6). With rare exception, organic molecules with two or more carbons contain ethane units in their backbone. Thus, in the remaining figures, the state point axes and locations will be marked in terms of reduced quantities $\hat{\xi}_{MI}$ and $\hat{\xi}_{\sigma E}$. For example, for glucose molecule, one has

$$\hat{\xi}_{MI} = \frac{0.717 \text{ bits}}{0.125 \text{ bits}} \approx 5.74 \tag{14.7}$$

$$\hat{\xi}_{\sigma E} = \frac{6.61 \text{ kilojoules/mole}}{4.53 \text{ kilojoules/mole}} \approx 1.46 \tag{14.8}$$

Among other things, the reduced coordinates confer a distance measure $\hat{L}_{\sigma E,MI}$ between any pair of state points. For instance, for 1-butanol ($C_4H_{10}O$) and 1-chlorobutane ($C_4H_9Cl$) of Figure 14.4,

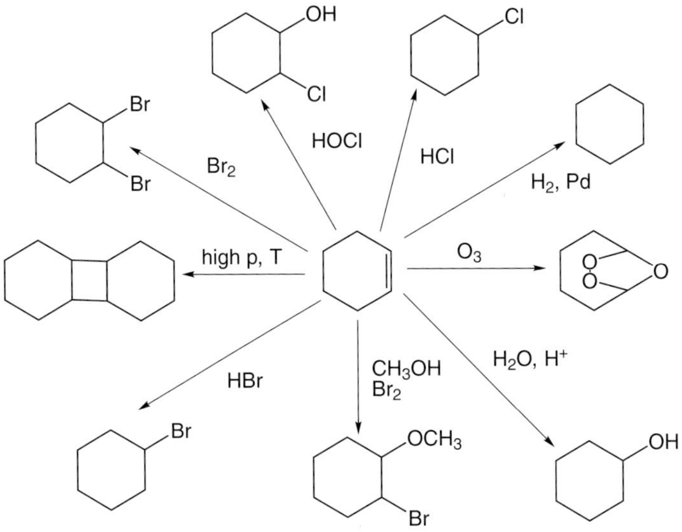


**Figure 14.7** Reactions of cyclohexene. A library of reactions representative of cyclohexene chemistry is shown in formula graph terms.

$$\hat{L}_{\sigma E,MI} = \sqrt{\left(\hat{\xi}_{MI}^{C_4H_{10}O} - \hat{\xi}_{MI}^{C_4H_9Cl}\right)^2 + \left(\hat{\xi}_{\sigma E}^{C_4H_{10}O} - \hat{\xi}_{\sigma E}^{C_4H_9Cl}\right)^2} \approx 2.45 \qquad (14.9)$$

A chemical reaction connects two or more points in state space. The greater $\hat{L}_{\sigma E,MI}$ is between the points, the more severe the differences regarding the electronic correlations and energy diversity. In contrast, reactants and products whose state points are separated by small $\hat{L}_{\sigma E,MI}$ do not diverge substantively in their correlations and diversity. This leads to several properties to note for reaction pathways.

The first deals with a single reactant and source of information, namely cyclohexene. This alkene can be transformed to various products using common stockroom reagents and standard procedures. A representative sampling is presented in Figure 14.7 by way of halogenations, oxidations, reductions, and condensations [19]. What state space plot accompanies the reaction library?

The answer appears in Figure 14.8. The open square marks the point for cyclohexene at $\hat{\xi}_{MI} = 2.11$, $\hat{\xi}_{\sigma E} = 4.29$. The filled squares mark the points for the product compounds of Figure 14.7. Each is labeled in terms of its functionality and distance $\hat{L}_{\sigma E,MI}$ from cyclohexene. The transition arrows (as in Figure 14.4) have been omitted for clarity.

One finds the point for the ozonide farthest from that of the starting material, as well as highest in correlation altitude. In contrast, the halogenated compounds present points clustered in the lower-left quadrant, not far from the starting-point. Not surprisingly, the state points for the saturated hydrocarbons lie the lowest. The molecules carrying hydroxyl and methoxy functionalities present points at altitudes higher than their alkene precursor.

As with the message transformations in computers, chemical reactions equate with jumps across state space – it is impossible for cyclohexene to be tuned

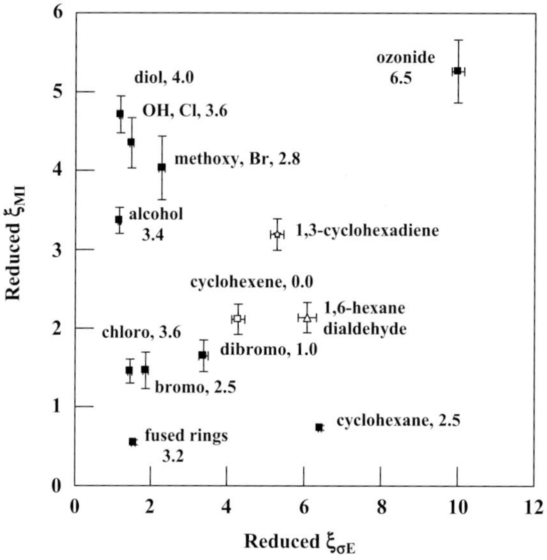


**Figure 14.8** State points allied with cyclohexene reactions. The open square marks the point for cyclohexene in the $\hat{\xi}_{MI}$ and $\hat{\xi}_{\sigma E}$ plane. The filled squares mark the points for the product compounds of Figure 14.7. Each is labeled in terms of its chemical functionality and distance $\hat{L}_{\sigma E,MI}$ from cyclohexene. The point for the 1,6-hexane dialdehyde that results from decomposition of the ozonide is marked with an open triangle. The point for 1,3-cyclohexadiene is marked with an open star.

monotonically to any of the product compounds. For a reaction library based on a single initial state, a normal distribution of $\hat{L}_{\sigma E,MI}$ is the general rule. This is crucial to understanding molecules in computational terms. Cyclohexene poses many more reactions than in Figure 14.7. Yet the $\hat{L}_{\sigma E,MI}$ average and standard deviation for the library, 3.24 and 1.41, respectively, offer guidelines for what nature can accomplish in a single transition. For example, the structure correlations in cyclohexene can be enhanced by simple conversion to alcohols and ozonides. The jump-distances level off, however, at $\hat{L}_{\sigma E,MI} \approx 6.5$. By the same token, message correlations in the molecule can be eradicated selectively; here the transition-lengths are limited to *c.* 3. For alkenes in general, the enhancements and diminutions of the message correlations transpire almost always at the expense of energy diversity. Only two of the products in Figure 14.7 demonstrate $\hat{\xi}_{\sigma E}$ greater than that of the starting compound. In Figure 14.8, the points are clustered in the left-hand portion of state space; there is a void on the right. Clearly, the possibilities for enhancing *both* the message correlations and energy diversity of the alkene are few and far between. When transforming cyclohexene and other unsaturated compounds, nature steers toward the product states with less energy diversity.

The most significant jump in state-space follows from ozonolysis. This reflects how the ring structure and unsaturated function are both mutated – the product of cyclohexene indeed presents a fused-ring, peroxide structure. In turn, $\hat{L}_{\sigma E,MI}$ allied with the transformation is about twice the library average. At such a remote location in the state plane, the ozonide converts readily not back to cyclohexene, but rather the 1,6-dialdehyde of *n*-hexane

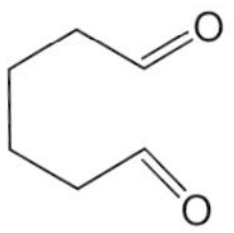

The state point for this molecule has been marked by the open triangle in Figure 14.8. The lowest-altitude molecule is a saturated hydrocarbon constructed of three fused rings. Its nearness to the state space origin reflects the lack of capacity for selective activation and message mutation. Moreover, its distance from cyclohexene, $\hat{L}_{\sigma E,MI} \approx 3.15$, reflects the significant functional differences. Perhaps surprisingly, the shortest leap incurs when cyclohexene is combined with bromine; the electronic states of 1,2-dibromocyclohexane and cyclohexene lie closest in terms of energy diversity and message correlations. In quantitative terms, one has:

$$\hat{L}_{\sigma E,MI} \approx 1.01$$

Figure 14.8 echoes a lesson of Figure 14.6, namely that the synthetically advantageous compounds are allied with high-altitude state points. These are the functionalized molecules which are dense with message correlations. It is the upper-left-quadrant molecules of Figure 14.8 that are gateway to the most numerous reaction pathways. For products of cyclohexene chemistry, the alcohol, diol, and methoxy compounds accommodate almost all the reactions developed for carbohydrates. Further, high-altitude compounds can be transformed so as to multiply the alkene functionality. The state point for 1,3-cyclohexadiene

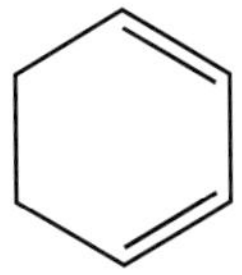

is located by the open star in Figure 14.8.

We next turn attention from a single reactant and source of information to a single *type* of transformation. The open squares of Figure 14.10 mark the state points for the reactants of Figure 14.9; the filled squares locate the product states formed via cyclization. The arrows indicate the state pairings while the indices match those of Figure 14.9. The distance $\hat{L}_{\sigma E,MI}$ separating each point-pair accompanies the label of the starting material. Briefly, Reaction 1 converts a tetraenic acetal to a tetracyclic compound [25]. Reaction 2 describes the thermally driven cyclization of glutaric acid to form an anhydride [26]. Reaction 3 portrays the combined cyclization, dehydration, and decarboxylation of pimelic acid [26]. Reaction 4 is an acid-catalyzed dehydration that yields a fused-ring unsaturated ketone [27]. Reaction 5 describes the dehydration of maleic acid to form the anhydride [26].

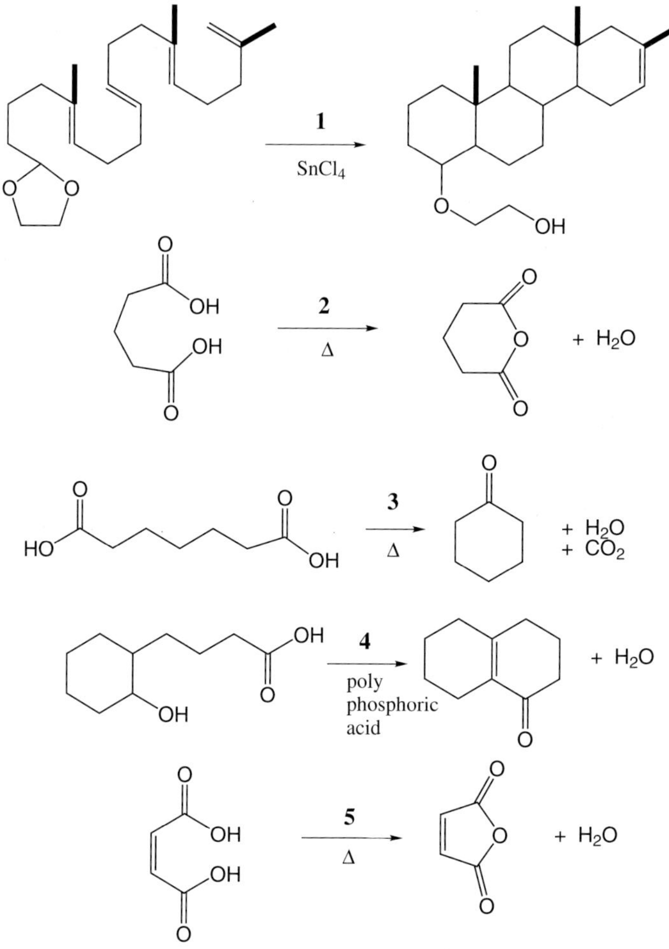


**Figure 14.9** Cyclization chemistry. A library of diverse cyclizations is illustrated via formula diagrams.

Reactions of the same genre share mechanisms and structure motifs [28]. In Figure 14.9, all the transformations center on ring construction; two are acid-initiated while three are heat-driven. Brownian computations contribute that the reactions of a genre – cyclizations, ring openings, cycloadditions, and so on – demonstrate characteristic jump-lengths and altitude adjustments in state space. This holds true, even for molecules of markedly disparate size, electronic structure and function. In the case of cyclizations – and, by extension, ring openings – typical $\hat{L}_{\sigma E,MI}$ is *c.* 2.4. Mechanism-wise, the rings always obtain at the expense of message correlations hosted by the reactant. Thus, all of the Figure 14.10 jumps are downward and terminate with lower-altitude states. In other words, message correlations are a necessary expense when nature fabricates rings at the Angstrom scale. Note that the

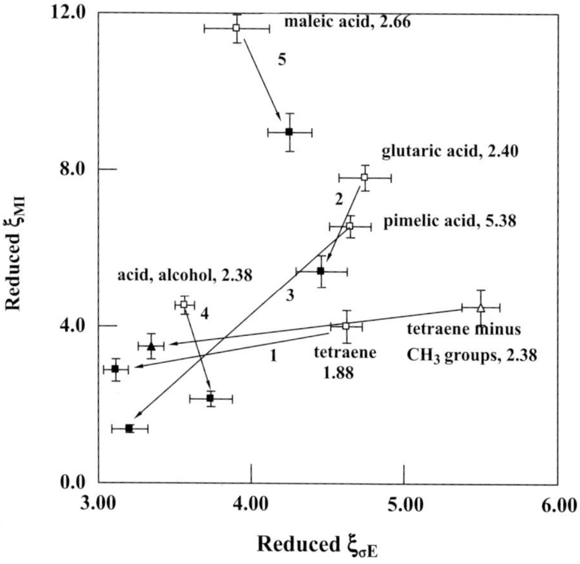


**Figure 14.10** State points allied with cyclization chemistry. The open and filled symbols mark the points for the reactants and products of Figure 14.9. The starting points are labeled in abbreviated terms and the distance $\hat{L}_{\sigma E,MI}$ from the terminal state. The integer labels correspond to those of Figure 14.9. The methyl groups of the tetraenic acetal of Reaction 1 are highlighted in boldface. The state points for the reactant and product minus the methyl groups are included in the plot.

altitude loss depends critically on the material jettisoned. There is no loss of water or other fragments in Reaction 1; here there is only a slight drop in the state plane. By contrast, Reaction 3 terminating in cyclohexanone demonstrates the greatest jump-length and vertical drop–$\hat{L}_{\sigma E,MI}$ is about twice the library average. This is because both dehydration and decarboxylation are incurred by the starting material. When this reaction is carried out in a lab, accommodation must be made for the absorption of water by a drying agent and the dissipation of carbon dioxide. The correlation losses run at about average if water molecules are the sole discard.

Reaction 1 is the most compelling in Figure 14.9. Four rings materialize with biogenic-type stereospecificity [25]. Yet such an electronic transformation is only slightly unusual from a computational standpoint. One finds $\hat{L}_{\sigma E,MI} \approx 1.88$; this is 30% less than the mean for a library of cyclizations. By comparison, the acid → anhydride transitions are much more severe. Importantly, the Brownian analysis sheds light on variations which can be contemplated for all cyclizations, including Reaction 1. One example is illustrated that involves the tetraene methyl groups; these have been indicated in bold in Figure 14.9. Classical principles hold these groups to enhance the stability of carbonium ions so formed during the covalent bond redistribution [29]. But there is more to learn by inspecting the energy diversity and correlations. The diamond symbols of Figure 14.10 mark the states of Reaction 1 compounds absent the methyl groups. One finds $\hat{L}_{\sigma E,MI} \approx 2.38$, a 30% increase in the jump-length. One learns that the role of the methyl groups includes mitigating the jump requirements across state space; the methyl groups render the alterations of the framework during the ring closures much less drastic. Note as well that the altitude loss is greater for the biogenic tetraene undergoing cyclization, not its methyl-absent analog. The correct orientation of methyl groups in the tetracycle product is more consuming of message correlations carried by the reactant.

Thus far, the transformations have been limited to a single reactant and product. The jumps in state space have accordingly been unidirectional. The product states posed either a gain or loss of correlation altitude, and an increase or decrease of the energy diversity. It is then interesting to view cases where the transit to a product state involves more than one direction. Figure 14.11 illustrates three transformations which terminate in the identical dicarboxylic methyl ester. Reactions 1, 2, and 3, respectively, describe an acylation using an acid chloride, a base-catalyzed alkyl halide

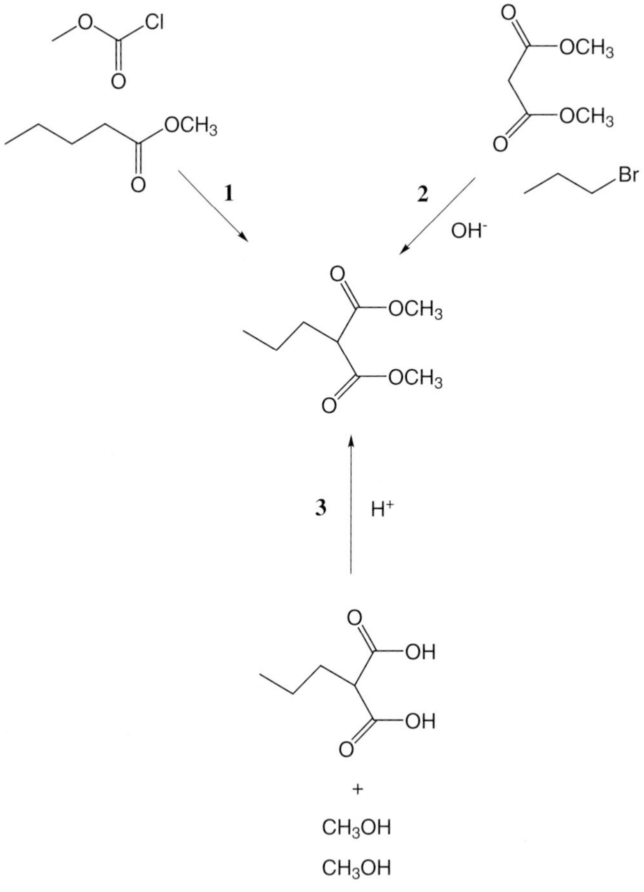


**Figure 14.11** Three reactions that share the same product. The starting materials are shown for three reactions that terminate in the identical substituted dicarboxylic ester.

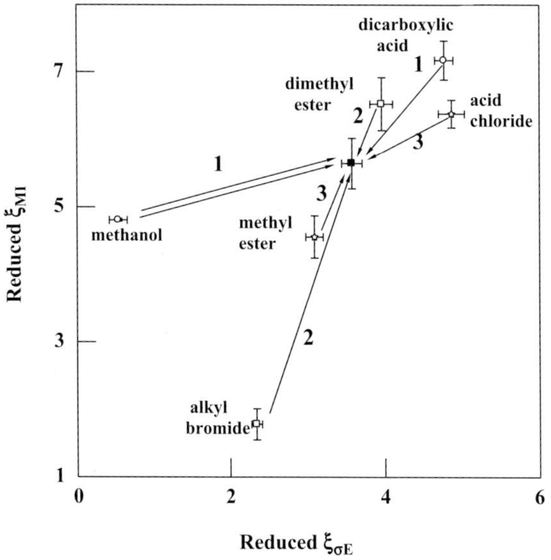


**Figure 14.12** State points for figure 14.11 reactants and products. The open and filled symbols are allied with reactants and products, respectively. The integer labels correspond to those of Figure 14.11.

addition, and an acid-catalyzed double esterification. Figure 14.12 presents the corresponding $\hat{\xi}_{MI}$ and $\hat{\xi}_{\sigma E}$ plot. The point for the product molecule has been marked by a filled square. Open symbols mark the reactant points while arrows indicate the jump transitions.

It is straightforward to piece together unidirectional reactions that begin, say, with acetic acid and terminate with the Figure 14.11 product. Figure 14.12 emphasizes their impracticality, however. Unidirectional reactions yield product state points away from, or toward the target compound – drifting further away is invariably as likely as moving closer. In the examples of Figure 14.11, the reactant states *bracket* the product. More to the point, each transformation features a reactant that gains altitude in the $\hat{\xi}_{MI}$ and $\hat{\xi}_{\sigma E}$ plane while the companion falls a little. In the strategic designs, the electronic correlations expended by one reactant are recovered in part by the companion. Reaction 2 featuring an alkyl bromide ($C_3H_7Br$) and dicarboxylic ester ($C_5H_8O$) is particularly instructive. Note how the change in correlation altitude with energy diversity of the ester is virtually canceled by the changes for the bromide. In quantitative terms, one has

$$\frac{\Delta\hat{\xi}_{MI}^{C_3H_7Br}}{\Delta\hat{\xi}_{\sigma E}^{C_3H_7Br}} + \frac{\Delta\hat{\xi}_{MI}^{C_5H_8O_4}}{\Delta\hat{\xi}_{\sigma E}^{C_5H_8O_4}} \approx 0 \tag{14.10}$$

As such, the transition arrows have slopes of approximately equal magnitude but opposite signs. The reactants and products lie along a single line in the state space.

Upon activation, the reactant states *pull* toward one another and land *cooperatively* at a point in-between.

The contrasts between unidirectional and bracketed reactions are most remarkable in natural product syntheses. In landmark chemistry conducted by Willstätter, cycloheptanone was converted to tropinone by a series of unidirectional reactions [30–32].

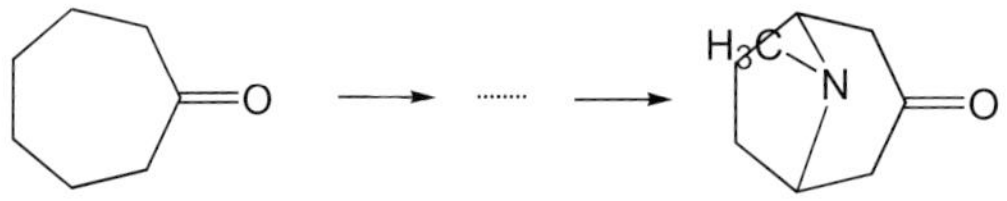

However brilliant, the design was ponderous as it required no less than 15 transformations executed in series. The chemistry is detailed not only in the original literature, but also in Fleming's *Selected Organic Syntheses* [33]. In Willstätter's design, the critical intermediate was a ring having strategically located bromine and amino substituents

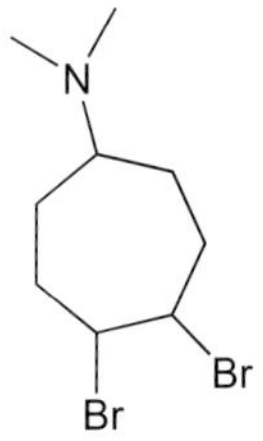

This enabled attainment of a 5,7-fused-ring structure which was isomorphous to the target. The above compound was late-stage, appearing only in the last third of the reaction series.

State space plots chart the chemistry in computational terms. The open squares and connecting lines of Figure 14.13 trace the pathway followed by Willstätter. Note how the synthesis was initiated using a low-altitude reagent in cycloheptanone. The inaugural step was necessarily to boost the correlation density by roughly a factor of five. This yielded a molecule with $\hat{\xi}_{MI}$ that exceeded the target. Only three transitions later, a molecule with a state point in the same neighborhood as tropinone was prepared, with $\hat{L}_{\sigma E,MI} \approx 1.14$. Figure 14.13 shows the remaining transitions to zig-zag through the neighborhood, eventually landing at the state point for tropinone. There are two lessons here. First is that Willstätter designed and executed what was, at root, a forced random walk of the molecules across the state space. Other series designs leading to tropinone offer zig-zag signatures as well. Secondly, from a computational point of view, the key intermediates were grasped early in the experiments. All of them located state points of high-correlation altitude close to the target, and offered multiple transformative options.

Willstätter's path to tropinone was not the only one. There was a beautiful alternative designed and materialized by Robinson in 1917 [34]. This entailed the Mannich-type fusion of three compounds

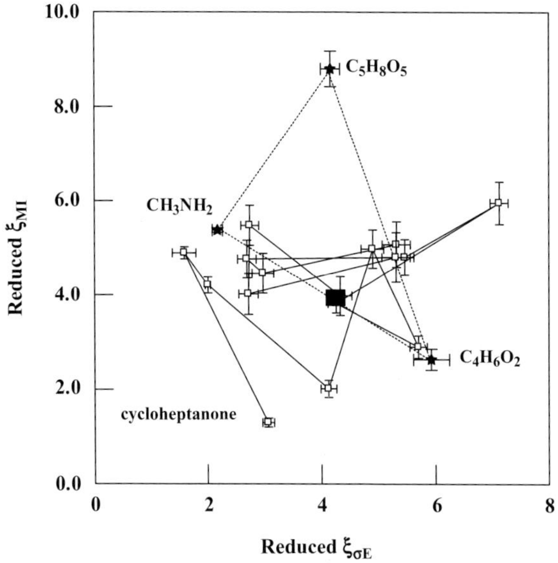


**Figure 14.13** State points and the synthesis of tropinone. The filled square marks the point for tropinone. The open squares and connecting lines mark the pathway followed by Willstätter. The stars mark the state points for reactants elected by Robinson.

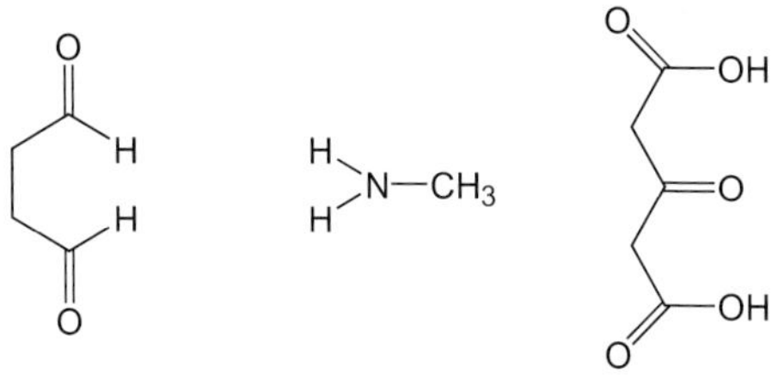


The state points for these molecules are marked by the stars in Figure 14.13 and connected by dotted lines. It is striking that the compounds bracket not only tropinone, but also the neighborhood sought, attained, and traversed by Willstätter. Comparisons of the $\hat{L}_{\sigma E,MI}$ sums for the two approaches are telling: 32.5 (Willstätter) and 9.47 (Robinson). The state space jumps in the reaction series were markedly more numerous and circuitous.

Equally illuminating is the progression of the molecular message correlations. The open squares of Figure 14.14 mark $\hat{\xi}_{MI}$ for Willstätter's pathway. The $\hat{\xi}_{MI}$ evolve more or less exponentially toward that of tropinone. This behavior is emblematic for organic reactions linked in series. In the typical case, there is a correlation gap between the starting material and target compound. Closing the gap by any single transformation occurs with a characteristic, low probability $P$; failing to narrow the gap happens with probability $1 - P$. Over the course of $N$ steps in series, failing to close the gap scales as $(1-P)^N \approx \exp(-Np)$. Correlations-wise, progress toward the target compound scales nominally as $1-\exp(-Np)$, that is, just as the exponential

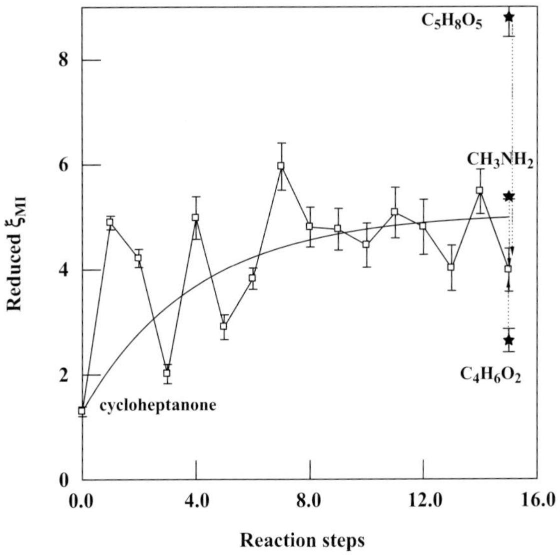


**Figure 14.14** Correlation intensity and tropinone synthesis. The open squares identify $\hat{\xi}_{MI}$ logged over the Willstätter pathway. The stars identify $\hat{\xi}_{MI}$ for reactants chosen by Robinson.

curve drawn in Figure 14.14. Bracket designs are superior because they are encumbered by less statistical baggage. The stars in Figure 14.14 mark $\hat{\xi}_{MI}$ for the three molecules engaged by Robinson. Note again how the $\hat{\xi}_{MI}$ surround the target above and below. This enables the correlations lost by one starting material to be recovered by another, and keeps the transitions to a minimum. As in computation and programming, it is the best-practice reaction designs that make economical use of message correlations.

The cases thus far have had three characteristics in common: (1) the reactants and products have been structurally well-defined, with nonoverlapping state points, (2) the transformations have been limited to one or a few functional groups, and (3) the product states have obtained by the breaking and formation of covalent bonds. These characteristics permeate chemistry research and development. At the same time, certain venues do not afford such luxuries. In pharmaceutical applications, a molecule such as an enzyme inhibitor is well characterized at the lab bench – its structure and synthesis offer a unique locus of state points. The precise nature of *in vivo* binding, however, is frequently unknown. This is because inhibitor docking at the active site of an enzyme rarely involves making and breaking covalent bonds. Rather, the intermolecular interactions predicate hydrogen and van der Waals bonds at multiple sites [35]. Insofar as covalent bonds are concerned, the number and identity of an inhibitor's degradation products formed by metabolism are generally unknown or ill-characterized.

Clearly, binding affinity alone does not determine the suitability of an inhibitor *in vivo*. It is conceivable that, given the chance, thousands of molecules would be able to deactivate a specific enzyme via binding. Moreover, a molecule that attaches to and inhibits one enzyme is likely to impact others *in vivo*. These are the reasons that a spectrum of properties – bioavailability, retention time, metabolites, resistance, toxicity, and side effects – must be weighed in pharmaceutical discovery and development [35].

The significance of electric charge inhomogeneities and message correlations cannot be overemphasized. Neither exists without the other, nor operates independently in Brownian environments. The inhomogeneities underpin a molecule's energy diversity and potential for site-selective activation. Just as crucial, the correlations regulate the entropy increases which accompany every spontaneous process. These principles apply as much to enzyme inhibitors *in vivo* as they do to the pedestrian molecules of this chapter.

The previous cases illustrated the synthetic utility of high $\hat{\xi}_{MI}$ compounds. Notably, molecules with therapeutic potential locate state points of ultrahigh $\hat{\xi}_{MI}$. For example, the following demonstrate high affinity toward aspartyl proteases necessary to retrovirus replication

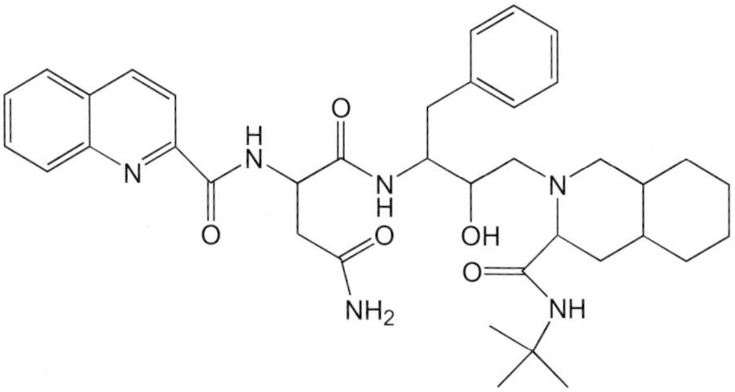


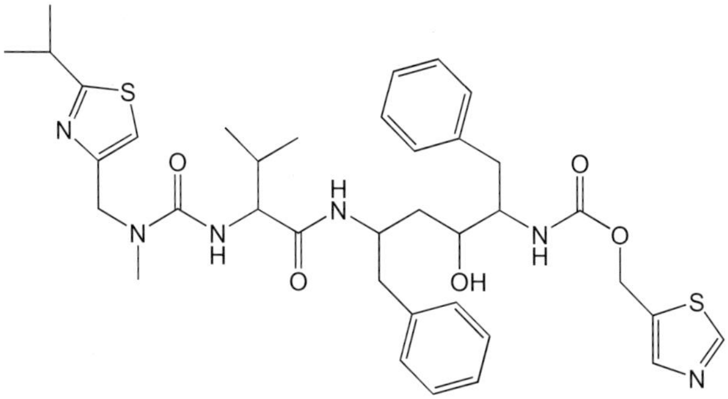


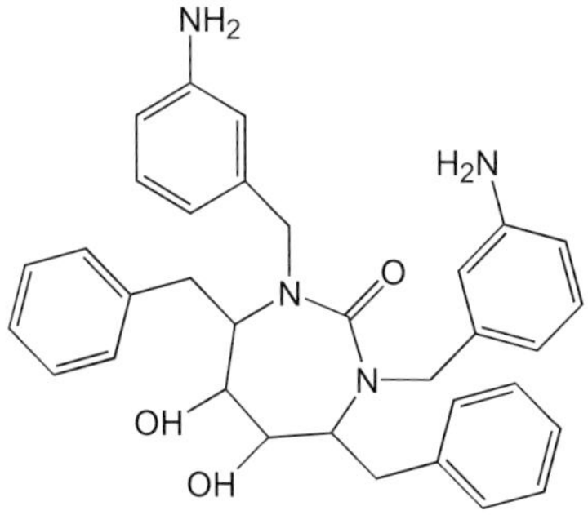

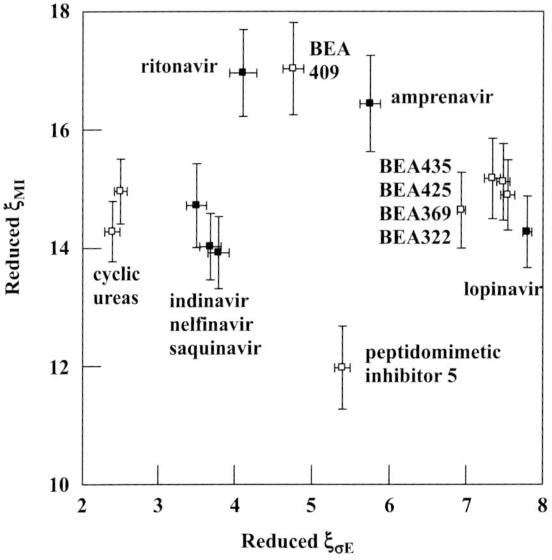


**Figure 14.15** State points and protease inhibitors. The filled squares mark state points for inhibitors approved for antiviral therapy. The open squares apply to compounds which are in the experimental stage, or have proven unsatisfactory for reasons such as low bioavailability.

Striking are the disparities of the three molecules regarding size, carbon framework, and functional groups. It is challenging, to say the least, to propose alternate systems which are just as diverse, and would miraculously target the same enzymes. The first two are well known, namely saquinavir and ritonavir [35]. The third is a cyclic urea with high specificity toward aspartyl proteases [36]. Saquinavir and ritonavir are long-approved antivirals, administered in combination with other compounds. The cyclic urea, along with the vast majority of inhibitors that have been synthesized and investigated, is not approved for therapy.

Figure 14.15 illustrates a state space plot for representative inhibitors of the same target enzymes. The points for therapeutically approved compounds are marked using filled squares. The open squares apply to high-affinity molecules that remain in the trial stage, or have proven unsuitable for reasons such as low bioavailability [37]. Notably, the state points are situated at high altitude – compare the vertical scale with previous figures. Note that the points for 10 inhibitors lie in the same correlation corridor: $14 \leq \hat{\xi}_{MI} \leq 16$. The points for three of the approved compounds prove clustered: indinavir, nelfinavir, and saquinavir. This is in spite of the marked differences in carbon scaffolding, heteroatom composition, and functional groups. Of the approved compounds, ritonavir lies highest in correlation altitude: $\hat{\xi}_{MI} \approx 17$. Its state point shares the diversity span occupied by indinavir, nelfinavir, and saquinavir: $3.5 \leq \hat{\xi}_{\sigma E} \leq 4.0$. The role of ritonavir includes inhibiting cytochromes which catalyze the degradation of protease inhibitors. In effect, ritonavir enhances the bioavailability of molecules like nelfinavir and significantly lowers their dose requirements.

Enzymes accelerate organic reactions with unparalleled efficacy and specificity. Their capabilities derive from the structure correlations encoded in the amino acid sequences and folded structures. Figure 14.15 shows that comparatively small organic molecules with extraordinarily high $\hat{\xi}_{MI}$ are able to deactivate biological function selectively. Thus, in sizing up candidates for antiviral therapy, one with $\hat{\xi}_{MI} < 12$ and/or $\hat{\xi}_{\sigma E} < 3$ would appear unlikely to meet the spectrum of requirements: bioavailability, low toxicity, tolerable side effects, and so on. Such an evaluation can be rendered on the basis of the formula diagram alone, the electronic blueprint so to speak. An antiviral candidate – or any other organic compound – need not be synthesized in order for its state point to be placed by Brownian computation. Note also that the $\hat{\xi}_{\sigma E}$ and $\hat{\xi}_{MI}$ coordinates for an inhibitor in free and enzyme-bound environments cannot lie too far from one another. This is on account of the absence of covalent bond breaking and making at protein active sites. Enzymes and inhibitors combine to make unique product complexes. This particular brand of chemistry, however, does not relocate the $\hat{\xi}_{\sigma E}$ and $\hat{\xi}_{MI}$ state point significantly, as with the other examples illustrated in this chapter.

Figure 14.15 shows that a variety of sizes and chemical functionalities yield state points in a given neighborhood of state space. Moreover, the potent therapy agents have state points that tend to cluster. Indinavir, nelfinavir, and saquinavir are very different structurally. At the same time, they pose virtually equal energy diversity and electronic correlations. This means that the activation and entropy-regulating facility of the three molecules are highly similar. Ritonovir locates the highest altitude state point. Its capacity for controlling entropy increases exceeds that of the other approved compounds in Figure 14.15. This may account for its inhibitory capacity that extends beyond a single enzyme.

## 14.4 Closing

Organic molecules are cataloged by what they are and what they can do. Functional groups and reaction classes provide the long-established organizational framework. Along the same lines, reactants and products are compared and contrasted by their structural overlap. Formula diagrams enable all these endeavors and it is not easy to imagine chemistry without them.

Brownian processing takes a different tack. Formula diagrams remain at the center stage. They are used, however, to construct probability functions, the descriptors of which connect with electronic communication in thermal environments. By traditional methods, it is difficult to categorize compounds that host multiple functional groups—see ritonavir and saquinavir for examples. Brownian processing is not hampered by such complexity. Every molecule predicates an electronic record tape, probability function, and state points in several coordinate systems. Reactants and products are related at once by the distance between points, altitude differences, and so forth. Note that this generality is not singular. Equilibrium systems, regardless of complexity, offer thermodynamic state points; likewise for computer messages. What

Brownian processing casts is a bridge between computation and thermodynamics, one accommodating of individual molecules.

State space plots do not capture the architectural beauty of molecules. But they do illuminate the transformations in information and diversity terms. They especially distinguish normative versus atypical states and chemistry for a single compound (e.g., cyclohexene), reaction genre (e.g., cyclizations), and synthetic design (e.g., series-versus-bracketed). Brownian processing is applicable when the reaction products are characterized incompletely, such as with enzyme inhibitors *in vivo*. A single chapter cannot portray the hundreds of reactions that can be posed for each of 20 million compounds registered in contemporary databases. In our studies, however, reactant/product $\hat{L}_{\sigma E,MI}$ distributions have proven highly normal with an average and standard deviation of 2.6 and 1.2, respectively. These statistical estimators provide the computational jump-lengths that would be anticipated for a typical reactant $\rightarrow$ product transformation.

## Acknowledgments

The author is grateful for the help of research students over the years, especially Matthew Schulmerich, Christopher Malarkey, Miriam Kim, Bea Chelsea, and Jessica Greminger. The author is equally appreciative to Donald May for discussion, comments, and criticism.

## References

1 Pauling, L. and Wilson, E.B. (1935) *Introduction to Quantum Mechanics with Applications to Chemistry*, McGraw-Hill, New York, Chapter 12.

2 Tribus, M. and McIrvine, E.C. (1971) Energy and information. *Scientific American*, **225**, 179.

3 Bonchev, D. and Trinajstic, N. (1978) Information theory, distance matrix, and molecular branching. *Journal of Chemical Physics*, **67**, 4517.

4 Basak, S.C., Grunwald, G.D., and Niemi, G.J. (1997) Use of graph-theoretic and geometrical molecular descriptors in structure-activity relationships, in *From Chemical Topology to Three-Dimensional Geometry* (ed A.T. Balaban), Plenum Press, New York, Chapter 4.

5 Dehmer, M., Varmuza, K., Borgert, S., and Emmert-Streib, F. (2009) On entropy-based molecular descriptors: Statistical analysis and synthetic chemical structures. *Journal of Chemical Information and Modeling*, **49**, 1655–1663.

6 Bonchev, D. and Rouvray, D.H. (1991) *Chemical Graph Theory. Introduction and Fundamentals*, Abacus Press, New York.

7 Trinajstić, N. (1992) *Chemical Graph Theory*, CRC Press, Boca Raton, FL.

8 Bertz, S.H. and Herndon, W.C. (1986) The similarity of graphs and molecules, in *Artificial Intelligence Applications in Chemistry* (eds. T.H. Pierce and B.A. Hohne), American Chemical Society, Chapter 15, pp. 169–175.

9 Bertz, S.H. (1981) The first general index of molecular complexity. *Journal of the American Chemical Society*, **103**, 3599.

10 Corey, E.J. (1967) General methods for the construction of complex molecules. *Pure and Applied Chemistry*, **14**, 19–37; See also Corey, E.J. and Wipke, W.T. (1969) Computer-assisted design of complex organic syntheses. *Science*, **166**, 178.

11 Ash, R.B. (ed.) (1990) Issues regarding error detection and correction during digital information processing are addressed, in *Information Theory*, Dover Publications, New York, Chapter 4.

12 Fermi, E. (1936) *Thermodynamics*, Dover, New York.

13 Pauling, L. (2001) *Selected Scientific Papers*, vol. I, World Scientific, New Jersey, pp. 10–18, Nobel Address.

14 Wannier, G.H. (1966) *Statistical Physics*, Dover, New York, Chapter 22.

15 Graham, D.J., Malarkey, C., and Schulmerich, M.V. (2004) Information content in organic molecules: Quantification and statistical structure via brownian processing. *Journal of Chemical Information and Computer Sciences*, **44**, 1601.

16 Graham, D.J. and Schulmerich, M.V. (2004) Information content in organic molecules: Reaction pathway analysis via Brownian processing. *Journal of Chemical Information and Computer Sciences*, **44**, 1612.

17 Graham, D.J. (2005) Information content and organic molecules: aggregation states and solvent effects. *Journal of Chemical Information and Modeling*, **45**, 1223.

18 Graham, D.J. (2007) Information content in organic molecules: brownian processing at low levels. *Journal of Chemical Information and Modeling*, **47**, 376.

19 Graham, D.J. and Greminger, J.L. (2009) On the information expressed in enzyme primary structure: Lessons from ribonuclease A. *Molecular Diversity*. doi: 10.1007/s11030-009-9211-3

20 Bennett, C.H. (1982) Thermodynamics of computation - a review. *International Journal of Theoretical Physics*, **21**, 905.

21 White, A., Handler, P., and Smith, E.I. (2011) *Principles of Biochemistry*, 5th edn, Mc-Graw Hill, New York, Chapter 28.

22 Shannon, C.E. (1948) A mathematical theory of communication. *Bell System Technical Journal*, **27**, 379–423. 623–656.

23 Kullback, S. (1997) *Information Theory and Statistics*, Dover, New York.

24 Pauling, L. (1970) *General Chemistry*, Dover, New York, Appendix VIII.

25 Johnson, W.S., Wiedhaup, K., Brady, S.F., and Olson, G.L. (1968) The nonenzymic, biogenic-like cyclization of a tetraenic acetal. *Journal of the American Chemical Society*, **90**, 5277–5279.

26 Roberts, J.D. and Caserio, M.C. (1965) *Basic Principles of Organic Chemistry*, W.A. Benjamin, New York, Chapter 16.

27 Kulkarni, S.B. and Dev, S. (1968) Organic reactions with polyphosphoric acid-VIII. *Tetrahedron*, **24**, 553–560.

28 House, H.O. (1972) *Modern Synthetic Reactions*, 2nd edn, W.A. Benjamin, New York.

29 March, J. (1977) *Advanced Organic Chemistry*, 2nd edn, McGraw-Hill, New York.

30 Willstatter, R. (1896) Ueber ψ-tropin. *Berichte der Deutschen Chemischen Gesellshaft*, **29**, 936–947.

31 Willstatter, R. (1901) Syntheses des tropidins. *Berichte der Deutschen Chemischen Gesellshaft*, **34**, 129–144.

32 Willstatter, R. (1901) Umwandlung von tropidin. *Berichte der Deutschen Chemischen Gesellshaft*, **34**, 3163–3165.

33 Fleming, I. (1973) *Selected Organic Syntheses*, John Wiley & Sons, Inc, New York.

34 Robinson, R. (1917) A synthesis of tropinone. *Journal of the Chemical Society*, 762–768.

35 Copeland, R.A. (2005) *Evaluation of Enzyme Inhibitors in Drug Discovery*, John Wiley & Sons, Inc, Hoboken, NJ.

36 Hodge, C.N., Aldrich, P.E., Bacheler, L.T., Chang, C.H., Eyermann, C.J., Garber, S., Grubb, M., Jackson, D.A., Jadhav, P.K., Korant, B., Lam, P.Y., Maurin, M.B., Meek, J.L., Otto, M.J., Rayner, M.M., Reid, C., Sharpe, T.R., Shum, L., Winslow, D.L., and Erickson-Viitanen, S. (1996) Improved cyclic urea inhibitors of the HIV-1 protease: Synthesis, potency, resistance profile, human pharmacokinetics and X-ray crystal structure of DMP 450. *Chemical Biology*, **3** 301–314.

37 Andersson, H.O., Fridborg, K., Lowgren, S., Alterman, M., Muhlman, A., Bjorsne, M., Garg, N., Kvarnstrom, I., Schaal, W., Classon, B., Danielsson, U.H., Ahlsen, G., Nillroth, U., Vrang, L., Oberg, B., Samuelsson, B., Hallberg, A., and Unge, T. (2003) Optimization of P1-P3 groups in symmetric and asymmetric HIV-1 protease inhibitors. *European Journal of Biochemistry/FEBS*, **270** 1746–1758.

# 15
# Generation of Chemical Transformations: Reaction Pathways Prediction and Synthesis Design

*Grażyna Nowak and Grzegorz Fic*

## 15.1
## Introduction

Computer-aided generation of chemical reactions consists of transforming one set of molecular species into another one according to a predefined set of reaction rules or reaction models formally expressed as graph transformations or graph rewritings. They describe all possible forms of structural changes, involving on general abstract level the bonds broken/formed or valence electrons redistributed in substrate–product molecular graphs. Both forward and backward directions of transform operations are possible. The first natural fashion corresponds to the reaction simulation, making it possible to generate all reaction products, given as input the graphs of starting materials. The second fashion, commonly known as the retrosynthetic transform strategy, allows us to obtain all possible reactants or precursor molecules starting from the desired target product. The iterative repetition of this process for successively generated species as molecular graphs enables the exhaustive search through the space of all formally possible transformations, including those candidates that do not have representants among the known chemical reactions. It creates fundamentals to build models of reaction networks or synthetic trees simulating the behavior of reacting molecules or interacting biological species, predicting the outcome and underlying mechanisms of organic reactions, tracing the formation of the specific products or undesirable by-products, as well as discovering new, unprecedented reactions and reaction pathways. There are several application areas including chemistry, engineering, biochemistry, and recently molecular biology in which these capabilities are of paramount importance.

At the very beginning, the idea of the generation and prediction of unknown reactions has been applied to computer-assisted organic synthesis (CAOS), an artificial chemistry discipline that dates back to the 1970s. It focuses on the generation and search for new routes of synthesizing organic compounds, simulation and predicting the reaction courses, and recently designing synthetically accessible virtual libraries. During the past decades, many significant advances have been made in this area, particularly addressed on the development of formal tools for the description,

*Statistical Modelling of Molecular Descriptors in QSAR/QSPR*. First Edition.
Edited by M. Dehmer, K. Varmuza, and D. Bonchev

classification and generation of chemical reactions, as well as inference strategies to guide the search during the reaction generation along potentially promising pathways. This task consists of the evaluation and selection of relevant transformations and species within a huge space of all combinatorially generated candidates. It is still one of the most difficult problems in research on reliable artificial chemistry models, whose resolving requires formalizing the extend of empirical chemical knowledge, as well as chemists predictive reasoning, creativity, and intuition. Some pioneering work on the logical formalisms of the synthesis design has been the Corey's retrosynthetic analysis for disconnections, a strategy recently advanced by using the tools of graph theory to be expressed mathematically [1, 2]. More important achievement was the Ugi and Dugundji mathematical concepts of reactions and a chemical distance that extended the search and prediction beyond the boundaries of the known empirical knowledge [3–5]. Original ideas for formal mathematical description of chemical reactions, the synthetic complexity, and the convergence were also proposed by Hendrickson that allowed building effective synthetic pathways for complex molecules and discovering novel strategies [6–8]. On this ground, new innovative and more sophisticated formalisms of the synthesis design have been developed related to the concept of "ideal synthesis" [9, 10]. An important contribution was empirical models developed by Gasteiger group to make predictions for chemical reaction courses in the context of the quantitative structure–reactivity relations. They are commonly used to extend the chemical and biological applications such as modeling the reaction mechanisms, establishing quantitative structure–properties or structure–activity relationships (QSAR, QSPR), or supporting the drug discovery and design process [11–15].

In spite of many improvements involved in the subsequent versions of the CAOS systems, their practical utility is still not fully satisfied, and generally limited by either low practicality or low creativity of generated solutions. There are two new trends in developing CAOS systems addressing these limitations. One is integration with chemical reaction databases and data mining tools for automatic deriving empirical transformation rules (generic reactions) that imitate "real chemistry". Their application into the reaction generation process gives the possibility of the exploitation of all the accumulated knowledge on chemical reactions (over several million stored in Beilstein and CASREACT databases) to guide the search toward more plausible, consistent with literature data results. In synthesis design applications this means the generating accessible and feasible synthetic pathways, while in the reaction simulation approach – developing more accurate, in the context of the regio- and stereoselectivity, models for the prediction of reaction courses and final products. In synthetic biology area, the similar approach is implemented in programs assisting in the reconstruction of an organism-specific metabolic networks using prior knowledge about the known enzymatic reactions from biological databases, the goal of which is finding feasible biosynthetic routes, enzymes, and product candidates. One can expect in future that unifying in one application the rule-based chemical pathways generation and biochemical pathways reconstruction will allow us to achieve new powerful possibilities by incorporating suitable enzymatic transformations into synthetic strategies, and search the space of chemicals that are accessible both by chemical synthesis and by employing microorganisms.

Additional contribution toward implementation of the "real chemistry" is the integration within the CAOS systems both in the qualitative and quantitative aspects of the reaction simulation. Consequently, different computational chemistry tools have been developed to direct the flow of the reaction generation toward reliable and feasible reaction routes. These include empirical models, molecular mechanics, and quantum chemistry calculations of physicochemical descriptors for the prediction of molecule reactivity, as well as thermodynamic and kinetic properties for estimating relative rate constants, ratios of products, and the effects of catalysts, or conditions. Similar approach has been applied earlier in the modeling and semiquantitative prediction mechanism of complex reacting systems (such as in petroleum refining, combustion, fermentation, and other industrial processes), where the key issue is to define the possible mechanisms and directionality of reactions in order to select only kinetically important elementary steps. Recently, the rate-based methods of reaction network modeling are used to predict and study the mechanisms of enzyme-catalyzed reactions offering a basis for the design of metabolic engineering strategies. In CAOS systems, quantitative estimation of thermodynamic and kinetic properties provides physicochemical criterion that allows us to eliminate less important or unreliable transformations offering simultaneously the possibility of exploring broader regions of a chemical space comparing to empirical heuristics. Although this approach determines in many respects practicality of generated results, large computational requirements make it difficult or ineffective to solve the problem of increased combinatorial complexity as result of structural targets increased in size.

Emergence of library-based drug discovery techniques and chemical genetic strategies associated with demands for the synthesis of more and more complex compounds with diverse structures and unusual properties served to focus the *attention* on alternative optimizing tools that require short computation times to span a much broader range of combinatorially conceivable chemical transformations and synthetic possibilities. This (second) new trend in developing the CAOS is related to unbiased and more abstract, in conceptual sense, methods for evaluating and selecting chemical transformations and species, in order to include those solutions that have no precedents in known chemistry. These are different topological metrics of the molecular complexity, connectivity, similarity, and concepts of symmetry, mainly derived from the field of information and graph theories. As the mathematical tools, independent of the state-of-the-art of organic chemistry, they can aid chemists in discovering novel types of reactions, original ideas for a synthesis or missing biotransformations or biomolecules in metabolic networks. But their exceptional value lies in possibilities of discovering or building the synthetic pathways in terms of both novelty and at the same time the practicality and feasibility. By tracking the changes of these topological measures during the generation of transformations and evolution of synthetic pathways, one can find the lowest required number of synthetic steps for construction of structurally complex molecules from simple starting materials. Strategies of this kind have emerged as one of the most promising approaches toward "ideal synthesis" and green chemistry [9, 10]. Allowing to achieve a rapid increase in structural complexity per step, they can be used for designing target-oriented complexity-based compound collections as well as in searching for synthetically accessible ligand candidates. The new approach in this area is diversity-oriented synthesis (DOS), proposed as a general strategy for generating skeletally diverse collections of molecules for chemical genetic studies [16]. Another innovative concept that extends the logical formalisms of synthesis planning beyond the traditional framework of structural complexity or diversity has been proposed by Weber. His idea of function-oriented synthesis (FOS) is to avoid complex reactions and guide the search for synthetic transformations by inferencing about physico-chemical or biological properties of a synthesized target [17]. It creates new perspective to integrate the CAOS systems and QSAR/QSPR programs.

The range of applications of the method and concepts originally developed in CAOS continuously grows, which means that the CAOS field still remains an area of concern. This chapter summarizes the most important achievements and new methods in solving the problem of prediction of reaction pathways and synthesis design, including other important applications such as drug design, combinatorial chemistry, modeling reaction mechanisms, reconstruction of metabolic pathways, and xenobiotic biotransformation prediction. Section 15.2 is devoted to models of graph transformations being fundamental starting point for reaction simulation and prediction. Two categories of the rules and corresponding concepts for the reaction generation are introduced with their applications. Automated construction of the reaction rules by learning and data mining tools are also presented in this section. Section 15.3 deals with the constructing hypothetical reaction pathways, trees, and networks for the prediction of reaction pathways and discovering retrosynthetic strategies. The problems associated with combinatorial nature of the space searched during the generation of transformations are highlighted along with strategies for the optimization and guiding the search.

## 15.2
## The Graph Transformation Rules for Generation of Chemical Reactions

The generation of chemical reactions is based on the mathematical formalism of graphs and graph transformations, in which labeled graphs represent molecules and graph transformations represent chemical reactions. This formalism allows simulating chemical reactions by generating all possible transformations on a set of initial graphs given a set of transformation or reaction rules. They describe, at a more or less general level, the topological changes in substrate–product molecular graphs that occur during the course of a reaction, for example, the changes in bonds, reaction centers, functional groups, and substructures. Graph transformation rules can be forward or reverse applied to an input molecular graph. A forward application gives a set of final graphs representing all possible reaction products, and in computer processing, corresponds to the reaction simulation. A reverse application leads to simpler graphs representing the reaction intermediates or substrates and creates fundamentals to the synthesis design.

Different types of reaction rules have been proposed in order to be applied for reaction simulation. In principle, they can be classified into two major categories, including the formal, graph-theoretic rules and empirical reaction ones. The former category refers to explicitly defined graph-theoretic tools to transform molecular graphs (usually encoded in double-pushout procedure or corresponding rewriting mathematical operations on molecular graphs) [18–21]. They enable the exhaustive search of the chemical space resulting from all possible chemical transformations and species generated combinatorially, creating a basis to the formal-logical concept of the reaction generation. The later one concerns manually created or automatically derived transformation patterns from the chemical reactions collected in chemical databases. In this case, the search is performed within the boundaries of the individual knowledge base and covers transformations that have analogs among the known, experimentally verified chemical reactions. The corresponding model of reaction generation is categorized as knowledge-based or empirical.

### 15.2.1
### The Graph-Theoretic Reaction Rules and Formal-Logical Approach for Reaction Generation

The graph-theoretic reaction rules describe with greater or lesser degree of generality, the way, the conversion of substrate molecular graphs into product molecular graph, by breaking, forming or changing the type of bonds represented by edges. These bond manipulations correspond to valence electrons redistribution that occurs during a reaction. Following this formalism, the reaction rule may be expressed as a symbolic equation, $G_B \Rightarrow G_E$, where the left-hand side graph $G_B$ represents the bonds changed in substrate(s), while the right-hand side graph $G_E$ specifies the new bonds formed in product(s). Each rule of this form implicitly defines the reactions that can be generated by condition applicability to input chemical species. It can be subgraphs isomorphic to the reactants graph $G_B$ for the reaction simulation or to the product graph $G_E$ for the synthesis design, and a new graph to be replaced in the generated products.

Figure 15.1 shows different types of rules for the example chemical reaction, defined within this general formalism. Considering only bond order changes or valence electron shifting, without writing atoms in explicit forms, corresponds to the rule at the medium level of generality (Figure 15.1c), that is usually called the reaction type. After deletion of edges that remain unchanged during the reaction, one can obtain much more general shape of the reaction rule. In this particular case, it describes, in an abstract sense, a class of reactions in which three bonds between arbitrary atoms are broken and formed in substrate–product molecular graphs (Figure 15.1d). If the left-hand side pattern in this rule matches the input graph then edges and nodes that are not in the right-hand side of the rule are deleted, while edges and nodes that are newly created in the right-hand side are placed into the graph. For applying this rule to different input graphs, it is possible to generate a range of chemical reactions of different types, for example, [2 + 4] cycloadditions, [3,3]-sigmatropic rearrangements, Darzens condensations, and other (Figure 15.2). Superimposition of the reactant graphs and the corresponding product graph of this rule gives a new graph that describes only the topology of the bond redistribution

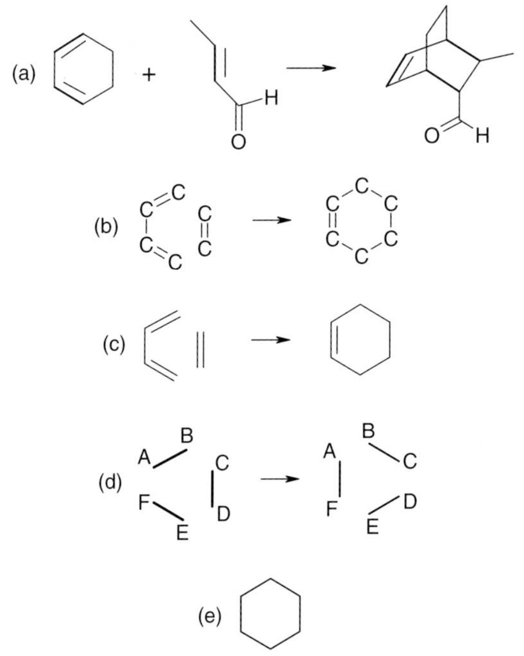


**Figure 15.1** The generalized graph transformation rules (A, B, C, . . ., F denote any atom symbol, double bonds are regarded as two single bonds).

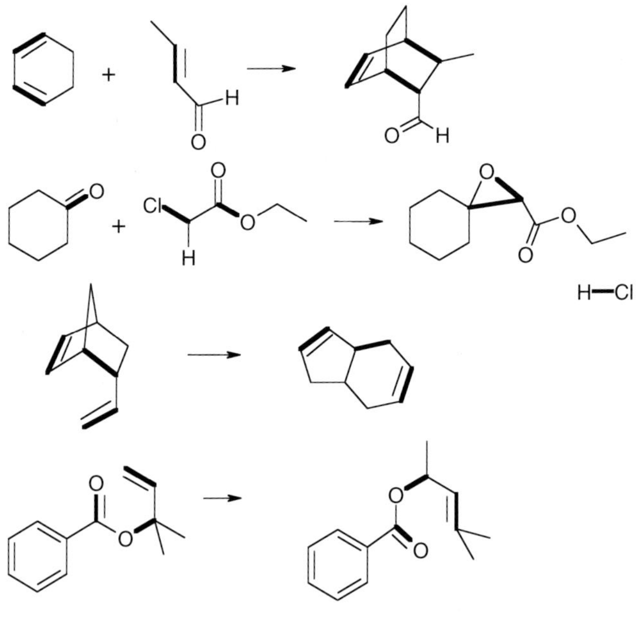


**Figure 15.2** Example reactions of different types that can be described with the transformation rule from Figure 15.1d.

(the relative position of bonds whose order changes during the reaction), called a topology identifying graph (Figure 15.1e). Two principles are imposed on the reaction rules, namely the principle of conservation of vertex labels and the principle of conservation of the total weight of edges that ensures that the transformations induced by the rule in analogy to chemical reactions are stoichiometrically balanced. Only if these conditions are satisfied, the graph transformation rules can model reaction mechanisms that the reactants can undergo.

Alternatively, additional structural information can be incorporated to the considered transformation types, for example, edge labels corresponding to the bond orders or length, node labels corresponding to the atom property attributes, such as chirality, aromaticity, hybridization state, and partial charge. The more detailed description of the transformation rule implies the lower number of isomorphic structures/substructures to which this rule is applicable, and in consequence, limits the search space. In the reaction simulation, it allows to discriminate between reactions involving the same reaction centers (in the sense bond breaking and forming).

#### 15.2.1.1 The Chemical Reaction Graph

Recently, much attention has been focused on a representation of reactions comprising only one graph containing information about both the starting as well as the product molecules. Following this line, Fujita has developed more compact description of chemical reactions as a reaction graph RG that is based on his concept of imaginary transition structure ITS [22, 23]. ITS is obtained by superimposition of the substrate and product molecular graphs. The method is an extension of the Vladutz idea to show the bond changes (as labeled edges) in a single structure from which the reactant (left-hand side of a transformation rule) and product graphs (right-hand side of a transformation rule) can be deduced [24]. Three kinds of bonds participating in a chemical reactions are distinguished in the reaction graph RG, namely the broken bonds that appear only in the starting materials (out-bonds marked with a double stroke), the formed bonds that appears only in products (in-bonds marked with a circle on the bond), and unchanged bonds that appear both in the starting materials and in products (par-bonds drawn as plain lines). This last type specifies which kinds of links are allowed to form new bonds during the generation of transformations. Given a reaction graph RG, the corresponding graph of substrates can be obtained by deleting in-bonds. On the other hand, an operation of deleting out-bonds affords the product graph (Figure 15.3).

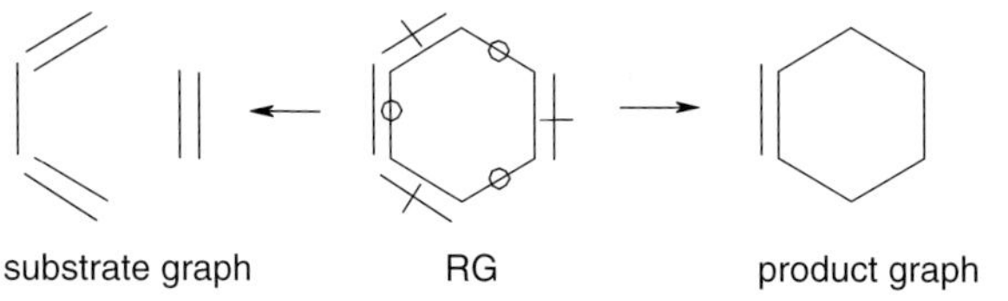


**Figure 15.3** Example reaction graph RG and the respective substrate and product graphs derived from RG.

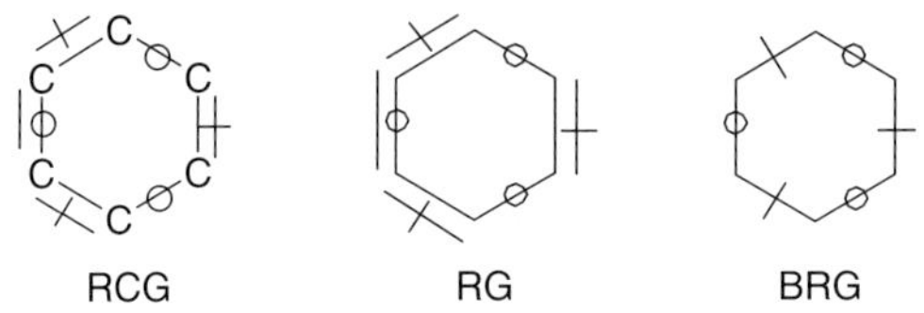


**Figure 15.4** Three categories of the reaction graphs according to Fujita's graph transformation formalism.

On the basis of Fujita's concept, two other reaction models can be derived from the reaction graph (RG). By substituting appropriate atoms for the vertices of a reaction graph RG one can obtain a reaction-center graph (RCG), whereas after deletion of par-bonds, the corresponding basic reaction graph (BRG) (or a simplified reaction graph) will be created. The last corresponds to the graph transformation rule from Figure 15.1 below are examples corresponding to these three types of reaction models (Figure 15.4).

This hierarchy from the basic reaction graph BRG to the more detailed reaction center graph RCG provides a flexible methodology for the generation and the prediction of reactions at different levels of abstraction. This formalism has been increasingly applied, especially in the field of molecular biology [25, 26]. More recently, the Fujita's imaginary transition structure (ITS) has been successively applied in program called the Toy Model for generating chemical mechanisms and studying properties of large-scale reaction networks [27, 28]. The similar concept has been adopted in MOLGEN-QSPR and MOLGEN-COMB, computer programs for applications in combinatorial chemistry, synthesis design, and mass spectroscopic structure elucidation [29].

#### 15.2.1.2 **Ugi and Dugundji Formal Theory for Reactions and Reaction Mechanisms**

The graph-theoretical reaction rules can be understood as an alternative formulation of the Ugi and Dugundji matrix model of reactions, introduced in 1970s, which has emerged as classical methodology for the representation and generation of chemical mechanisms [3–5]. It is based on the encoding of the bond order changes during a transformation with the use of the reaction matrix (R-matrix). Negative numbers in the reaction matrix correspond to the cleavage of bonds and positive numbers correspond to the formation of bonds. A direct representation of the reactants and products molecular graphs are the respective B- and E-matrices, in which nondiagonal entries correspond to the order of bonds, and diagonal entries represent the number of free electrons. The R-matrix is equivalent representation to the graph transformation rule of type defined in Figure 15.1d. The example reaction matrix encoding this rule is shown in Figure 15.5. According to the fundamental equation $B + R = E$, it is possible to deduce the corresponding molecular graphs $G_B$ and $G_E$ from B- and E-matrices of the educts and the products. When the R-matrix is added to the matrix representing the substrate ensemble, the resulting matrices specify products of all the mathematically possible transformations (within the predefined R-matrix or the

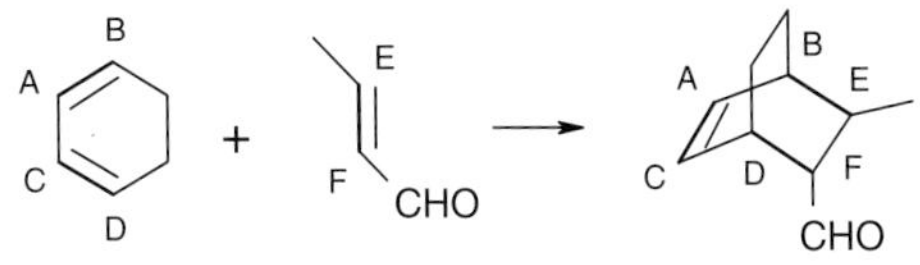


A-B + C-D + E-F ⇒ A-C + B-E + D-F

**B**

| | A | B | C | D | E | F |
|---|---|---|---|---|---|---|
| A | | 2 | | | | |
| B | 2 | | | | | |
| C | | | | 2 | | |
| D | | | 2 | | | |
| E | | | | | | 2 |
| F | | | | | 2 | |

+

**R**

| | A | B | C | D | E | F |
|---|---|---|---|---|---|---|
| A | | −1 | +1 | | | |
| B | −1 | | | | +1 | |
| C | +1 | | | −1 | | |
| D | | | −1 | | | +1 |
| E | | +1 | | | | −1 |
| F | | | | +1 | −1 | |

=

**E**

| | A | B | C | D | E | F |
|---|---|---|---|---|---|---|
| A | | 1 | 1 | | | |
| B | 1 | | | | 1 | |
| C | 1 | | | 1 | | |
| D | | 1 | | | | 1 |
| E | | 1 | | | | 1 |
| F | | | | 1 | 1 | |

B + R = E

**Figure 15.5** Representation of a graph transformation rule by R-matrix (the number of induced transformations results from the different permutations within the given scheme of bond changes and application of this scheme to all bonds with appropriate valence state in the input molecular graph).

scheme of the bond changes); analogical operation on the product matrix specifies all the reaction substrates.

The Ugi–Dugundji theory of R- and BE-matrices was designed for the formal description of reactions that involve only localized single, double, or triple bonds. The R-matrices can be extended for embedding more information on reaction centers, as atomic charges, valences, topological information, aromaticity, and others. Following this line, Stein has proposed an extension capable of dealing with delocalized electrons [30]. It can encompass ∏-electron systems, organometallic compounds, and electron donor–acceptor complexes, being useful for the investigation of reaction mechanisms, especially catalytic processes. Kvasnicka and Koca have developed a synthon model that is a graph-theoretical extension of the Ugi model. The "synthon" defines additional types of reaction centers along with other components like lone-pairs electrons and free radicals, so that the reactions in which they participate are still isomerization processes [31]. The idea of the graph-based representation of R-matrices is also implemented in the CSB system developed by Hippe group, which includes 54 reaction models with participation of free electrons, electron gap, and radicals (Table 15.1) [32]. Such reaction models can be used for the simulation of complex mechanisms of multicentered and multicomponent reactions by means of one execution stage. Reisen *et al.* have proposed the extension of R-matrices for the description of transformations of functional groups [33]. A similar concept has been used for the description and classification of the enzyme functions and generation of metabolic transformations [34].

A classical application area of the Ugi model was CAOS, including the first program IGOR developed by Ugi group [35]. Their second program RAIN was designed for reaction network generation and study of the reaction mechanisms [36]. Another examples in which fundamental elements of this model or its extension have been implemented are CICLOPS, SYNCHEM, EROS, TOSCA, CASINO, SCAN-MAT, and CSB. The idea of R-matrices combined with kinetic parameters estimation

**Table 15.1** Graph transformation schemes encoded in the CSB library of reaction generator.

| No. | Graph of transformation | Number of schemes |
|---|---|---|
| 1 | A–B + C–D → A–C + B–D | 2 |
| 2 | A–B + C–D + E–F → A–C + B–E + D–F | 8 |
| 3 | :A + B–C → A–B + :C | 2 |
| 4 | :A + B–C → B–A–C | 1 |
| 5 | A–B + C–D + E=F → A–E–C + B–F–D | 6 |
| 6 | A–B + C–D → A + B–C +:D | 8 |
| 7 | A–B + D–C–E → A–C–B + D–E | 1 |
| 8 | :A + B–C + D–E → B–D + A–E +:C | 8 |
| 9 | A–B + D–C–E → A–E + B–D +:C | 2 |
| 10 | A=B + D–C–E → D–A–E + B=C | 2 |
| 11 | A + B–C → A–B + C | 2 |
| 12 | B–A–C → B–C + :A | 1 |
| 13 | A–B → A + :B | 2 |
| 14 | :A + B → A–B | 1 |
| 15 | :A + B + C–D → A–C + B–D | 2 |
| 16 | A–B + C–D + E–F + G–H → A–D + B–H + C–E + F–G | 60 |
| 17 | A–B + C–D +:E–F → A–E–D + B–C: + F | 8 |
| 18 | A–B + C=D + F–E–G → E=D + A–C–F + B–G | 8 |
| 19 | :A +:B → A=B | 1 |
| 20 | A +:B–C → A–B +:C | 1 |
| 21 | A=B → :A +:B | 1 |
| 22 | :A + B=C → A=B +:C | 2 |
| 23 | :A +:B + C=D → A=C +:D–B | 4 |
| 24 | :A +:B–C → A=B +:C | 1 |
| 25 | :A + B=C + D–E → D–B–E +:C–A | 2 |
| 26 | A–B–C + D–E–F → B=E + A–F + C–D | 3 |
| 27 | A=B + C=D → A=C + B=D | 2 |
| 28 | :A + B–C–D → A–C: + B–D | 1 |
| 29 | A–B +:C–D → A–C–B +:D | 1 |
| 30 | A–B + C–D +:E–F → A–E–D + B–C +:F | 4 |
| 31 | A–B–C + D–E–F + G–H → A–F + B=E + C–H + D–G | 12 |
| 32 | :A + B → A +:B | 1 |
| 33 | A–B–C + D–E + F–G → A–E + G–B–F + C–D | 4 |
| 34 | A–B → A* + B* | 1 |
| 35 | A* + B* → A–B | 1 |
| 36 | A–B + C* → A–C + B* | 2 |
| 37 | A–B + C–D → A–C + B* + D* | 4 |
| 38 | A=B + C–D–E → B=D + C–E +:A | 2 |
| 39 | A–B + C=D → A–C–B +:D | 2 |
| 40 | A–B + C* + D* → A–C + B–D | 2 |
| 41 | A=B + D–C:–E → A=C=B + D–E | 1 |
| 42 | A–B + C–D + E–F + H=G=I → A–G:–B + C–H–D + E–I–F | 6 |
| 43 | A: + B–C + D–E + F=G → B–F–A–E + C–G–D | 16 |
| 44 | A–B + C=D + F–E–G +:H–I → A–C–H=E + I–G + B–D–F | 8 |
| 45 | A=B +:C–D + G–E–F + I–H–K → K–E–A–C=H + F–B–I + D–G | 16 |

**Table 15.1** (*Continued*)

| No. | Graph of transformation | Number of schemes |
|---|---|---|
| 46 | A: + B=C + E–D–F– + H–G–I → I–D–B–A–G–F + E–C–H | 16 |
| 47 | A: + B–C + D=E +:F–G + I–H–K → B–D–F=H + A–G + C–E–I + K: | 8 |
| 48 | A–B + C–D + E=F +:G–H +:I–K → A–E–G=I + B–F–K + C: + D–H | 32 |
| 49 | A–B + C–D + E–F + G–H: + I=K → A–E + B–F + C–I–H=K + D–G | 24 |
| 50 | A–B + C=D + F–E–G +:H–I → A–C–H=E + I–B + F–D–G | 4 |
| 51 | A=B + F–E–D–C: → A=E + B=C +:D–F | 2 |
| 52 | A–B + C–D + E–F + G–H + I=K → A–G + B–H + C–I–E + D–K–F | 48 |
| 53 | A–B + C–D + E=F + G=H + K–I–L → A–E–I–G–C + B–F–K + D–H–L | 128 |
| 54 | A: + B: + C–D → A–B–C + D: | 4 |

has recently been applied for modeling the reaction mixtures and for discovering novel metabolic pathways [37].

The graph-theoretic transformation rules provide the general formalism for the description and simulation of any type of chemical reaction in organic chemistry, inorganic chemistry, and biochemistry. These include reactions at the level of elementary steps, complex multicentered multibond forming reaction mechanisms, reactions with participation of enzymes, catalytic systems, and a variety of biochemical reactions such as transcription, translation, and binding. They enable the exhaustive combinatorial search by generating all possible transformations and chemical species, offering the possibility of discovering unknown reactions and reaction mechanisms. Such capabilities are essential in searching for new ideas of synthesizing complex organic compounds, designing of novel metabolic pathways, suggesting novel biochemical species, and elucidating yet unknown reaction mechanisms. As an example, one can mention the pyrolysis of α-formyloxy ketones discovered with the IGOR program, the reaction that was later confirmed in the laboratory [38, 39]. Also Herges demonstrated the discovery of novel butadiene syntheses by logical formal reaction generation with the SYNGEN [40]. CONAN (connectivity analysis) was capable of discovering new reaction types that provide a basis to the introduction by Wender group of novel metal catalyzed [4 + 2], [5 + 2], [4 + 4], [6 + 2], and [5 + 2 + 1] cycloadditions for 6-, 7-, and 8-membered rings [10]. However, the potential in the ability of these models to generate a whole chemical space is limited by combinatorial problem, what means that they do not encompass the breadth of chemical species required to fully describe complex networks, especially those found in biological systems or complex reacting mixtures. In practice, usually only two to four levels of intermediates (or successors) can be generated and analyzed, without consideration of the vast majority of other formally possible transformations and species. A variety of techniques have been developed to overcome this obstacle capable of detecting good candidates and rejecting less probable or unreal solutions. Among them are explicitly defined or extracted automatically heuristic rules, quantitative computational chemistry tools, and formal graph theoretic concepts of the reaction similarity, distance, and complexity. Some of more important solutions are discussed in Section 15.3.

### 15.2.2
### The Empirical Reaction Rules and Knowledge-Based Approach for Reaction Generation. Automated Creation of Rules by Learning and Reaction Database Mining

Conceptually different is the category of empirical transformation rules that contain domain-specific knowledge to model chemical reactions. They are generally regarded as generalized descriptions of known chemical reactions in terms of a reaction topology (graph transformation patterns) as well as chemical conditions, requirements, and constraints that control the direction and extend of the reaction. A set of reaction rules encoded in the knowledge base determines the size of the space to be searched. As a result, transformations that have analogs among the known chemical reactions are generated.

This category of reaction models includes both the transformation rules manually encoded by expert chemists as well as automatically extracted from the reaction databases by perceiving common transformation patterns, clustering, and generalization. The former attempts to replicate the detailed knowledge of experienced chemists what ensure generating more accurate and realistic results, although within limited knowledge domains. The most prominent examples are the LHASA (Logic and Heuristics Applied to Synthetic Analysis) and CAMEO programs [41–43]. The later have not the perfect quality level as the manually created rules but give the possibility of utilizing almost all the accumulated knowledge on chemical reactions and search within the much larger region of chemistry space for the reaction prediction.

#### 15.2.2.1 **Automatically Derived Reaction Rules**

The rapid increase of chemical information collected in different databases was a strong incentive for developing and integrating techniques to mine the reaction data and extract relevant knowledge, with programs for the rule-based reaction generation. Currently, several CAOS programs are capable to build the rule knowledgebases automatically, including the first SynChem and AIPHOS, built in the early 1990s, and some newer as EROS, SOPHIA, KOSP, Route Designer, RETROSYN, CSB, and COSYMA [44–46].

There are a number of approaches to formalize and describe transformation rules within a framework of the empirical or knowledge-based model of chemistry. One may say that every program of this category in order to conduct the heuristic search and construct the reaction pathways applies its own representation method. In general, the rule contains two elements, the reaction topology and reaction chemical context. The former consists of the description of topological structural changes from reactants to products, with respect to the reaction centers (the core of a reaction), extended reaction centers on surrounding bonds and atoms (within defined distance), functional groups, or substructures. The later represents the reaction conditions, requirements, and constrains, for example, activating groups, interfering functional groups, leaving groups, steric hindrance, ring strain, and stereochemistry.

Commonly used fundamental representation, especially with regards to automatically induced rules, takes a form of balanced atom-mapped transformations of reaction centers in substrate and product molecular graphs. The reaction centers, that is, those molecular subgraphs that are modified in a reaction, are identified by maximal common subgraph technique used to find atom–atom correspondences between the reactant and reaction products. The corresponding algorithm was developed by Willet in 1980 [47, 48]. These kinds of rules considered as the patterns of the reaction center changes can be used to modeling reactions at the level of elementary steps for generating networks of reaction mechanisms as well as the overall reactions from starting materials to final products for simulating the synthetic routes to complex target products. In narrow domain-specific applications, the use is made of manually created rules of these types, usually representing reactions of a certain classes. Examples are knowledge-based systems for modeling the reaction mechanisms of the specific catalytic system, combustion of a particular hydrocarbon, in which a limited set of rules representing hypothesized elementary steps, is used to generate and determine a relative importance of a specific reaction within the network of reaction mechanisms [49, 50].

To include more details and limit applicability of the transformation rule, the description of reaction centers is extended taking into account their nearer and further structural environment that influence the reaction (Figure 15.6). In the bond-centered approach by Wilcox and Levinson, the adjacent unchanging bonds are added to the reaction center bonds, providing a basis for hierarchical distinction of the transformation rules and reaction classes [51]. In some programs (such as the

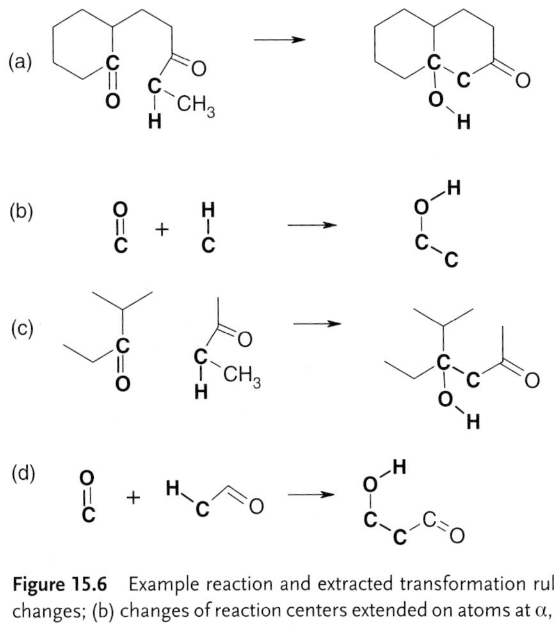


**Figure 15.6** Example reaction and extracted transformation rules describing (a) reaction center changes; (b) changes of reaction centers extended on atoms at α, β (c); changes of reaction centers extended with activating group (d).

InfoChem Reaction Classification Program), the extension is defined by the bond distance from the reaction center [52]. Such programs also include SOPHIA and KOSP (AIPHOS), in which the structural and stereo characteristic of the reaction sites (bonds broken/formed) and their environments on atoms at the α, β, and γ positions are described [53–55]. These automatically derived rules also include the information on type of reaction conditions classified on general classes and subclasses. Another concept implemented in RETROSYN and Route Designer programs is to extend the reaction center only with important neighboring functionality (regardless of distance), including leaving groups and activating functionality [56, 57].

In the PathPred web-based server for the reconstruction and prediction of biodegradation and biosynthesis pathways (based on information extracted from different biological databases integrated within KEGG service), the use is made of biotransformation patterns (called RDM patterns) that describe atom type changes for a pair of reactant–product structures, taking into account the reaction center atom (R), and its neighboring atoms located in the unmatched differencing substructure (D) and matched substructure (M) [58]. In contrast to other systems for the prediction of metabolic pathways that mostly use the transformation patterns encoding changes in functional groups, the RDM patterns provides unified model for the description of any enzymatic reaction and enzyme function.

#### 15.2.2.2 **Functional Group Transformations**

A concept of transformation rules based on changes of functional groups is a convenient representation for reactions of specific class or domain. It is implemented in particular into the programs for the prediction of xenobiotic metabolism, the most known are TIMES, METEOR, META, and MetabolExpert [59]. TIMES whose unique feature is evaluating the generated biotransformations and metabolites in quantitative manner uses a set of 341 principal enzyme reaction rules simulating metabolism in mammals, classified into transformations of highly reactive groups (non-rate-determining), and rate-determining transformations (for predicting and explaining reactivity, RESYN-ASSISTANT).

#### 15.2.2.3 **Substructure-Based Transformations**

The methods presented above are based on the smallest substructure related to the structural change between reactants and products of the reaction. Diversely, Corey *et al.* have proposed a concept of a hypothetical presumed substructure called "retron" that constitutes a basis of the retrosynthetic analysis of simplifying disconnections and generating retrosynthetic transformations (the inverse of chemical reactions) [60]. A retron is minimum molecular substructure that covers all required functionalities along with stereochemistry for the determination of bonds being broken in the product structure. A pair of the retron substructure and the corresponding product substructure creates so-called goal or simplifying transforms that along with other transformation types, including the functional group interchanges FGI and functional group addition FGA, are employed in the knowledge base of the LHASA program for synthesis design [61]. When the retron pattern only partially matches a target substructure, the LHASA is able to perform a sequence of changes

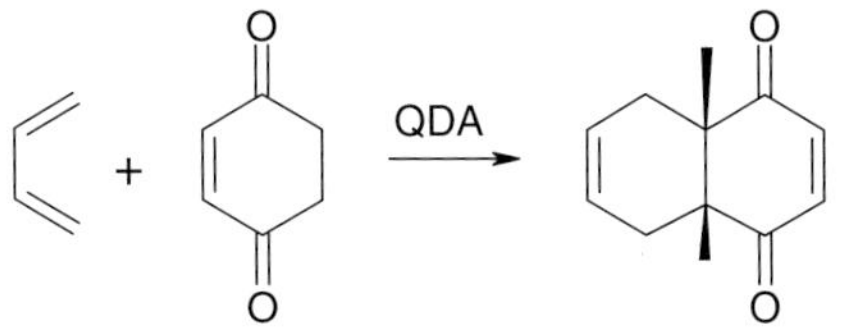


the retron for the quinone Diels–Alder transform QDA

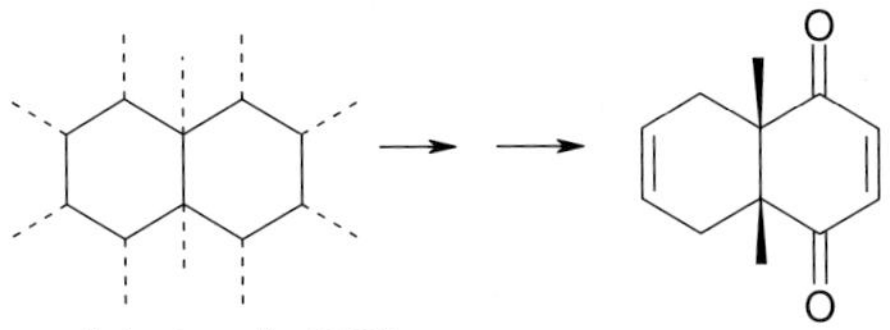


partial retron for QDA

**Figure 15.7** The quinone Diels–Alder transformation (on left, the quinone retron QDA) and cis-decalin partial retron below and its modification. The description of this transform contains its molecular context with stereochemical requirements, steric constraints, and other heuristics to assess its suitability.

in terms of functional modifications needed for the application of the given simplifying transform (Figure 15.7).

The idea of substructure-based transformations can also be implemented in programs for the reaction prediction. In this case, the substructure(s) describing the reactive site in reactant molecules consists of the applicability condition, while the corresponding product substructure determines the result of application of the transformation rule. The rules of these types are automatically created using one of the learning methods embodied within the CSB system for the reaction prediction and synthesis design. They describe different structural changes of so-called active subgraphs of reactants and associated products in unified form. These subgraphs differently as in other learning methods are recognizing on-the-fly simulation with the CSB of the instances from the reaction database. More details are given in [62]. Figure 15.8 shows example reaction and the transformation extracted as a result of the CSB learning.

The CAMEO and EROS programs for the modeling and the prediction of the reaction courses use the substructure-based transformations generalized and arranged for each reaction type or category. They require the user to specify them for reaction prediction. Reaction categories and subcategories in CAMEO are associated with the reaction mechanisms and the type of reagent and conditions [42, 43]. The following categories are involved: nucleophilic and electrophilic, pericyclic, redox, radical, reactions of carbenoid species, reduction reactions by hydrides, and others. Each category is represented by a set of substructure transformations with heuristics (including stereochemistry, steric constraints, reactive nucleophilic, and electrophilic sites) modeling elementary reaction steps. Furthermore, the evaluation rules, including assessment of $pK_a$, $\Delta H_f$, the HOMO–LUMO energies, and heats of reactions, are taken into account.

The EROS belongs to the category of formal-logical programs that can generate new reactions and reaction mechanisms. It uses a rule knowledge base as a tool

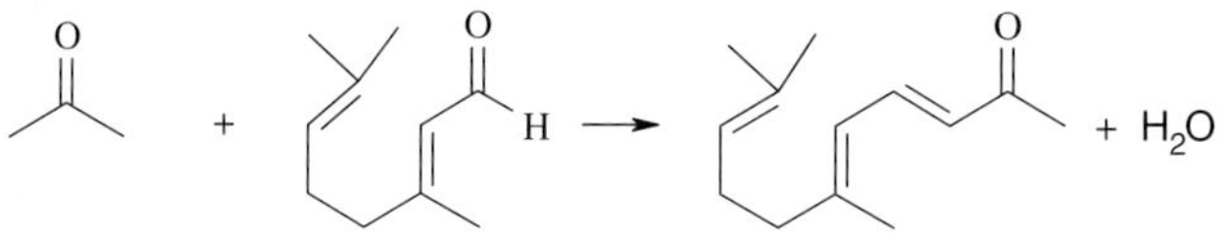


**RA:**
if
M1={AS_236} and M2={AS_107}
then
RM=2 and GT_10_2{(236:1,2),(107:1,3),(107:1,4)} and basic

**RB:**
if
M1={AS_236, AS_107}
then
RM=1 and GT_10_2{(236:1,2),(107:1,3),(107:1,4)} and basic

active subgraphs:

AS_236 AS_107

$X_1$ = {C,H} $X_2$ = {C,H}

graph of transformation: GT_10, scheme: 2

A=B + D-C-E ⟶ A=C + D-B-E

medium: basic

**Figure 15.8** Example reaction (at the top) and transformation rules (RA and RB) derived with the CSB program. Cause (part *if*): list of identifiers of active subgraphs recognized in molecular graphs of substrates. Consequence (part *then*): transformation vector describing the course of the transformation process for active subgraphs (comprises, among others, the list of the broken edges in active subgraphs and the transformation scheme from Reaction Generator Library). The active subgraphs in the rule RA have to be placed in two different reactants M1 and M2, and in the rule RB, in one reactant.

controlling and guiding the process of the generation of chemical reactions. The task is realized in twofold manner, by the identification of the reactive sites and quantitative prediction of chemical reactivity [44–46]. Automatically deduced rules in the knowledge base of the EROS contain apart from topological transformation the ranges of allowed values of physicochemical descriptors for the reactive sites, such as partial atomic charges, electronegativities, delocalization measure, polarizability effect, and hyperconjugation effect. Independent module consists of empirical method calculating electronic and energy parameters and reactivity for the entered molecules and intermediates generated, allowing discriminating and selecting between probable reaction courses. This approach combining the generalized knowledge derived from chemical reaction data with quantitative model of chemical reactivity provides a comprehensive method for analyzing chemo- and regio- and stereoselectivity and predicting the outcome of chemical reactions.

The empirical transformation rules can serve as tools for classifying, searching, and retrieving reactions described in explicit way in the databases as well as provide reasoning tool for the reaction generation allowing us to predict the reaction product or the reaction substrates and build models of hypothetical reaction pathways. The introduction of the learning and data mining tools for the automated extraction of empirical transformation patterns have inspired the further development of the rule-based (empirical) systems in the CAOS area. Hierarchically organized process of the reaction classification and generalization enables to create the respective rules at different levels abstraction-resolution. Thus, depending on the knowledge level involved in the generating process, different models of the reaction pathways can be constructed and analyzed and tested for a given input. This approach extends the prediction capacity of rule-based systems and simultaneously offers the possibility of achieving accurate, experimentally verified results.

## 15.3
## Combinatorial Complexity Problem: Strategies for the Directed Reaction Generation

A computer generation of chemical transformations is a technique to build models of hypothetical reaction pathways that can serve for many applications. The most important are the simulating and predicting of the reaction courses, modeling the reaction mechanisms, and designing synthetic pathways and synthetically accessible compound collections. Such models can have a form of reaction trees or complex reaction networks generated by the iterative application of a transformation rule (or a series of rules) to the input graph and successive intermediates. However, the exhaustive search within the general framework of graph transformations involves combinatorial problem even for medium-sized molecules, producing large number of less important or unreal reactions and species. This limitation still remains a significant challenge, particularly with respect to the mathematical or graph-theory rules employed in this process to enable the search for novel chemical reactions or synthetic pathways and reveal previously undiscovered reaction mechanisms. This phenomenon also arises when a large rule set (usually extracted automatically) is involved in the knowledge-based reaction generation, although this search type is implicitly constrained by additional information, beyond the connectivity, encoded in empirical rules (e.g., ranges of allowed values of attributes for atoms and bonds and physicochemical descriptors) to limit their applicability.

A variety of approaches have been developed to effectively conduct the search within the combinatorial space of possible solutions. They can be grouped into three categories including heuristics, calculation models, and formal approaches (mostly in terms of graph theory) imposed on the reaction generation. But their categorization should also be considered from the standpoint of the application area and functionality of the computer program. In different applications, different optimization or selection criteria emerge as being important. In the programs designed to predict reaction courses, the particular attention is devoted to achieving an accurate reaction prediction; therefore, theoretically grounded heuristics and quantitative physicochemical criteria play a significant role for the optimization of the search. Additionally, in order to select the most favorable or reliable route, it is necessary to consider and evaluate all the possible solutions generated taking into account different reaction sites, stereoisomers, the effects of catalysts, or conditions. Similarly, in reaction mechanisms generation, it is essential to have precise evaluation criteria for discrimination between all alternative variants of elementary steps. But, for large-scale reaction networks, one of the main limitations of this type search is still its computational complexity. Differently, the programs for synthesis design aim to propose alternative solutions for novel efficient routes, synthetic strategies, or "creative" ideas for a synthesis of a given complex target structure. Thus, different approaches for the optimization of the search can be involved and different alternatives for solutions would be possible. In this case, it is reasonable to use formal methods for the selection of transformations and generated intermediates. These may be theoretical semiquantitative measures of topological complexity, similarity, connectivity, or even intuitive measures of synthetic complexity that have no physical justification. As unbiased, independent of domain knowledge tools, they allow us to span a broader range and diversity of chemical reaction space to explore little investigated, unconventional or unknown chemical transformations. Perceiving and applying such transformations to build synthetic pathways provides the opportunity to find innovative strategies or original ideas for the synthesis of complex molecules.

### 15.3.1
### Retrosynthetic Generation of Chemical Transformations: Computer-Assisted Synthesis Design

Most computer programs for synthesis design use the concept of a retrosynthetic search that starts from a target product and applies transformation rules in reverse. A uniform logical basis for leading the search in this fashion has been formulated by Corey in the framework of retrosynthetic analysis of strategic disconnections [1, 2]. According to its main principles, only those bonds are disconnected for converting the target into its constituent components that provide a significant decrease in topological or synthetic complexity. Such a bonds formed in synthetic direction will equivalently increase complexity of the assembled product structure along with the difficulty of the corresponding reactions. The search for the best bond for disconnections or retrosynthetic simplifying transformations can be conducted in two ways. The first rely on the recognition of guiding patterns inside the target structure, corresponding to potentially available starting materials, building blocks, or functional substructures. The second is based on the recognition and ranking of bonds giving the greatest simplification of the target structure. A variety of the disconnective strategies have been developed taking as a basis these general principles. They include both the heuristically guided search for simplifying disconnections and new formal approaches, for example, topology-based disconnective strategy, graph-based complexity analysis, identification of isomorphic substructures, or detection of molecular symmetry.

#### 15.3.1.1 Recognition of Guiding Patterns, Molecular Symmetry, or Isomorphic Substructures

The identification of the patterns of starting materials within the target structure is routinely used approach in many CAOS systems to find the shortest accessible pathways. Some typical examples are the LHASA, SECS, WODCA, and SYNCHEM using their original starting material libraries and different approaches to estimate the synthetic accessibility. LHASA uses the quantitative idea of synthetic proximity between two structures, associated with the number of atoms or bonds needed to be changed to convert one into another [63]. WODCA employs different structural similarity criteria [64]. The SST program involves hierarchical search and abstract subgraph definition to select the candidates from starting material library. Here, the identity or similarity relationship is carried out at three levels, that is, at carbon skeleton level, functionality, and stereochemistry. It also uses a function to evaluate the appropriateness of the functionality of the substrate candidates [65]. The CHIRON program is capable to recognize inside the target structure the chiral templates of starting materials based on the library of available or easily accessible enantiopure starting materials. Like the SST program, it uses some kind of the hierarchical search (carbon skeleton, functionality, and stereochemistry) to find apparent or hidden structural relations [66, 67]. SESAM program also searches for nonobvious or hidden structural identities between the starting materials and complex targets, but only at the level of their carbon skeletons. It can locate any restructured starting material inside the complex synthetic target without the use of the starting materials database [68]. CAESA (Computer-Assisted Estimation of Synthetic Accessibility) selects starting materials based on their total coverage of the target compound, so that the starting material that covers the most area will be the most favored [69].

The original idea of combining synthetic accessibility and convergence criterion to search for the shortest routes for constructing complicated molecular skeletons was introduced by Hendrickson's and implemented in the SYNGEN program [70]. It generates synthetic routes by dividing the target and successive intermediates into two fragments, which are matched with starting materials, involving their skeletons. The important extension of this concept was a symmetry-based strategy proposed by Bertz and Sommer to guide the search for reflexive or two-directional routes [71]. Such synthetic routes allow utilizing the same starting material (or intermediates) to construct a target molecule. Their approach involves recognizing disjoint isomorphic substructures in a target structure and intermediates by enumerating all possible substructures method introduced for measuring topological similarity and complexity [72, 73]. The same concept of detecting the potential symmetry in synthetic targets has been implemented in RESYN_Assistant, a program for computer-aided understanding of synthesis problems, developed by Vismara *et al.* The authors have proposed another original solution to find isomorphic subgraphs that lies in disconnecting a minimum number of edges and vertices in a graph of the target structure, so that the resulting subgraph will consists two isomorphic connected components [74].

#### 15.3.1.2 Complexity-Based Disconnective Strategies

The second way to optimize the search space of retrosynthetic transformations consists of the systematic reduction of molecular complexity by disconnecting only strategically important bonds, that is, generating precursors that are easier to synthesize. In order to identify such bonds in a target structure, Corey devised in 1971 a series of heuristic rules based on detailed analysis and recognition of all complexity elements effecting difficulty of reactions (e.g., appendages, chiral centers, rings of medium and larger size, and bridged rings) [75]. However, as Bertz stated, recognizing strategic disconnections taking into account the synthetic complexity will depend on the current state of chemistry knowledge that evolves rapidly along with the progress in synthetic methodologies, effective catalyst, and reagents to control the reaction selectivity and to increase the reaction scope [76, 77].

An important extension of the heuristic retrosynthetic analysis has become formal strategy to optimize the search, derived from the fields of information and graph theories, based on assessment and ordering alternative disconnections in terms of the molecular complexity, connectivity, or diversity. They quantitatively define the retrosynthetically best bonds, as those that give the largest decrease in topological complexity resulting from disconnections. By perceiving and successively disconnecting such bonds in a complex target structure we can find the shortest sequence of the retrosynthetic transformations to afford a simple starting material. Going in a reverse, synthetic direction, and evaluating the increase in complexity during the bond formation, we can build the complex structures in the minimal number of steps. The advantage of this approach is that the search for retrosynthetic transformations is not heuristically confined to the "known chemistry," what ensures a larger degree of novelty. The second benefit is practicality of solutions, from the viewpoint of the minimization of the number of steps, the maximization of product complexity, or minimization of excess complexity, the principles of "ideal synthesis" [9, 10, 77].

The leading approach to quantify molecular complexity in the context of chemical transformations relies on complexity metrics, introduced by Bertz and Sommer. They are based on the concept that a molecular graph is more complex when it contains more subgraphs. The representative examples of the subgraph-based complexity measures are the number of kinds of subgraphs, $N_S$, and the total number of subgraphs, $N_T$, based on constructing and counting subgraphs of a molecular graph [72, 73, 78]. Rucker and Rucker have proposed an idea to measuring the complexity by counting all walks of all lengths in a molecular graph. These computationally simple graph invariants are, for example, the total walk count and the walk complexity [79–81]. Like the subgraph count, the total walk count has proven an adequate topological index to discriminate between chemical structures, as well as between the alternative bond disconnections. These measures have been applied in the program MASA (Mathematics Applied to Synthetic Analysis) to calculate the complexity change and identify the retrosynthetically best bonds. Comparing the results based on the MASA pure mathematics and those obtained with the LHASA heuristics, the authors have found some degree of correspondence [82, 83]. In the CONAN program, strategies for retrosynthetic disconnections of complex molecules are based on connectivity analysis described by Wender and Miller [10]. The program ranks the bonds and provides a variety of options for different disconnection types. These include the disconnection of rings according to [x + y] formal scheme representing reaction types [2 + 2], [2 + 4], [3 + 2], and so on, disconnections of polycyclic ring systems to obtain linear precursors, convergent disconnections, and other [84]. However, as Bertz concluded, using rigorously mathematical approach for calculation molecular complexity enables to analyze and compare different strategies for constructing complex molecules, without accounting for synthetic complexity/feasibility. On the other hand, empirical heuristics assume structural features that affect the synthetic difficulty, but depend on the state-of-the-art of organic synthesis, which is constantly developing.

Empirical metric of structural complexity corresponding with chemical concept has been proposed by Whitlock [85]. It is simply calculated from complexity values of arbitrary defined structural features such as the number of heteroatoms, unsaturations, chiral centers, and rings. According to the author, this chemical intuition measure was devised to provide a simple way of analyzing and comparing complex multistep chemical transformations by measuring complexity intermediates and constructing plots of complexity versus step. Barone and Chanon have proposed an extension of this complexity measure by considering the connectivity of atoms and the size of the rings [86, 87]. Clearly, whatever approach is taken, the complexity remains a relative concept expressing a measure that depends on the components selected for examination.

#### 15.3.1.3 **Concept of the Strategic Bond Tree for Disconnections**

The disconnective strategies based on topological indices require disconnection of each bond in the target and analyzing all possible precursor candidates in order to calculate the corresponding changes of molecular complexities. We have proposed another approach for recognizing and ordering all possible strategic bond sets in single procedure, constructing a strategic bond tree [88]. Identification of bonds and generation of the strategic bond tree is based on hierarchical analysis of molecular connectivity in order to partition the atoms and bonds of a target molecule. The paths on the bond tree represent the alternative bond sets for the disconnection of the target (Figure 15.9). The bond in the root has a highest priority, and the last bond on the path has a lowest priority. After selecting the bond set and deleting the bonds from the target structure, the resulting structural units can be tested for reassembly of the target by means of biomimetic transformation patterns, describing various one-pot-multistep reaction cascades [89].

Figure 15.10 presents the application of different disconnective strategies and CAOS programs to the problem of the taxane core synthesis concept. Strategies of this kind are applied to plan target-oriented synthetic pathways dedicated to a particular product or combinatorial libraries of analogues with specific structural and biological properties. Their development is still continuing, driven by increasing

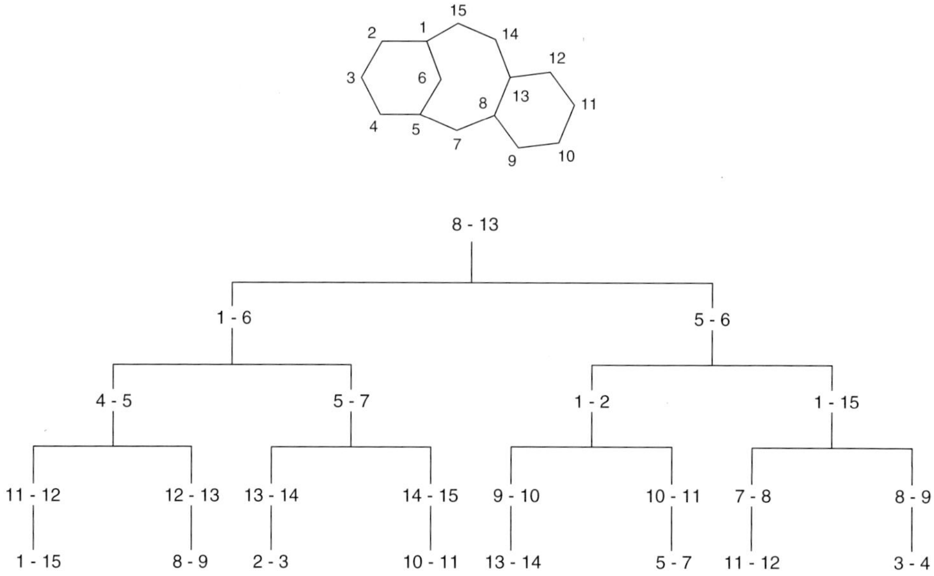

Figure 15.9 The strategic bond tree for the taxane core.

demands to synthesize compounds having more and more complex scaffolds and by needs to produce them more quickly and more cheaply. Besides the CAOS programs, this same sort of strategy is involved in drug discovery process for filtering libraries of candidate compounds that could be synthetically accessible from simple available building blocks.

### 15.3.2
### Forward Generation of Chemical Transformations: Computer-Assisted Reaction Prediction

In computer programs developed to assist in the prediction of the reaction pathways, the reactions are simulated by exhaustive application of transformation rules in a forward direction, that is, for the reactants and iteratively generated intermediates. The search within the entire reaction space is optimized with strategies controlling the bond formations between reactivity centers identified in the reacting molecules. They include heuristic rules describing reactivity of the bond disconnections/formations, mathematical models for quantitative prediction of the chemical reactivity, or abstract graph-based evaluation criteria describing chemical reactions in the context of topological distance, similarity, or complexity.

The use of heuristic rules describing transformation of reaction centers is inherent feature of empirical systems. They comprise both structural and stereochemical requirements, as well as the reaction conditions that influence the reaction or a class of similar reactions having the same reaction centers (see Section 15.2.2). Such rules combined with formal graph-theoretic models for reaction generation give the

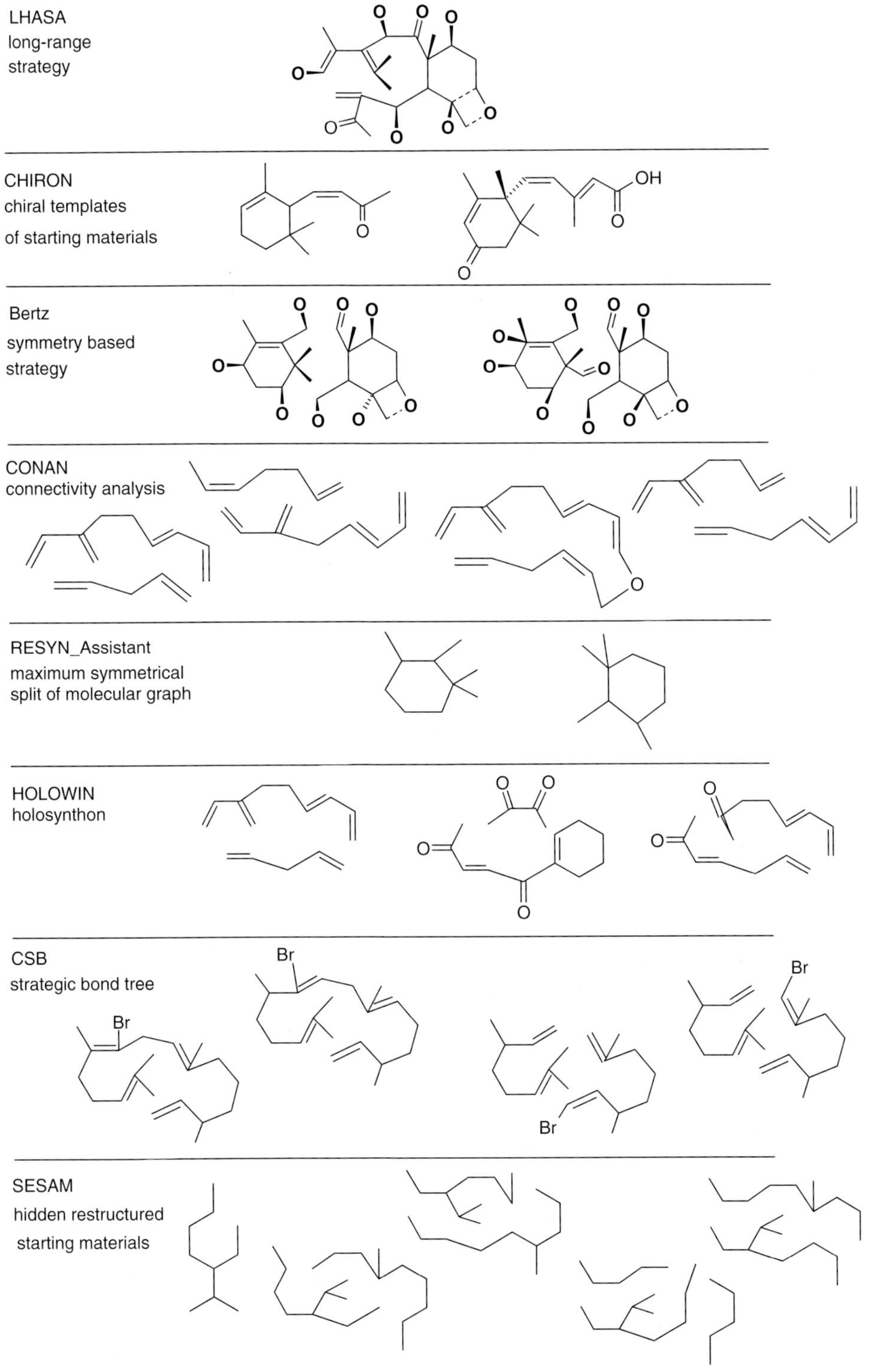


**Figure 15.10** Application of different disconnective strategies to the taxol and taxane ring core.

possibilities to explore (optionally) different regions of the reaction space with the guidance of the search along reliable paths. Examples include EROS, CAMEO, SOPHIA, and CSB hybrids systems that integrate knowledgebases, databases, and learning tools within a framework of formal mathematical reaction models. The application of heuristic rules to the graphs of starting materials involves the recognition of all possible reactive sites (or all substructures that are isomorphic to the left-hand side of transformation rules) and the generation of the respective product candidates. The use of more details on reaction centers allows the better discrimination between different reaction courses for obtaining more reliable and accurate solutions. However, the heuristic rules, prepared by expert or automatically derived from reaction databases, are not enough for the detailed description of the chemical reactivity and selectivity. Additionally, the heuristic search in the reaction space is constrained by literature precedents and do not encompasses the vast majority of possible chemical reactions that are not yet discovered.

#### 15.3.2.1 Quantitative Models for Reactivity Prediction

In most of the CAOS programs, besides the qualitative prediction of the reactivity sites, the use is made of general quantitative models for determining the chemical reactivity and studying the reactions in terms of directionality and feasibility. In order to use them effectively in the reaction simulation process requiring consideration of hundred of alternatives, a balance between a suitable level of accuracy and resource requirements in time and memory is still of importance. Therefore, they are mostly based on fast empirical calculations of global molecular properties and physico-chemical effects of bonds and atoms involved in the reactions. Some newest applications use more sophisticated quantum calculations for molecular and atomic properties, reactivity descriptors, and thermodynamic and kinetic parameters, but in highly approximated forms. Important contributions in quantitative modeling of chemical reactivity are various empirical methods developed by Gastaiger's group. Based on the partial equalization of orbital electronegativities and modified Hückel molecular orbital methods, they enable calculating the most important molecular, atomic, and bond properties, such as $\sigma$-charge and $\pi$-charge distribution, electronegativities, delocalization energies, polarizability, bond dissociation energies, and heats of formations [11–15]. Their main advantage is short computational time and usefulness in processing large data sets of chemical structures and reactions. These methods are implemented in the early-mentioned EROS program for calculating reactivity and predicting regioselectivity of reactions. They are also known as application software PETRA (Parameter Estimation for the Treatment of Reactivity Applications), commonly used to extend the chemical and biological problems such as establishing quantitative structure properties or structure activity relationships (QSPR, QSAR), assessing similarity and classification of chemical and enzymatic reactions, assessing similarity/diversity of chemical libraries, modeling and predicting metabolism, lead discovery, and lead optimization [90].

In another program for the generation and study of the reaction networks, known in the literature as the Toy Model of Chemistry, the reactions are simulated by applying the graph rewriting rules to the starting materials, represented by orbital graphs [27]. Their vertices represent atom orbitals, labeled by atom hybridization states (from VSEPR theory), while edges dente overlaps adjacent orbitals, tabulated according to the atom's hybridization. The molecular physical properties of reacting molecules (electronic energy and charge density) are considered taking into account the energy function derived in a simplified shape from the extended Hückel theory [91]. For measuring regioselectivities, the reactivity index is calculated based on the *Klopman-Salem* concept of charge and orbital control [92]. The related equation expresses the driving force of a chemical reaction by the summation of the electrostatic and frontier orbital terms.

The methods implemented in the Beppe program to evaluate the reactivity are based on the calculations of atomic and molecular properties of reactants, defined in the frame of the density functional theory, including electronic energies, electronegativities, hardness, and other related reactivity descriptors [93, 94]. They are used to select the reactive sites in the reactants and evaluate the imaginary intermediate, interpreted as an approximated model of the transition state. Combining the molecular electronic energy, electrostatic interaction energy, and hardness, in the respective equation, the program calculates the imaginary reaction rate and the quantity related to the imaginary intermediate, and in this context used as the useful controlling mechanism during the reaction simulation [95, 96]. The interesting feature of this program is the capability of modeling the reaction conditions involving the solvent, catalyst, and temperature. The selection of the respective solvent for the principal reaction (with the highest value of the imaginary reaction rate) is based on the concept of the similarity between the solvent and the imaginary intermediate, calculated in terms of the electronegativities of the atoms and ionic or covalent radii of the atoms. Accordingly, considering different aspects of chemical reactivity, the Beppe can simulate the behavior of the reacting molecules by comparing feasibility, predicting regioselectivity, and modeling reaction conditions.

Differently, the approach in the ROBIA (Reaction Outcome by Informatics Analysis) combines rule-based techniques with molecular mechanics and quantum chemistry calculations by using shell scripts developed for launching the popular computational chemistry software packages, such as Maestro, MacroModel, Jaguar, GAMESS, and MOPAC in the reaction prediction [97]. The program applies a series of heuristic rules to direct the reaction courses, by recognizing and locating the reactive sites in reacting molecules and by selecting the most probable product candidates. It allows to limit the set of structures before being entered to molecular modeling packages. Its important feature is capability of assigning a configuration of chiral centers and generating all possible stereoisomer structures. After the rule-based selection of the respective product candidates, the 2D structures are transformed into 3D forms and optimized by energy minimization and conformational search with the Maestro. Finally, molecular and quantum mechanics energy calculations are performed on the all products, including their stereoisomers and conformationally stable forms. Calculations of the respective energies of all competing transition state structures and corresponding product candidates make the simulation of the kinetically or thermodynamically controlled reactions possible. The program considering all possible reaction courses (generated within a set of several transformation rules) enables to predict in a reasonable time and with sufficient accuracy the most feasible alternatives in terms of regio- and stereoselectivity of reaction pathways.

The generation of chemical transformations combined with the quantitative prediction reactivity enables to limit the reaction space and guiding the search along the most favorable pathways. The use of a variety of electronic and energetic properties and reactivity descriptors provides reliable and unbiased by experimental knowledge estimates of the reaction selectivity and feasibility. In contrast to the heuristic tools controlling the reaction simulation, they enable to predict unknown reactions and reaction mechanisms. However, large computational requirements have made difficult at the present time the exploitation of their full potential in applications to the reaction simulation, even at the semiempirical level.

#### 15.3.2.2 Formal-Logical Approach to the Search Space of Possible Chemical Transformations

Another approach to guide the search in the space of combinatorially generated chemical reactions concerns the more formal measures, describing chemical reactions in the context of topological distance, similarity, or complexity. According to the Ugi principle, the generation of chemical transformations should be controlled by the requirement of decreasing chemical distances going through reaction path from reactants to the product via intermediates. This measure in terms of the matrix model is defined as the sum of the absolute values of all entries of the R-matrix, which is the number of broken/formed bonds or twice the number of relocated valence electrons during the reaction [98]. Accordingly, only those intermediates have to be generated for obtaining feasible reaction routes that are closer (in terms of the structural changes) to the product graph. The analogous concept of the reaction distance is employed in the FORWARD program developed by Hendrickson for the generation of the shortest synthetic routes to the given target compound [99]. This measure takes into account the degree of the similarity between any mapped pair of compounds, in terms of the refunctionalization steps needed to convert one into the other. It allows to guide computer generation toward the shortest synthetic pathways by selecting those starting materials or generated intermediates that are functionally nearer to the target compound. Following the Hendrickson's concept, the ideal routes to the target are those containing exclusively sequences of construction reactions with no refunctionalization steps. According to the Kwasnicka definition, the reaction distance is the minimum length of a path between two molecular graphs in the reaction network [100, 101]. All these measures are closely related and can be regarded as quantitative expression of the classical principle of minimal structural changes used for the explanation of reaction mechanisms. Applying these measures for the reaction prediction will result in generating feasible pathways involving the minimum number of edges that should be broken and formed or redistributions of valence electrons. Thus, this approach can be valuable to propose mechanisms and explain the courses of reactions at the level of electron flows in elementary reaction steps. However, the application of these measures is associated with combinatorial algorithms that may become ineffective for larger molecular graphs and for large-scale reaction networks.

Bertz approach to the evaluation of the structural changes that take place during the course of synthetic reactions is based on calculation of complexity indices with the use of the graph and information theory. The respective criterion for complexity-based construction of the reaction pathways involves a large increase of molecular complexity in order to rapidly build molecular skeletons. It corresponds to Wender's principle of "maximization of target-relevant complexity," allowing to find a minimum number of synthetic transformations (or steps) to rich a complex target [10]. As concluded by Berz, the use of rigorously mathematical complexity measures enables to "compare the power of various reactions to increase molecular or topological complexity," however, without consideration of the structural features affecting the synthetic difficulty [77]. Alternatively, empirical metrics can be a useful to estimate synthetic complexity of chemical tractions, but from viewpoint of currently available synthetic methods. The former as an unbiased and objective tool can aid chemists in searching for novel unusual reactions or original ideas for synthesis of complex molecules. The latter can become a useful fast method for evaluating large sets of virtual (combinatorial) libraries and high throughput screening hits from the viewpoint of synthetic feasibility and accessibility.

Different approach to efficiently explore the space of synthetically accessible and structurally complex chemicals consists of the use of the multibond-forming/braking transformations that imitate cascade reactions. They are automatically created using one of the combinatorial learning methods implemented in the CSB program, by combining several elementary transformations to be generated in one iteration [88]. Applying them in the reaction generation enables us to form several bonds and construct considerable part of the product structure in one transformation (iteration), without the necessity of generating intermediates. This option used independently or in combination with a backward search for strategic disconnections can be used for discovering the shortest reaction pathways for complex molecules as well as synthetically feasible and accessible libraries.

All the mentioned strategies aim at controlling the generation of chemical transformations in terms of structural or topological features. Another innovative concept that extends the logical formalisms of synthesis planning beyond the traditional framework of structural complexity or diversity has been proposed by Weber. His idea of FOS is to avoid complex reactions and guide the search for synthetic transformations by inferences about physicochemical or biological properties of a synthesized target [17]. It creates new perspective to integrate the CAOS systems and drug design and drug discovery programs.

## 15.4
## Conclusion

Recent years have seen many new developments in the area of CAOS, to provide creative problem-solving tools to chemists with the potential of predicting the outcome of chemical reactions, proposing mechanisms to explain the reaction courses, discovering new strategies for the synthesis of valuable synthetic targets, and designing synthetically accessible combinatorial libraries. The most important achievements concern integration of an artificial chemistry with chemical databases and tools to mine the reaction data and extract relevant knowledge, as well as computational models of chemical reactivity, reaction kinetics, and thermodynamics for simulating chemical reactions under controlled conditions. In combination, they offer unprecedented power in reaction prediction and synthesis design with respect to reliability, accuracy, and practicality of the generated solutions. This trend in the development of the CAOS systems will be continued along with increasing computing performance and storage capacity to fully exploit their potential. Another increasingly important focus of research is on logical and more sophisticated formalism for describing molecular transformations and building synthetic routes. This approach involves graph theory tools and measures of the structural changes, the reaction distance, molecular connectivity, and complexity allowing to explore all combinatorially conceivable chemical transformations and extending the prediction beyond the boundaries of the known empirical knowledge. On this ground, novel creative ideas for synthesis design of complex molecules will be further developed, as early-mentioned DOS and FOS. The progress toward greater abstraction will also provide new opportunities for possible biological applications of the CAOS systems, for example, designing novel metabolic pathways and biochemical species and discovering missing biotransformations or biomolecules in metabolic networks.

## References


1 Corey, E.J. and Cheng, X.M. (1995) *The Logic of Chemical Synthesis*, Wiley, New York.

2 Corey, E.J. (1988) Retrosynthetic thinking – essentials and examples. *Chemical Society Reviews*, **17**, 111–133.

3 Dugundji, J. and Ugi, I. (1973) An algebraic model of constitutional chemistry as a basis for chemical computer programs. *Topics in Current Chemistry*, **39**, 19–64.

4 Ugi, I., Bauer, J., Bley, K., Dengler, A., Dietz, A., Fontain, E., Gruber, B., Herges, R., Knauer, M., Reitsam, K., and Stein, N. (1993) Computer-assisted solution of chemical problems – the historical development and the present state of the art of a new discipline of chemistry. *Angewandte Chemie (International Edition in English)*, **32**, 201–227.

5 Ugi, I., Bauer, J., Blomberger, C., Brandt, J., Dietz, A., Fontain, E., Gruber, B., Scholley-Pfab, A.v., Senff, A., and Stein, N. (1994) Models, concepts, theories, and formal languages in chemistry and their use as a basic for computer assistance in chemistry. *Journal of Chemical Information and Computer Sciences*, **34**, 3–16.

6 Hendrickson, J.B. (1971) A systematic characterization of structures and reactions for use in organic synthesis. *Journal of the American Chemical Society*, **93**, 6847–6854.

7 Hendrickson, J.B. (1975) Systematic synthesis design. IV. Numerical codification of construction reactions. *Journal of the American Chemical Society*, **97**, 5784–5800.

8 Hendrickson, J.B., Grier, D.L., and Toczko, A.G. (1985) A logic-based program for synthesis design. *Journal of the American Chemical Society*, **107**, 5228–5238.

9 Wender, P.A., Handy, S.T., and Wright, D.L. (1997) Towards the ideal synthesis. *Chemistry and Industry (London, UK)*, 765–769.

10 Wender, P.A. and Miller, B.L. (1993) Organic synthesis: theory and

applications, in *Toward the Ideal Synthesis: Connectivity Analysis and Multi-bond Forming Processes*, vol. **2** (ed. T. Hudlicky), JAI Press, Greenwich, pp. 27–66.

11 Gasteiger, J. and Marsili, M. (1978) A new model for calculating atomic charges in molecules. *Tetrahedron Letters*, 3181–3184.

12 Gasteiger, J. and Marsili, M. (1980) Iterative partial equalization of orbital electronegativity – a rapid access to atomic charges. *Tetrahedron*, **36**, 3219–3228.

13 Hutchings, M.G. and Gasteiger, J. (1983) Residual electronegativity – an empirical quantification of polar influences and its application to the proton affinity of amines. *Tetrahedron Letters*, **24**, 2541–2544.

14 Gasteiger, J. and Saller, H. (1985) Calculation of the charge distribution in conjugated systems by a quantification of the resonance concept. *Angewandte Chemie-International Edition*, **24**, 687–689.

15 Gasteiger, J. and Hutchings, M.G. (1984) Quantification of effective polarisability. Applications to studies of X-ray photoelectron spectroscopy and alkylamine protonation. *Journal of the Chemical Society – Perkin Transactions 2*, 559–564.

16 Schreiber, S.L. (2000) Target-oriented and diversity-oriented organic synthesis in drug discovery design. *Science*, **287**, 1964–1969.

17 Wender, P.A., Verma, V.A., Paxton, T.J., and Pillow, T.H. (2008) Function-oriented synthesis, step economy, and drug design. *Accounts of Chemical Research*, **41**, 40–49.

18 Ehrig, H. (1979) Introduction to the algebraic theory of graph grammars (A survey). *Lecture notes in Computer Science*, **73**, 1–69.

19 Ehrig, K., Heckel, R., and Lajios, G. (2006) Molecular analysis of metabolic pathways with graph transformation. *Lecture Notes in Computer Science*, **4178**, 107–121.

20 Rosselló, F. and Valiente, G. (2005) Chemical graphs, chemical reaction graphs, and chemical graph transformation. *Electronic Notes in Theoretical Computer Science*, **127**, 157–166.

21 Félix, L., Rosselló, F., and Valiente, G. (2009) Efficient reconstruction of metabolic pathways by bidirectional chemical search. *Bulletin of Mathematical Biology*, **71**, 750–769.

22 Fujita, S. (1986) Description of organic reactions based on imaginary transition structures. Part 1–5. *Journal of Chemical Information and Computer Sciences*, **26**, 205–242.

23 Fujita, S. (1987) Description of organic reactions based on imaginary transition structures. Part 6–9. *Journal of Chemical Information and Computer Sciences*, **27**, 99–120.

24 Vladutz, G. (1986) Do we still need a classification of reactions? in *Modern Approaches to Chemical Reaction Searching* (ed. P. Willett), Gower, Aldershot, pp. 202–220.

25 Rosello, F. and Valiente, G. (2005) Graph transformation in molecular biology. *Lecture Notes in Computer Science*, **3393**, 116–133.

26 Rosello, F. and Valiente, G. (2004) Analysis of metabolic pathways by graph transformation. *Lecture Notes in Computer Science*, **3256**, 70–82.

27 Benko, G., Flamm, C., and Stadler, P.F. (2003) A graph-based toy model of chemistry. *Journal of Chemical Information and Computer Sciences*, **43**, 1085–1093.

28 Benko, G., Flamm, C., and Stadler, P.F. (2005) Explicit collision simulation of chemical reactions in a graph based artificial chemistry. *Lecture Notes in Computer Science*, **3630**, 725–733.

29 Kerber, A., Laue, R., Meringer, M., and Rücker, C. (2007) Molecules in silico. A graph description of chemical reactions. *Journal of Chemical Information and Modeling*, **47**, 805–817.

30 Stein, N. (1995) New perspectives in computer-assisted formal synthesis design-treatment of delocalized electrons. *Journal of Chemical Information and Computer Sciences*, **35**, 305–309.

31 Hladka, E., Koca, J., Kratochvil, M., Kvasnicka, V., Matyska, L., Pospichal, J.,

and Potucek, V. (1992) The synthon model and the program PEGAS for computer assisted organic synthesis. *Topics in Current Chemistry*, **166**, 121–197.

**32** Fic, G. and Nowak, G. (1998) Extended concept of reaction generators in organic reaction prediction system CSB. *Computers and Chemistry*, **22**, 141–146.

**33** Reisen, F.H., Schneider, G., and Proschak, E. (2009) Reaction-MQL: Line notation for functional transformation. *Journal of Chemical Information and Modeling*, **49**, 6–12.

**34** Li, C., Henry, C.S., Jankowski, M.D., Ionita, J.A., Hatzimanikatis, V., and Broadbelt, L.J. (2004) Computational discovery of biochemical routes to speciality chemicals. *Chemical Engineering Science*, **59**, 5051–5060.

**35** Bauer, J. (1989) IGOR2: A PC-program for generating new reactions and molecular structures. *Tetrahedron Computer Methodology*, **2**, 269–280.

**36** Fontann, E. and Reitsam, K. (1991) The generation of reaction networks with RAIN. 1. The reaction generator. *Journal of Chemical Information and Computer Sciences*, **31**, 96–101.

**37** Hatzimanikatis, V., Li, C., Ionita, J.A., Henry, C.S., Jankowski, M.D., and Broadbelt, L.J. (2005) Exploring the diversity of complex metabolic networks. *Bioinformatics*, **21**, 1603–1609.

**38** Broadelt, L.J., Stark, S.M., and Klein, M.T. (1994) Computer generated pyrolyzis modeling: On-the-fly generation of species, reactions, and rates. *Industrial and Engineering Chemistry Research*, **33**, 790–799.

**39** Bauer, J., Fontain, E., and Ugi, I. (1988) Computer-assisted bilateral solution of chemical problems and generation of reaction networks. *Analytica Chimica Acta*, **210**, 123–134.

**40** Herges, R. and Hoock, C. (1992) Reaction planning: Computer-aided discovery of a novel elimination reaction. *Science*, **255**, 711–713.

**41** Corey, E.J., Long, A.K., and Rubenstein, S.D. (1985) Computer-assisted analysis in organic synthesis. *Science*, **228**, 4698. 408–418.

**42** Jorgensen, W.L., Laird, E.R., Gushurst, A.J., Fleischer, J.M., Gothe, S.A., Helson, H.E., Paderes, G.D., and Sinclair, S. (1990) CAMEO: A program for the logical prediction of the products of organic reactions. *Pure and Applied Chemistry*, **62**, 1921–1932.

**43** Ellen, R., Laird, E.R., and Jorgensen, W.L. (1990) Computer assisted analysis of reactions involving organic free radicals and diradicals. *Journal of Chemical Information and Computer Sciences*, **30**, 458–466.

**44** Gasteiger, J. and Jochum, C. (1978) EROS – a computer program for generating sequences of reactions. *Topics in Current Chemistry*, **74**, 93–126.

**45** Röse, P. and Gasteiger, J. (1990) Automatic derivation of reaction rules for the EROS 6.0 system for reaction prediction. *Analytica Chimica Acta*, **235**, 163–168.

**46** Röse, P. and Gasteiger, J. (1990) in *Software-Development in Chemistry 4* (ed. J. Gasteiger), Springer, Heidelberg, pp. 275–288.

**47** Lynch, M.F. and Willett, P. (1978) The automatic detection of chemical reaction sites. *Journal of Chemical Information and Computer Sciences*, **18**, 154–159.

**48** Willett, P. (1980) The evaluation of an automatically indexed, machine-readable chemical reactions file. *Journal of Chemical Information and Computer Sciences*, **20**, 93–96.

**49** Ratkiewicz, A. and Truong, T.N. (2003) Application of chemical graph theory for automated mechanism generation. *Journal of Chemical Information and Computer Sciences*, **43**, 36–44.

**50** Hsu, S.H., Krishnamurthy, B., Rao, P., Zhao, C., Jagannathan, S., and Venkatasubramanian, V. (2008) A domain-specific compiler theory based framework for automated reaction network generation. *Computers and Chemical Engineering*, **32**, 2455–2470.

**51** Wilcox, C.S. and Levinson, R.A. (1986) Self-organized knowledge base for recall, design and discovery in organic chemistry. *ACS Symposium Series*, **306**, 209–230.

**52** http://infochem.de/content/downloads/classify.pdf.

53 Satoh, H. and Funatsu, K. (1995) SOPHIA, a knowledge base-guided reaction prediction system-utilization of a knowledge base derived from a reaction database. *Journal of Chemical Information and Computer Sciences*, **35**, 34–44.

54 Satoh, H. and Funatsu, K. (1996) Further development of areaction generator in the SOPHIA system for organic reaction prediction, knowledge-guided addition of suitable atoms and/or atomic groups to product skeleton. *Journal of Chemical Information and Computer Sciences*, **36**, 173–184.

55 Satoh, H. and Funatsu, K. (1999) A novel approach to retrosynthetic analysis using knowledge bases derived from reaction databases. *Journal of Chemical Information and Computer Sciences*, **39**, 316–325.

56 Blurock, E.S. (1990) Computer-aided synthesis design at risc-linz: automatic extraction and use of reaction classes. *Journal of Chemical Information and Computer Sciences*, **30**, 505–510.

57 Law, J., Zsoldos, Z., Simon, A., Reid, D., Liu, Y., Khew, S.Y., Johnson, A.P., Major, S., Wade, A.W., and Ando, H.Y. (2009) Route designer: A retrosynthetic analysis tool utilizing automated retrosynthetic rule generation. *Journal of Chemical Information and Modeling*, **49**, 593–602.

58 Kotera, M., Okuno, Y., Hattori, M., Goto, S., and Kanehisa, M. (2004) Computational assignment of the EC numbers for genomic-scale analysis of enzymatic reactions. *Journal of the American Chemical Society*, **126**, 16487–16498.

59 Kulkarni, S.A., Zhu, J., and Blechinger, S. (2005) In silico techniques for the study and prediction of xebobiotic metabolism: A review. *Xenobiotica*, **35**, 955–973.

60 Ott, M. and Noordik, J.H. (1997) Long-range strategies in the LHASA program: the quinone Diels–Alder transform. *Journal of Chemical Information and Computer Sciences*, **37**, 98–108.

61 Long, A.K., Kappos, J.C., Rubenstein, S.D., and Walker, G.E. (1994) Computer-assisted synthetic analysis. A generalized procedure for subgoal transform selection based on a two-dimensional pattern language. *Journal of Chemical Information and Computer Sciences*, **34**, 922–933.

62 Fic, G. and Nowak, G. (2001) Implementation of similarity model in the CSB system for chemical reaction prediction. *Computers and Chemistry*, **25**, 177–186.

63 Johnson, A.P., Marshall, C., and Judson, P.N. (1992) Starting material oriented retrosynthetic analysis in the LHASA program. 1. General description. *Journal of Chemical Information and Computer Sciences*, **32**, 411–417.

64 Gasteiger, J., Ihlenfeldt, W.D., Fick, R., and Rose, J.R. (1992) Similarity concept for the planning of organic reactions and syntheses. *Journal of Chemical Information and Computer Sciences*, **32**, 700–712.

65 Wipke, W.T. and Rogers, D. (1984) Artificial intelligence in organic synthesis. SST: starting material selection strategies. An application of superstructure search. *Journal of Chemical Information and Computer Sciences*, **24**, 71–81.

66 Hanessian, S., Franko, J., and Larouche, B. (1990) The psychobiological basis of heuristic synthesis planning – man, machine and the chiron approach. *Pure and Applied Chemistry*, **62**, 1887–1910.

67 Hanessian, S., Botta, M., Larouche, B., and Boyaroglu, A. (1992) Computer-assisted perception of similarity using the chiron program: A powerful tool for the analysis and prediction of biogenetic patterns. *Journal of Chemical Information and Computer Sciences*, **32**, 718–722.

68 Mehta, G., Barone, R., and Chanon, M. (1998) computer-aided organic synthesis – SESAM: A simple program to unravel "hidden" restructured starting materials skeleta in complex targets. *European Journal of Organic Chemistry*, 1409–1412.

69 Gillet, V.J., Myatt, G., Zsoldos, Z., and Johnson, A.P. (1995) SPROUT, HIPPO and CAESA: Tools for de novo structure generation and estimation of synthetic accessibility. *Perspectives in Drug Discovery and Design*, **3**, 34–50.

70 Hendrickson, J.B. (1977) Systematic synthesis design. 6. yield analysis and

convergency. *Journal of the American Chemical Society*, **99**, 5439–5450.

71 Bertz, S.H. and Sommer, T.J. (2003) The role of isomorphism in synthetic analysis. Pruning the search tree by finding disjoint isomorphic substructures. *Chemical Communications*, 1000–1001.

72 Bertz, S.H. (2001) The bond cover approach to chemical topology. *Chemical Communications*, 2516–2517.

73 Bertz, S.H. and Sommer, T.J. (1997) Rigorous mathematical approaches to strategic bonds and synthetic analysis based on conceptually simple new complexity indices. *Chemical Communications*, 2409–2410.

74 Vismara, P., Tognetti, Y., and Laurenco, C. (2005) Maximum symmetrical split of molecular graphs. Application to organic synthesis design. *Journal of Chemical Information and Modeling*, **45**, 685–695.

75 Corey, E.J., Howe, W.J., Orf, H.W., Pensak, D.A., and Petersson, G. (1975) General methods of synthetic analysis. Strategic bond disconnections for bridged polycyclic structures. *Journal of the American Chemical Society*, **97**, 6116–6124.

76 Bertz, S.H. and Wright, W.F. (1998) The graph theory approach to synthetic analysis: Definition and application of molecular complexity and synthetic complexity. *Graph Theory Notes of New York*, **35**, 32–48.

77 Bertz, S.H. (2003) Complexity of synthetic routes: Linear, convergent and reflexive syntheses. *New Journal of Chemistry*, **27**, 870–879.

78 Bertz, S.H. (1981) The first general index of molecular complexity. *Journal of the American Chemical Society*, **103**, 3599–3601.

79 Rücker, G. and Rücker, C. (2001) Substructure, subgraph, and walk counts as measures of the complexity of graphs and molecules. *Journal of Chemical Information and Computer Sciences*, **41**, 1457–1462.

80 Rücker, G. and Rücker, C. (1999) On topological indices, boiling points, and cycloalkanes. *Journal of Chemical Information and Computer Sciences*, **39**, 788–802.

81 Rücker, G. and Rücker, C. (2000) Walk counts, labyrinthicity, and complexity of acyclic and cyclic graphs and molecules. *Journal of Chemical Information and Computer Sciences*, **40**, 99–106.

82 Bertz, S.H., Rücker, C., Rücker, G., and Sommer, T.J. (2003) Simplification in synthesis. *European Journal of Organic Chemistry*, 4737–4740.

83 Rücker, C., Rücker, G., and Bertz, S.H. (2004) Organic synthesis – art or science? *Journal of Chemical Information and Computer Sciences*, **44**, 378–386.

84 Barone, R. and Chanon, M. (2005) Search for strategies by computer: the CONAN approach. Application to steroid and taxane frameworks. *Tetrahedron*, **61**, 8916–8923.

85 Whitlock, H.W. (1998) On the structure of total synthesis of complex natural products. *Journal of Organic Chemistry*, **63**, 7982–7989.

86 Barone, R. and Chanon, M. (2001) A new and simple approach to chemical complexity. Application to the synthesis of natural products. *Journal of Chemical Information and Computer Sciences*, **41**, 269–272.

87 Barone, R., Petitjean, M., Baralotto, C., Piras, P., and Chanon, M. (2003) Information theory description of synthetic strategies. A new similarity index. *Journal of Physical Organic Chemistry*, **16**, 9–15.

88 Nowak, G. and Fic, G. (2010) Search for complexity generating chemical transformations by combining connectivity analysis and cascade transformation patterns. *Journal of Chemical Information and Modeling*, **50**, 1369–1377.

89 Nowak, G. and Fic, G. (2009) Machine learning approach to discovering cascade reaction patterns. Application to reaction pathways prediction. *Journal of Chemical Information and Modeling*, **49**, 1321–1329.

90 PETRA - Parameter Estimation for the Treatment of Reactivity Applications http://www2.chemie.uni-erlangen.de/software/petra/.

91 Hoffman, R. (1963) An extended hückel theory. I. Hydrocarbons. *Journal of Chemical Physics*, **39**, 1397–1412.

92 Klopman, G. (1968) Chemical reactivity and the concept of charge- and frontier-

controlled reactions. *Journal of the American Chemical Society*, **90**, 223–243.

93 Baumer, L., Sala, G., and Sello, G. (1990) A new metod for the calculation of atomic and local hardness. *Journal of Computational Chemistry*, **11**, 694–699.

94 Baumer, L. and Sello, G. (1992) New method for the calculation of bond native polarity using molecular electronic energy. *Journal of Chemical Information and Computer Sciences*, **32**, 125–130.

95 Sello, G. (1992) Reaction prediction: The suggestions of the beppe program. *Journal of Computational Chemistry*, **32**, 713–717.

96 Durante, M. and Sello, G. (2000) Prediction of organic reaction products: Determining the best reaction condition. *Journal of Chemical Information and Computer Sciences*, **40**, 221–235.

97 Socorro, I.M. and Goodman, J.M. (2006) The ROBIA program for predicting organic reactivity. *Journal of Chemical Information and Modeling*, **46**, 606–614.

98 Ugi, I., Bauer, R., Baumgartner, E., Fontain, E., Forstmeyer, D., and Lohberger, S. (1988) Computer assistance in the design of syntheses and a new generation of computer programs for the solution of chemical problems by molecular logic. *Pure and Applied Chemistry*, **60**, 1573–1586.

99 Hendrickson, J. and Parks, C.A. (1992) A program for the forward generation of synthetic routes. *Journal of Chemical Information and Computer Sciences*, **32**, 209–215.

100 Kvasnička, V. and Pospichal, J. (1989) Two metrics for a graph-theoretical model of organic chemistry. *Journal of Mathematical Chemistry*, **3**, 161–191.

101 Kvasnicka, V. and Pospichal, J. (1991) Chemical and reaction metrics for graph-theoretical model of organic chemistry. *Journal of Molecular Structure: THEOCHEM*, **227**, 17–42.

# Index



*Statistical Modelling of Molecular Descriptors in QSAR/QSPR*. First Edition.
Edited by M. Dehmer, K. Varmuza, and D. Bonchev

***n***